National 5 MATHEMATICS
STUDENT BOOK

N5 MATHS STUDENT BOOK

Craig Lowther • Judith Barron • Robin Christie
Brenda Harden • Andrew Thompson
John Ward • Stuart Welsh

001/27052013

10 9 8 7 6 5 4 3 2 1

ISBN 9780007504626

Published by
Leckie & Leckie Ltd
An imprint of HarperCollins*Publishers*
Westerhill Road, Bishopbriggs, Glasgow, G64 2QT
T: 0844 576 8126   F: 0844 576 8131
leckieandleckie@harpercollins.co.uk   www.leckieandleckie.co.uk

Special thanks to
Jouve (layout); Ink Tank (cover design); Lucy Hadfield, John Hodgson, Peter Lindsay, Laura Rigon, Paul Cruickshank (answers); Laurice Suess (proofread); Caleb O'Loan (proofread); Alistair Coats (proofread)

A CIP Catalogue record for this book is available from the British Library.

Acknowledgements
Whilst every effort has been made to trace the copyright holders, in cases where this has been unsuccessful, or if any have inadvertently been overlooked, the Publishers would gladly receive any information enabling them to rectify any error or omission at the first opportunity.

Leckie & Leckie would like to thank the following copyright holders for permission to reproduce their material:

Cover: matabum; Kellis; Nuttapong Wongcheronkit; photos in book: Shutterstock

UNIT 1 – EXPRESSIONS AND FORMULAE

UNIT 3 – APPLICATIONS

Introduction

About this book

This book provides a resource to practise and assess your understanding of the mathematics covered for the National 5 qualification. There is a separate chapter for each of the operational skills specified in the Expressions and Formulae, Relationships and Applications units, and most of the chapters use the same features to help you progress. You will find a range of worked examples to show you how to tackle problems, and an extensive set of exercises to help you develop the whole range of operational and reasoning skills needed for the National 5 assessment.

You should not work through the book from page 1 to page xxx. Your teacher will choose a range of topics throughout the school year and teach them in the order they think works best for your class, so you will use different parts of the book at different parts of the year.

Features

CHAPTER TITLE

The chapter title shows the operational skill covered in the chapter.

9 Calculating the length of an arc or the area of a sector of a circle

THIS CHAPTER WILL SHOW YOU HOW TO:

Each chapter opens with a list of topics covered in the chapter, and tells you what you should be able to do when you have worked your way through the chapter.

This chapter will show you how to:

- calculate the length of an **arc** of a circle
- calculate the area of a **sector** of a circle

YOU SHOULD ALREADY KNOW:

After the list of topics covered in the chapter, there is a list of topics you should already know before you start the chapter. Some of these topics will have been covered at National 4, and others will depend on preceding chapters in this book.

You should already know

- the vocabulary associated with circles
- how to calculate the circumference of a circle given radius or diameter
- how to calculate the area of a circle given radius
- how to round answers to an appropriate degree of accuracy.

EXAMPLE

Each new topic is demonstrated with at least one worked Example, which shows how to go about tackling the questions in the following Exercise. Each Example breaks the question and solution down into steps, so you can see what calculations are involved, what kind of rearrangements are needed and how to work out the best way of answering the question. Most Examples have comments, which help explain the processes.

Example 9.3

Calculate the length of the minor arc AB.

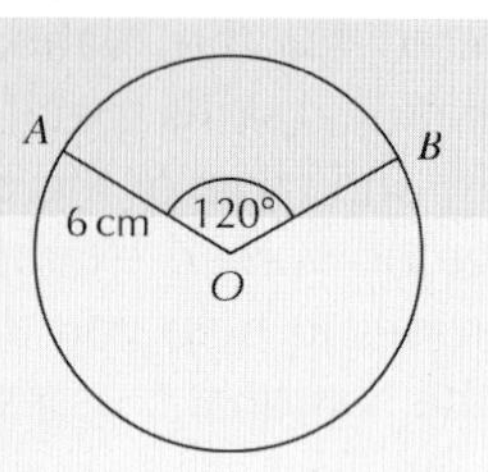

$$\text{arc length} = \frac{\theta}{360} \times \pi d$$

Remember to use diameter (2 × radius).

$$= \frac{120}{360} \times \pi \times 12$$

EXERCISE

The most important parts of the book are the Exercises. The questions in the Exercises are carefully graded in difficulty, so you should be developing your skills as you work through an Exercise. If you find the questions difficult, look back at the Example for ideas on what to do. Use Key questions, marked with a star ★, to assess how confident you can feel about a topic. Questions which require reasoning skills are marked with a cog icon ⚙.

Exercise 9C

You may use a calculator for Questions 1 and 4.

★ 1 For each of the circles below, calculate θ, the sector angle.

Round your answers to the nearest whole degree.

ANSWERS

Answers to all Exercise questions are provided online at www.leckieandleckie.co.uk/N5maths.

ACTIVITIES

Most chapters have Activities which help extend your understanding of the topic and get you thinking more deeply about the topic.

GO! Activity

Eratosthenes was a Greek mathematician who lived in the 3rd century BCE. He worked in the Great Library of Alexandria. Eratosthenes studied geometry and, having observed several lunar eclipses, believed the Earth to be shaped like a sphere. He came up with a

HINTS

Where appropriate, Hints are given in the text and in Examples, to help give extra support.

The index number tells you the number of times the base number is multiplied by itself.

LINK TO HIGHER

N5 Mathematics is a stepping stone to Higher Mathematics. Where appropriate, Link to Higher topics show the progression from N5 to Higher.

Link to Higher

A **rate of change** is a measure of how much one variable changes relative to a corresponding change in another. When calculating gradient, a large change in y relative to a small change in x gives a larger gradient, whereas a small change in y relative to a large change in x gives a smaller gradient. When we talk about a rate of change we are really talking about gradient.

END-OF-CHAPTER SUMMARY

Each chapter closes with a summary of learning statements showing what you should be able to do when you complete the chapter. The summary identifies the Key questions for each learning statement. You can use the End-of-chapter summary and the Key questions to check you have a good understanding of the topics covered in the chapter. Traffic light icons can be used for your own self-assessment.

- I can calculate the length of an arc or the area of a sector. ★ Exercise 9A Q2
- I can work with arc lengths and sector areas expressed in terms of π. ★ Exercise 9B Q3

ASSESSMENTS

End-of-unit assessments are provided for each of the Expressions and Formulae, Relationships and Applications units. These assessments cover the minimum competence for the unit content and are a good preparation for your unit assessment.

The Preparation for Assessment provides examples of examination style questions (sometimes called the 'added value' unit). In this section there are examples to show how you should write your answers and some added value questions to practise on.

NATIONAL 5 NUMERACY – PROBABILITY

Two extra chapters and appropriate assessment material on probability are available online. These chapters complete the essential coverage for National 5 Numeracy, and students working towards National 5 Numeracy should download these chapters from

www.leckieandleckie.co.uk/N5maths

These chapters cover:

- Making and explaining decisions based on the interpretation of data
- Making and explaining decisions based on probability

TEACHER NOTES

Teacher notes, giving guidance and suggestions for teaching the topics covered in each chapter are provided online at www.leckieandleckie.co.uk/N5maths.

1 Working with surds

In this chapter you will learn how to:

- identify and simplify **surds**
- multiply and divide surds
- calculate and manipulate surds
- **rationalise** a denominator.

You should already know:

- how to calculate squares of numbers
- how to calculate square roots of numbers
- how to calculate cubes of numbers
- how to calculate cube roots of numbers.

Sets of numbers

In mathematics different sets of numbers are used.

The set of **natural numbers** is $\mathbb{N} = \{1, 2, 3, 4, 5, \ldots\}$. These are the numbers used for counting.

The set of **whole numbers** is $\mathbb{W} = \{0, 1, 2, 3, 4, \ldots\}$. The set of whole numbers is the same as the set of natural numbers with the addition of 0.

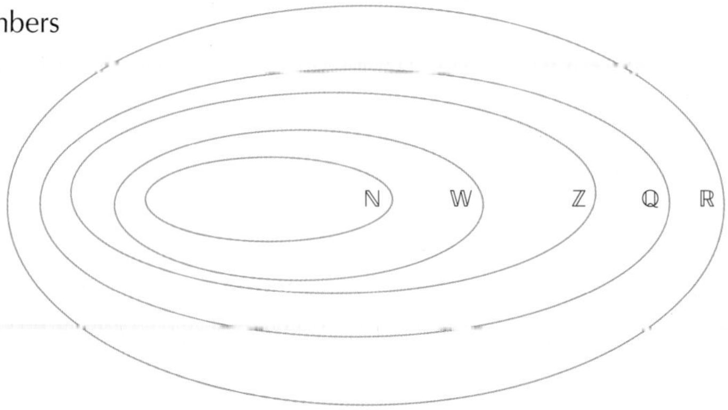

$\mathbb{N}$ $\mathbb{W}$ $\mathbb{Z}$ $\mathbb{Q}$ $\mathbb{R}$

The set of **integers** is $\mathbb{Z} = \{\ldots -3, -2, -1, 0, 1, 2, 3, \ldots\}$. The set of integers is the same as the set of whole numbers with the addition of negative numbers.

The set of **rational numbers** is $\mathbb{Q}$ and this set consists of all the numbers which can be expressed in the form of a fraction $\frac{a}{b}$, where a and b are integers and the denominator b does not equal 0.

Examples of rational numbers are 3, 4.7, $\frac{11}{5}$, $7\frac{2}{5}$, -8, $\sqrt{9} = 3$, $\sqrt[3]{64} = 4$.

An **irrational number** is any number that **cannot** be expressed as a fraction $\frac{a}{b}$.

Examples of irrational numbers include:

- **surds** such as $\sqrt{5}$, $\sqrt[3]{15}$, $\sqrt[5]{18}$. Surds are numbers which when left in square root form (or a higher root form) cannot be simplified. $\sqrt{5}$ cannot be simplified as $\sqrt{5} = 2.236\,067\,977\ldots$ The decimal equivalent is never ending and there are no repeated patterns. Similarly, $\sqrt[3]{6} = 1.817\,120\,593\ldots$

- the ratio of a circle's circumference to its diameter is the mathematical constant $\pi = 3.141\,592\,65\ldots$ π is also known as a **transcendental number**.

The set of **real numbers** is $\mathbb{R}$, made up of all rational and irrational numbers. All integers are rational numbers.

Identifying and simplifying surds

In situations where an accurate answer is needed, surds can be used in complex calculations rather than a rounded decimal number. A decimal approximation can lead to inaccuracies which can be made worse when used in further calculations, especially when large numbers are involved.

The numerical value of $\sqrt{2} = 1.414\,213\,562\ \ldots$ has no exact value. However, when $\sqrt{2}$ is left as $\sqrt{2}$ this is referred to as the **exact** value.

Example 1.1

Which of these numbers are surds?

a $\sqrt{9}$ b $\sqrt{12}$ c $\sqrt[3]{8}$ d $\sqrt{0.4}$ e $\sqrt{0.16}$ f $\sqrt[3]{33}$

a $\sqrt{9} = 3$ $\sqrt{9}$ has an **exact** value so **is not** a surd. ($3 \times 3 = 9$)

b $\sqrt{12}$ has **no exact** value, so it **is** a surd.

c $\sqrt[3]{8} = 2$ $\sqrt[3]{8}$ has an **exact** value so **is not** a surd. ($2 \times 2 \times 2 = 8$)

d $\sqrt{0.4}$ has **no exact** value, so it **is** a surd.

e $\sqrt{0.16} = 0.4$ $\sqrt{0.16}$ has an **exact** value so **is not** a surd. ($0.4 \times 0.4 = 0.16$)

f $\sqrt[3]{33}$ has **no exact** value so it **is** a surd.

Simplifying surds and expressing surds in the simplest form

Simplifying surds is similar to simplifying algebraic expressions. Only **like terms** can be added or subtracted:

$3y + 2y = 5y$ and $3\sqrt{y} + 2\sqrt{y} = 5\sqrt{y}$

$2a + 3b - 6a = 3b - 4a$ and $2\sqrt{a} + 3\sqrt{b} - 6\sqrt{a} = 3\sqrt{b} - 4\sqrt{a}$

Surds can be simplified numerically by identifying factors which are square or cube numbers. It is useful to learn some square numbers and cube numbers, as shown in the table:

	1	2	3	4	5	6	7	8	9	10	11	12
square number (n^2)	1	4	9	16	25	36	49	64	81	100	121	144
cube number (n^3)	1	8	27	64	125	216	343	512	729	1000	1331	1728

To simplify $\sqrt{44}$, express 44 as the product of two numbers, where one of the numbers is a squared number:

$44 = 4 \times 11$ so $\sqrt{44} = \sqrt{4 \times 11}$

$= \sqrt{4} \times \sqrt{11}$

$= 2\sqrt{11}$

Rule 1 $\sqrt{ab} = \sqrt{(a \times b)} = \sqrt{a} \times \sqrt{b}$

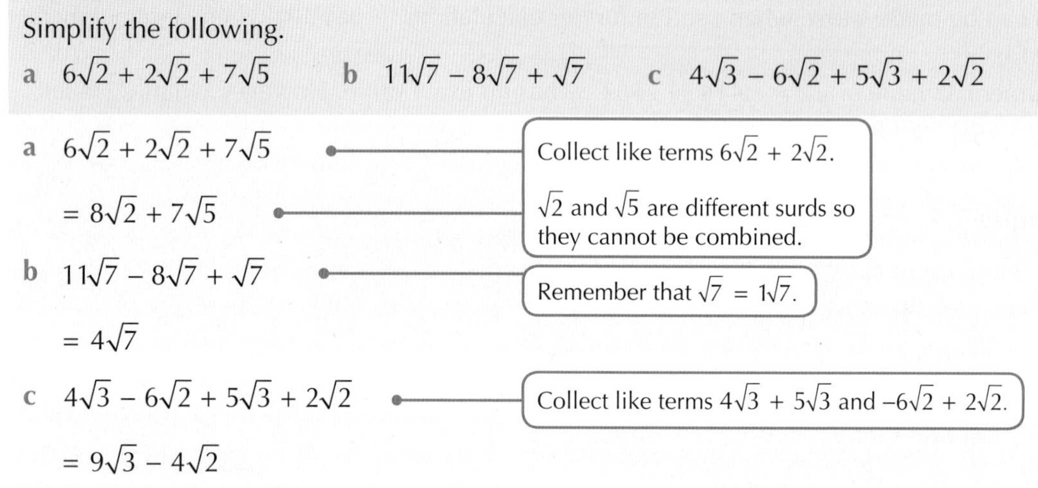

Example 1.2

Simplify the following.

a $6\sqrt{2} + 2\sqrt{2} + 7\sqrt{5}$ b $11\sqrt{7} - 8\sqrt{7} + \sqrt{7}$ c $4\sqrt{3} - 6\sqrt{2} + 5\sqrt{3} + 2\sqrt{2}$

a $6\sqrt{2} + 2\sqrt{2} + 7\sqrt{5}$ — Collect like terms $6\sqrt{2} + 2\sqrt{2}$.

$= 8\sqrt{2} + 7\sqrt{5}$ — $\sqrt{2}$ and $\sqrt{5}$ are different surds so they cannot be combined.

b $11\sqrt{7} - 8\sqrt{7} + \sqrt{7}$ — Remember that $\sqrt{7} = 1\sqrt{7}$.

$= 4\sqrt{7}$

c $4\sqrt{3} - 6\sqrt{2} + 5\sqrt{3} + 2\sqrt{2}$ — Collect like terms $4\sqrt{3} + 5\sqrt{3}$ and $-6\sqrt{2} + 2\sqrt{2}$.

$= 9\sqrt{3} - 4\sqrt{2}$

Example 1.3

Express each of the following in its simplest form.

a $\sqrt{12}$ b $\sqrt{72}$ c $5\sqrt{18}$ d $\sqrt{18} + 5\sqrt{2}$

a $\sqrt{12} = \sqrt{4 \times 3}$ — Express 12 as a product of two numbers where one number is a square number.

$= \sqrt{4} \times \sqrt{3}$

$= 2\sqrt{3}$

b $\sqrt{72} = \sqrt{36} \times \sqrt{2}$ — 36 is the greatest square number which is a factor.

$= 6\sqrt{2}$

or

$\sqrt{72} = \sqrt{9} \times \sqrt{8}$ — Check to see if the surds can be simplified again. $\sqrt{8} = \sqrt{4} \times \sqrt{2}$

$= 3 \times \sqrt{4} \times \sqrt{2}$

$= 3 \times 2 \times \sqrt{2}$

$= 6\sqrt{2}$ — In **b**, 72 can be factorised in different ways. Always try to choose the highest square number which is a factor.

c $5\sqrt{18} = 5 \times \sqrt{9} \times \sqrt{2}$

$= 5 \times 3 \times \sqrt{2}$

$= 15\sqrt{2}$

d $\sqrt{18} + 5\sqrt{2} = (\sqrt{9} \times \sqrt{2}) + 5\sqrt{2}$ — $\sqrt{18}$ can be simplified.

$= 3\sqrt{2} + 5\sqrt{2}$

$= 8\sqrt{2}$

Exercise 1A

★ 1 Identify the surds and evaluate the ones which are not surds.

a $\sqrt{8}$ b $\sqrt{81}$ c $\sqrt[3]{27}$ d $\sqrt{5}$ e $\sqrt[3]{1}$ f $\sqrt{900}$

g $\sqrt{2.5}$ h $\sqrt{0.25}$ i $\sqrt[3]{52}$ j $\sqrt{0.04}$ k $\sqrt{63}$ l $\sqrt{10}$

2 Simplify each of the following.

a $3\sqrt{5} + 7\sqrt{5}$ b $6\sqrt{2} - 5\sqrt{2}$ c $9\sqrt{7} - 4\sqrt{7}$

d $\sqrt{3} + 8\sqrt{3}$ e $3\sqrt{11} - 5\sqrt{11}$ f $\sqrt{2} + 4\sqrt{3} - 5\sqrt{2}$

g $16\sqrt{5} - 3\sqrt{10} - 7\sqrt{5}$ h $4\sqrt{3} + \sqrt{3} - 6\sqrt{3}$ i $5\sqrt{2} + 3\sqrt{3} - 3\sqrt{2} + 8\sqrt{3}$

3 Express each of the following in its simplest form.

a $\sqrt{24}$ b $\sqrt{500}$ c $\sqrt{32}$ d $\sqrt{75}$

e $\sqrt{1000}$ f $3\sqrt{8}$ g $6\sqrt{12}$ h $5\sqrt{50}$

★ 4 Simplify each of the following.

a $5\sqrt{2} + \sqrt{12}$ b $\sqrt{50} - 6\sqrt{2}$ c $3\sqrt{7} + \sqrt{98}$

d $\sqrt{27} - 4\sqrt{3}$ e $\sqrt{125} + 3\sqrt{5}$ f $\sqrt{112} - \sqrt{28}$

g $\sqrt{8} - 3\sqrt{32}$ h $3\sqrt{48} + 2\sqrt{75}$ i $6\sqrt{4} - 4\sqrt{9}$

Multiplying and dividing surds

In the construction industry many calculations require exact values and we need to be able to multiply and divide surds so that we can arrive at an exact answer. Measurements need to be exact so that material is not wasted by cutting lengths too short.

There are four general rules for multiplying and dividing surds. Rule 1 was used in Examples 1.2 and 1.3, and in Exercise 1A.

Rule 1 $\sqrt{a} \times \sqrt{b} = \sqrt{ab}$ **Rule 2** $C\sqrt{a} \times D\sqrt{b} = CD\sqrt{ab}$

Rule 3 $\dfrac{\sqrt{a}}{\sqrt{b}} = \sqrt{\dfrac{a}{b}}$ **Rule 4** $\dfrac{C\sqrt{a}}{D\sqrt{b}} = \dfrac{C}{D}\sqrt{\dfrac{a}{b}}$

An extension of Rule 1 shows that $\sqrt{a} \times \sqrt{a} = \sqrt{a^2} = a$.

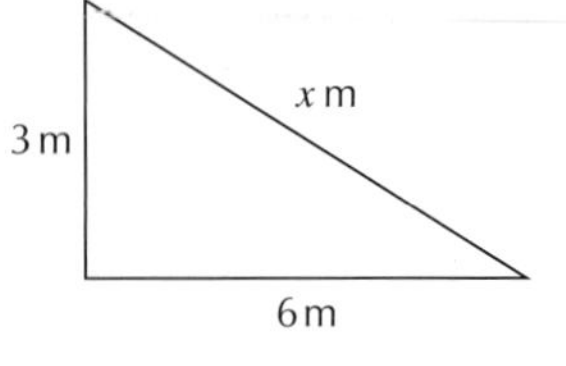

The diagram shows a cross-section of a roofing strut in a building. Pythagoras' theorem is used to find the **exact** length of wood for the longest side in the triangle.

$$x^2 = 6^2 + 3^2$$
$$= 36 + 9$$
$$x = \sqrt{45}$$
$$= \sqrt{9} \times \sqrt{5}$$
$$= 3\sqrt{5} \text{ metres}$$

Example 1.4

Simplify each of the following, leaving your answer in surd form where necessary.

a $\sqrt{3} \times \sqrt{5}$ b $\sqrt{2} \times \sqrt{18}$ c $\dfrac{\sqrt{21}}{\sqrt{3}}$

d $\sqrt{54} \div \sqrt{6}$ e $5\sqrt{2} \times 6\sqrt{3}$ f $\dfrac{4\sqrt{6} \times 2\sqrt{5}}{3\sqrt{15}}$

a $\sqrt{3} \times \sqrt{5}$
$= \sqrt{3 \times 5}$
$= \sqrt{15}$

b $\sqrt{2} \times \sqrt{18}$
$= \sqrt{2 \times 18}$
$= \sqrt{36}$
$= 6$

c $\dfrac{\sqrt{21}}{\sqrt{3}}$
$= \sqrt{21 \div 3}$
$= \sqrt{7}$

d $\sqrt{54} \div \sqrt{6}$
$= \sqrt{54 \div 6}$
$= \sqrt{9}$
$= 3$

e $5\sqrt{2} \times 6\sqrt{3}$
$= 5 \times 6 \times \sqrt{(2 \times 3)}$
$= 30\sqrt{6}$

f $\dfrac{4\sqrt{6} \times 2\sqrt{5}}{3\sqrt{15}}$
$= \dfrac{8\sqrt{30}}{3\sqrt{15}}$
$= \dfrac{8}{3}\sqrt{2}$

Exercise 1B

1 Simplify each of the following, leaving your answer in surd form where necessary.

a $\sqrt{3} \times \sqrt{2}$ b $\sqrt{5} \times \sqrt{5}$ c $\sqrt{16} \times \sqrt{9}$

d $\sqrt{6} \times \sqrt{3}$ e $\sqrt{10} \times \sqrt{40}$ f $3\sqrt{18} \times 4\sqrt{2}$

g $5\sqrt{6} \times 3\sqrt{2}$ h $4\sqrt{32} \times \sqrt{2}$ i $7\sqrt{5} \times 7\sqrt{5}$

2 Simplify each of the following, leaving your answer in surd form where necessary.

a $\sqrt{8} \div \sqrt{2}$ b $\sqrt{32} \times \sqrt{\dfrac{9}{16}}$ c $\dfrac{\sqrt{30}}{\sqrt{10}}$

d $\sqrt{5} \div \sqrt{5}$ e $\sqrt{48} \div \sqrt{3}$ f $\dfrac{10\sqrt{50}}{2\sqrt{5}}$

g $\dfrac{6\sqrt{28}}{3\sqrt{7}}$ h $16\sqrt{20} \div 2\sqrt{2}$ i $9\sqrt{7} \div 3\sqrt{7}$

3 Simplify each of the following, leaving your answer in surd form where necessary.

a $8\sqrt{5} \times 2\sqrt{6} \div 4\sqrt{10}$ b $12\sqrt{21} \div 2\sqrt{3} \times 3\sqrt{2}$ c $4\sqrt{15} \div 2\sqrt{5} \times 3\sqrt{3}$

d $\dfrac{10\sqrt{2} \times 3\sqrt{8}}{5\sqrt{2}}$ e $\left(\dfrac{2}{\sqrt{3}}\right)^2$ f $\left(\dfrac{\sqrt{7}}{5}\right)^2$

★ 4 Simplify each of the following, leaving your answer in surd form where necessary.

a $\sqrt{125}$ b $\sqrt{54}$ c $\sqrt{288}$

d $6\sqrt{3} + \sqrt{27}$ e $10\sqrt{7} - \sqrt{98}$ f $\sqrt{7} \times \sqrt{8}$

g $\sqrt{20} \times \sqrt{10}$ h $\sqrt{56} \div \sqrt{8}$ i $\dfrac{\sqrt{54}}{\sqrt{18}}$

j $3\sqrt{6} \times 5\sqrt{2} \times 4\sqrt{3}$ k $7\sqrt{6} \times 6\sqrt{12} \div 2\sqrt{8}$

5 What value of x makes these statements true?

a $\sqrt{6} \times \sqrt{x} = \sqrt{30}$ b $3\sqrt{x} \times \sqrt{10} = 30$

c $4\sqrt{x} \times \sqrt{x} = 20$ d $6\sqrt{x} \times 2\sqrt{x} \div 3\sqrt{x} = \sqrt{32}$

Calculating and manipulating surds

Manufacturers need to deal with exact values when making parts. They need to calculate precisely how many parts can be made so that this will lead to the maximum profit for the company when selling the goods.

For example, when manufacturing jewellery, circular discs of diameter $\sqrt{5}$ cm have to be cut from a strip of gold with dimensions $\sqrt{5}$ cm by 100 cm. How many discs can be cut?

$$\sqrt{5} = 2.236\ 067\ \ldots$$

- Rounding $\sqrt{5}$ to the nearest cm gives 2. Dividing 100 cm by 2 gives 50 discs.
- Rounding $\sqrt{5}$ to 1 decimal place gives 2.2. Dividing 100 cm by 2.2 gives 45 discs.
- Using the exact value $\sqrt{5}$ gives $100 \div \sqrt{5} = 44.721$. Only whole numbers of discs can be used, so this is rounded to 44 discs.

Different answers are worked out depending on the rounding and this has further cost implications if used in subsequent calculations. Note also that if an accurate value of $\sqrt{5}$ is used, it is impossible to get 50 discs of diameter $\sqrt{5}$ cm from a strip 100 cm long, so rounding can give very misleading results.

When accurate calculations are needed, it is often useful not to round midway through a calculation, but to leave your answer as a surd and then use this exact value in following calculations.

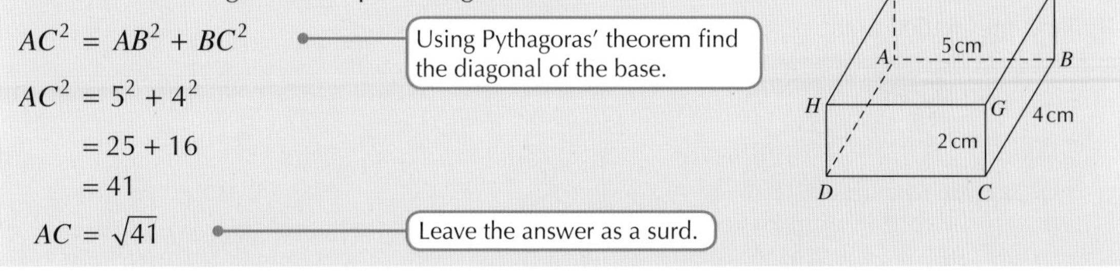
Example 1.5
Find the exact length of the space diagonal AG.
AC² = AB² + BC²
Using Pythagoras' theorem find the diagonal of the base.
AC² = 5² + 4²
= 25 + 16
= 41
AC = √41
Leave the answer as a surd.
E
F
A
5 cm
B
H
G
4 cm
2 cm
D
C

(continued)

Space diagonal $AG^2 = AC^2 + CG^2$

$AG^2 = 41 + 2^2$ — $\sqrt{41}^2 = 41$

$= 41 + 4 = 45$

$AG = \sqrt{45} = \sqrt{9} \times \sqrt{5}$

$= 3\sqrt{5}$ cm

Exercise 1C

1 **a** Using a calculator:

  i calculate $1000 \div \sqrt{7}$ when $\sqrt{7}$ is rounded to the nearest unit

  ii calculate $1000 \div \sqrt{7}$ when $\sqrt{7}$ is rounded to 1 decimal place

  iii calculate $1000 \div \sqrt{7}$ when $\sqrt{7}$ is rounded to 2 decimal places.

b Comment on your answers in part **a**.

2 **a** Using a calculator:

  i calculate $\sqrt{77}$ rounding your answer to the nearest unit

  ii calculate $\sqrt{3}$ rounding your answer to the nearest unit

  iii use your rounded answers to calculate $\sqrt{77} \div \sqrt{3}$.

b Repeat **a** rounding each answer to 1 decimal place.

c Using a calculator calculate $\sqrt{77} \div \sqrt{3}$.

d Comment on your answers.

3 Find the area of the following rectangles, leaving your answer as a surd in its simplest form where necessary.

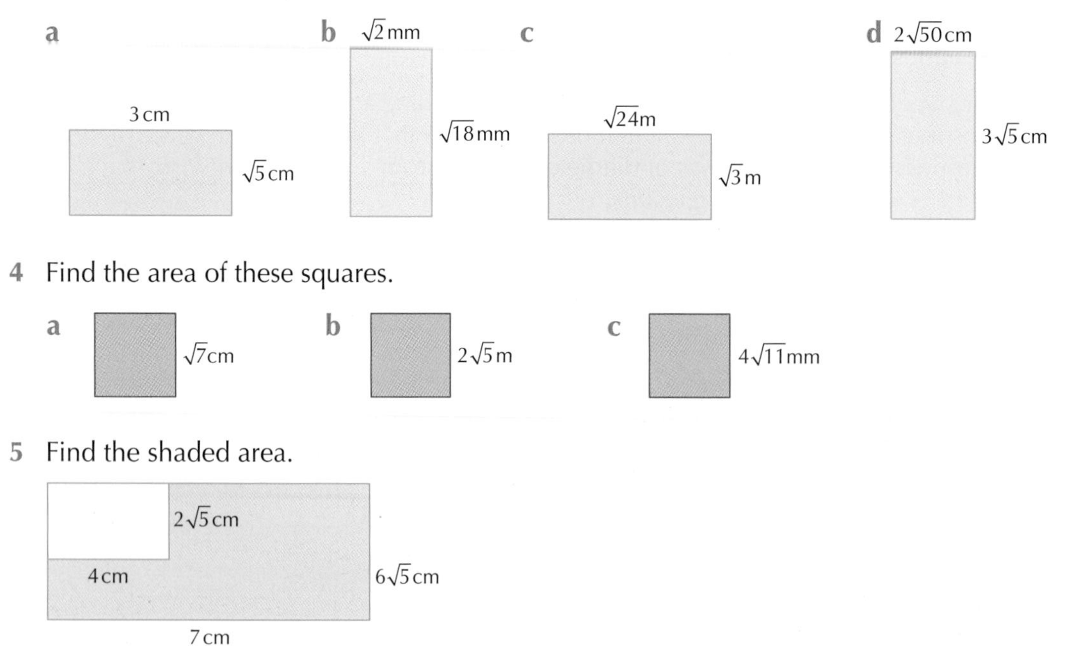

4 Find the area of these squares.

a $\sqrt{7}$ cm **b** $2\sqrt{5}$ m **c** $4\sqrt{11}$ mm

5 Find the shaded area.

6 A garden has dimensions $8\sqrt{3}$ m by 5 m. Jack the gardener decides to create a 1 metre grass area surrounding the flower bed (pink shaded area).

a What are the **exact** dimensions of the flower bed?

b What is the **exact** area of the flower bed?

c What is the **exact** area of grass?

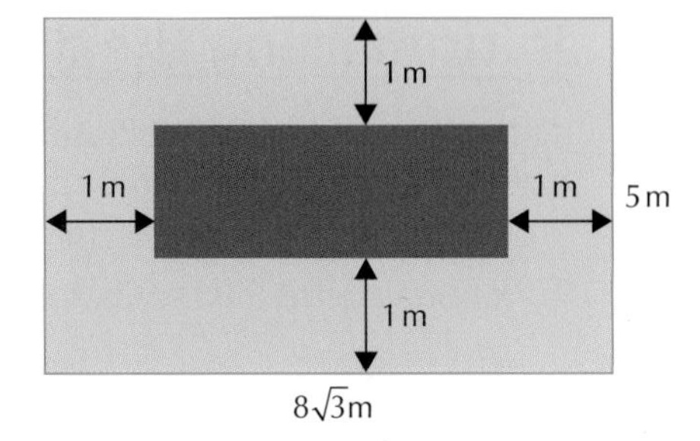

7 Find the length of side x in the following triangles, leaving your answer as a surd in its simplest form where necessary.

8 Find the **exact** length of PS.

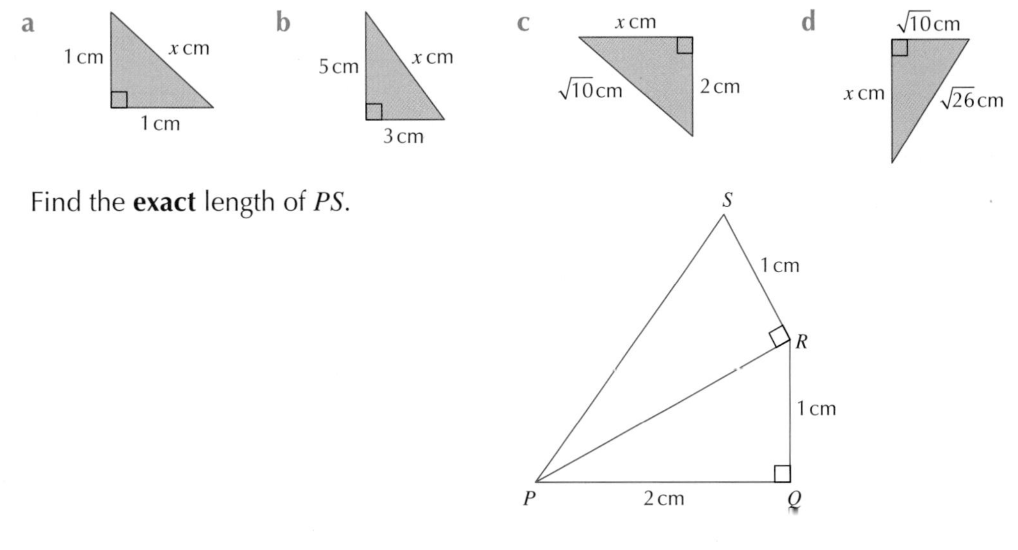

9 Find the **exact** length of the space diagonal leaving your answer as a surd in its simplest form.

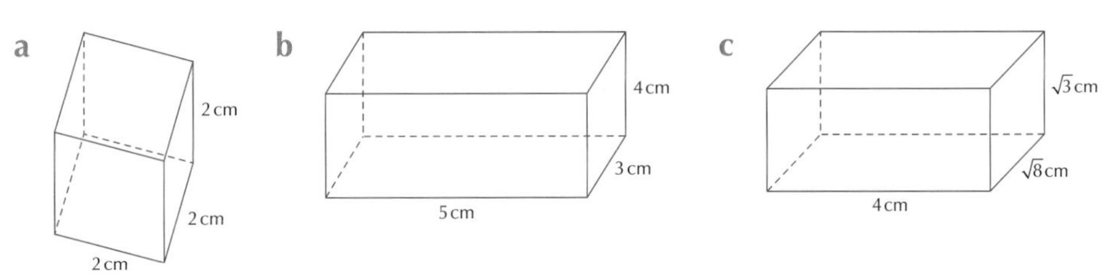

10 Find the **exact** length of side x.

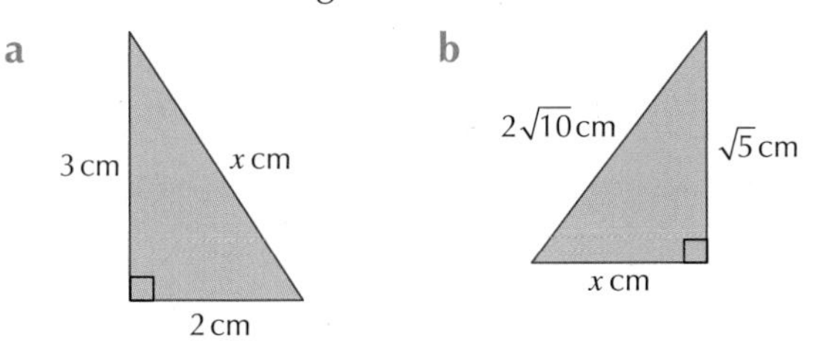

11 a Find the **exact** length of the space diagonal of a cube of side 4 cm.

b Find the **exact** length of the space diagonal of a cuboid with dimensions $2\sqrt{3}$ cm, 3 cm, $2\sqrt{5}$ cm.

Rationalising the denominator of a surd

At times we do not want to have an irrational number as the denominator of a fraction as this can make values difficult to compare. Also a surd is not fully simplified if the answer is left with an irrational denominator. Removing irrational denominators can also help you solve equations.

To rationalise the denominator we need to multiply by 1, where for example:

$$1 = \frac{\sqrt{2}}{\sqrt{2}} \qquad 1 = \frac{\sqrt{3}}{\sqrt{3}} \qquad 1 = \frac{\sqrt{7}}{\sqrt{7}} \text{ etc.}$$

Example 1.6

Rationalise the denominator of: **a** $\frac{5}{\sqrt{3}}$ **b** $\frac{\sqrt{7}}{\sqrt{8}}$

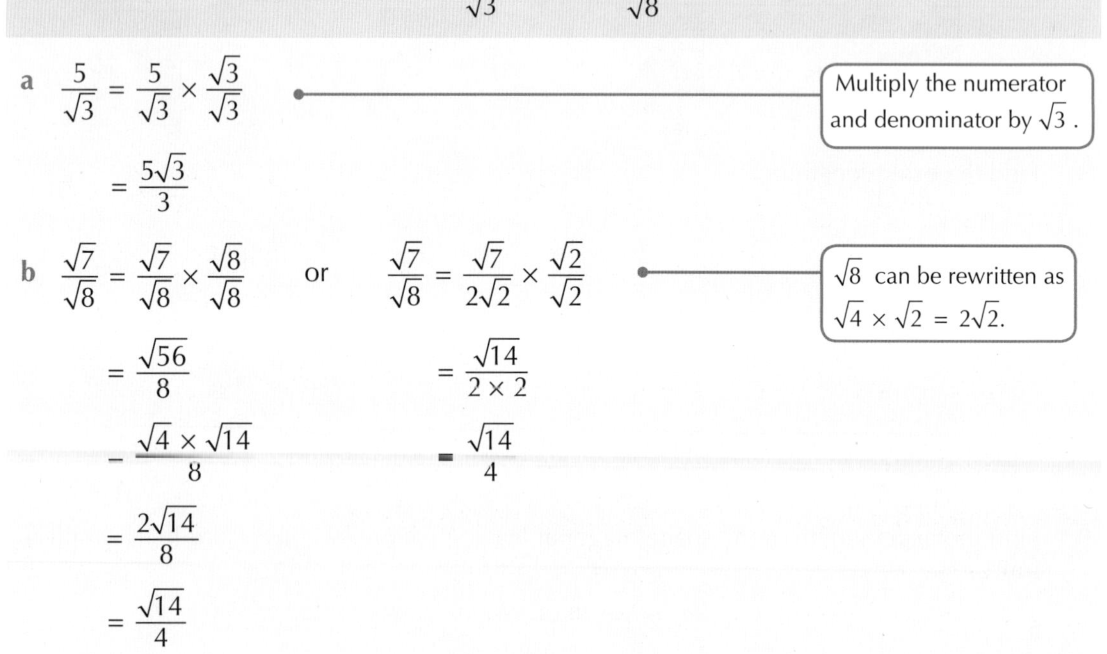

a $\frac{5}{\sqrt{3}} = \frac{5}{\sqrt{3}} \times \frac{\sqrt{3}}{\sqrt{3}}$

$= \frac{5\sqrt{3}}{3}$

Multiply the numerator and denominator by $\sqrt{3}$.

b $\frac{\sqrt{7}}{\sqrt{8}} = \frac{\sqrt{7}}{\sqrt{8}} \times \frac{\sqrt{8}}{\sqrt{8}}$

$= \frac{\sqrt{56}}{8}$

$= \frac{\sqrt{4} \times \sqrt{14}}{8}$

$= \frac{2\sqrt{14}}{8}$

$= \frac{\sqrt{14}}{4}$

or $\frac{\sqrt{7}}{\sqrt{8}} = \frac{\sqrt{7}}{2\sqrt{2}} \times \frac{\sqrt{2}}{\sqrt{2}}$

$= \frac{\sqrt{14}}{2 \times 2}$

$= \frac{\sqrt{14}}{4}$

$\sqrt{8}$ can be rewritten as $\sqrt{4} \times \sqrt{2} = 2\sqrt{2}$.

Exercise 1D

1 Rationalise the denominators of these expressions.

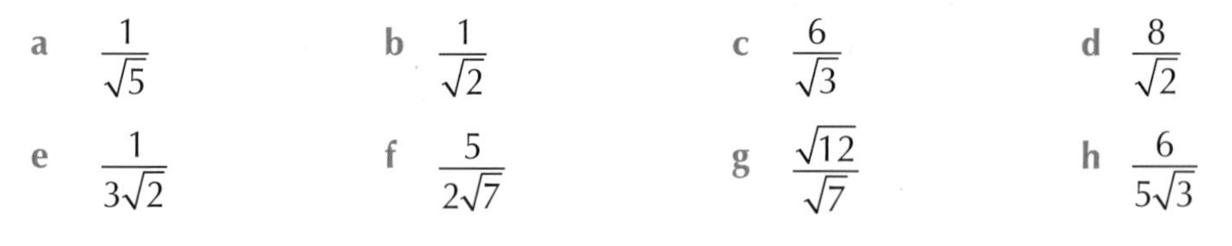

a $\frac{1}{\sqrt{5}}$ **b** $\frac{1}{\sqrt{2}}$ **c** $\frac{6}{\sqrt{3}}$ **d** $\frac{8}{\sqrt{2}}$

e $\frac{1}{3\sqrt{2}}$ **f** $\frac{5}{2\sqrt{7}}$ **g** $\frac{\sqrt{12}}{\sqrt{7}}$ **h** $\frac{6}{5\sqrt{3}}$

★ **2** Express each of the following in its simplest form with a rational denominator.

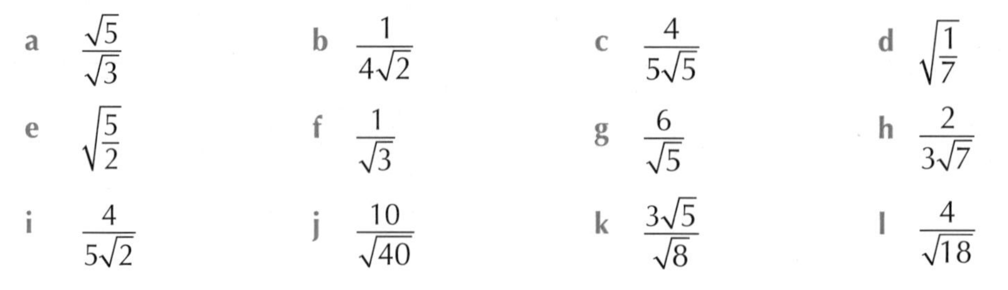

a $\frac{\sqrt{5}}{\sqrt{3}}$ **b** $\frac{1}{4\sqrt{2}}$ **c** $\frac{4}{5\sqrt{5}}$ **d** $\sqrt{\frac{1}{7}}$

e $\sqrt{\frac{5}{2}}$ **f** $\frac{1}{\sqrt{3}}$ **g** $\frac{6}{\sqrt{5}}$ **h** $\frac{2}{3\sqrt{7}}$

i $\frac{4}{5\sqrt{2}}$ **j** $\frac{10}{\sqrt{40}}$ **k** $\frac{3\sqrt{5}}{\sqrt{8}}$ **l** $\frac{4}{\sqrt{18}}$

Example 1.7

Multiply the brackets and simplify $\sqrt{5}(2+\sqrt{8})$.

$$\sqrt{5}(2+\sqrt{8}) = 2\sqrt{5} + \sqrt{5} \times \sqrt{8}$$
$$= 2\sqrt{5} + \sqrt{40}$$
$$= 2\sqrt{5} + \sqrt{(4 \times 10)}$$
$$= 2\sqrt{5} + 2\sqrt{10}$$

Exercise 1E

1 Multiply the brackets and simplify.

a $\sqrt{2}(3+\sqrt{2})$ b $\sqrt{5}(\sqrt{5}-1)$ c $\sqrt{7}(5-\sqrt{7})$ d $2\sqrt{3}(6-\sqrt{3})$

e $3(\sqrt{2}-\sqrt{5})$ f $7(\sqrt{3}-8)$ g $\sqrt{8}(\sqrt{2}-\sqrt{3})$ h $\sqrt{6}(3-2\sqrt{2})$

2 Multiply the brackets and simplify.

a $(\sqrt{2}+1)(\sqrt{2}-3)$ b $(\sqrt{3}-2)(\sqrt{3}+4)$ c $(\sqrt{3}-5)(\sqrt{3}+5)$

d $(\sqrt{2}-\sqrt{7})(\sqrt{2}+\sqrt{7})$ e $(\sqrt{5}+1)^2$ f $(\sqrt{3}-\sqrt{2})^2$

g $(\sqrt{7}-5)^2$ h $3(\sqrt{5}+1)(\sqrt{5}+4)$

★ 3 A rectangle has sides of length $(3+\sqrt{2})$ cm and $(3-\sqrt{2})$ cm.

Calculate:

a the area of the rectangle b the length of a diagonal.

It can help to draw a diagram.

Activity

Surds are used in the science and industry, in art and also occur in nature. A famous occurrence of surds is in the Golden Ratio. The Golden Ratio describes a ratio of dimensions and numbers which appear in a host of applications and natural functions such as the Golden Rectangle in maths, the Fibonacci sequence in patterns in flowers, trees and beehives, and patterns in music. The Golden Ratio is also known as the Divine Proportion, because artists use it to produce paintings which have a pleasing balance.

The Golden Ratio is defined as $\Phi = \dfrac{1+\sqrt{5}}{2}$.

In measurements, this is the ratio of length to breadth, and is known to be 'pleasing to the eye'. The measurements of many Classical Roman and Greek buildings illustrate the Golden Ratio. One of the best known is the Parthenon in Athens.

(continued)

The lines below are in the Golden Ratio:

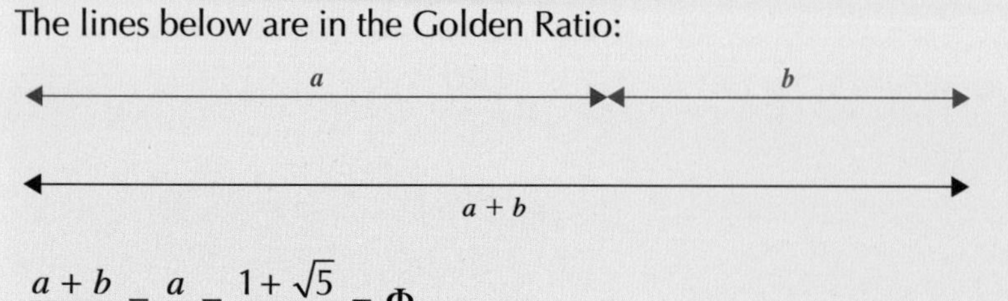

$$\frac{a+b}{a} = \frac{a}{b} = \frac{1+\sqrt{5}}{2} = \Phi$$

The ratio of $a + b$ to a is the same ratio as a to b.

a Using the equation for Φ above, evaluate the Golden Ratio to 5 decimal places.
b Look at a Fibonacci sequence 1, 1, 2, 3, 5, 8, 13, 21, . . . etc.
Divide term 2 by term 1, then term 3 by term 2, then term 4 by term 3 and so on.
Do you see a link to the Golden Ratio?
c Investigate the Golden Ratio in art and nature and music.

- I can identify surds. ★ Exercise 1A Q1
- I can simplify surds and leave my answer as a surd in its simplest form. ★ Exercise 1A Q4
- I can add, subtract, multiply and divide surds. ★ Exercise 1B Q4
- I can use my knowledge of surds in problem questions. ★ Exercise 1C Q10, Q11 ★ Exercise 1E Q3
- I can express surds with a rational denominator. ★ Exercise 1D Q2

For further assessment opportunities, see the Preparation for Assessment for Unit 1 on pages 89–92.

2 Simplifying expressions using the laws of indices

In this chapter you will learn how to:

- write and use **index notation** for **positive** and **negative indices**
- simplify expressions using the laws of indices where there are:
  - » positive and negative indices
  - » fractional indices
- write numbers using **scientific notation**
- solve problems for **very large** or **small numbers** using scientific notation.

You should already know:

- how to find the square, cube and square root of a number
- how to add, subtract, multiply and divide decimals
- how to add, subtract, multiply and divide fractions.

Writing and using index notation

Index notation is an important part of mathematics and science, which is used in different ways to show and calculate:

- repeated multiplication and division, using positive and negative indices

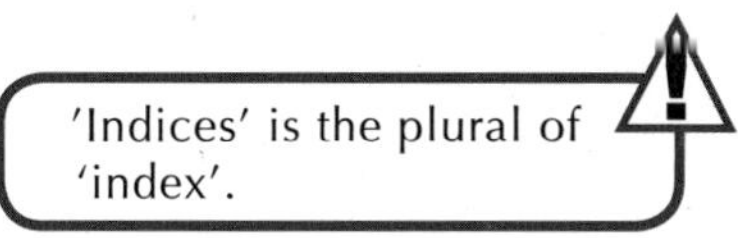
'Indices' is the plural of 'index'.

- roots of numbers, using fractional indices.

Index notation is also used in **scientific notation** to accurately show very large and very small numbers, and in systems of measurement to show combinations of units.

Positive indices

A useful way of writing repeated multiplications is to use **index notation**. You should already know that 4^2 is the same as 4×4, and that 4^3 is the same as $4 \times 4 \times 4$.

- $\mathbf{4^2}$ is the mathematical way of writing **4 to the power of 2**, or **4 squared**.
- $\mathbf{4^3}$ is the mathematical way of writing **4 to the power of 3**, or **4 cubed**.

You can extend repeated multiplication as many times as you want. For example 4^5 (4 to the power of 5) means $4 \times 4 \times 4 \times 4 \times 4$; where 4 is the **base** and 5 is the **index** (also known as the **exponent**).

The first rule for indices is that any number (or base) with an index of zero is equal to 1.

The index number tells you the number of times the base number is multiplied by itself.

Rule 1 $\quad a^0 = 1$

The exception to this rule is when zero is the base number. Zero raised to any power is always equal to zero (i.e. $0^3 = 0$), except in the special case of 0^0. 0^0 is undefined and has no value.

For example:

$9^0 = 1 \quad 4.5^0 = 1 \quad 123\,456^0 = 1$

You can use a calculator to calculate powers. Check your calculator to see what the powers key looks like. It will probably be something like this: x^y

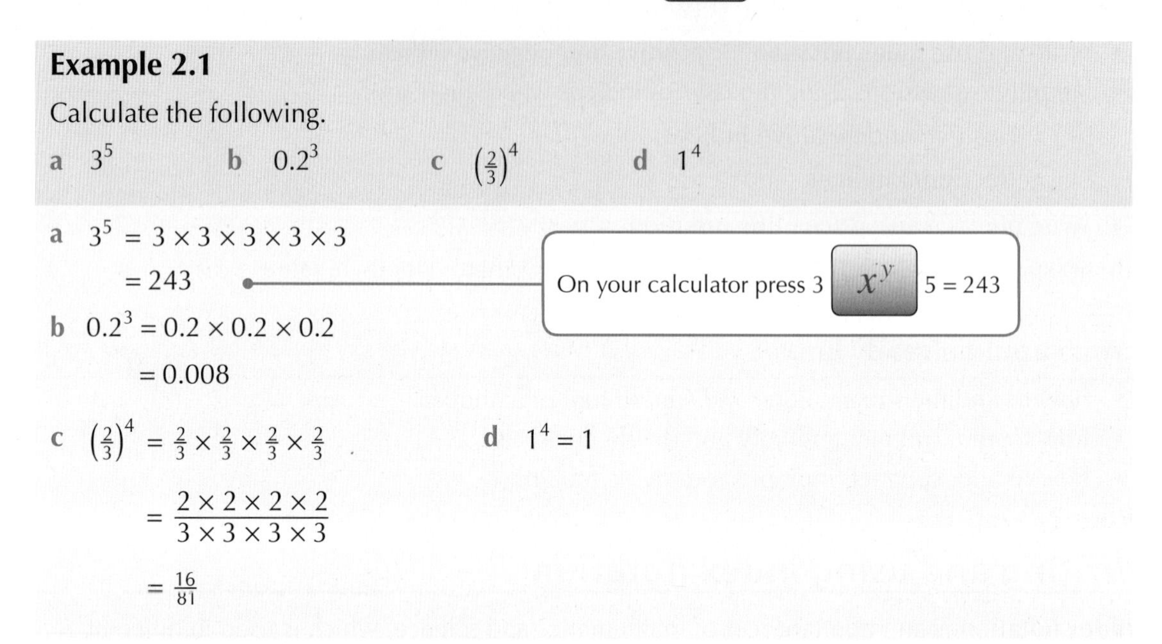

Example 2.1

Calculate the following.

a 3^5 b 0.2^3 c $\left(\frac{2}{3}\right)^4$ d 1^4

a $3^5 = 3 \times 3 \times 3 \times 3 \times 3$

$= 243$

On your calculator press 3 x^y 5 = 243

b $0.2^3 = 0.2 \times 0.2 \times 0.2$

$= 0.008$

c $\left(\frac{2}{3}\right)^4 = \frac{2}{3} \times \frac{2}{3} \times \frac{2}{3} \times \frac{2}{3}$

$= \dfrac{2 \times 2 \times 2 \times 2}{3 \times 3 \times 3 \times 3}$

$= \frac{16}{81}$

d $1^4 = 1$

Example 2.2

Write these expressions using index notation.

a $2 \times 2 \times 2 \times 2 \times 2$ b $f \times f \times f \times f$

a $2 \times 2 \times 2 \times 2 \times 2 = 2^5$

There are five 2s so the answer is 2 raised to the power of 5. (This can also be calculated numerically: $2^5 = 32$. However, the question asked for index notation, so you would leave your answer as 2^5).

b $f \times f \times f \times f = f^4$

We don't know the value of f so we cannot evaluate f^4.

Exercise 2A

★ **1** Calculate the value of the following without a calculator.

a 2^4 b 3^6 c 4^3 d 10^6

e 23^1 f 0.3^3 g $\left(\frac{1}{4}\right)^4$ h 3^0

★ **2** Use a calculator with a power key x^y to check your answers for Question 1.

3 Write the following expressions using index notation.

a 7×7 b $30 \times 30 \times 30$ c $0.3 \times 0.3 \times 0.3 \times 0.3$

d $10 \times 10 \times 10 \times 10$ e $g \times g \times g$ f $a \times a \times a \times b \times b$

4 Use a calculator with a power key to find, where possible, the value of your answers to Question 3.

Activity

The number 16 can be written in index notation as 2^4 or 4^2. How many different ways can you find to write the following numbers in index form?

a 64 b 81 c 256 d 729

Writing and calculating negative indices

Reciprocals are widely used for algorithms in computer science. For many machines, the process of division is slower than multiplication. Changing a division into a multiplication by using indices can speed up calculations. This is essential in modern computing, when millions of calculations can be needed in single processes such as weather forecasting and financial analysis.

The reciprocal of 5 is $\frac{1}{5}$. If you multiply 5 by its reciprocal $\frac{1}{5}$ you get the answer 1: $5 \times \frac{1}{5} = 1$

If you multiply any term or number by its reciprocal you always get the answer 1. As before, zero is a special case. There is no number that you can multiply zero by to get 1, and any number multiplied by 0 will always give 0.

The reciprocal of a is $\frac{1}{a}$. This is written in index form as a^{-1}.

Rule 2 $a^{-m} = \frac{1}{a^m}$

Example 2.3

Write the following expressions with a positive exponent.

a 4^{-3} b x^{-2} c $6n^{-3}$ d $\frac{2}{3}a^{-4}$

a $4^{-3} = \frac{1}{4^3}$ b $x^{-2} = \frac{1}{x^2}$ c $6n^{-3} = \frac{6}{n^3}$

d $\frac{2}{3}a^{-4} = \frac{2}{3a^4}$

Note that the constant 6 remains on top of the fraction. The reciprocal of n^{-3} moves to the bottom of the fraction to give $\frac{6}{n^3}$.

You should become confident in moving numbers with a power from the numerator to the denominator of a fraction.

Example 2.4

Evaluate 4^{-2}, leaving your solution as a fraction.

$4^{-2} = \frac{1}{4^2} = \frac{1}{16}$

Remember that a negative power means the reciprocal, not that the answer is negative.

Exercise 2B

1 Evaluate.

a 6^{-3} b 2^{-5} c 3^{-4} d $(3a)^{-2}$

★ 2 Write with positive indices.

a x^{-3} b y^{-8} c $3t^{-4}$ d $7y^{-6}$ e $\frac{2}{7}t^{-5}$ f $\frac{1}{2}y^{-3}$

3 Write in index form.

a $\frac{6}{x^3}$ b $\frac{9}{t^5}$ c $\frac{3}{m^4}$ d $\frac{10}{a^8}$

4 Find the value of each of the following:

a when $m = 4$

i m^3 ii m^{-2} iii $5m^{-1}$

b when $a = 2$

i a^5 ii a^{-3} iii $8a^{-4}$

5 Consider these terms in m: m^0 m^3 m^{-2}

a Arrange from smallest to largest if m is a positive whole number greater than 1.

b Arrange from smallest to largest if m is a negative whole number less than –1.

GO! Activity

The standard prefix for a factor of 1000 in the SI system (International System of Units) is **kilo**. For example 1000 metres is 1 **kilo**metre or 1 **k**m. If the factor is one million times, the prefix is **mega** so a million tonnes is a **mega**tonne or 1 **M**t. Notice that although you write the word **mega** with a lower case **m**, you use a capital **M** when you use it as an abbreviation. It is important not to confuse a capital **M** for **mega** and lower case **m** for **milli** – they are quite different.

a Try to find all the standard prefixes for the SI system.

b Which is bigger; a yotta (Y) or a zetta (Z)?

Simplifying expressions using the laws of indices

The laws of indices can be used to simplify a wide range of expressions used in algebra, science and engineering. You will use laws of indices to:

- multiply and divide numerical expressions
- raise a power of a number or value to a further power
- find roots of numbers or expressions
- simplify products when sets of brackets are multiplied out.

Rules for multiplying and dividing numbers and expressions in index form

The great mathematician and inventor **Archimedes of Syracuse**, who lived in Greece in the third century BCE, was the first person to discover and prove the law of indices for multiplying and dividing numbers in index form. He showed that $10^m \times 10^n = 10^{(m+n)}$.

This law can be applied to any base number. For example, simplify the following:

$4^5 \times 4^2 = (4 \times 4 \times 4 \times 4 \times 4) \times (4 \times 4)$

$= 4^{(5+2)}$

$= 4^7$

You can check with your calculator that $4^5 \times 4^2$ has the same value as 4^7.

When **multiplying** powers of the same variable we **add** the indices.

Rule 3 $a^m \times a^n = a^{m+n}$

A similar rule is developed for division of numbers that are expressed in index form.

Simplify $6^5 \div 6^2$:

$$\frac{6^5}{6^2} = \frac{\not{6} \times \not{6} \times 6 \times 6 \times 6}{\not{6} \times \not{6}} = 6^3$$

Cancel out equal numbers of 6s in the numerator and denominator.

or:

$$= 6^{(5-2)} = 6^3$$

When **dividing** powers of the same variable we **subtract** the indices.

Rule 4 $\boldsymbol{a^m \div a^n = a^{m-n}}$

Example 2.5

Simplify the following. Write your answer in index form with a positive exponent.

a $3^4 \times 3^5$ b $7^2 \times 7^{-5}$ c $5x^8 \times 6x^4$ d $10y^3 \times 4y^{-5}$

a $3^4 \times 3^5 = 3^{4+5}$
$= 3^9$

b $7^2 \times 7^{-5} = 7^{(2+-5)}$
$= 7^{-3}$
$= \frac{1}{7^3}$

Remember **Rule 2:** $a^{-m} = \frac{1}{a^m}$

c $5x^8 \times 6x^4 = (5 \times 6)x^{8+4}$
$= 30x^{12}$

d $10y^3 \times 4y^{-5} = (10 \times 4)\, y^{3-5}$
$= 40y^{-2}$
$= \frac{40}{y^2}$

Example 2.6

Simplify the following. Write your answer in index form with a positive exponent.

a $8^7 \div 8^4$ b $2^3 \div 2^9$ c $5^3 \div 5^{-7}$ d $18a^9 \div 6a^2$

a $8^7 \div 8^4 = \frac{8^7}{8^4}$
$= 8^{(7-4)}$
$= 8^3$

b $2^3 \div 2^9 = 2^{3-9}$
$= 2^{-6}$
$= \frac{1}{2^6}$

Rewrite the answer with a positive exponent using **Rule 2**.

c $5^3 \div 5^{-7} = 5^{3-(-7)}$
$= 5^{10}$

d $18a^9 \div 6a^2 = (18 \div 6)a^{9-2}$
$= 3a^7$

Exercise 2C

★ 1 Simplify these expressions. Write your answer in index form with a positive exponent.

a $4^5 \times 4^3$ b $7^4 \times 7$ c $x^{10} \times x^2$ d $t^2 \times t^3 \times t^4$

e $3^2 \times 3^{-7}$ f $c^3 \times c^{-9}$ g $a^8 \times a^{-8}$ h $4y^3 \times 5y^6$

i $c \times 4c^2 \times 2c^3$ j $8c^2 \times 3c^{-7}$ k $10a^7 \times 3a^{-20}$ l $4t^3 \times 3t^{-8} \times 2t^2$

★ **2** Simplify these expressions leaving your answer in index form.

a $3^7 \div 3^2$ b $6 \div 6^3$ c $x^8 \div x^5$ d $t^3 \div t$

e $p^3 \div p^{-2}$ f $y^{-3} \div y^{-3}$ g $12y^{10} \div 3y^3$ h $24y^3 \div 12y^8$

i $15x^2 \div 3x^{-4}$ j $42p^6 \div (-7p)^{-2}$ k $\frac{4t^5 \times -7t^3}{14t^{-4}}$ l $\frac{5y^2 \times 4y^{-6}}{2y^3}$

3 Simplify these expressions.

a $3x^2y \times 5x^3y^2$ b $3a^2b^3 \times 7ab^4$ c $30x^3y \div 6x^2y^4$

Raising a power to a further power

What happens when you want to square a number that already has an exponent?

Consider $(4^3)^2$ (we say this as **four cubed all squared**).

$(4^3)^2 = (4^3) \times (4^3) = (4 \times 4 \times 4) \times (4 \times 4 \times 4) = 4^6$

So $(4^3)^2 = 4^{3 \times 2} = 4^6$

We can extend this to any power: $(4^5)^3 = 4^{5 \times 3} = 4^{15}$

When **raising** a power to a further power you **multiply** the indices.

Rule 5 $(a^m)^n = a^{m \times n} = a^{mn}$

Example 2.7

Simplify the following.

a $(x^7)^4$ b $(7^6)^{-2}$ c $(y^{-3})^{-4}$ d $(4a)^3$

a $(x^7)^4 = x^{7 \times 4}$
$= x^{28}$

b $(7^6)^{-2} = 7^{6 \times -2}$
$= 7^{-12}$
$= \frac{1}{7^{12}}$

c $(y^{-3})^{-4} = y^{12}$

Remember that $(-3) \times (-4) = 12$.

d $(4a)^3 = 4a \times 4a \times 4a$
$= 4^3a^3$
$= 64a^3$

From Example 2.7 part d, you should see that $(ab)^n = a^nb^n$.

For example:

$(2a)^5 = 2^5 \times a^5 = 32a^5$ and $(2x^4y^3)^2 = 2^2(x^4)^2(y^3)^2 = 4x^8y^6$

Exercise 2D

1 Simplify the following.

a $(3^4)^5$ b $(2^3)^4$ c $(10^5)^3$ d $(t^3)^{-4}$ e $(a^7)^3$

2 Simplify the following.

a $(3y)^2$ b $(x^3y^4)^5$ c $(ab^3)^4$

d $(3p^4q^2)^3$ e $(2t^3u^{-2})^4$ f $(10u^{-5}v^{-2})^3$

★ 3 Simplify the following.

a $(6^4)^3$ b $(2^7)^4$ c $(a^5)^6$

d $(t^{-3})^7$ e $(x^{-2})^{-5}$ f $(6a^3b^4)^2$

g $(2x^{-3}y^5)^4$ h $(3a^6b^{-3})^5$ i $(x^4y^{-2}z^3)^3$

4 State true or false for each of the following, giving your reasons.

a $3^3 \times 3^4 = 3^{12}$ b $2^3 = 3^2$ c $8^9 \div 8^5 = 8^4$

d $(4^2)^3 = (4^3)^2$ e $12^3 \div 6^3 = 2^3$ f $5^5 \times 5^5 = 5^{10}$

g $7^6 \div 7^6 = 7^0$ h $(3a^5)^3 = 27a^8$ i $\dfrac{2x^6 \times 3x^2}{6x^{-2}} = x^6$

Fractional indices

Indices can be fractions as well as integers. The two cases to investigate are those with the numerator = 1, and those with the numerator > 1.

> Remember that $3 = 3^1$. See Chapter 1 for more about multiplying surds.

Consider $3^x \times 3^x = 3$.

We know, using Rule 3, that $3^x \times 3^x = 3^{x+x}$.

So $3^x \times 3^x = 3^{2x}$.

If $3^{2x} = 3^1$, we can equate the index numbers, so $2x = 1$, and so $x = \frac{1}{2}$.

Now we have $3^{\frac{1}{2}} \times 3^{\frac{1}{2}} = 3$. We also know that $\sqrt{3} \times \sqrt{3} = 3$.

So $3^{\frac{1}{2}} = \sqrt{3}$.

We can express this as:

Rule 6 $a^{\frac{1}{n}} = \sqrt[n]{a}$

The table shows how to write indices which are unit fractions.

$a^{\frac{1}{2}}$	$\sqrt{a}$	square root
$a^{\frac{1}{3}}$	$\sqrt[3]{a}$	cube root
$a^{\frac{1}{4}}$	$\sqrt[4]{a}$	fourth root
$a^{\frac{1}{5}}$	$\sqrt[5]{a}$	fifth root

If the numerator of a fractional index is greater than 1, there are two operations involved. For example, in a term such as $8^{\frac{2}{3}}$, the index notation is telling you to square *and* find the cube root of 8.

The two operations can be done in any order, so look for the easiest path to the solution.

To simplify $8^{\frac{2}{3}}$, we could find $\sqrt[3]{8^2}$ which is $\sqrt[3]{64} = 4$.
Or we could find $\left(\sqrt[3]{8}\right)^2$ which is $2^2 = 4$.
You get the same answer either way.
So for fractional indices:

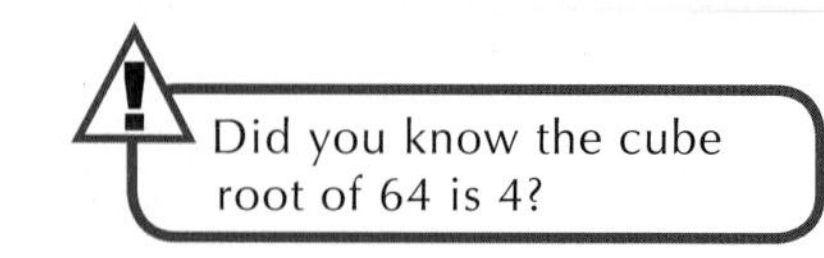

Did you know the cube root of 64 is 4?

Rule 7 $a^{\frac{m}{n}} = \sqrt[n]{a^m} = \left(\sqrt[n]{a}\right)^m$

Example 2.8

Write in index form.

a $\sqrt[4]{a^3}$ b $\sqrt{m^5}$

a $\sqrt[4]{a^3} = a^{\frac{3}{4}}$ b $\sqrt{m^5} = m^{\frac{5}{2}}$

Example 2.9

Evaluate.

a $49^{\frac{1}{2}}$ b $32^{\frac{3}{5}}$ c $32^{-\frac{3}{5}}$

a $49^{\frac{1}{2}} = \sqrt{49} = 7$

b $32^{\frac{3}{5}} = \left(\sqrt[5]{32}\right)^3$
$= 2^3$
$= 8$

c $32^{-\frac{3}{5}} = \dfrac{1}{\left(\sqrt[5]{32}\right)^3}$
$= \dfrac{1}{2^3}$
$= \dfrac{1}{8}$

Remember that a negative index is a reciprocal, not a negative value.

Exercise 2E

1 Use the rules to express the following with root signs of the form $\sqrt[n]{a^m}$.

a $a^{\frac{1}{3}}$ b $a^{\frac{1}{5}}$ c $t^{\frac{1}{2}}$ d $a^{\frac{2}{3}}$ e $a^{\frac{3}{5}}$

f $t^{\frac{5}{2}}$ g $x^{\frac{4}{3}}$ h $y^{\frac{2}{5}}$ i $p^{\frac{1}{4}}$ j $m^{\frac{3}{4}}$

2 Write in index form.

a $\sqrt{t^5}$ b $\sqrt[4]{a^3}$ c $\sqrt[5]{x^3}$ d $\sqrt[7]{m^4}$ e $\sqrt[3]{a^{12}}$

3 Evaluate.

a $9^{\frac{1}{2}}$ b $16^{\frac{1}{4}}$ c $8^{\frac{2}{3}}$ d $49^{\frac{3}{2}}$ e $25^{-\frac{1}{2}}$

f $81^{-\frac{3}{4}}$ g $100^{-\frac{3}{2}}$ h $\left(\frac{1}{27}\right)^{\frac{2}{3}}$ i $\left(\frac{49}{81}\right)^{\frac{1}{2}}$ j $\left(\frac{16}{25}\right)^{\frac{3}{2}}$

4 Simplify by applying the appropriate rules.

a $\left(x^{6}\right)^{\frac{1}{2}}$ b $\left(y^{3}\right)^{\frac{2}{3}}$ c $\left(a^{-12}\right)^{\frac{1}{4}}$ d $\left(t^{\frac{4}{3}}\right)^{0}$ e $3t^{\frac{1}{2}} \times 6t^{-\frac{1}{2}}$

f $5a^{\frac{2}{3}} \times 3a^{\frac{1}{3}}$ g $12x^{\frac{4}{3}} \div 6x^{-\frac{2}{3}}$ h $5y^{\frac{2}{5}} \times \left(-3y^{\frac{7}{5}}\right)$ i $10t^{\frac{3}{2}} \div 5t^{\frac{5}{2}}$

5 Arrange the following from smallest to largest.

5^{0} 20^{-1} $8^{\frac{2}{3}}$ $4^{\frac{1}{2}}$

★ **6** **a** Evaluate.

i $81^{\frac{1}{2}}$ ii $27^{\frac{2}{3}}$ iii $36^{-\frac{1}{2}}$

iv $100^{-\frac{3}{2}}$ v $\left(\frac{25}{49}\right)^{\frac{1}{2}}$ vi $\left(\frac{8}{27}\right)^{-\frac{1}{3}}$

b Simplify and leave your answer in positive index form where necessary.

i $\left(y^{3}\right)^{\frac{1}{3}}$ ii $\left(x^{\frac{1}{2}}\right)^{-5}$ iii $\sqrt[3]{a^{7}}$

iv $\sqrt[4]{y^{3}}$ v $4t^{\frac{1}{2}} \times 3t^{-\frac{3}{2}}$ vi $6t^{\frac{1}{3}} \div 2t^{-\frac{2}{3}}$

GO! **Activity**

Use your understanding of the rules for indices to solve these equations for x.

a $4^{x} = 16$ b $3^{x} = 81$ c $\sqrt[x]{64} = 4$ d $\sqrt[x]{32} = 2$ e $125^{x} = 5$

f $x^{4} = 1$ g $x^{6} = 64$ h $x^{\frac{2}{3}} = 4$ i $16^{\frac{3}{4}} = x$ j $x^{\frac{1}{3}} = 3$

Multiplying out brackets

In an assessment you may be asked to combine your understanding of multiplying out brackets with the rules for indices. You need to be able to use all the rules:

Rule 1 $a^{0} = 1$ **Rule 2** $a^{-m} = \frac{1}{a^{m}}$ **Rule 3** $a^{m} \times a^{n} = a^{m+n}$

Rule 4 $a^{m} \div a^{n} = a^{m-n}$ **Rule 5** $(a^{m})^{n} = a^{(m \times n)} = a^{mn}$ **Rule 6** $a^{\frac{1}{n}} = \sqrt[n]{a}$

Rule 7 $a^{\frac{m}{n}} = \sqrt[n]{a^{m}} = \left(\sqrt[n]{a}\right)^{m}$

See Chapter 3 for more on multiplying out brackets.

Example 2.10

Multiply the bracket and simplify $x^{\frac{1}{2}}\left(2x^{2} + x^{-\frac{3}{2}}\right)$

$$x^{\frac{1}{2}}(2x^{2} + x^{-\frac{3}{2}}) = x^{\frac{1}{2}} \times 2x^{2} + x^{\frac{1}{2}} \times x^{-\frac{3}{2}}$$ Multiply out the bracket.

$$= 2x^{\left(\frac{1}{2}+2\right)} + x^{\left(\frac{1}{2}+\left(-\frac{3}{2}\right)\right)}$$ Use **Rule 3**.

$$= 2x^{\frac{5}{2}} + x^{-1} \text{ or } 2x^{\frac{5}{2}} + \frac{1}{x}$$ Use **Rule 2**.

Exercise 2F

1 Multiply the bracket and simplify where necessary.

a $a^2(a^3 + 1)$ b $x^{-4}(x^2 + x^{-1})$ c $y^3(y^{-2} + y^{-3})$ d $m^3(m^{-5} - 4)$ e $5a^2(2a^{-2} - 7a^3)$

★ 2 Expand.

a $p^{\frac{1}{2}}(p + 3)$ b $t^{\frac{1}{2}}\left(t + t^{-\frac{1}{2}}\right)$ c $3z^{\frac{1}{3}}\left(z^{\frac{5}{3}} + 2\right)$

d $b^{-\frac{1}{4}}\left(b^{\frac{3}{4}} - b^{-\frac{1}{2}}\right)$ e $y^{\frac{4}{5}}\left(y^{-\frac{4}{5}} + 3y^{\frac{1}{5}}\right)$ f $c^{-\frac{1}{2}}\left(c^{\frac{1}{2}} - c^{\frac{1}{3}}\right)$

3 For each of your answers in Questions 1 and 2, use your calculator to find the exact value of each expression if $a = 3$, $b = 16$, $c = 64$, $m = 5$, $p = 4$, $t = 9$, $x = 2$, $y = 10$ and $z = 27$.

4 Expand and simplify.

a $\left(t^2 - 4\right)^2$ b $\left(y^5 + 2\right)\left(y^{-3} - 1\right)$ c $\left(x^{\frac{1}{2}} + 3\right)^2$

d $(m^3 + 1)(m^3 - 1)$ e $\left(c^{\frac{2}{3}} + 3\right)\left(c^{\frac{2}{3}} - 3\right)$ f $\left(5 - m^{\frac{1}{4}}\right)\left(m^{\frac{1}{2}} + 2\right)$

Scientific notation (Standard form)

Using powers and scientific notation, we can write very large or very small numbers quickly and concisely. Scientific notation is used to simplify recording and to avoid errors by accidentally adding or missing out a digit in a long chain of numbers. This is the common format for scientists, engineers, mathematicians and in commerce.

Writing numbers using scientific notation

The Earth has an elliptical orbit around the Sun. When the Earth is at its closest point to the Sun the distance between them is 147 million kilometres – 147 000 000 km. Astronomers call this the **perihelion**. At its furthest point from the Sun, the distance is 152 000 000 kilometres, This is called the **aphelion**.

147 000 000 km can be written as 1.47×10^8 km and 152 000 000 km as 1.52×10^8 km.

The mass of an electron is a very small number. It is about

0.000 000 000 000 000 000 000 000 000 000 910 938 22 kg.

You can see that it is very easy to make a mistake when writing this number. We can write it more clearly using scientific notation as 9.1×10^{-31} kg.

In scientific notation all numbers are written in the form $\boldsymbol{a} \times \mathbf{10}^{\boldsymbol{b}}$ (we say ***a* times 10 to the power of *b*** or ***a* times 10 to the *b***). In this form $1 \leqslant a < 10$ and b is an integer.

Example 2.11

Write the following numbers using scientific notation.

a 34 b 569 c 35 807 d 0.8 e 0.345 f 0.0078

a $34 = 3.4 \times 10$
$= 3.4 \times 10^1$

b $569 = 5.69 \times 100$
$= 5.69 \times 10^2$

c $35\,807 = 3.5807 \times 10\,000$
$= 3.5807 \times 10^4$

d $0.8 = 8 \times 0.1$
$= 8 \times 10^{-1}$

e $0.345 = 3.45 \times 0.1$
$= 3.45 \times 10^{-1}$

f $0.0078 = 7.8 \times 0.001$
$= 7.8 \times 10^{-3}$

Example 2.12

Write the following numbers in full.

a 2.8×10^1	b 1.342×10^4	c 9.11×10^2
d 3.6×10^{-1}	e 1.78×10^{-2}	f 5.54×10^{-3}
a 28	b 13 420	c 911
d 0.36	e 0.0178	f 0.005 54

Exercise 2G

★ 1 Write in scientific notation.

a 2340 000	b 1070	c 35 000 000	d 27
e $3\frac{1}{2}$million	f 712 000 000 000	g 0.000 56	h 0.0312
i 0.000 000 408	j 0.78	k 0.006 04	l 0.000 005 100

★ 2 Write the following in full.

a 5×10^6	b 6.32×10^4	c 7.01×10^8
d 4.7×10^{-5}	e 8.04×10^{-7}	f 8.89×10^{-8}

Calculations using scientific notation

Calculations with very large or very small numbers can be completed more easily if you use scientific notation. To further simplify your calculations you should also be able to enter numbers in scientific notation onto your calculator. It's a useful idea to spend some time learning how to use scientific notation on your calculator. In particular, you must know how to enter numbers as positive and negative powers. Use Example 2.13 and Exercise 2H to practise.

Example 2.13

Proxima Centauri is the closest star to our Solar System. It is approximately 40 000 000 000 000 km away. If the human race could build a star ship that could travel at the speed of light, how long would it take the ship to get there? (The speed of light is approximately 300 000 000 ms^{-1}.)

The distance to travel is 40 000 000 000 000 km or 4×10^{13} km.

The speed is 300 000 000 ms^{-1} or 300 000 $kms^{-1} = 3 \times 10^5\ kms^{-1}$.

Time taken is Distance ÷ Speed.

Time $= (4 \times 10^{13}) \div (3 \times 10^5) = (4 \div 3) \times (10^{13} \div 10^5)$ — **Rule 4** – subtract the powers.

$= 1.\dot{3} \times 10^8$ seconds — You can also use your calculator but be careful if and when you round.

We can convert our answer to minutes, hours, days or even years to help us understand the answer.

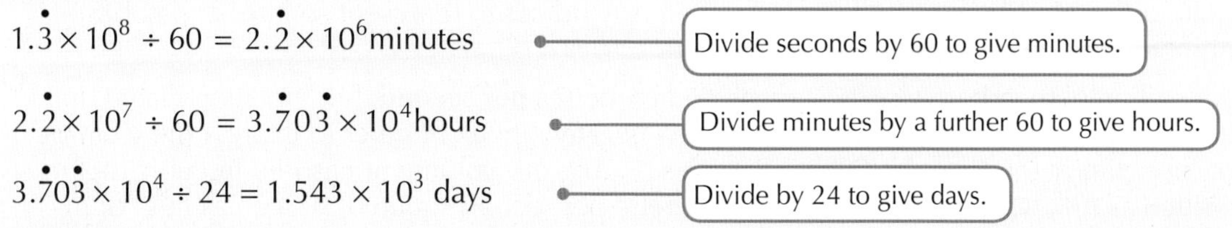
$1.\dot{3} \times 10^8 \div 60 = 2.\dot{2} \times 10^6$minutes
Divide seconds by 60 to give minutes.
$2.\dot{2} \times 10^7 \div 60 = 3.\dot{7}0\dot{3} \times 10^4$hours
Divide minutes by a further 60 to give hours.
$3.\dot{7}0\dot{3} \times 10^4 \div 24 = 1.543 \times 10^3$ days
Divide by 24 to give days.

(continued)

$1.543 \times 10^3 \div 365.25 = 4.23$ years (rounded to 3 s.f.)

Divide by 365.25 to give years. (Don't forget leap years.)

So a star ship that can travel at the speed of light would take a bit more than 4 years to travel to the nearest star outside our Solar System.

See chapter 11 for information about significant figures (s.f.).

Exercise 2H

1 Calculate and express your answer in scientific notation.

a $(4.2 \times 10^7) \times (2 \times 10^5)$ b $(6.34 \times 10^8) \times (3 \times 10^{-3})$ c $(8.4 \times 10^7) \div (4 \times 10^5)$

d $(4.2 \times 10^7) \div (6 \times 10^{-2})$ e $\dfrac{(8.4 \times 10^4) \times (4 \times 10^7)}{2 \times 10^5}$ f $\dfrac{(9.4 \times 10^5) \times (4 \times 10^{-3})}{8 \times 10^{-6}}$

For Questions 2–9 express your final answer in scientific notation.

2 The speed of light is approximately 3×10^8 ms^{-1}. How many metres does light travel in one hour?

3 The mass of one atom of oxygen is 2.7×10^{-23} grams. What is the mass of 5×10^{30} atoms of oxygen?

4 The distance the Earth travels in one orbit is 558 000 000 miles. How far will the Earth travel in 75 orbits?

5 The average distance between the Earth and Mars is 225 million kilometres. Light travels at a speed of 3×10^5 kilometres per second. How long does it take for light to travel from Earth to Mars?

★ 6 A light year is the distance a ray of light travels in one year. This distance is 5.88×10^{12} miles.

a How far does a ray of light travel in a millennium?

b How far does a ray of light travel in a minute? (Round to 3 significant figures.)

7 The volume of the supergiant star Betelgeuse is 2.75×10^{35} m^3 and the volume of the Earth is 1×10^{21} m^3. How many Earths would fit into Betelgeuse?

8 A pixel on a computer screen is 2×10^{-2} cm long and 7×10^{-3} cm wide. What is the area of a pixel?

9 An adult has approximately 2×10^{13} red blood cells. The red blood cell has an approximate mass of 10^{-10} g. Find the total mass of red blood cells in an adult.

Using indices in units of measurement

The International System of Units (abbreviated as **SI**) is the metric system of measurement. It is the system most used in science and increasingly in all forms of commerce. There are many quantities that are a combination of units and the way we *say* and *write* these units can be quite different. When we write units we can use indices to represent squares and cubes of numbers and negative indices to show 'per'.

For example, **speed** or velocity is measured in **metres per second**, but the abbreviated unit of this is **m/s** or **ms^{-1}**. Notice that we can use a negative index number to express the reciprocal or 'per' part of the formula. Other examples of units are volume (measured in cubic metres, written as m^3), acceleration (measured in metre per second squared, written as m/s^2 or ms^{-2}), and the joule (the unit for energy, work or amount of heat), written as m^2.kg/s^2 or m^2.kg.s^{-2}.

The SI system has seven base units as shown in the table.

Base quantity	Name	Symbol
length	metre	m
mass	kilogram	kg
time	second	s
electric current	ampere	A
thermodynamic temperature	kelvin	K
amount of substance	mole	mol
luminous intensity	candela	cd

Exercise 2I

1 The following is a list of SI Derived Quantities. These are made up of the base units shown in the table above. For each quantity, change the units so that they are expressed with positive index numbers (the first one has been done for you).

Derived quantity	Name	Units
a speed, velocity	metre per second	$ms^{-1} = m/s$
b acceleration	metre per second squared	ms^{-2}
c wave number	reciprocal metre	m^{-1}
d frequency	hertz	s^{-1}
e force	newton	$m.kg.s^{-2}$
f pressure	pascal	$m^{-1}.kg.s^{-2}$
g energy	joule	$m^2.kg.s^{-2}$
h power	watt	$m^2.kg.s^{-3}$
i electric potential difference	volt	$m^2.kg.s^{-3}.A^{-1}$
j electric resistance	ohm	$m^2.kg.s^{-3}.A^{-2}$
k magnetic flux density	tesla	$kg.s^{-2}.A^{-1}$
l inductance	henry	$m^2.kg.s^{-2}.A^{-2}$

2 For the following derived quantities rewrite the units so that they appear only as a numerator (the first one has been done for you).

a mass density: $kg/m^3 = kg.m^{-3}$

b specific volume: m^3/kg

c magnetic field strength: A/m

d absorbed radioactive dose: m^2/s^2

e catalytic activity: mol/s

f magnetic flux: $m^2.kg/s^2A$

g electric conductance: $s^3A^2/m^2.kg$

h capacitance: $s^4.A^2/m^2.kg$

3 Try to find out the names of the special units for the derived quantities in Question 2 parts **d** to **h**.

Activity

Consider the planets in our solar system: Mercury, Venus, Earth, Mars, Jupiter, Saturn, Uranus, Neptune.

- Investigate the following topics, using scientific notation where appropriate:
  - » the distance of each planet from the Sun
  - » the mass of each planet
  - » the circumference of each planet
  - » the surface area of each planet.
- Create a presentation to present your findings to a group of friends.

- I know that any term (except zero) raised to the power of zero is equal to 1. ★ Exercise 2A Q1
- I can enter index numbers onto a calculator. ★ Exercise 2A Q2
- I can read, evaluate and write numbers using index notation. ★ Exercise 2A Q1 ★ Exercise 2B Q2
- I can simplify expressions that are multiplied together by adding the indices. ★ Exercise 2C Q1
- I can simplify expressions that are divided by subtracting the indices. ★ Exercise 2C Q2
- I can simplify expressions where there are powers of powers. ★ Exercise 2D Q3
- I can simplify expressions when there are fractional indices. ★ Exercise 2E Q6
- I can multiply out brackets that have terms with powers. ★ Exercise 2F Q2
- I can convert numbers to scientific notation and back. ★ Exercise 2G Q1, Q2
- I can solve problems for very large or small numbers using scientific notation. ★ Exercise 2H Q6

For further assessment opportunities, see the Preparation for Assessment for Unit 1 on pages 89–92.

3 Working with algebraic expressions involving expansion of brackets

In this chapter you will learn how to:

- use the **distributive law** to expand brackets
- multiply brackets of the form $ax(bx + c)$
- **multiply** pairs of brackets
- **expand** a binomial bracket and a trinomial bracket
- **solve problems** by multiplying brackets.

You should already know:

- how to collect like terms
- how to add and subtract algebraic expressions
- how to multiply and divide algebraic expressions
- how to write expressions using symbols.

Collecting like terms

Algebraic expressions involve the use of **variables** to represent quantities and properties. Algebraic expressions will often include the same variables in different parts of the expressions. In order to simplify the expressions to make them easier to calculate, it is necessary to:

- **identify** the same variables, or terms, in different parts of the expression
- **collect** like terms together.

Like terms contain the same variable. If a letter is used for the variable, the like terms are all those which are of the same letter and power.

- $3a$, a, $7a$, are like terms – they all have the variable a. The **coefficients** 3, 1 and 7 are different.

Remember $a = 1a$.

- $7t^2$ and $10t^2$ are like terms – they both have the same power of the variable, t^2. The coefficients 7 and 10 are different.
- $7t^2$ and $7t$ are not like terms – they have different powers of the variable, t and t^2.
- $5c$ and $2f$ are not like terms – they have different variables, c and f.

We can collect like terms by **adding** and **subtracting** like terms. This allows us to combine and simplify expressions.

Example 3.1

Collect the like terms in the expression $5a + 2b + 3a - 6b$.

$5a + 2b + 3a - 6b = 5a + 3a + 2b - 6b$

Collect terms in a and b and rearrange.

$= 8a - 4b$

Exercise 3A

1 Simplify the expressions by collecting like terms.

a $3t + 4t$ b $6y + 2y - 10y$ c $5x^2 - 3x^2$

d $4ab + 7ab - ab$ e $8x + 9y - 3x + 7y$ f $4a - 6a^2 + 7a^2 + 2a$

★ 2 Simplify the following where possible.

a $8p + 2p - 5p$ b $9t - 4r$ c $6c + 7c - 8$

d $-3pq + 7qp$ e $4rt + 3rt - 7rt$ f $5a^2 - 6a - 8a^2$

g $-2 + 3a - 4a^2$ h $4m - 7n + 3m + 2n$ i $8rs + 7s - 12sr - 2s$

Using the distributive law to expand brackets

Algebraic expressions may be presented as combinations of expressions within brackets. **Expanding brackets** is the process of multiplying terms inside the brackets. To do this we multiply every term inside the bracket by the term outside the bracket to get two or more terms and finish with an equivalent expression without any brackets.

Every number inside the bracket is multiplied by the number outside the bracket.

The **distributive law** is used when multiplying a bracket.

In general:

$a(b + c) = ab + ac$

Geometrically, this can be shown in a diagram.

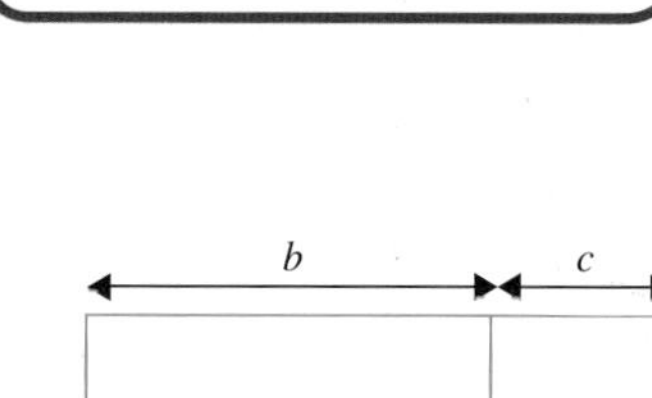

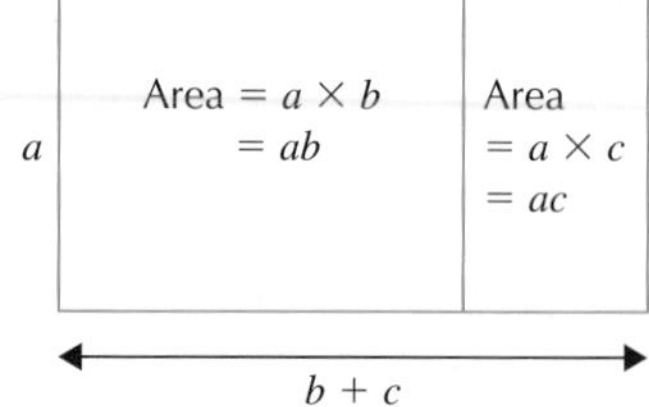

The area of the two small rectangles is $ab + ac$.

The area of the large rectangle is $a(b + c) = ab + ac$.

When expanding brackets, always be systematic. Multiply the first term in the bracket by the term outside the bracket, then the second term inside the bracket by the term outside the bracket, and so on.

Normally the number is written first and if there are 2 or more letters they are written alphabetically.

Remember: $y \times 5 = 5y$

$a \times 3 \times b = 3ab$

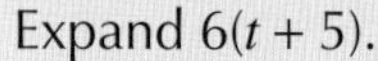

Example 3.2

Expand $6(t + 5)$.

$6(t + 5) = (6 \times t) + (6 \times 5)$

$= 6t + 30$

Multiply each term inside the bracket by the number outside.

Or, using a diagram, find the area of a rectangle with sides $(t + 5)$ cm and 6 cm, express your answer in two ways.

t cm | 5 cm | 6 cm | Area $= 6 \times t = 6t$ cm^2 | Area $= 6 \times 5 = 30$ cm^2 | $(t + 5)$ cm

Area of the complete rectangle
= area of the left-hand rectangle
+ the area of the right-hand rectangle.

Total area $= 6t + 30$

So: $6(t + 5) = 6t + 30$

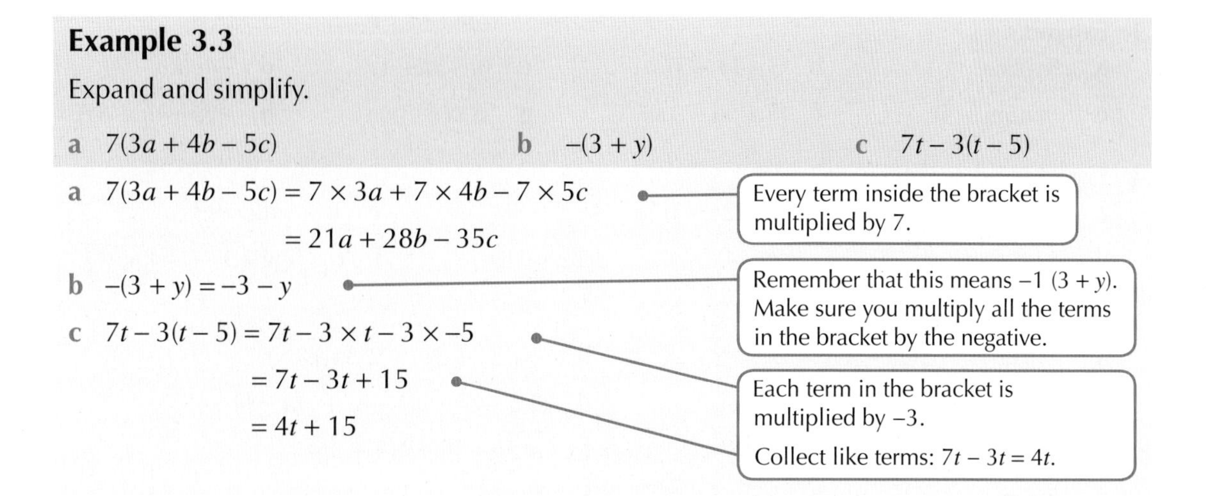

Example 3.3

Expand and simplify.

a $7(3a + 4b - 5c)$ b $-(3 + y)$ c $7t - 3(t - 5)$

a $7(3a + 4b - 5c) = 7 \times 3a + 7 \times 4b - 7 \times 5c$
$= 21a + 28b - 35c$

Every term inside the bracket is multiplied by 7.

b $-(3 + y) = -3 - y$

Remember that this means $-1\ (3 + y)$. Make sure you multiply all the terms in the bracket by the negative.

c $7t - 3(t - 5) = 7t - 3 \times t - 3 \times -5$
$= 7t - 3t + 15$
$= 4t + 15$

Each term in the bracket is multiplied by -3.

Collect like terms: $7t - 3t = 4t$.

Example 3.4

Expand and simplify.

a $7(3x - 2) - 5(4x - 3)$ b $6(4y + 1) - (3y + 11)$

a $7(3x - 2) - 5(4x - 3) = 21x - 14 - 20x + 15$
$= 21x - 20x - 14 + 15$
$= x + 1$

b $6(4y + 1) - (3y + 11) = 24y + 6 - 3y - 11$
$= 24y - 3y + 6 - 11$
$= 21y - 5$

Example 3.5

Expand.

a $x(x + 4)$ b $5x(2t - 3u)$ c $7y(4y + 3t)$ d $5t(2t - 1) + 3t(-4t + 1)$

a $x(x + 4) = x^2 + 4x$

b $5x(2t - 3u)$
$= 10tx - 15ux$

$xt = tx$ but it is conventional to write the answer alphabetically.

c $7y(4y + 3t) = 28y^2 + 21ty$

d $5t(2t - 1) + 3t(-4t + 1) = 10t^2 - 5t - 12t^2 + 3t$
$= -2t^2 - 2t$

Exercise 3B

1 Expand.

a $2(t + 4)$ b $5(m - 3)$ c $-6(2a + 1)$ d $-10(11 - 9y)$

e $8(2t + 3y + 1)$ f $-5(-4m + 2n - 7r)$ g $a(4 + c)$ h $2a(8 - c)$

i $5x(3y - 4)$ j $y(y - 4)$ k $-b(b - c)$ l $a(b - c + a)$

★ 2 Expand and simplify.

a $3(2x + 7) - 12$ b $10(3y - 7) + 8y$ c $6 + 3(2 + y)$ d $12 + 4(2t - 3)$

e $8p - 5(4 - p)$ f $7 + 6(-3 + 2y)$ g $7 - (2p + 3)$ h $2t - (9 + 2t)$

★ **3** Expand.

a $5x(2x + 3)$ b $3y(4y - 5)$ c $6t(-5t + 1)$ d $-4c(2c - 7)$

e $9m(5m + 4)$ f $8w(2m - 3w)$ g $x(-x + 7y)$ h $-9s(-4u + 3s)$

4 Expand and simplify.

a $5x^2 + 3x(x + 2)$ b $2y - y(5y - 4)$ c $8 - 5x(2x + 3)$

d $11t^2 - t(t + 3)$ e $4x(x + 7) + 3(2x - 1)$ f $6w(2w + 1) - 4w(w + 1)$

5 Expand and simplify.

a $2(5x - 4) - 3(2x - 1)$ b $10(2x + 3y) + 2(3x + 5y)$

c $-7(2t - 3w) - 11(t - 1)$ d $a(b - c) + b(c - a) + c(a - b)$

e $x(3 + y) + y(4 + x)$ f $a(2b + 7) + b(5 - 3a)$

g $5t(2m + 3) - 3m(2t - 7)$ h $6p(3n - 1) + 2n(3p + 5)$

6 Write down an expression for the area of each of the following parallelograms and simplify where necessary.

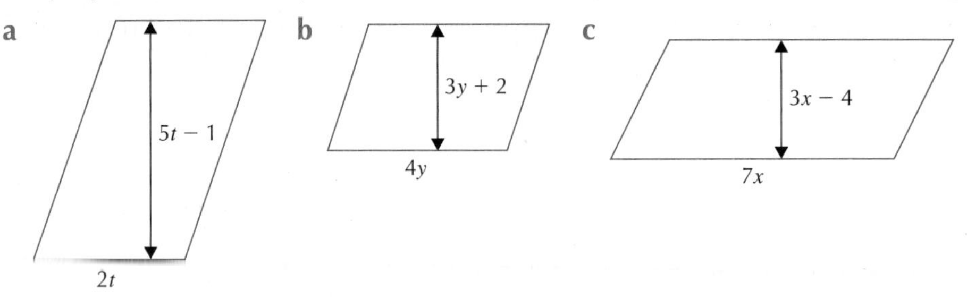

7 Write down an expression for the area of each of the following triangles and simplify where necessary.

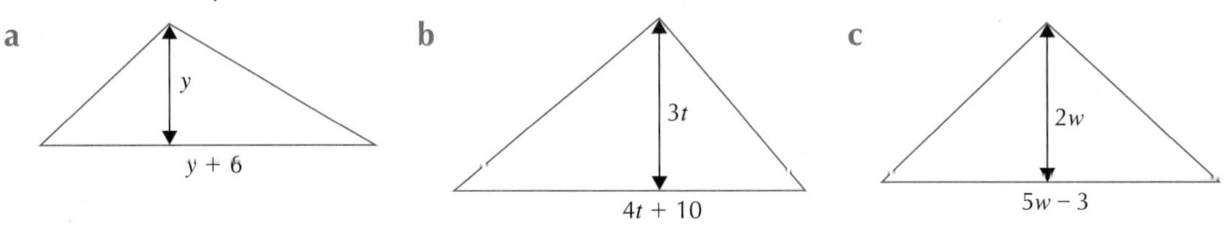

8 The dimensions of the base of a house are:

breadth = $(4x + 5)$ metres and length = $6x$ metres.

A conservatory is going to be erected along the breadth of the house and the length of the house will increase by 3 metres. What is the new area of the base of the house?

9

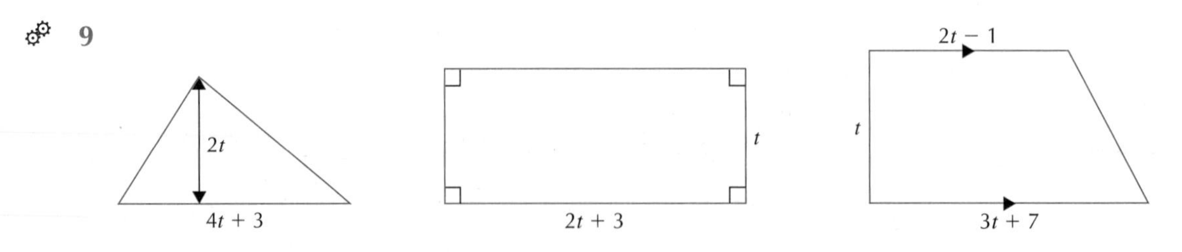

a Which shape has the largest area?

b Which shape has the smallest area?

Multiplying pairs of brackets

So far we have looked at multiplying the terms inside one bracket by a single term outside the bracket. Now we will multiply pairs of brackets. To do this, we multiply every term in the second bracket by every term in the first bracket. There are different methods, shown in the following examples.

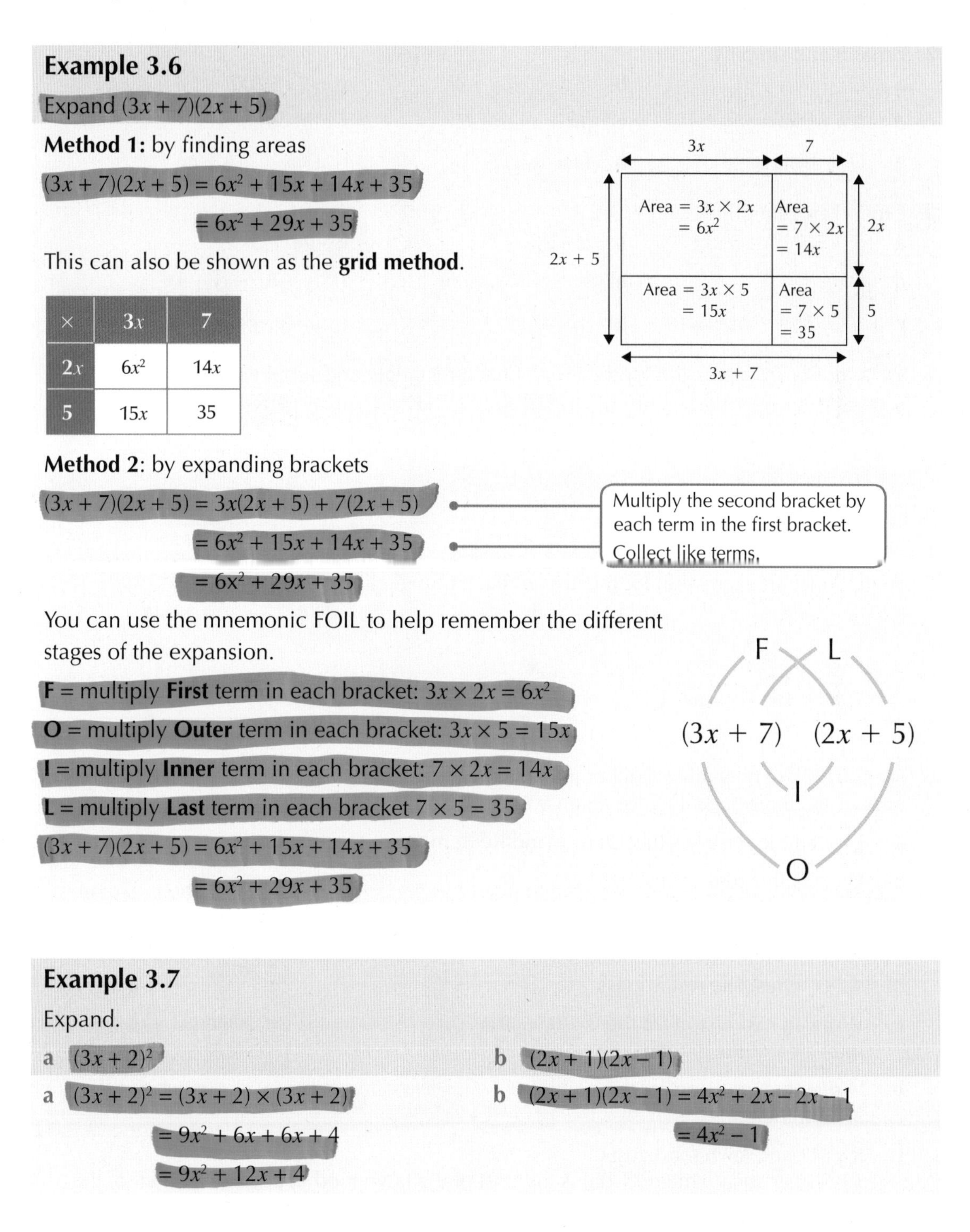

Example 3.6

Expand $(3x + 7)(2x + 5)$

Method 1: by finding areas

$(3x + 7)(2x + 5) = 6x^2 + 15x + 14x + 35$

$= 6x^2 + 29x + 35$

This can also be shown as the **grid method**.

×	$3x$	7
$2x$	$6x^2$	$14x$
5	$15x$	35

Method 2: by expanding brackets

$(3x + 7)(2x + 5) = 3x(2x + 5) + 7(2x + 5)$ — Multiply the second bracket by each term in the first bracket.

$= 6x^2 + 15x + 14x + 35$ — Collect like terms.

$= 6x^2 + 29x + 35$

You can use the mnemonic FOIL to help remember the different stages of the expansion.

F = multiply **First** term in each bracket: $3x \times 2x = 6x^2$

O = multiply **Outer** term in each bracket: $3x \times 5 = 15x$

I = multiply **Inner** term in each bracket: $7 \times 2x = 14x$

L = multiply **Last** term in each bracket $7 \times 5 = 35$

$(3x + 7)(2x + 5) = 6x^2 + 15x + 14x + 35$

$= 6x^2 + 29x + 35$

Example 3.7

Expand.

a $(3x + 2)^2$

b $(2x + 1)(2x - 1)$

a $(3x + 2)^2 = (3x + 2) \times (3x + 2)$

$= 9x^2 + 6x + 6x + 4$

$= 9x^2 + 12x + 4$

b $(2x + 1)(2x - 1) = 4x^2 + 2x - 2x - 1$

$= 4x^2 - 1$

Exercise 3C

1 Expand and simplify.

a $(x + 3)(x + 2)$ b $(y + 7)(y + 4)$ c $(t + 4)(t + 8)$

d $(a - 7)(a - 3)$ e $(w - 2)(w - 9)$ f $(z - 10)(z - 8)$

g $(r - 3)(r + 10)$ h $(t + 11)(t - 4)$ i $(a - 9)(a + 7)$

★ **2** Expand and simplify.

a $(2x + 1)(x - 3)$ b $(5y + 7)(3y - 4)$ c $(8u - 3)(u + 6)$

d $(7a - 2)(a + 5)$ e $(6t - 5)(3t - 2)$ f $(b - 8)(8b - 3)$

g $(7 + 3w)(2w - 5)$ h $(4 + 3s)(6 - 7s)$ i $(3m - 2)(4 + m)$

3 Expand and simplify.

a $(x + 5)^2$ b $(t - 2)^2$ c $(4a - 3)^2$

d $(3x + 1)^2$ e $(x + 2)^2 + (x + 6)^2$ f $(y + 8)^2 - (y - 3)^2$

g $(x^2 + 7)^2$ h $(t^2 - 4)^2$ i $(a^2 + 9)^2 + (7 - a)^2$

4 Write down an expression for the area of each rectangle and simplify.

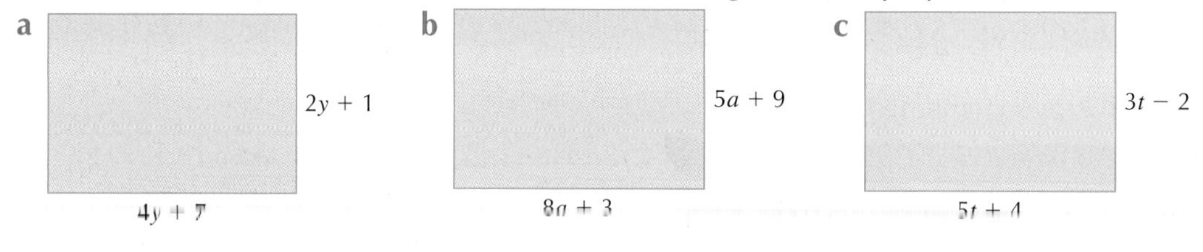

5 Write down an expression for the area of each triangle and simplify.

a
3y − 1
4y + 7

b
2t − 1
6t − 3

c
3y + 2
5y + 3

6 Mr Gray the mathematics teacher asks his class to expand $(5x + 1)^2$. Denise gave an answer of $25x^2 + 1$. Keith gave an answer of $5x^2 + 10x + 1$.

a Explain the mistakes that Denise and Keith made.

b What is the correct answer?

★ **7** A garden measuring $6y + 5$ metres by $2y - 3$ metres has two square flower beds of side y metres. The rest is grass.

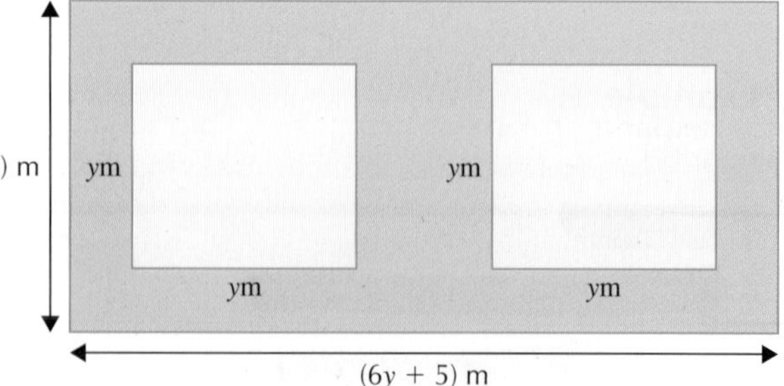

a Write an expression for the total area of the garden.

b Write an expression for the area of the grass.

c If $y = 2$ and lawn food costs £1.50 per square metre, what is the cost of the lawn food needed to feed all of the grass?

d If lawn food costs £x per square metre, what is the cost of the lawn food?

Multiplying a binomial expression by a trinomial expression

Binomial expressions contain two terms inside the brackets. **Trinomial expressions** contain three terms inside the brackets. The rules for multiplying binomials and trinomials are just the same, so we multiply each term in the second bracket by each term in the first bracket. It is important to be systematic in your working, because it's easy to miss a calculation if you try to do too much in your head.

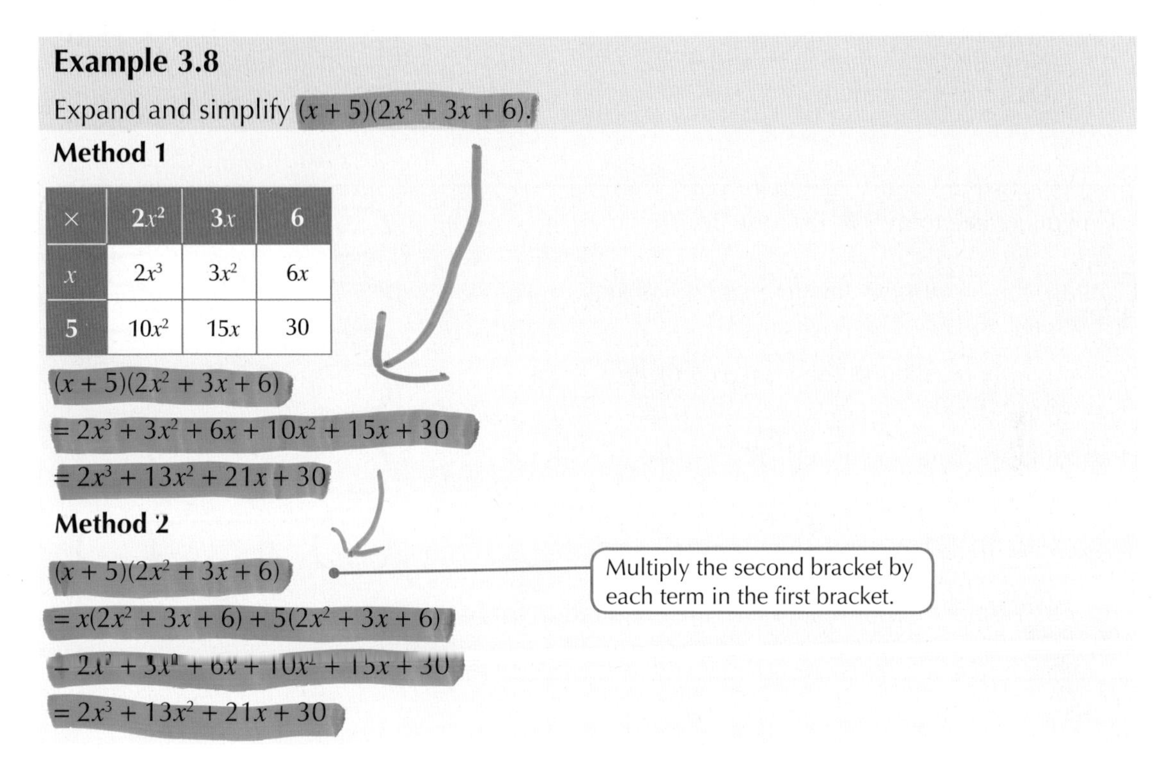

Example 3.8

Expand and simplify $(x + 5)(2x^2 + 3x + 6)$.

Method 1

×	$2x^2$	$3x$	6
x	$2x^3$	$3x^2$	$6x$
5	$10x^2$	$15x$	30

$(x + 5)(2x^2 + 3x + 6)$

$= 2x^3 + 3x^2 + 6x + 10x^2 + 15x + 30$

$= 2x^3 + 13x^2 + 21x + 30$

Method 2

$(x + 5)(2x^2 + 3x + 6)$ — Multiply the second bracket by each term in the first bracket.

$= x(2x^2 + 3x + 6) + 5(2x^2 + 3x + 6)$

$= 2x^3 + 3x^2 + 6x + 10x^2 + 15x + 30$

$= 2x^3 + 13x^2 + 21x + 30$

Exercise 3D

1 Expand and simplify.

a $(x + 1)(3x^2 + 2x + 7)$	**b** $(y + 4)(2y^2 - 5y + 2)$	**c** $(x + 3)(5x^2 - x - 1)$
d $(t - 2)(3t^2 + 6t - 1)$	**e** $(w - 5)(w^2 - 4w - 2)$	**f** $(5 + a)(4a^2 - 2a + 5)$

2 Expand and simplify.

a $(3x - 2)(4x^2 + 3x + 1)$	**b** $(6y + 1)(2y^2 - 3y - 2)$	**c** $(7a + 4)(2a^2 - 5a + 3)$
d $(4w - 5)(w^2 - 3w + 4)$	**e** $(8b - 7)(2b^2 + 7b + 9)$	**f** $(6x^2 + 2x - 3)(5x - 1)$

★ **3** Expand and simplify.

a $(x + 7)(2x^2 + 9x + 5)$	**b** $(a - 2)(3a^2 - 7a + 4)$	**c** $(6 - a)(5a^2 + 6a - 1)$
d $(9u + 5)(3u^2 - 8u + 7)$	**e** $(b + 5)(6b^2 - 2b + 5)$	**f** $(4w^2 - 5w + 3)(8w - 1)$

Activity

Investigate how expanding brackets can help when squaring numbers. For example, consider the square of 1.5.

$$
\begin{aligned}
1.5^2 &= (1 + 0.5)^2 \\
&= (1 + 0.5)(1 + 0.5) \\
&= 1 + 0.5 + 0.5 + 0.25 \\
&= 2.25
\end{aligned}
$$

Now find 2.5^2, then 3.5^2, and 4.5^2.
Describe the pattern.

- I can collect like terms. ★ Exercise 3A Q2
- I can use the distributive law to expand brackets. ★ Exercise 3B Q2
- I can multiply brackets of the form $ax(bx + c)$. ★ Exercise 3B Q3
- I can multiply pairs of brackets. ★ Exercise 3C Q2, Q7
- I can expand a binomial bracket and a trinomial bracket. ★ Exercise 3D Q3

For further assessment opportunities, see the Preparation for Assessment for Unit 1 on pages 89–92.

4 Factorising an algebraic expression

In this chapter you will learn how to:

- factorise by finding a **common factor**
- factorise a **difference of two squares** expression
- factorise by finding a common factor followed by difference of two squares
- factorise **trinomials** with a unitary x^2 coefficient
- factorise trinomials with non-unitary x^2 coefficient.

You should already know:

- how to find the highest common factor (HCF)
- how to add, subtract, multiply and divide integers
- how to multiply and divide term in index notation.

Factorise by finding a common factor

Factorisation is the process used to write an expression to include brackets. It is therefore the reverse process of **expanding brackets**. Factorising is an essential skill as it simplifies expressions and makes them easier to evaluate. It is also useful when solving quadratic equations.

Use the distributive law to expand:

$$a(b + c) = ab + ac$$

See Chapter 3 for a reminder about expanding.

So when we factorise $ab + ac$, we divide both terms ab and ac by the common factor a to give get $a(b + c)$. So, factorising $ab + ac$ gives $a(b + c)$.

Always try to find the **highest common factor** (HCF) when you factorise. In complex expressions, there may be more than one common factor, so be sure to check that there are no further common factors left within the bracket after one factorisation. If there is, your factorisation is not yet complete. A common factor can be a number, a letter, or a number and a letter.

Example 4.1

Factorise.

a $9m + 15n$ b $4tx + 9xy$ c $24ab + 6bc$ d $49x^2 - 7xy$ e $4ab + 16ac - 10ad$

a $9m + 15n = 3(3m + 5n)$ — The HCF = 3 so divide each term in the original expression by 3.

b $4tx + 9xy = x(4t + 9y)$ — The HCF = x so divide each term in the original expression by x.

c $24ab + 6bc = 6b(4a + c)$ — The HCF = $6b$ so divide each term in the original expression by $6b$. This could be done in two stages, first diving each term by 6 and then dividing again by b.

d $49x^2 - 7xy = 7x(7x - y)$ — The HCF = $7x$.

e $4ab + 16ac - 10ad$
$= 2a(2b + 8c - 5d)$ — The HCF = $2a$.

Check your final answers by expanding the brackets. If you have factorised correctly, you should get back to the original expression.

Exercise 4A

1 Factorise the following expressions.

a $3b + 3c$ b $2a + 10b$ c $4x + 14y$ d $at + ar$
e $12x - 8y$ f $ab - bc$ g $cy^2 - cy$ h $24ab - 12bc$
i $14y - 35z$ j $4t^2 - 6at$ k $4p - 5pr$ l $20b - 20b^2$

2 Factorise the following where possible.

a $pq - qr$ b $5xt - 10ay$ c $2\pi r^2 - 6\pi rh$ d $8a^2b - 20ab^2$
e $4m^2n - 5t$ f $12t^2 - 6u$ g $3t^2 - 5ty + 4t$ h $24xy - 16xz$
i $ab + bc - bd$ j $m^4 + m^3 + m^2$ k $r(p + q) + s(p + q)$ l $6qp - 8rs$

3 By using factorisation carry out the following calculations.

a $(53 \times 48) + (53 \times 52)$ b $(74 \times 63) - (74 \times 53)$
c $(2.7 \times 8.6) + (1.4 \times 2.7)$ d $(3.9 \times 6.75) + (3.9 \times 3.25)$
e $(63 \times 24) + (39 \times 63) + (63 \times 37)$ f $(0.17 \times 7.9) + (2.8 \times 0.17) - (0.7 \times 0.17)$

★ 4 Factorise the following where possible.

a $9t + 12r$ b $ay - by$ c $24 - 6t$ d $p^2 - p$
e $8yz - 9ut$ f $abc + bcd$ g $35m - 14n$ h $25t^2 - 15tx + 20tz$
i $5x^2 - 4y$ j $15x - 9y + 6z$ k $6rt + 3ts - 12ty$ l $t^6 + t^4 - t^3$

Factorise an expression with a difference of two squares

A difference of two squares is a squared term minus another squared term, such as:

$x^2 - y^2$ $49 - 36 = 7^2 - 6^2$ $9a^2 - 25 = (3a)^2 - 5^2$

A difference of two squares requires a special type of factorisation. Consider what happens when you multiply $(x + y)$ by $(x - y)$:

$$(x + y)(x - y) = x^2 + xy - xy - y^2$$
$$= x^2 - y^2$$

So we know that factorising $x^2 - y^2$ will give $(x - y)(x + y)$ or $(x + y)(x - y)$.

One bracket has a positive y term and the other has a negative y term, so when each of these is multiplied by x, the resultant terms add to give 0 $(-xy + xy = 0)$.

When factorising a difference of two squares, you should find a solution in which the pairs of brackets contain the same two terms but have opposite signs between the terms ('+' and '−').

The method of the difference of two squares can be a useful tool for multiplication calculations. Consider for example 38×42. This is the same as $(40 - 2) \times (40 + 2)$. In the terminology we have been using, this would be written as $(40 - 2)(40 + 2)$.

From the expansion above, we know that:

$$(40 - 2)(40 + 2) = 40^2 - 2^2$$
$$= 1600 - 4$$
$$= 1596$$

Simple examples for factorising the difference of two squares

Example 4.2

Factorise:

a $t^2 - 25$ b $4y^2 - 9z^2$

a $t^2 - 25 = (t - 5)(t + 5)$

> You should spot that $25 = 5^2$, so this can be rewritten as $t^2 - 5^2$.

b $4y^2 - 9z^2 = (2y - 3z)(2y + 3z)$

> You should spot that $4y^2 = (2y)^2$ and that $9z^2 = (3z)^2$ so this can be rewritten as $(2y)^2 - (3z)^2$.

Exercise 4B

1 Factorise the following.

a $x^2 - 7^2$ b $a^2 - b^2$ c $36 - a^2$ d $1 - t^2$ e $25a^2 - b^2$

f $9t^2 - 25s^2$ g $64c^2 - 49d^2$ h $36a^2 - 1$ i $c^2 - 16d^2$ j $100x^2 - 49y^2$

2 Factorise completely.

a $t^4 - 1$ b $16 - a^4$ c $256 - p^4$

d $t^4 - 81$ e $(m - n)^2 - (m + n)^2$ f $x^2 - (y + z)^2$

★ 3 Factorise the following.

a $t^2 - 9^2$ b $s^2 - t^2$ c $81 - p^2$ d $m^2 - 1$ e $a^2 - 4b^2$

f $25p^2 - q^2$ g $25a^2 - 36b^2$ h $100 - 49^2$ i $9x^2 - 25y^2$ j $a^2 - 4b^2c^2$

★ 4 Find an expression in factorised form for the shaded area.

a The outer square has length t metres and the inner square has length 1 metre.

b A button has radius R mm. Each small circle has radius r mm.

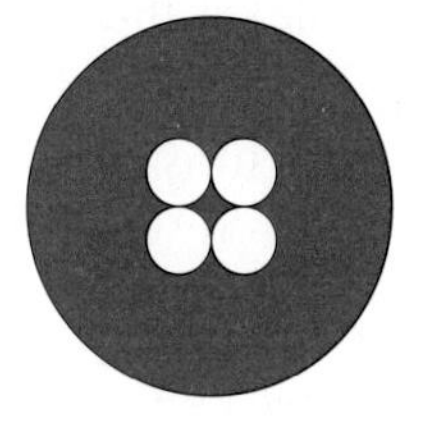

5 Find the length of side x, leaving your answer in factorised form.

a

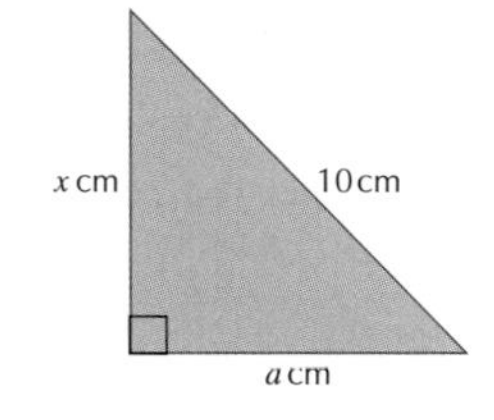

b

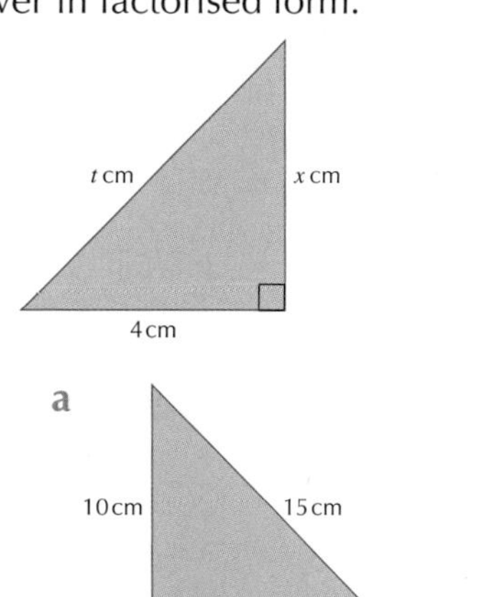

★ 6 Find the **exact** length of side x, using factorising. Write your answer as a simplified surd.

a

b

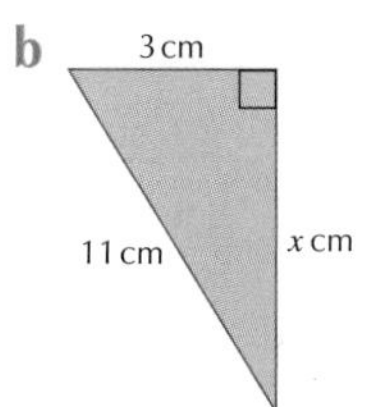

Factorisation by finding the HCF and then factorising a difference of two squares

When factorising look for the highest common factor (HCF) first. After removing the HCF you should always check to see if there is a difference of two squares in the brackets. If there is, factorise the difference of two squares. The HCF remains in front of the first bracket.

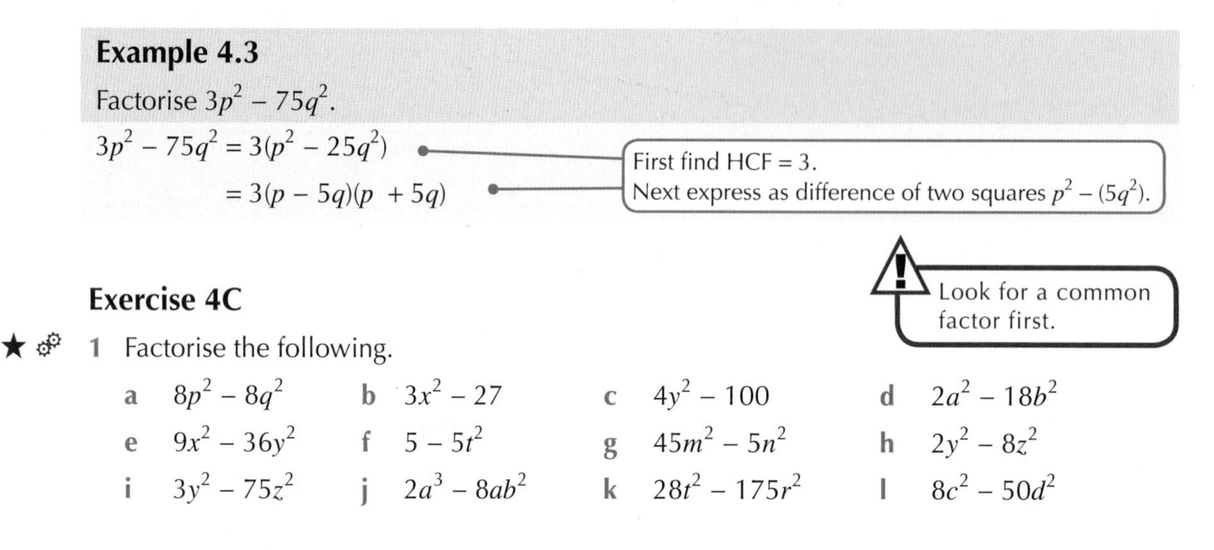

Example 4.3

Factorise $3p^2 - 75q^2$.

$3p^2 - 75q^2 = 3(p^2 - 25q^2)$

$= 3(p - 5q)(p + 5q)$

First find HCF = 3.
Next express as difference of two squares $p^2 - (5q^2)$.

Exercise 4C

Look for a common factor first.

★ 1 Factorise the following.

a	$8p^2 - 8q^2$	b	$3x^2 - 27$	c	$4y^2 - 100$	d	$2a^2 - 18b^2$
e	$9x^2 - 36y^2$	f	$5 - 5t^2$	g	$45m^2 - 5n^2$	h	$2y^2 - 8z^2$
i	$3y^2 - 75z^2$	j	$2a^3 - 8ab^2$	k	$28t^2 - 175r^2$	l	$8c^2 - 50d^2$

Factorising trinomials

A **trinomial** is an expression with three terms, such as $6x^3 + 3x^2 - 2$ or $4x^2 + 2x - 1$.

In this section trinomials of the form $ax^2 + bx + c$ will be factorised.

Factorise trinomials with a unitary x^2 coefficient

A **unitary x^2 coefficient** means the number in front of the x^2 term is 1.

To factorise an expression such as $x^2 + 5x + 6$, notice that both the middle term and the last term are positive, so the factorisation will be of the form:

$(x + \square)(x + \square)$

Look at the last number and find pairs of factors of 6:

$6 = 1 \times 6$ and $6 = 2 \times 3$

Identify the pair of factors which add to give 5.

$5 = 2 + 3$

So:

$x^2 + 5x + 6 = (x + 2)(x + 3)$ or $(x + 3)(x + 2)$

Example 4.4

Factorise $x^2 - 8x + 15$.

The last term is positive and the middle term is negative so the factorisation is of the form shown.

$x^2 - 8x + 15 = (x - \square)(x - \square)$

$= (x - 3)(x - 5)$ or $(x - 5)(x - 3)$

Look at the last number and find pairs of factors of 15 which add to give −8: $-5 \times -3 = 15$, $-5 + -3 = -8$.

Example 4.5

Factorise $x^2 - 2x - 8$.

$x^2 - 2x - 8 = (x - \square)(x + \square)$

$= (x - 4)(x + 2)$ or $(x + 2)(x - 4)$

The last term is negative and the middle term is negative so the factorisation is of the form shown.

Find pairs of factors of −8 which add to give −2: $-4 \times 2 = -8$, $-4 + 2 = -2$.

Example 4.6

Factorise $x^2 + 3x - 10$

$x^2 + 3x - 10 = (x - \square)(x + \square)$

$= (x - 2)(x + 5)$ or $(x + 5)(x - 2)$

The last term is negative and the middle term is positive so the factorisation is of the form shown.

$5 \times -2 = 10$, $-2 + 5 = 3$.

Use these general rules to factorise $x^2 + bx + c$:

- If c is **positive**, then the factors are either **both positive** or **both negative**.
  - » If b is **positive**, then the factors are **positive**
  - » If b is **negative**, then the factors are **negative**.

  Look for factors of c that **add** to b.
- If c is **negative**, then **one** factor is **positive** and the **other** factor is **negative**.
  - » If b is **positive**, then the **larger** factor is **positive**.
  - » If b is **negative**, then the **larger** factor is **negative**.

  Look for factors of c that have a **difference** of b.

Example 4.7

Factorise fully $3y^2 - 18y - 21$.

$3y^2 - 18y - 21 = 3(y^2 - 6y - 7)$

$= 3(y + \square)(y - \square)$

$= 3(y + 1)(y - 7)$

Find the HCF first.

The last term is negative so one bracket has a positive last term and other bracket has a negative last term.

Exercise 4D

1 Factorise.

a $a^2 + 12a + 11$ b $x^2 - 9x + 20$ c $w^2 - 11w + 28$

d $b^2 - 10b + 24$ e $p^2 + 24p + 63$ f $x^2 - 11x + 18$

g $19 - 20t + t^2$ h $34 + 19y + y^2$ i $t^2 + t - 12$

j $y^2 - 3y - 18$ k $x^2 - 2x - 63$ l $y^2 - 5y - 36$

2 Factorise.

a $7x^2 - 7y^2$ b $2t^2 + 4t + 2$ c $3a^2 - 18a + 24$ d $4x^2 + 20x + 24$

e $2t^2 + 22t + 48$ f $2y^2 + 30y + 100$ g $5m^2 - 10m - 40$ h $6t^2 + 12t - 48$

★ 3 Factorise.

a $t^2 - t - 6$ b $m^2 + 7m - 8$ c $x^2 + 6x - 7$ d $y^2 + 4y + 4$
e $u^2 + 2u - 3$ f $c^2 - c - 20$ g $y^2 - 5y - 24$ h $m^2 - 7m - 8$
i $2p^2 + 4p - 30$ j $3y^2 + 18y + 54$ k $2x^2 - 6x - 20$ l $4a^2 - 32a + 60$

Factorise more complex trinomials with non-unitary x^2 coefficient

In this section expressions of the form $ax^2 + bx + c$ where $a \neq 1$ will be factorised. In these expressions, you need to identify factors of the coefficient a and of the last term c.

Example 4.8

Factorise $4x^2 + 21x - 18$.

$4x^2 + 21x - 18 = (\square x + \square)(\square x - \square)$ — The last term is negative.

Identify factors of 4 and factors of 18:

Factors of 4 are: 1×4 -1×-4 2×2 -2×-2

Factors of -18 are: 1×-18 -1×18 2×-9 -2×9 3×-6 -3×6

Identify the combination of the factors of 4 and 18 which will add to give 21:

$(4 \times 6) + (1 \times -3) = 24 - 3 = 21$

You can use the grid method to carry out the separate calculations:

×	x	6
$4x$	$4x^2$	$24x$
-3	$-3x$	-18

So: $4x^2 + 21x - 18 = (x + 6)(4x - 3)$

Exercise 4E

★ 1 Factorise.

a $2x^2 + 5x + 3$ b $3t^2 + t - 2$ c $12m^2 - 8m + 1$
d $4y^2 + 7y - 2$ e $8u^2 + 10u - 3$ f $4p^2 + 3p - 7$
g $4t^2 + 12t + 9$ h $6m^2 + 17m - 3$ i $-8y^2 - 2y + 3$

2 Factorise.

a $4t^2 + 14t + 6$ b $6m^2 - 15m + 6$ c $15x^2 - 10x - 40$
d $45y^2 + 36y - 9$ e $8u^2 + 4u - 4$ f $18c^2 - 12c - 48$
g $x^2 + 8xy + 12y^2$ h $4m^2 - 7mn - 2n^2$
i $3x^2 - 5x - 2$ j $2b^2 - 11b - 21$

! Look for a common factor first.

★ 3 Factorise.

a $8y - 24$ b $m^2 - 36$ c $x^2 + 4x + 4$
d $6y^2 + 4y$ e $2u^2 + 3u - 5$ f $14c^2 - 56d^2$
g $2m^2 - 7m - 15$ h $16p^2 - 8p + 1$ i $10y^2 - y - 3$

Activity

Take two consecutive odd numbers, square each of them and then subtract the smaller square from the larger square.

Investigate for different pairs of consecutive odd numbers.

Can you see a connection?

If the first odd number is $2x - 1$, what is the next odd number?

Square each of these and subtract smaller from larger. Does this answer agree with the connection you found?

- I can factorise by finding a common factor. ★ Exercise 4A Q4
- I can factorise an expression with a difference of two squares. ★ Exercise 4B Q3
- I can factorise by finding a common factor followed by difference of two squares. ★ Exercise 4C Q1
- I can factorise trinomials with a unitary x^2 coefficient. ★ Exercise 4D Q3
- I can factorise trinomials with non-unitary x^2 coefficient. ★ Exercise 4E Q1
- I can factorise mixed examples and apply to problems. ★ Exercise 4B Q4, Q6 ★ Exercise 4E Q3

For further assessment opportunities, see the Preparation for Assessment for Unit 1 on pages 89–92.

5 Completing the square in a quadratic expression with unitary x^2 coefficient

In this chapter you will learn how to:

- **complete the square** of a quadratic expression with unitary x^2 coefficient
- solve equations by completing the square
- complete the square of a quadratic expression with non-unitary x^2 coefficient.

You should already know:

- how to add, subtract, multiply and divide integers
- how to expand squared brackets.

Completing the square

Completing the square is a useful technique for solving quadratic equations and graphing quadratic expressions. (See Chapters 16–18.)

A **quadratic expression** is a polynomial expression of the form $ax^2 + bx + c$ where $a \neq 0$. Quadratic expressions create parabolas and are frequently found and used in mathematics, science and engineering. The shapes of telescopes, satellite dishes, lenses are all defined by quadratic curves.

Completing the square is a useful technique for analysing parabolas. It is also used to determine the coordinates of the maximum or minimum turning point and the equation of the line of symmetry for the curve.

A parabola is symmetrical about the line of symmetry through the turning point. Completing the square allows you to find the maximum or minimum turning point of a parabola.

- The y-intercept of a parabola is the point at which $x = 0$.
- The x-intercepts of a parabola are the points at which $y = 0$.

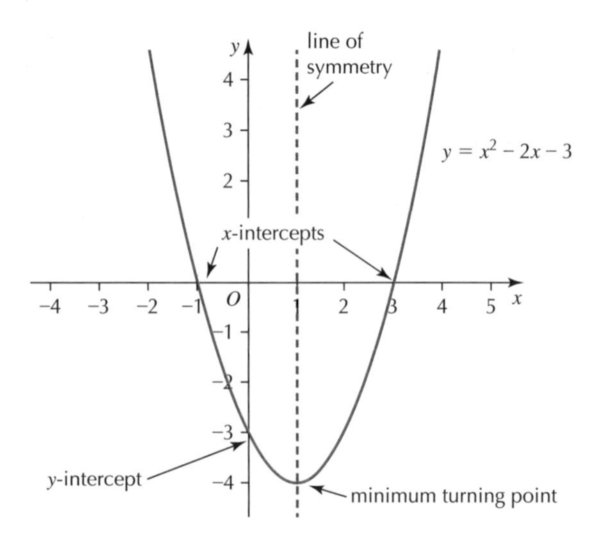

The equation of this curve is:

$y = x^2 - 2x - 3$

By completing the square the following is obtained:

$y = (x - 1)^2 - 4$

The **minimum value** of y is found when $(x - 1)^2 = 0$, so the minimum value occurs when $x = 1$.

Substituting the value $x = 1$ into the equation of the curve gives the y-value of the minimum point.

The minimum turning point is (1, –4). This can be confimed by a visual check of the graph.

Complete the square with unitary x^2 coefficient

In Chapter 3, brackets were expanded to give individual components. For example:

$(x + a)^2 = x^2 + 2ax + a^2$

This can be rearranged to give:

$x^2 + 2ax = (x + a)^2 - a^2$

This gives a way of solving equations of the form $x^2 + 2ax$.

Example 5.1

Complete the square.

a $x^2 + 10x$ b $y^2 - 16y$ c $m^2 + 8m + 2$

a $x^2 + 10x = (x + 5)^2 - 25$

Divide the x-coefficient 10 by 2, giving $10 \div 2 = 5$. Multiply out $(x + 5)^2$ to give $x^2 + 10x + 25$. To keep the value of the expression correct, we need to **subtract 25:** $x^2 + 10x + 25 - 25$

b $y^2 - 16y = (y - 8)^2 - 64$

Divide the y-coefficient –16 by 2, giving –8. Multiply out $(y - 8)^2$ to give $y^2 - 16y + 64$. To maintain the value of the expression, **subtract 64**: $y^2 - 16y + 64 - 64$.

c $m^2 + 8m + 2 = (m + 4)^2 - 14$

Divide the m-coefficient 8 by 2, giving 4. Multiply out $(m + 4)^2$ and then **subtract 16**: $m^2 + 8m + 16 + 2 - 16$.

Exercise 5A

1 Complete the square.

a $x^2 + 6x$ b $x^2 + 14x$ c $y^2 + 20y$ d $m^2 - 2m$ e $t^2 - 8t$ f $a^2 - 12a$

g $y^2 - 6y$ h $w^2 - w$ i $x^2 + 5x$ j $y^2 + 4y$ k $t^2 - 30t$ l $x^2 + 7x$

2 Write in the form $(x + a)^2 + b$ and state the values of a and b.

a $x^2 + 10x + 3$ b $y^2 - 4y + 6$ c $t^2 + 14t - 9$ d $m^2 - 6m + 4$

e $w^2 - 20w + 10$ f $x^2 + 12x - 3$ g $x^2 + 8x + 1$ h $m^2 + 7m + 3$

i $x^2 + 3x - 1$ j $a^2 - 4a - 2$ k $w^2 - 18w + 5$ l $t^2 + 9t - 3$

★ 3 Complete the square.

a $m^2 + 2m$ b $t^2 - 10t$ c $x^2 + 12x$ d $y^2 - 8y + 4$

e $a^2 - 4a - 3$ f $t^2 + 22t - 15$ g $p^2 + 16p - 7$ h $m^2 + 2m + 7$

i $y^2 + 10y - 5$ j $y^2 - 5y + 3$ k $a^2 - a + 4$ l $x^2 + 7x - 2$

Solving equations by completing the square

Numerical solutions to quadratic equations can be found by factorising or by using the quadratic formula. It is also possible to use the method of completing the square to find the numerical solutions.

Example 5.2

Solve $x^2 - 14x + 3 = 0$ by completing the square. Round your answers to 1 decimal place.

$x^2 - 14x + 3 = 0$

$(x - 7)^2 + 3 - 49 = 0$ — Complete the square.

$(x - 7)^2 - 46 = 0$

$(x - 7)^2 = 46$

$(x - 7) = \pm\sqrt{46}$ — Take the square root of both sides.

$(x - 7) = +\sqrt{46}$ or $(x - 7) = -\sqrt{46}$

$x = 7 + \sqrt{46}$ or $x = 7 - \sqrt{46}$

$x = 7 + 6.7823$ or $x = 7 - 6.7823$

$x = 13.7823$ or $x = 0.2177$

$x = 13.8$ or $x = 0.2$ (to 1 d.p.)

Example 5.3

Solve $x^2 + 6x - 5 = 0$ by completing the square. Leave your answers in surd form.

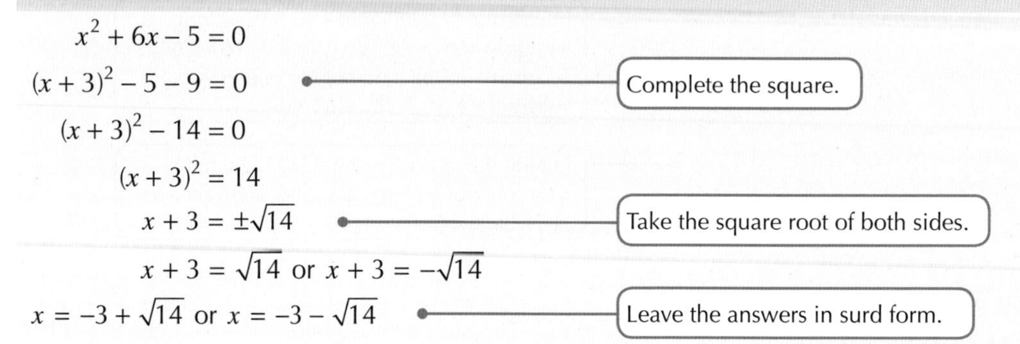
Complete the square.
Take the square root of both sides.
Leave the answers in surd form.

$x^2 + 6x - 5 = 0$

$(x + 3)^2 - 5 - 9 = 0$

$(x + 3)^2 - 14 = 0$

$(x + 3)^2 = 14$

$x + 3 = \pm\sqrt{14}$

$x + 3 = \sqrt{14}$ or $x + 3 = -\sqrt{14}$

$x = -3 + \sqrt{14}$ or $x = -3 - \sqrt{14}$

Exercise 5B

★ **1** Solve these equations by completing the square. Leave the answers in surd form.

a $y^2 + 2y - 7 = 0$ **b** $t^2 - 6t - 11 = 0$ **c** $x^2 - 10x + 4 = 0$

d $a^2 + 4a + 1 = 0$ **e** $y^2 + 8y + 13 = 0$ **f** $t^2 - 3t - 6 = 0$

g $x^2 + 14x - 5 = 0$ **h** $x^2 - 6x + 3 = 0$ **i** $x^2 + 6x + 3 = 0$

★ **2** Solve these equations by completing the square. Round the answers to 1 decimal place.

a $m^2 + 8m + 3 = 0$ **b** $x^2 - 12x + 4 = 0$ **c** $w^2 - 5w - 10 = 0$

d $t^2 - 4t + 1 = 0$ **e** $a^2 + 14a + 13 = 0$ **f** $x^2 + 3x - 7 = 0$

3 Solve the following equations by completing the square.

a $x^2 + 2x - 5 = 0$ **b** $x^2 - 4x - 7 = 0$ **c** $x^2 + 2x - 9 = 0$

Complete the square of a quadratic expression with non-unitary x^2 coefficient

Not all quadratic expressions have a unitary x^2 coefficient. The following extension Exercise is an introduction to solving quadratic equations with a non-unitary x^2 coefficient. This will be taken further at Higher.

Example 5.4

Complete the square $3x^2 + 6x + 4$.

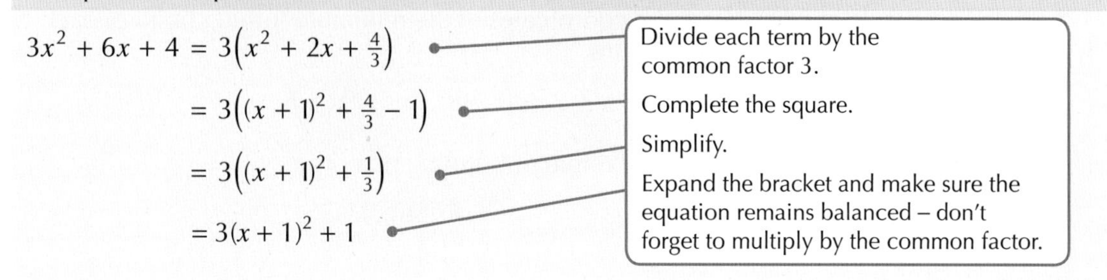

$$3x^2 + 6x + 4 = 3\left(x^2 + 2x + \tfrac{4}{3}\right)$$
$$= 3\left((x+1)^2 + \tfrac{4}{3} - 1\right)$$
$$= 3\left((x+1)^2 + \tfrac{1}{3}\right)$$
$$= 3(x+1)^2 + 1$$

Divide each term by the common factor 3.

Complete the square.

Simplify.

Expand the bracket and make sure the equation remains balanced – don't forget to multiply by the common factor.

Example 5.5

Write the expression $-x^2 + 3x - 2$ in the form $a(x + p)^2 + q$.

$$-x^2 + 3x - 2 = -(x^2 - 3x + 2)$$
$$= -\left(\left(x - \tfrac{3}{2}\right)^2 + 2 - \tfrac{9}{4}\right)$$
$$= -\left(\left(x - \tfrac{3}{2}\right)^2 - \tfrac{1}{4}\right)$$
$$= -\left(x - \tfrac{3}{2}\right)^2 + \tfrac{1}{4}$$

Divide each term by the common factor –1.

Complete the square.

Simplify.

Expand the bracket and remember to multiply by the common factor.

Exercise 5C

★ **1** Complete the square, leaving your answers in the form $a(x + p)^2 + q$.

a $4x^2 + 16x + 3$ b $2y^2 + 12y - 3$ c $5t^2 - 30t - 8$

d $-m^2 + 6m + 2$ e $6w^2 + 12w - 4$ f $3t^2 + 12t - 3$

GO! Activity

Part 1

Use a graph-drawing programme or graphic calculator to investigate different parabolas and find the connection between the x^2 coefficient and whether the parabola has a maximum or minimum turning point.

A quadratic equation takes the form $ax^2 + bx + c$.

Choose some positive values for a, b, and c. Input the expression into your graph-drawing programme and make a sketch of the parabola, noting the following:

i the equation

ii where it crosses the x- and y-axes

iii the nature of the curve; is it a maximum or minimum turning point?

Now try some different values for a, b and c. And make notes for **i**, **ii** and **iii**.

What happens if a is a negative number?

Can you predict the shape of the curve for any values of a, b and c?

Part 2

For any quadratic expression $ax^2 + bx + c$, if $a > 0$, the shape of the parabola gives a minimum.

a Find by completing the square the coordinates of the minimum turning point of each of the following quadratic expressions.

i $x^2 + 6x + 2$ ii $x^2 - 4x + 2$ iii $x^2 + 9x - 5$

b Find the quadratic expression if the minimum turning point is at:

i (–3, 4) ii (5, 2) iii (6, –1)

- I can complete the square of a quadratic expression with unitary x^2 coefficient. ★ Exercise 5A Q3
- I can solve equations by completing the square. ★ Exercise 5B Q1, Q2
- I can complete the square of a quadratic expression with non-unitary x^2 coefficient. ★ Exercise 5C Q1

For further assessment opportunities, see the Preparation for Assessment for Unit 1 on pages 89–92.

6 Reducing an algebraic fraction to its simplest form

In this chapter you will learn how to:

- use **factorising** to simplify algebraic fractions
- use multiplication to **simplify** complex algebraic fractions
- use a **negative common factor** to simplify algebraic fractions.

You should already know:

- what is meant by the highest common factor (HCF)
- how the HCF is used when simplifying fractions
- how to factorise an algebraic expression using a common factor
- how to factorise an algebraic expression using the difference of two squares
- how to factorise an algebraic expression in the form of a trinomial
- how to factorise an algebraic expression using a combination of the above.

Using factorising to simplify an algebraic fraction

Simplifying algebraic fractions is an important mathematical skill used to help solve equations, rearrange formulae and simplify complex expressions. To do this you will need to be able to factorise algebraic expressions (see Chapter 4). Understanding the concept of the **highest common factor** (HCF) is also useful and you will see that the method used to simplify numerical fractions can also be applied to algebraic fractions.

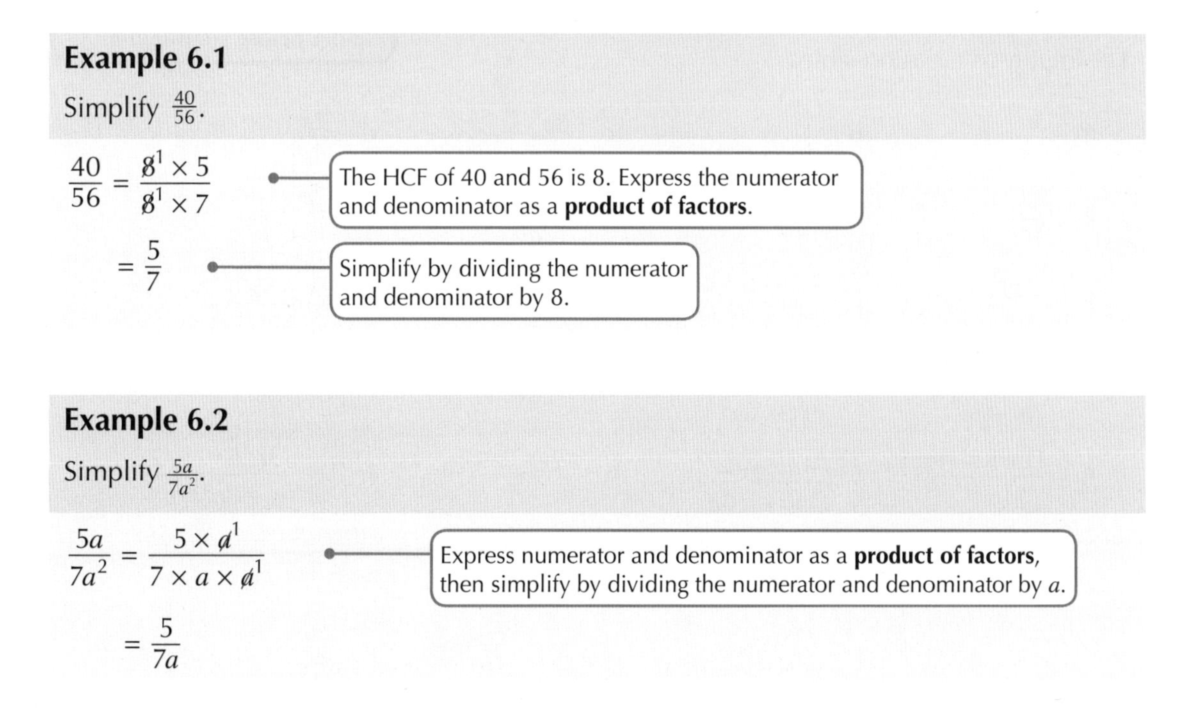

Example 6.1

Simplify $\frac{40}{56}$.

$\frac{40}{56} = \frac{8^1 \times 5}{8^1 \times 7}$ — The HCF of 40 and 56 is 8. Express the numerator and denominator as a **product of factors**.

$= \frac{5}{7}$ — Simplify by dividing the numerator and denominator by 8.

Example 6.2

Simplify $\frac{5a}{7a^2}$.

$\frac{5a}{7a^2} = \frac{5 \times a^1}{7 \times a \times a^1}$ — Express numerator and denominator as a **product of factors**, then simplify by dividing the numerator and denominator by a.

$= \frac{5}{7a}$

Example 6.3

Simplify.

a $\dfrac{3b^2+12b}{5b^2+20b}$ b $\dfrac{4a^2-b^2}{2a-b}$ c $\dfrac{x^2-4}{x^2+3x+2}$ d $\dfrac{12x^2-3}{2x^2+5x-3}$

a $\dfrac{3b^2+12b}{5b^2+20b}=\dfrac{3b(b+4)^1}{5b(b+4)^1}$

$=\dfrac{3}{5}$

Factorise to give the common factor $b(b+4)$, then simplify.

See Chapter 4 for a reminder about the difference of two squares.

b $\dfrac{4a^2-b^2}{2a-b}=\dfrac{(2a+b)(2a-b)^1}{(2a-b)^1}$

$=2a+b$

You should recognise the numerator as being a difference of two squares as it is of the form a^2-b^2. Factorise to give the common factor $2a-b$, then simplify.

c $\dfrac{x^2-4}{x^2+3x+2}=\dfrac{(x+2)^1(x-2)}{(x+2)^1(x+1)}$

$=\dfrac{x-2}{x+1}$

The numerator is a difference of two squares. The denominator is a trinomial. Factorise and simplify.

See Chapter 4 for a reminder about trinomials.

d $\dfrac{12x^2-3}{2x^2+5x-3}=\dfrac{3(4x^2-1)}{(2x-1)(x+3)}$

$=\dfrac{3(2x+1)(2x-1)^1}{(2x-1)^1(x+3)}$

$=\dfrac{3(2x+1)}{x+3}$

Take the common factor 3 from the numerator to give a difference of two squares.
Factorise and simplify.

Exercise 6A

Remember to factorise fully before simplifying.

1 Simplify as far as possible.

a $\dfrac{5a^2}{a}$	b $\dfrac{b}{3b}$	c $\dfrac{8c^2}{2c^2}$	d $\dfrac{2x}{4y}$
e $\dfrac{6e}{3e}$	f $\dfrac{7xy}{14y}$	g $\dfrac{9ab^2}{12ab}$	h $\dfrac{5xy^2}{15x^2y}$
i $\dfrac{(3p)^2}{6p}$	j $\dfrac{7y+3y^2}{3y}$	k $\dfrac{8a+2}{2a}$	l $\dfrac{2x+3x^2}{3x}$
m $\dfrac{5m}{7m-m^2}$	n $\dfrac{3xy}{9x+6x^2}$	o $\dfrac{8x+4}{4y}$	p $\dfrac{5p+10q}{15pq}$
q $\dfrac{12x-4xy}{8x^2}$	r $\dfrac{4ab+8a^2}{5ab+10a^2}$	s $\dfrac{(2r)^2-12r}{4r}$	t $\dfrac{3a^2bc}{3a^3bc+3a^2b^2c}$

2 Simplify as far as possible.

a $\frac{a^2 - 4}{3a - 6}$ b $\frac{3b + 9}{b^2 - 9}$ c $\frac{c^2 - 1}{5c - 5}$ d $\frac{d^2 - 3d}{d^2 - 2d - 3}$

e $\frac{e^2 + 4e}{2e^2 + 8e}$ f $\frac{x^2 - 16}{x^2 + 2x - 8}$ g $\frac{x^2 + 6x - 40}{x^2 - 100}$ h $\frac{x^2 + 6x + 5}{x^2 - x - 2}$

i $\frac{x^2 - 4x - 21}{x^2 - 5x - 14}$ j $\frac{2x^2 + 3x - 2}{x^2 + 5x + 6}$ k $\frac{4x^2 + 4x - 3}{6x^2 + x - 2}$ l $\frac{6x^2 - x - 12}{6x^2 + 5x - 4}$

★ 3 Simplify as far as possible.

a $\frac{x^2 + 3x + 2}{4x^2 - 4}$ b $\frac{2x^2 + 4x}{2x^2 - 8}$ c $\frac{3x^2 - 3x - 6}{6x^2 - 6}$

d $\frac{x^3 - 4x}{x^2 - 5x + 6}$ e $\frac{(x + 2)^3}{\left(x^2 - 4\right)(x + 2)}$ f $\frac{12x^3 - 27x}{12x^3 + 6x^2 - 18x}$

Activity

Work with a partner. Create an algebraic fraction that will simplify to:

a 5 b $2x$ c $x + 1$ d $3xy$ e $\frac{2x}{y}$ f $\frac{x + 1}{x - 1}$

Simplifying complex algebraic fractions using multiplication

In a complex fraction the numerator, or the denominator, or sometimes both, contain a fraction, such as $\frac{\frac{2}{3}}{5}$ or $\frac{x - \frac{4}{x}}{x - 2}$.

When dealing with complex algebraic fractions, use multiplication to remove any fractions in the numerator or denominator to create an equivalent fraction that is easier to work with.

Example 6.4

Simplify $\frac{2x - \frac{1}{x}}{\frac{1}{x}}$

$$\frac{2x - \frac{1}{x}}{\frac{1}{x}} = \frac{2x - \frac{1}{x}}{\frac{1}{x}} \times \frac{x}{x}$$

Remove the fractions in the numerator and denominator by multiplying by $\frac{x}{x}$. This doesn't change the overall value of the fraction because $\frac{x}{x} = 1$.

$$= \frac{x\left(2x - \frac{1}{x}\right)}{x\left(\frac{1}{x}\right)}$$

$$= \frac{2x^2 - \frac{x}{x}}{\frac{x}{x}}$$

$$= \frac{2x^2 - 1}{1} = 2x^2 - 1$$

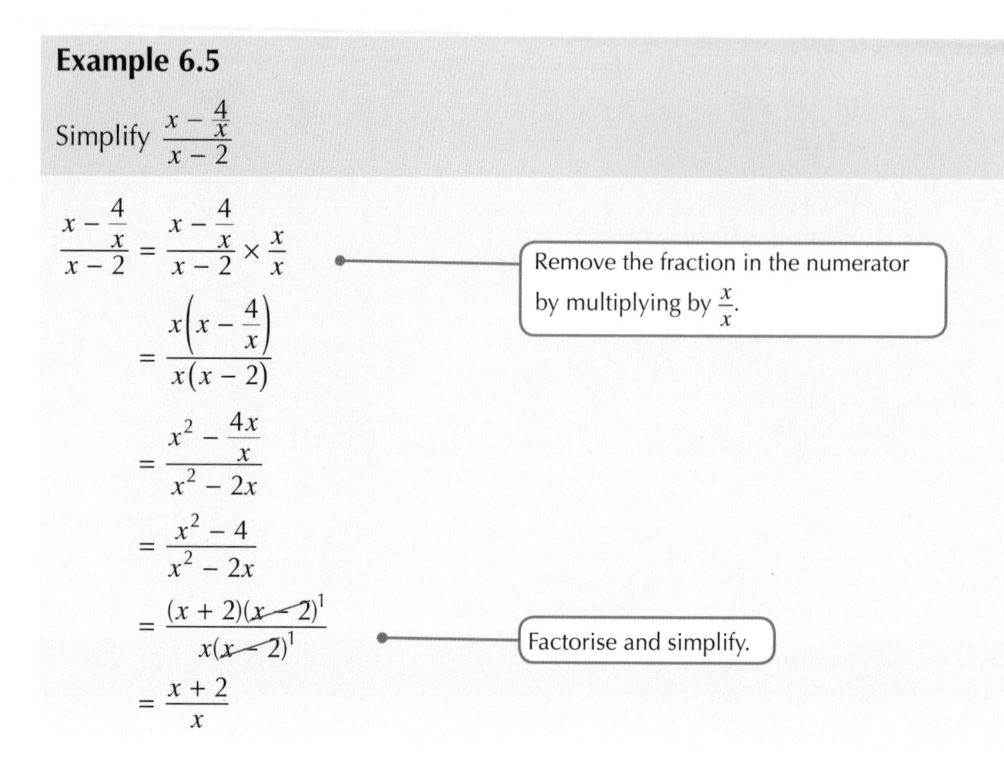

Example 6.5

Simplify $\dfrac{x - \frac{4}{x}}{x - 2}$

$$\frac{x - \frac{4}{x}}{x - 2} = \frac{x - \frac{4}{x}}{x - 2} \times \frac{x}{x}$$

Remove the fraction in the numerator by multiplying by $\frac{x}{x}$.

$$= \frac{x\left(x - \frac{4}{x}\right)}{x(x - 2)}$$

$$= \frac{x^2 - \frac{4x}{x}}{x^2 - 2x}$$

$$= \frac{x^2 - 4}{x^2 - 2x}$$

$$= \frac{(x + 2)\cancel{(x - 2)}^1}{x\cancel{(x - 2)}^1}$$

Factorise and simplify.

$$= \frac{x + 2}{x}$$

Exercise 6B

★ **1** Simplify as far as possible, writing your answers without fractions.

a $\dfrac{x + \frac{1}{x}}{x}$ **b** $\dfrac{3x + \frac{1}{4}}{\frac{1}{4}}$ **c** $\dfrac{3x - \frac{1}{x}}{2}$ **d** $\dfrac{x + \frac{1}{3}}{2x - \frac{1}{2}}$ **e** $\dfrac{3x + \frac{1}{x}}{x + \frac{2}{x}}$ **f** $\dfrac{x^2 - \frac{1}{4}}{2x + 1}$

2 Simplify as far as possible.

a $\dfrac{\frac{1}{x} + \frac{1}{2}}{\frac{1}{x} - \frac{1}{2}}$ **b** $\dfrac{\frac{a}{b} + \frac{a}{c}}{\frac{a}{b} - \frac{a}{c}}$

Activity

Work with a partner to create five algebraic fractions which can be simplified. Swap your fractions with another pair and try to simplify their fractions before they can simplify yours.

Simplifying complex algebraic fractions using a negative common factor

Sometimes a little manipulation using negatives is required to fully simplify an algebraic fraction. It is important to note that 'forcing' out a common factor of –1 does not change the overall value of the fraction but it does allow you to see clearly how your simplification should proceed.

Example 6.6

Simplify $\dfrac{4 - 4x + x^2}{x - 2}$

$$\frac{4 - 4x + x^2}{x - 2} = \frac{(2 - x)(2 - x)}{x - 2}$$

Take out a common factor of –1 from the denominator, so $(x - 2)$ becomes $-(2 - x)$, and the simplification is easier to see.

$$= \frac{(2 - x)\cancel{(2 - x)}^1}{-\cancel{(2 - x)}^1}$$

$$= \frac{(2 - x)}{-1}$$

$$= -(2 - x)$$

The division by –1 makes the whole expression negative.

$$= x - 2$$

The last step is to multiply the –1 back in.

Exercise 6C

★ **1** Simplify as far as possible.

a $\dfrac{x - 3}{3 - x}$ **b** $\dfrac{9x - 3y}{2y - 6x}$ **c** $\dfrac{3x^2 - 2xy}{4xy - 6x^2}$ **d** $\dfrac{x^2 - 9}{6 - 2x}$

e $\dfrac{1 - x^2}{x^3 \quad x}$ **f** $\dfrac{x^2 - a^2}{a - x}$ **g** $\dfrac{6 - x - x^2}{x^2 - 9}$ **h** $\dfrac{x^2 - 2xy + y^2}{y^2 - x^2}$

GO! Activity

Work with a partner to create five algebraic fractions which require a negative common factor to allow them to be simplified. Swap your fractions with another pair and try to simplify their fractions before they can simplify yours.

- I can use factorising to write the numerator and/or denominator of a fraction as a product of factors to help write an algebraic fraction in its simplest form. ★ Exercise 6A Q3
- I can use multiplication to simplify complex algebraic fractions. ★ Exercise 6B Q1
- I can use a negative common factor to help simplify algebraic fractions. ★ Exercise 6C Q1

For further assessment opportunities, see the Preparation for Assessment for Unit 1 on pages 89–92.

7 Applying one of the four operations to algebraic fractions

In this chapter you will learn how to:

- add and subtract algebraic fractions with one term in the numerator or denominator
- add and subtract algebraic fractions with more than one term in the numerator or denominator
- multiply algebraic fractions
- divide algebraic fractions.

You should already know:

- what is meant by the lowest common multiple (LCM)
- how to use the LCM when adding and subtracting fractions with numbers
- how to multiply and divide numerical fractions
- how to factorise an algebraic expression using a common factor
- how to factorise an algebraic expression using the difference of two squares
- how to factorise an algebraic expression in the form of a trinomial.

Adding and subtracting algebraic fractions with one term in the numerator or denominator

Adding or subtracting fractions containing algebraic terms requires the same technique as adding or subtracting numerical fractions. The first step is to express each fraction with a common denominator. To avoid having to simplify later on, the common denominator should be the **lowest common multiple** (LCM) of the denominators given.

Example 7.1

Express $\frac{2}{3} + \frac{1}{4}$ as a single fraction in its simplest form

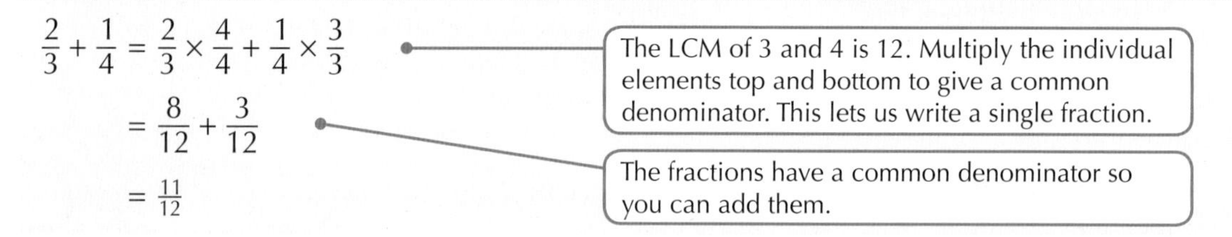

The LCM of 3 and 4 is 12. Multiply the individual elements top and bottom to give a common denominator. This lets us write a single fraction.
The fractions have a common denominator so you can add them.

$$\frac{2}{3} + \frac{1}{4} = \frac{2}{3} \times \frac{4}{4} + \frac{1}{4} \times \frac{3}{3}$$

$$= \frac{8}{12} + \frac{3}{12}$$

$$= \frac{11}{12}$$

Example 7.2

Express $\frac{2}{y} + \frac{3}{x}$ as a single fraction in its simplest form.

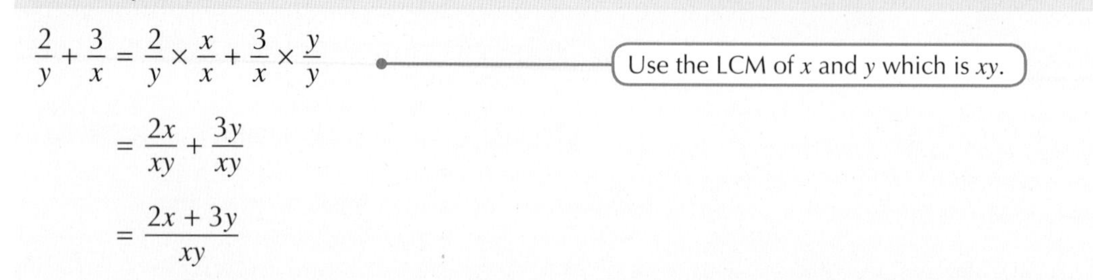

Use the LCM of x and y which is xy.

$$\frac{2}{y} + \frac{3}{x} = \frac{2}{y} \times \frac{x}{x} + \frac{3}{x} \times \frac{y}{y}$$

$$= \frac{2x}{xy} + \frac{3y}{xy}$$

$$= \frac{2x + 3y}{xy}$$

Example 7.3

Express $\frac{3}{2x} - \frac{1}{4x}$ as a single fraction in its simplest form.

$$\frac{3}{2x} - \frac{1}{4x} = \frac{3}{2x} \times \frac{2}{2} - \frac{1}{4x}$$

Use the LCM of $2x$ and $4x$ which is $4x$.

$$= \frac{6}{4x} - \frac{1}{4x}$$

$$= \frac{6-1}{4x}$$

$$= \frac{5}{4x}$$

Example 7.4

Express $2 - \frac{1}{x}$ as a single fraction in its simplest form.

$$2 - \frac{1}{x} = \frac{2}{1} - \frac{1}{x}$$

You do not have to show this line of working but it can help to think of non-fraction terms like 2 as $\frac{2}{1}$. Then multiply $\frac{2}{1}$ by $\frac{x}{x}$ to get a common denominator.

$$= \frac{2x}{x} - \frac{1}{x}$$

$$= \frac{2x-1}{x}$$

Exercise 7A

1 Express each of the following as a single fraction in its simplest form.

a $\frac{5x}{8} - \frac{x}{8}$ b $\frac{7x}{8} + \frac{x}{4}$ c $\frac{3x}{4} - \frac{x}{5}$ d $\frac{x}{3} + \frac{5x}{9}$

e $\frac{x}{2} + \frac{x}{3} + \frac{x}{6}$ f $\frac{x}{3} - \frac{x}{4} + \frac{5x}{6}$ g $\frac{x}{2} + \frac{2x}{3} - \frac{3x}{4}$ h $\frac{5x}{3} - \frac{x}{4} - \frac{x}{5}$

★ 2 Express each of the following as a single fraction in its simplest form.

a $\frac{5}{x} - \frac{2}{y}$ b $\frac{a}{c} + \frac{b}{d}$ c $\frac{3}{4x} + \frac{2}{3x}$ d $\frac{7}{4x} - \frac{3}{5x^2}$

e $\frac{1}{x} + \frac{2}{y} + \frac{3}{z}$ f $\frac{1}{pr} - \frac{3}{qr}$ g $\frac{2}{s^2t} - \frac{3}{st^2}$ h $\frac{a}{d} - \frac{b}{e} + \frac{c}{f}$

3 Express each of the following as a single fraction in its simplest form.

a $7 + \frac{3}{x}$ b $5 - \frac{2}{3x}$ c $x - \frac{2}{3}$ d $\frac{2}{3x} + 1$

e $x + \frac{5}{x}$ f $7 + \frac{x-y}{x}$ g $3 - \frac{x-y}{x}$ h $x^2 - \frac{5}{x}$

4 In physics the net resistance (R) of two resistors R_1 and R_2 connected in parallel can be calculated using the relationship $\frac{1}{R} = \frac{1}{R_1} + \frac{1}{R_2}$.

By expressing the right-hand side of this relationship as a single fraction, show that the net resistance (R) can be written as $\frac{R_1R_2}{R_1 + R_2}$.

Adding and subtracting algebraic fractions with more than one term in the numerator or denominator

The same method is used to add and subtract algebraic fractions where the numerators and/or denominators contain more than one term but you need to remove brackets and simplify. Some denominators must be factorised before you can find the common denominator.

Example 7.5

Express $\frac{3}{x-1} + \frac{4}{x-2}$ as a single fraction in its simplest form.

$$\frac{3}{x-1} + \frac{4}{x-2} = \frac{3(x-2)}{(x-1)(x-2)} + \frac{4(x-1)}{(x-1)(x-2)}$$

$$= \frac{3x - 6 + 4x - 4}{(x-1)(x-2)}$$

$$= \frac{7x - 10}{(x-1)(x-2)}$$

Multiply the first fraction by $\frac{x-2}{x-2}$ and the second fraction by $\frac{x-1}{x-1}$ to give a common denominator of $(x-1)(x-2)$.

Example 7.6

Express $\frac{x-1}{3} + \frac{x-3}{4}$ as a single fraction in its simplest form.

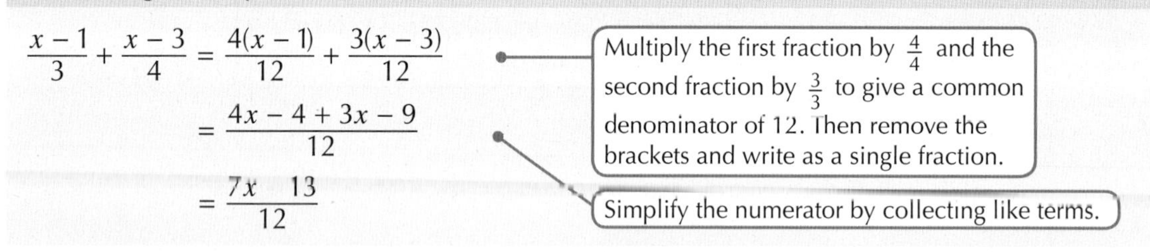

$$\frac{x-1}{3} + \frac{x-3}{4} = \frac{4(x-1)}{12} + \frac{3(x-3)}{12}$$

$$= \frac{4x - 4 + 3x - 9}{12}$$

$$= \frac{7x - 13}{12}$$

Multiply the first fraction by $\frac{4}{4}$ and the second fraction by $\frac{3}{3}$ to give a common denominator of 12. Then remove the brackets and write as a single fraction.

Simplify the numerator by collecting like terms.

Example 7.7

Express $\frac{3x-1}{3} + \frac{x+3}{5}$ as a single fraction in its simplest form.

$$\frac{3x-1}{3} - \frac{x+3}{5} = \frac{5(3x-1)}{15} - \frac{3(x+3)}{15}$$

$$= \frac{15x - 5 - 3x - 9}{15}$$

$$= \frac{12x - 14}{15}$$

Take care with the signs here.

Factorising a trinomial is explained in Chapter 4.

Example 7.8

Express $\frac{1}{x+2} + \frac{1}{x^2 + 3x + 2}$ as a single fraction in its simplest form.

$$\frac{1}{x+2} + \frac{1}{x^2 + 3x + 2} = \frac{1}{x+2} + \frac{1}{(x+1)(x+2)}$$

$$= \frac{x+1}{(x+1)(x+2)} + \frac{1}{(x+1)(x+2)}$$

$$= \frac{\cancel{x+2}^{\,1}}{(x+1)\cancel{(x+2)}^{\,1}}$$

$$= \frac{1}{x+1}$$

Factorising the denominator makes the LCM easier to find.

Multiply the first fraction by $\frac{x+1}{x+1}$ to give a common denominator. The second fraction doesn't change.

Write as a single fraction then simplify by dividing the numerator and the denominator by $x + 2$.

Exercise 7B

★ **1** Express each of the following as a single fraction in its simplest form.

a $\frac{x}{3} + \frac{x-2}{6}$ b $\frac{x+1}{4} + \frac{x-3}{8}$ c $\frac{2x-1}{3} - \frac{x}{4}$ d $\frac{x-3}{3} + \frac{x-2}{5}$

e $\frac{2x-2}{3} - \frac{x+1}{2}$ f $\frac{2x-1}{3} - \frac{x-3}{4}$ g $\frac{2x-1}{3} + \frac{x-3}{4} - \frac{2x-3}{6}$

★ **2** Express each of the following as a single fraction in its simplest form.

a $\frac{3}{x+1} - \frac{2}{x}$ b $\frac{4}{x-2} + \frac{3}{x}$ c $\frac{5}{x-2} + \frac{3}{x+3}$

d $\frac{3}{x+1} - \frac{2}{1-x}$ e $\frac{2}{2x+1} + \frac{3}{x-1}$ f $\frac{7}{3x-1} - \frac{2}{x+1}$

g $\frac{x+1}{x-2} + \frac{3}{x-1}$ h $\frac{x^2+1}{x+3} - \frac{x-1}{x+1}$ i $\frac{x+3}{2x+1} - \frac{1-2x^2}{x-1}$

★ **3** Express each of the following as a single fraction in its simplest form.

a $\frac{1}{x^2-16} - \frac{1}{x+4}$ b $\frac{1}{3x^2-3} + \frac{1}{x+1}$ c $\frac{2}{x+2} - \frac{5}{x^2-x-6}$

d $\frac{x+1}{x^2-4} + \frac{3}{x^2+3x+2}$ e $\frac{4x+12}{x^2-9} - \frac{3}{x+3}$ f $\frac{1}{x^2+x-12} - \frac{1}{x^2+3x-4}$

4 A man paddles his kayak a distance of 500 m up a river. Fighting against the flow of water he averages a speed of x metres per second. When he returns downstream to his starting point his average speed is $x + 3$ metres per second.

a Write down an expression for the time he takes to paddle upstream.

b Write down an expression for the time he takes to return to his starting point.

c Show that the total time for his journey can be expressed as $\frac{100(10x+15)}{x^2+3x}$

Multiplying algebraic fractions

Multiplying algebraic fractions follows the same process as multiplying numerical fractions. Simply multiply the numerators then multiply the denominators. There is no need to find a common denominator. If one of the numerators has a factor in common with one of the denominators then it can be cancelled before the multiplication. Sometimes you have to factorise numerators and denominators before you can see if cancelling is possible.

Example 7.9

Express $\frac{3}{x} \times \frac{2}{x}$ as a single fraction in its simplest form.

$$\frac{3}{x} \times \frac{2}{x} = \frac{6}{x^2}$$

Example 7.10

Express $\frac{3}{x^2} \times \frac{x}{6}$ as a single fraction in its simplest form.

$$\frac{3}{x^2} \times \frac{x}{6} = \frac{\cancel{3}^1}{x^{\cancel{2}1}} \times \frac{\cancel{x}^1}{\cancel{6}^2}$$

3 and 6 have a common factor of 3. x^2 and x have a common factor of x. Dividing by the common factors simplifies the multiplication. Notice that cancelling reduces the x^2 on the denominator of the first fraction to x^1 which is written as x.

$$= \frac{1}{x} \times \frac{1}{2}$$

This line of working shows what remains after cancelling. You don't have to include it but doing so can help avoid errors when writing the final answer.

$$= \frac{1}{2x}$$

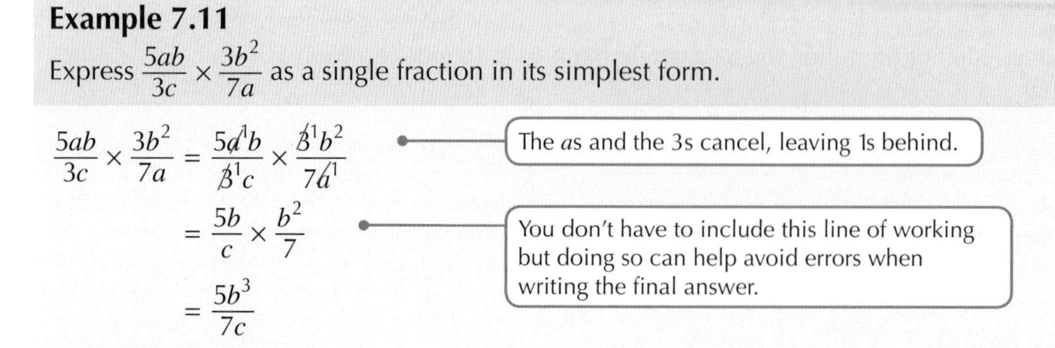

Example 7.11

Express $\frac{5ab}{3c} \times \frac{3b^2}{7a}$ as a single fraction in its simplest form.

$$\frac{5ab}{3c} \times \frac{3b^2}{7a} = \frac{5\not{a}^1 b}{\not{3}^1 c} \times \frac{\not{3}^1 b^2}{7\not{a}^1}$$

The *a*s and the 3s cancel, leaving 1s behind.

$$= \frac{5b}{c} \times \frac{b^2}{7}$$

You don't have to include this line of working but doing so can help avoid errors when writing the final answer.

$$= \frac{5b^3}{7c}$$

Example 7.12

Express $\frac{2a^2}{3} \times 7a^2$ as a single fraction in its simplest form.

$$\frac{2a^2}{3} \times 7a^2 = \frac{2a^2}{3} \times \frac{7a^2}{1}$$

You do not have to show this line of working but it can help to think of non-fraction terms such as $7a^2$ as $\frac{7a^2}{1}$, then multiply numerators and denominators as before.

$$= \frac{14a^4}{3}$$

Example 7.13

Express $\frac{a^2-4}{2a+6} \times \frac{a^2+5a+6}{a-2}$ as a single fraction in its simplest form.

$$\frac{a^2-4}{2a+6} \times \frac{a^2+5a+6}{a-2} = \frac{(a+2)\cancel{(a-2)}^1}{2\cancel{(a+3)}^1} \times \frac{(a+2)\cancel{(a+3)}^1}{\cancel{(a-2)}^1}$$

Factorising then cancelling can make the problem much simpler.

$$= \frac{(a+2)}{2} \times \frac{(a+2)}{1}$$

$$= \frac{a^2+4a+4}{2}$$

Exercise 7C

1 Express each of the following as a single fraction in its simplest form.

a $\frac{7}{x} \times \frac{2}{x}$ b $\frac{4}{x} \times \frac{x}{2y}$ c $\frac{5}{x} \times \frac{3}{10} \times \frac{xy}{3}$

d $\frac{2xy}{z} \times \frac{5}{4x^2}$ e $\frac{3}{2x^5} \times \frac{x^4}{9}$ f $\frac{(x+5)(x+2)}{x+3} \times \frac{2(x+3)}{x+2}$

★ 2 Express each of the following as a single fraction in its simplest form.

a $\frac{5}{x} \times 2y$ b $x \times \frac{5x}{3}$ c $\frac{4x^2}{5} \times 3x^2$

d $(x+5) \times \frac{x-5}{3}$ e $\frac{5}{4x^2-4} \times (x+1)$ f $x+3 \times \frac{x^2+9}{x^2-9}$

g $\frac{3x}{x^2+3x-28} \times \frac{x^2-49}{x^2}$ h $\frac{x+3}{x^2+3x+2} \times \frac{x^2-4}{x^2+4x+3}$

i $\frac{x^2+x-12}{x^2-x-6} \times \frac{x^2-2x-8}{x^2-16}$

Dividing algebraic fractions

Two numbers are **reciprocals** of each other if their product is 1, such as $\frac{1}{3}$ and 3 or $\frac{3}{4}$ and $\frac{4}{3}$. To divide by a fraction, multiply by the reciprocal:

$6 \div \frac{3}{4} = 6 \times \frac{4}{3}$

The same technique is applied to algebraic fractions. Sometimes the division might be written as a complex fraction, such as $\dfrac{x}{\frac{y}{z}}$. If this is the case it can help to first re-write as $x \div \frac{y}{z}$.

Example 7.14

Express $x^3 \div \dfrac{3}{x}$ as a single fraction in its simplest form.

$$x^3 \div \frac{3}{x} = x^3 \times \frac{x}{3}$$
$$= \frac{x^4}{3}$$

Remember $x^3 = \frac{x^3}{1}$.

Example 7.15

Express $\dfrac{3}{x^2} \div \dfrac{6}{x}$ as a single fraction in its simplest form.

$$\frac{3}{x^2} \div \frac{6}{x} = \frac{\cancel{3}^1}{x^{\cancel{2}^1}} \times \frac{\cancel{x}^1}{\cancel{6}^2}$$
$$= \frac{1}{2x}$$

Notice that cancelling reduces the x^2 on the denominator of the first fraction to x^1 which is written as x.

Example 7.16

Express $\dfrac{2x}{5y^2} \div \dfrac{x^2}{y} \times \dfrac{xy}{4}$ as a single fraction in its simplest form.

$$\frac{2x}{5y^2} \div \frac{x^2}{y} \times \frac{xy}{4} = \frac{\cancel{2}^1 \cancel{x}^1}{5y^{\cancel{2}^1}} \times \frac{\cancel{y}^1}{x^{\cancel{2}^1}} \times \frac{\cancel{x}^1\cancel{y}^1}{\cancel{4}^2}$$
$$= \frac{1}{5} \times \frac{1}{1} \times \frac{1}{2}$$
$$= \frac{1}{10}$$

There is a lot of cancelling here. Notice how single x terms from two different numerators combine to cancel the x^2. The same happens with the y terms giving a surprisingly simple answer.

Example 7.17

Express $\dfrac{x^2 - 9}{x^5} \div \dfrac{x + 3}{x^3}$ as a single fraction in its simplest form.

Factorising a difference of two squares is covered in Chapter 4.

$$\frac{x^2 - 9}{x^5} \div \frac{x + 3}{x^3} = \frac{x^2 - 9}{x^5} \times \frac{x^3}{x + 3}$$
$$= \frac{\cancel{(x+3)}^1(x - 3)}{x^{\cancel{5}^2}} \times \frac{\cancel{x^3}^1}{\cancel{(x+3)}^1}$$
$$= \frac{x - 3}{x^2} \times \frac{1}{1}$$
$$= \frac{x - 3}{x^2}$$

$x^2 - 9$ is a difference of two squares and it factorises to give $(x + 3)(x - 3)$.

Notice that cancelling reduces the x^5 on the denominator of the first fraction to x^2.

Exercise 7D

★ **1** Express each of the following as a single fraction in its simplest form.

a $3x^4 \div \frac{x^2}{5}$ b $x^3y \div \frac{3x^2}{y}$ c $\frac{4x^3}{7} \div 2x^2$

d $\frac{2xy^2}{3} \div 6x^2y$ e $\frac{2x^2}{3} \div \frac{5x^3}{6}$ f $-\frac{2x^2y^3}{5} \div \frac{4x^3y^2}{5}$

2 Express each of the following as a single fraction in its simplest form.

a $\frac{3}{5x} \div \frac{6}{x^2} \times \frac{5}{4x}$ b $\frac{yz^2}{4x} \times \frac{x^2}{y} \div \frac{(xz)^2}{2}$ c $\frac{3x-3}{x+2} \div \frac{x^2-1}{x^2+4x+4}$

d $\frac{x^3+7x^2+12x}{y^2-9} \div \frac{x^2+4x}{y-3}$ e $-\frac{b}{x^2+bx} \div \frac{b}{x+b}$ f $\frac{3x+3}{x^2+4x+4} \div \frac{3x^2+6x+3}{x^2-4}$

3 Express each of the following as a single fraction in its simplest form.

a $1 \div \frac{x}{y}$ b $\frac{1}{\frac{x}{x+1}}$ c $\frac{1}{\frac{1}{x-2}}$ d $\frac{\frac{a}{b}}{\frac{c}{d}}$

Activity

In a magic square the totals of the rows, columns and diagonals are the same. The total is three times the value of the centre square. Make a copy of the table and insert the algebraic fractions given to create a magic square.

$\frac{2(x-1)}{3}$ $\frac{2(x+2)}{3}$ $\frac{x+3}{2}$ $\frac{x+2}{3}$ $\frac{x+5}{3}$ $\frac{x-1}{3}$

	$\frac{x+1}{2}$	
$\frac{2x+1}{3}$		$\frac{x-1}{2}$

- I know how to add and subtract simple algebraic fractions where the numerators and denominators contain one number or term. ★ Exercise 7A Q2
- I know how to add and subtract more complicated algebraic fractions where the numerators contain more than one term. ★ Exercise 7B Q1
- I know how to add and subtract more complicated algebraic fractions where the denominators contain more than one term and can use factorising where necessary to help find the LCM. ★ Exercise 7B Q2, Q3
- I know how to multiply algebraic fractions and can use cancelling and factorising where appropriate to simplify my answers. ★ Exercise 7C Q2
- I know how to divide algebraic fractions and can use cancelling and factorising where appropriate to simplify my answers. ★ Exercise 7D Q1

For further assessment opportunities, see the Preparation for Assessment for Unit 1 on pages 89–92.

8 Determining the gradient of a straight line, given two points

In this chapter you will learn how to:

- use the **gradient formula** $m = \frac{y_2 - y_1}{x_2 - x_1}$
- solve problems involving gradient
- use gradient to determine if lines are **parallel**
- calculate gradient using the angle a line makes with the x-axis
- calculate **perpendicular gradients**
- solve problems involving perpendicular gradients.

You should already know:

- how to calculate the gradient of a straight line from horizontal and vertical distances
- that horizontal lines have a gradient of zero
- that vertical lines have undefined gradient
- that lines sloping up from left to right have a positive gradient
- that lines sloping down from left to right have a negative gradient.

Gradient and the gradient formula

The gradient (m) of a line is a measure of its steepness. A basic definition of gradient is:

change in vertical distance ÷ change in horizontal distance

This is expressed mathematically as:

$$\textbf{gradient } (m) = \frac{\textbf{vertical change}}{\textbf{horizontal change}}$$

We can use this definition of gradient to derive a formula that will allow us to calculate the gradient of any straight line if we are given two points on the line.

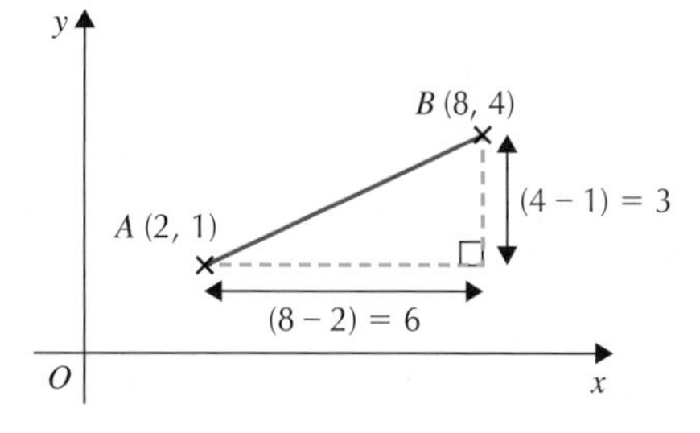

The diagram shows a straight line joining the points A (2, 1) and B (8, 4).

Creating a right-angled triangle allows us to determine the vertical and horizontal distances from A to B.

The gradient of the line joining points A and B is given by:

$$m_{AB} = \frac{\text{vertical change}}{\text{horizontal change}} = \frac{4-1}{8-2} = \frac{3}{6} = \frac{1}{2}$$

Gradients should be left as fractions in their simplest form. Applying this process to the general points A (x_1, y_1) and B (x_2, y_2) we get a formula that lets us determine the gradient of any straight line if we know two points on the line:

$$m_{AB} = \frac{y_2 - y_1}{x_2 - x_1}$$

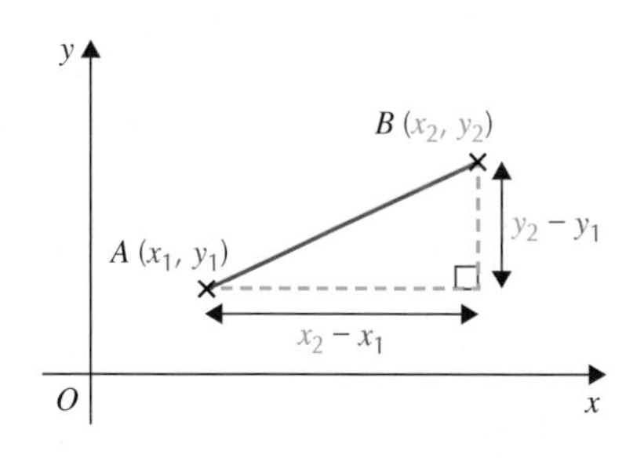

Notice that y values are in the numerator and the x values are in the denominator.

This makes sense since gradient is $\frac{\text{vertical change}}{\text{horizontal change}}$ and any change in y is a vertical change and any change in x is a horizontal change.

Link to Higher

A **rate of change** is a measure of how much one variable changes relative to a corresponding change in another. When calculating gradient, a large change in y relative to a small change in x gives a larger gradient, whereas a small change in y relative to a large change in x gives a smaller gradient. When we talk about a rate of change we are really talking about gradient.

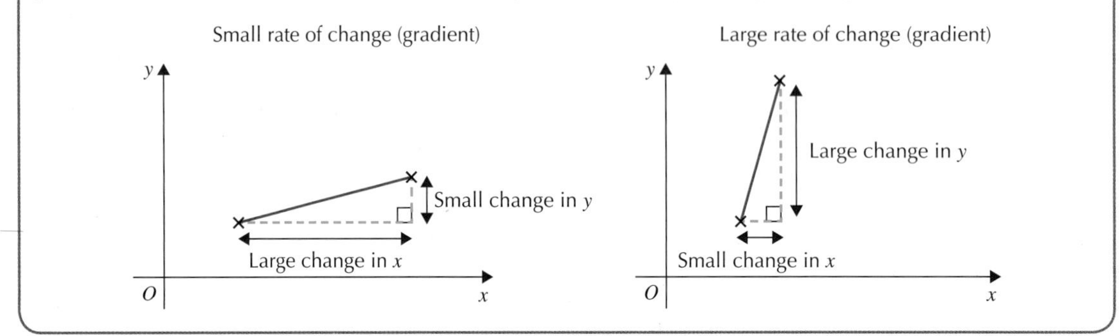

Example 8.1

Calculate the gradient of the line joining A (–3, 2) and B (2, 12).

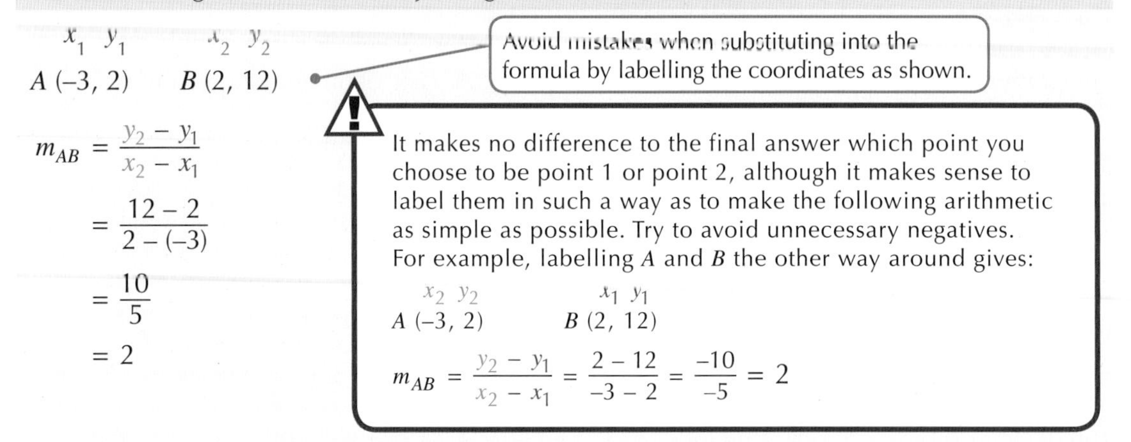

x_1 y_1 $\qquad$ x_2 y_2

A (–3, 2) $\qquad$ B (2, 12)

Avoid mistakes when substituting into the formula by labelling the coordinates as shown.

$$m_{AB} = \frac{y_2 - y_1}{x_2 - x_1}$$

$$= \frac{12 - 2}{2 - (-3)}$$

$$= \frac{10}{5}$$

$$= 2$$

It makes no difference to the final answer which point you choose to be point 1 or point 2, although it makes sense to label them in such a way as to make the following arithmetic as simple as possible. Try to avoid unnecessary negatives. For example, labelling A and B the other way around gives:

x_2 y_2 $\qquad$ x_1 y_1

A (–3, 2) $\qquad$ B (2, 12)

$$m_{AB} = \frac{y_2 - y_1}{x_2 - x_1} = \frac{2 - 12}{-3 - 2} = \frac{-10}{-5} = 2$$

Example 8.2

The line joining C (4, –2) and D (0, p) has a gradient of –3. Calculate the value of p.

$$\frac{y_2 - y_1}{x_2 - x_1} = m_{CD}$$

$$\frac{p - (-2)}{0 - 4} = -3$$

Substitute the values and produce an equation which can be solved to find p.

$$\frac{p + 2}{-4} = -3$$

Remove the fraction by multiplying both sides by –4.

$$p + 2 = 12$$

$$p = 10$$

Example 8.3

Calculate the gradient of the line joining $E\left(\frac{5}{2}, \frac{3}{2}\right)$ and $F\left(\frac{9}{2}, \frac{15}{2}\right)$.

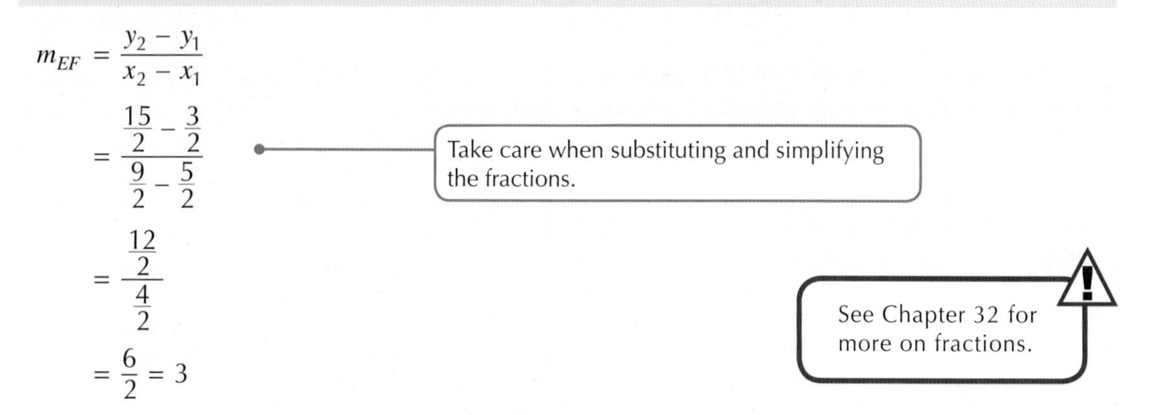

$$m_{EF} = \frac{y_2 - y_1}{x_2 - x_1}$$

$$= \frac{\frac{15}{2} - \frac{3}{2}}{\frac{9}{2} - \frac{5}{2}}$$

Take care when substituting and simplifying the fractions.

$$= \frac{\frac{12}{2}}{\frac{4}{2}}$$

See Chapter 32 for more on fractions.

$$= \frac{6}{2} = 3$$

Exercise 8A

★ **1** Calculate the gradient of the straight line joining the following pairs of points.

a A (2, 1) and B (3, 4) **b** C (–1, 6) and D (0, 4) **c** E (–1, –3) and F (1, 5)

d G (7, 2) and H (–4, –1) **e** J (–2, 5) and K (1, 7) **f** M (–7, 3) and N (–3, –2)

g P (–7, –8) and Q (–3, –2) **h** R (–11, 4) and S (–2, –8) **i** T (–9, –10) and U (–3, 5)

2 a Calculate the gradient of the line joining A (–3, 5) and B (7, 5).

b The line in part **a** is horizontal. Which axis is it parallel to?

c How could you have known from the coordinates that the line in **a** is horizontal?

3 a Calculate the gradient of the line joining C (5, 4) and D (5, –7).

b The line in part **a** is vertical, its gradient is undefined. Which axis is it parallel to?

c How could you have known from the coordinates that the line is **a** is vertical?

4 Calculate the gradient of the straight line joining the following pairs of points.

a $T\left(-\frac{1}{2}, -\frac{3}{2}\right)$ and $U\left(-\frac{9}{4}, \frac{15}{4}\right)$ **b** $V\left(\frac{1}{3}, \frac{1}{5}\right)$ and $W\left(-\frac{16}{9}, \frac{41}{15}\right)$

5 A is the point (5, 1) and B is the point (8, y). Find y if $m_{AB} = 2$.

6 C is the point (–1, –7) and D is the point (x, –5). Find x if $m_{CD} = -\frac{2}{5}$.

7 E is the point $\left(\frac{3}{2}, \frac{5}{3}\right)$ and F is the point $\left(\frac{11}{2}, y\right)$. Find y if $m_{EF} = -\frac{3}{4}$.

8 Find and simplify an expression for the gradient of the line joining $G\ (a, a^2)$ and H (–2, 4).

See Chapter 6 for help on simplifying expressions.

9 The graph below shows the acceleration, constant speeds and deceleration of a rider in a bike race.

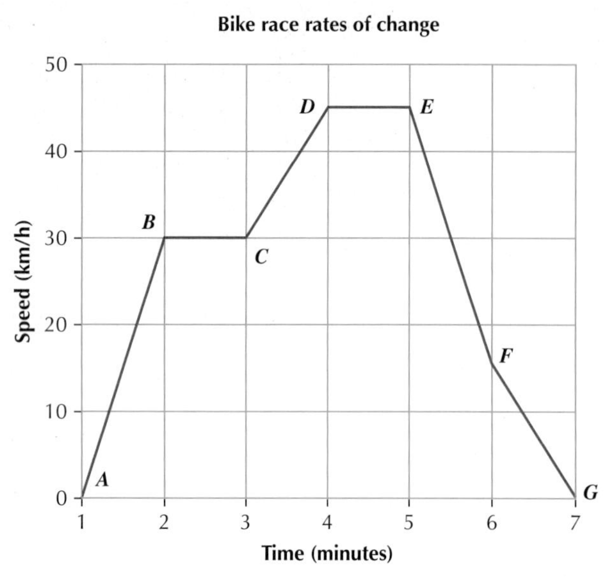

a At which point do you think the race finishes?

b Identify the parts of the race that show the greatest rate of change.

c Calculate the maximum acceleration of the bike.

d The rate of acceleration between points A and B is the same as the rate of deceleration between points E and F. True or false? Explain your answer.

e What does a gradient of zero represent on the graph?

> Notice that the speed is shown in kilometres per hour and the time is shown in minutes.

10 The graph shows the temperature variations on a winter's day in Crieff.

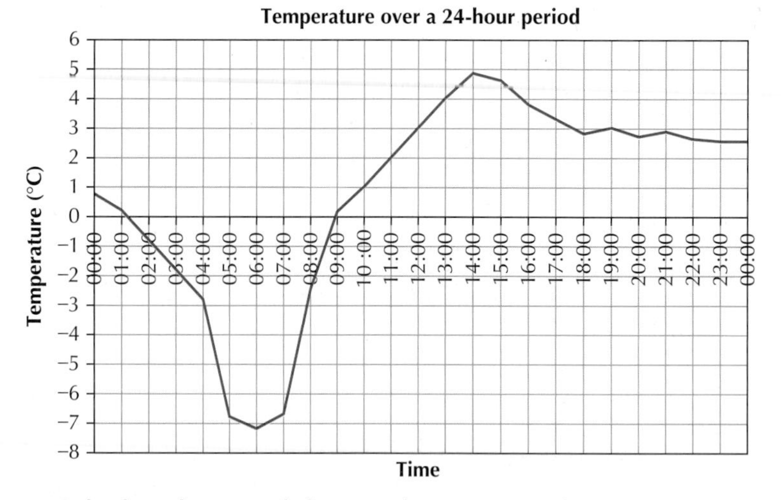

a Calculate the rate of change of temperature between 09:00 and 13:00.

b What is the maximum rate of change of temperature, and when does is take place?

c When does the rate of change show little variation?

Parallel lines

Parallel lines never meet, no matter how far they are extended. Parallel lines have **equal** gradients.

Are the grey lines in this image parallel?

Use your ruler to check if the distance between any two lines remains constant from one side to the other.

Example 8.4

Consider the points $P\ (3, 4)$, $Q\ (6, 8)$, $R\ (-2, -5)$ and $S\ (4, 3)$. Show that PQ and RS are parallel.

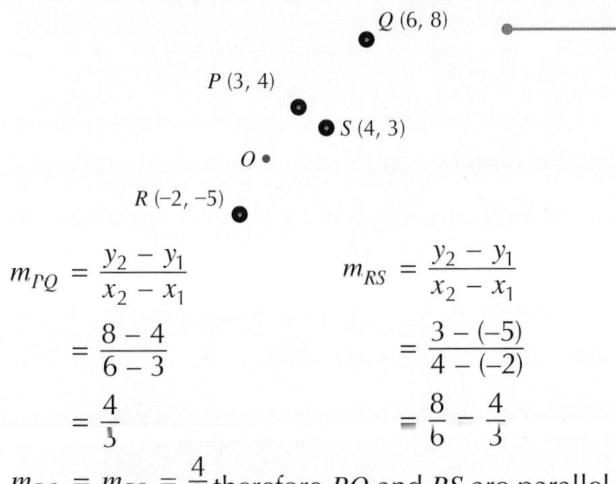

It is helpful to make a quick sketch for questions like this. You don't need to draw axes and accurately plot the points; just make a small dot on your page for the origin (shown here in red), then estimate where the points would be.

$$m_{PQ} = \frac{y_2 - y_1}{x_2 - x_1} = \frac{8-4}{6-3} = \frac{4}{3} \qquad m_{RS} = \frac{y_2 - y_1}{x_2 - x_1} = \frac{3-(-5)}{4-(-2)} = \frac{8}{6} = \frac{4}{3}$$

$m_{PQ} = m_{RS} = \frac{4}{3}$ therefore PQ and RS are parallel.

The mathematical way to indicate that two lines are parallel is to write $PQ \parallel RS$. If they are not parallel we would write $PQ \nparallel RS$.

Exercise 8B

1 Consider the points $P\ (1, 4)$, $Q\ (6, 6)$, $R\ (-5, -3)$ and $S\ (0, -1)$.

a Find m_{PQ}.

b Find m_{RS}.

c Explain why lines PQ and RS are parallel.

★ 2 Consider the points $T\left(-\frac{3}{2}, -5\right)$, $U\left(\frac{5}{2}, 3\right)$, $V\left(\frac{7}{2}, \frac{2}{5}\right)$ and $W\left(\frac{15}{2}, \frac{42}{5}\right)$.
Show that TU and VW are parallel.

★ 3 The points $P\ (-8, 4)$, $Q\ (-5, 8)$, $R\ (-1, 5)$ and $S\ (-4, 1)$ are the vertices of a quadrilateral. Show that $PQRS$ is a rhombus.

You should first make a sketch here. Then you will need to consider the properties of a rhombus: opposite sides are parallel *and* all sides are equal in length.

Link to Higher

Extending the line AB we can see that it meets the positive direction of the x-axis at an angle θ.
This corresponds to an angle of θ inside the triangle (using the fact that corresponding angles are equal).
From the definition of the tangent ratio we can say that

$$\tan\theta = \frac{\text{opposite}}{\text{adjacent}} = \frac{y_2 - y_1}{x_2 - x_1}$$

Since $m = \frac{y_2 - y_1}{x_2 - x_1}$ this allows us to say that the gradient of the line equals the tangent of the angle the line makes with the positive direction of the x-axis. This is represented by:

$m = \tan\theta$

The following Examples and Exercise 8C use $m = \tan\theta$ and are best attempted if appropriate trigonometry work in Chapter 24 has already been covered.

Example 8.5

Calculate (to 1 decimal place) the gradient of the straight line ST which makes an angle of 60° with the positive direction of the x-axis.

$$\begin{aligned} m_{ST} &= \tan\theta \\ &= \tan 60° \\ &= 1.7 \text{ (to 1 d.p.)} \end{aligned}$$

Example 8.6

Calculate (to 1 decimal place) the angle θ that the line GH with gradient $-\frac{5}{2}$ makes with the positive direction of the x-axis.

$$\begin{aligned} \tan\theta &= m_{GH} \\ &= -\frac{5}{2} \end{aligned}$$

Note that the gradient is negative, so you should expect an obtuse angle.

$\theta = 180° - 68.2° = 111.8°$ (to 1 d.p.)

Exercise 8C

You may use a calculator for this Exercise and you should round your answers to 1 decimal place.

★ 1 Calculate the gradient of the line which meets the positive direction of the x-axis at the following angles. Give your answers to 2 decimal places.

a 30° b 45° c 120° d 135° e 150° f 180°

2 Calculate the angle (θ) which is made with the positive direction of the x-axis by the lines with the following gradients. Give your answers to 3 significant figures.

a $m = 2$ b $m = \frac{1}{2}$ c $m = -3$ d $m = -\frac{3}{5}$ e $m = \frac{7}{2}$ f $m = -\frac{13}{4}$

3 Find the size of the angle that the line joining the points A (–4, 2) and B (5, 5) makes with the positive direction of the x-axis. Give your answer to 1 decimal place.

4 Find the size of the angle that the line joining the points C (–5, 1) and D (3, –3) makes with the positive direction of the x-axis. Give your answer to 3 decimal places.

5 The line OA connects the origin to point A (3, 7).
The line OB connects the origin to point B (5, 2).

a Calculate the size of angle $\angle AOB$.

b Use this to calculate the gradient of AB.

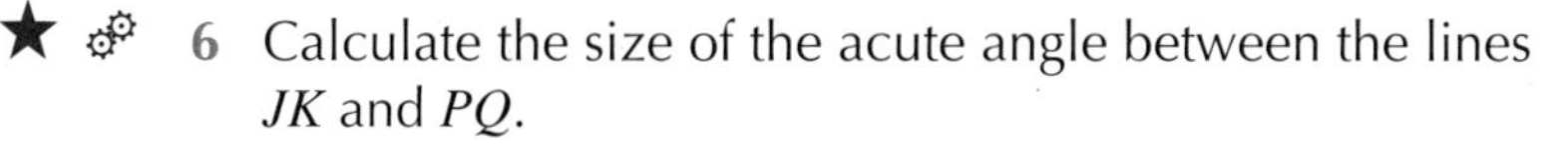

★ 6 Calculate the size of the acute angle between the lines JK and PQ.

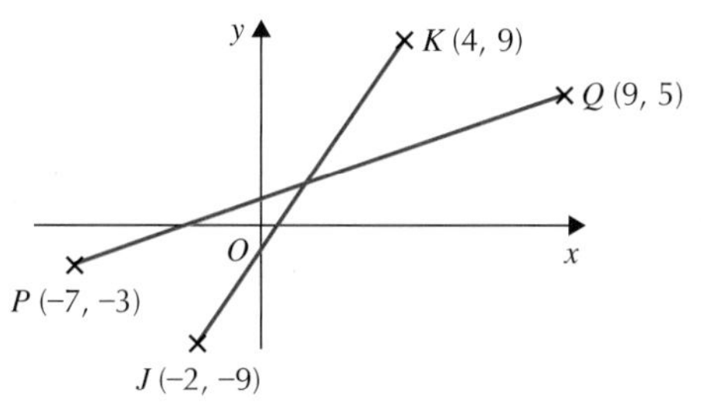

Activity

On squared paper plot the points A (−5, 2), B (−2, 6) and C (2, 3).

Add a fourth point (D) so that the points form the vertices of a square. Join the vertices together.

Calculate m_{AB} and m_{CD}. Since opposite sides of a square are parallel then, assuming you have plotted D correctly, you should find that $m_{AB} = m_{CD}$.

Check that the other pair of sides are parallel by calculating and comparing m_{AD} and m_{BC}.

Any two lines which meet at right angles are said to be **perpendicular** to one another.

One of the properties of a square is that all the angles are right angles. This means that, for example, side AB meets side BC at 90°. There is a connection between the gradient of side AB and side BC. Can you see it? What is the product of these two gradients?

Repeat the above process for the points P (−4, 5), Q (2, 8) and R (5, 2), this time adding a fourth point (S) so that the points form the vertices of a square. Calculate $m_{PQ} \times m_{QR}$.

In mathematics, a **conjecture** is an opinion or theory which hasn't yet been proven. Make a conjecture about the product of the gradients of perpendicular lines.

Link to Higher

Any two lines which meet at right angles are said to be **perpendicular** to one another. $m_\perp$ is the notation for a perpendicular gradient.

If two lines are perpendicular then their gradients will multiply to give −1, that is, $m \times m_\perp = -1$.

Note that the converse is also true, that is, if two gradients multiply to give −1 then the lines are perpendicular.

If we divide −1 by the gradient of a line we can find the gradient of any line perpendicular to it.

(continued)

For example:

a If $m = \frac{3}{4}$ then $m_{\perp} = -1 \div \frac{3}{4} = -\frac{4}{3}$

b If $m = -\frac{5}{2}$ then $m_{\perp} = -1 \div \frac{5}{2} = -\frac{2}{5}$

c If $m = -4$ then $m_{\perp} = -1 \div -4 = \frac{1}{4}$

d If $m = \frac{1}{3}$ then $m_{\perp} = -1 \div \frac{1}{3} = -3$

Dividing –1 like this can be a little tricky. There is an easier way to get a perpendicular gradient. Have you noticed it? Simply take the negative reciprocal of the first gradient. Reciprocals are covered in Chapter 7 (swapping numerator and denominator creates a reciprocal). When finding a reciprocal it can help to think of whole numbers such as 4 as $\frac{4}{1}$.

Example 8.7

Consider the points A (1, 0), B (–3, 5) and C (6, 4). Show that the line AB is perpendicular to AC.

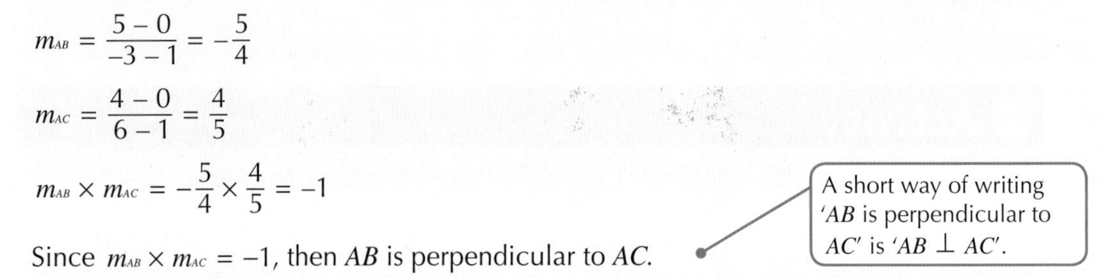

A short way of writing 'AB is perpendicular to AC' is 'AB ⊥ AC'.

$$m_{AB} = \frac{5-0}{-3-1} = -\frac{5}{4}$$

$$m_{AC} = \frac{4-0}{6-1} = \frac{4}{5}$$

$$m_{AB} \times m_{AC} = -\frac{5}{4} \times \frac{4}{5} = -1$$

Since $m_{AB} \times m_{AC} = -1$, then AB is perpendicular to AC.

Example 8.8

Find the gradient of a line perpendicular to the line with gradient $\frac{2}{3}$.

$$m \times m_{\perp} = -1$$

$$\frac{2}{3} \times m_{\perp} = -1$$

$$m_{\perp} = -\frac{3}{2}$$

Since $\frac{2}{3} \times -\frac{3}{2} = -1$.

Example 8.9

Find the gradient of a line perpendicular to the line with gradient –7.

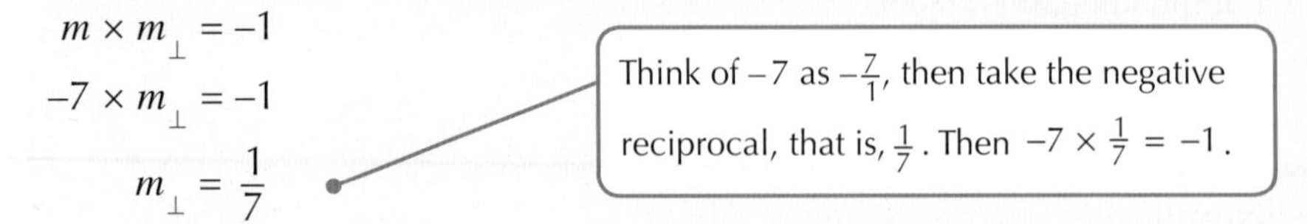

Think of –7 as $-\frac{7}{1}$, then take the negative reciprocal, that is, $\frac{1}{7}$. Then $-7 \times \frac{1}{7} = -1$.

$$m \times m_{\perp} = -1$$

$$-7 \times m_{\perp} = -1$$

$$m_{\perp} = \frac{1}{7}$$

Exercise 8D

1 Write down the gradient of a line that is perpendicular to each of these lines with the following gradients.

a $m = \frac{3}{4}$ b $m = -\frac{5}{2}$ c $m = \frac{1}{3}$ d $m = 5$

e $m = -\frac{4}{5}$ f $m = -1$ g $m = -\frac{1}{2}$ h $m = 1$

★ 2 Find the gradient of a line perpendicular to the line joining the points A (–5, –4) and B (–3, 11).

3 Make a sketch showing the points C (–2, –3), D (–5, 7) and E (8, 0). Show that the line CD is perpendicular to CE.

★ 4 Make a sketch showing the points C (1, 1), D (8, 2) and E (5, –2). Show that CDE is a right-angled triangle.

★ 5 The points P (–7, 1), Q (–2, 13), R (10, 8) and S (5, –4) are the vertices of a quadrilateral. Show that $PQRS$ is a square.

> You should first make a sketch then you will need to consider the properties of a square: opposite sides are parallel, all sides are equal in length and all sides meet at 90°.

- I can use the gradient formula to determine gradient when given two points. ★ Exercise 8A Q1
- I can use gradient to determine if lines are parallel. ★ Exercise 8B Q2
- I can solve problems involving gradient. ★ Exercise 8B Q3 ★ Exercise 8C Q6 ★ Exercise 8D Q5
- I can calculate gradient using the angle a line makes with the x-axis. ★ Exercise 8C Q1
- I can calculate the gradients of perpendicular lines. ★ Exercise 8D Q2
- I can solve problems involving the gradients of perpendicular lines. ★ Exercise 8D Q4

For further assessment opportunities, see the Preparation for Assessment for Unit 1 on pages 89–92.

9 Calculating the length of an arc or the area of a sector of a circle

This chapter will show you how to:

- calculate the length of an **arc** of a circle
- calculate the area of a **sector** of a circle
- express arc lengths and sector areas in terms of π
- calculate the sector angle
- solve problems involving arcs and sectors.

You should already know

- the vocabulary associated with circles
- how to calculate the circumference of a circle given radius or diameter
- how to calculate the area of a circle given radius
- how to round answers to an appropriate degree of accuracy.

Calculating arc lengths and sector areas

An **arc** of a circle is part of its **circumference**.

The **minor arc** AB is the shorter distance around the circle from point A to point B.

The **major arc** AB is the longer distance.

The **sector** AOB is a part of the circle enclosed by the arc AB and the radii OA and OB where O is the centre of the circle.

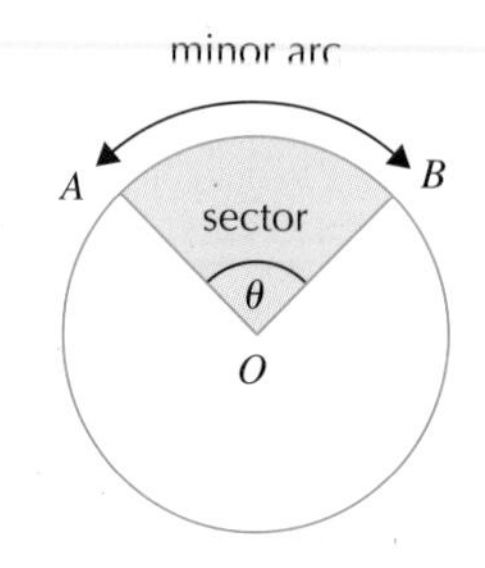

$\angle ABC$ is called the **sector angle** and is often denoted by the Greek letter θ (theta). In geometry we say that an arc length *subtends* an angle (θ); the word 'subtend' is of Latin origin and means 'stretches under'.

This circle has been divided into four identical sectors.

The sector angle (θ) is $\frac{1}{4}$ of a full turn, that is, $360 \div 4 = 90°$.

The arc length (AB) is $\frac{1}{4}$ of the full circumference.

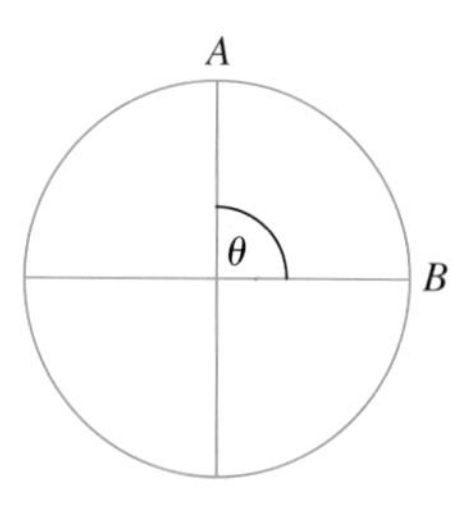

This circle has been divided into five identical sectors.

The sector angle (θ) is $\frac{1}{5}$ of a full turn, that is, $360 \div 5 = 72°$.

The arc length (CD) is $\frac{1}{5}$ of the full circumference.

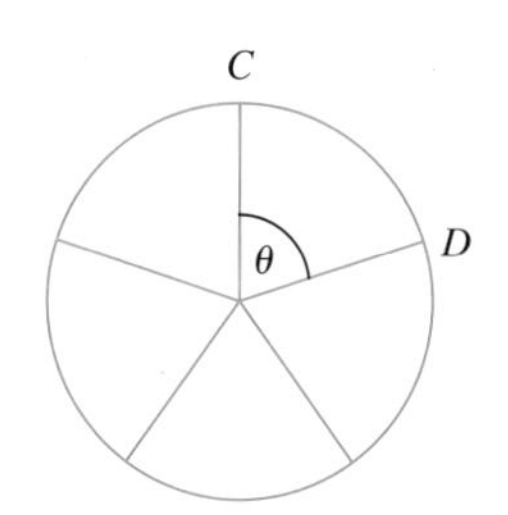

This circle has been divided into six identical sectors.

The sector angle (θ) is $\frac{1}{6}$ of a full turn, that is, $360 \div 6 = 60°$.

The arc length *(EF)* is $\frac{1}{6}$ of the full circumference.

The length of the arc is a fraction of the circumference and depends on the size of the sector angle θ, as shown in these examples.

In fact, the size of the arc is $\frac{\theta}{360}$ of the circumference. This gives us the relationship:

$$\textbf{arc length} = \frac{\theta}{360} \times \pi d$$

where d is the diameter.

The area of the sector is a fraction of the total area of the circle and again depends on θ.
In fact the area of the sector is $\frac{\theta}{360}$ of the total area. This leads to the relationship:

$$\textbf{sector area} = \frac{\theta}{360} \times \pi r^2$$

where r is the radius.

Example 9.1

Calculate (to 1 decimal place) the length of the minor arc AB.

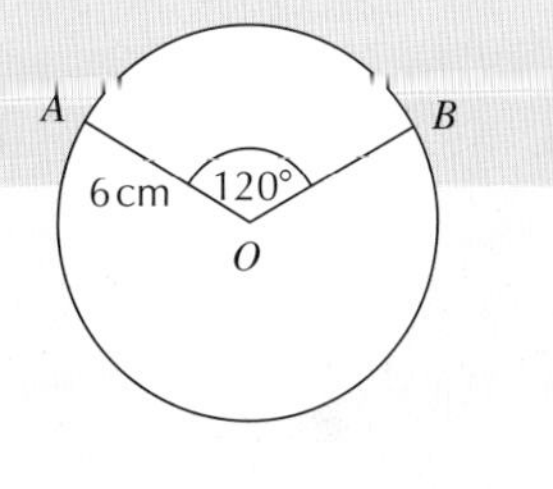

$$\text{arc length} = \frac{\theta}{360} \times \pi d$$

$$= \frac{120}{360} \times \pi \times 12$$

Remember to use diameter (2 × radius).

$$= \frac{1}{3} \times \pi \times 12$$

$$= 12.6 \text{ cm (to 1 d.p.)}$$

Example 9.2

Calculate (to 1 decimal place) the area of sector AOB.

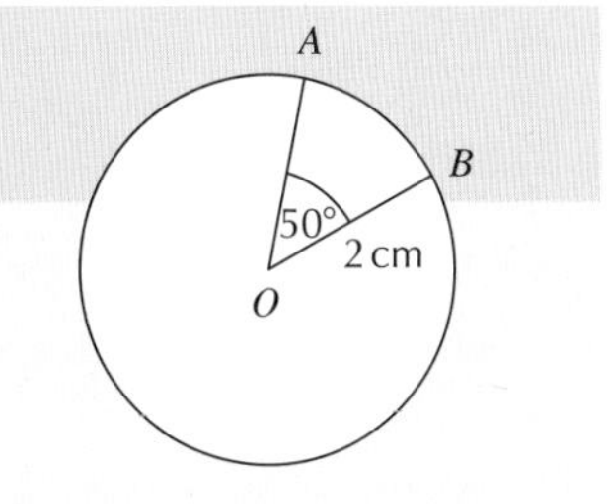

$$\text{sector area} = \frac{\theta}{360} \times \pi r^2$$

$$= \frac{50}{360} \times \pi \times r^2$$

$$= \frac{5}{36} \times \pi \times 4$$

$$= 1.7 \text{ cm}^2 \text{ (to 1 d.p.)}$$

Exercise 9A

You may use a calculator for this Exercise and you should round your answers to 1 decimal place unless otherwise stated.

1 For each of the circles below, calculate:

i the length of the minor arc **ii** the area of the sector.

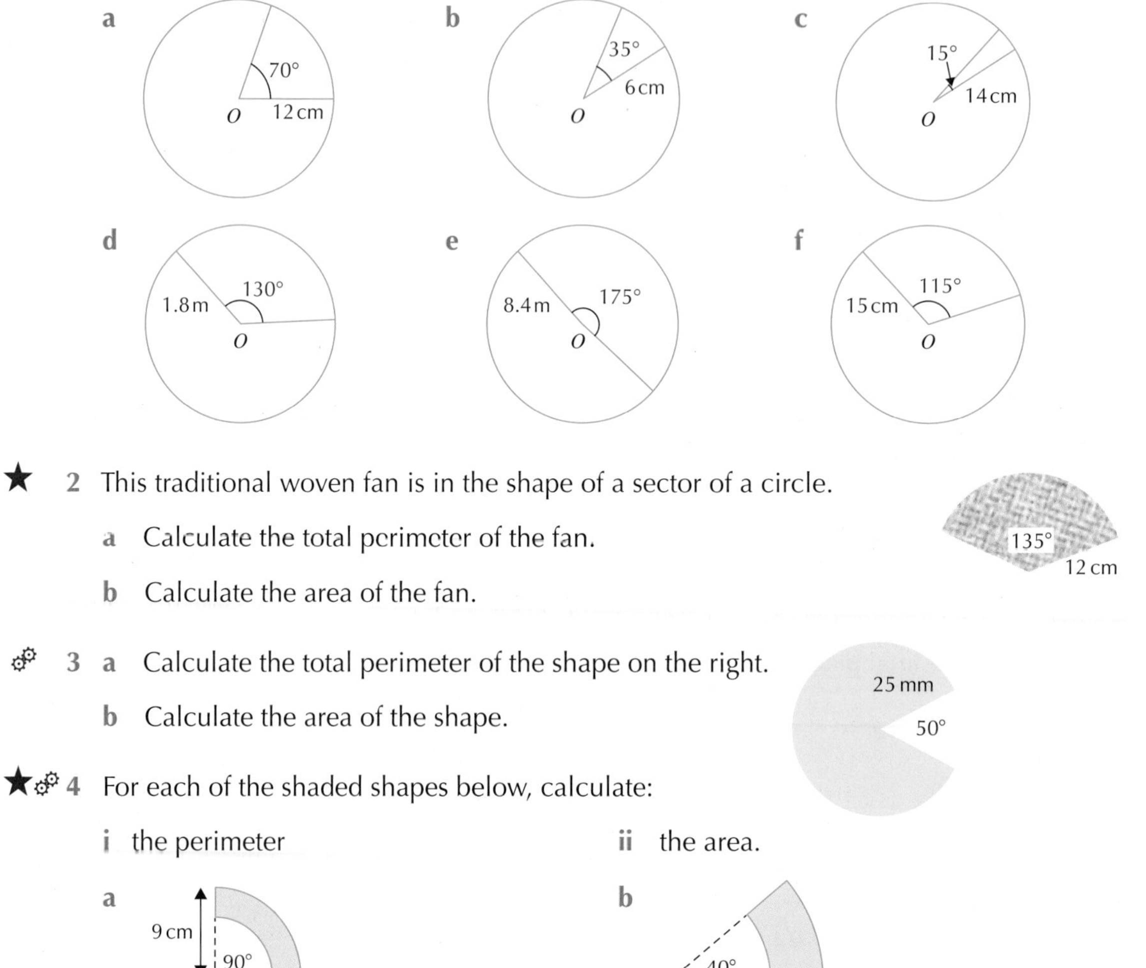

★ **2** This traditional woven fan is in the shape of a sector of a circle.

a Calculate the total perimeter of the fan.

b Calculate the area of the fan.

135°
12 cm

3 a Calculate the total perimeter of the shape on the right.

b Calculate the area of the shape.

25 mm
50°

★ **4** For each of the shaded shapes below, calculate:

i the perimeter **ii** the area.

a

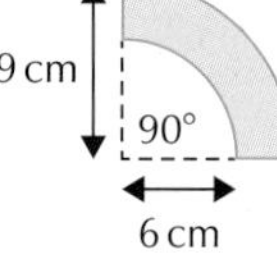

b

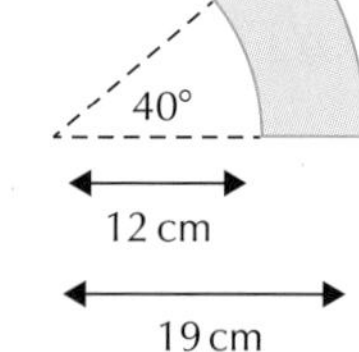

5 Calculate the area of the shaded segment. Give your answer to 3 significant figures.

90°
15 cm

Significant figures are covered in Chapter 11.

6 Kai and Maria are making decorations. Their decorations involve sectors cut from circles. They have four fabric circles of diameter 20 cm and want to make 40 identical sectors.

a Calculate the arc length of one of the sectors.

Kai and Maria want to trim the edges of their fabric sectors with ribbon.

b Calculate the minimum amount of ribbon they need.

7 Calculate the area the minute hand of a clock sweeps through in 25 minutes if the diameter of the clock is 30 cm.

Expressing arc lengths and sector areas in terms of π

To write the result of any calculation involving π as a decimal fraction will require rounding to some degree of accuracy. To avoid losing accuracy through rounding, especially if you need to use the value in a further calculation, you can express arc lengths and areas of sectors in terms of π.

Example 9.3

Calculate the length of the minor arc AB.

$$\text{arc length} = \frac{\theta}{360} \times \pi d$$

Remember to use diameter (2 × radius).

$$= \frac{120}{360} \times \pi \times 12$$

$$= \frac{1}{3} \times \pi \times 12 = 4\pi \text{ cm}$$

A, B, O, 6 cm, 120°

Example 9.4

Find the length of the radius in this circle where the sector angle is 120° and the sector area is $12\pi \text{ cm}^2$.

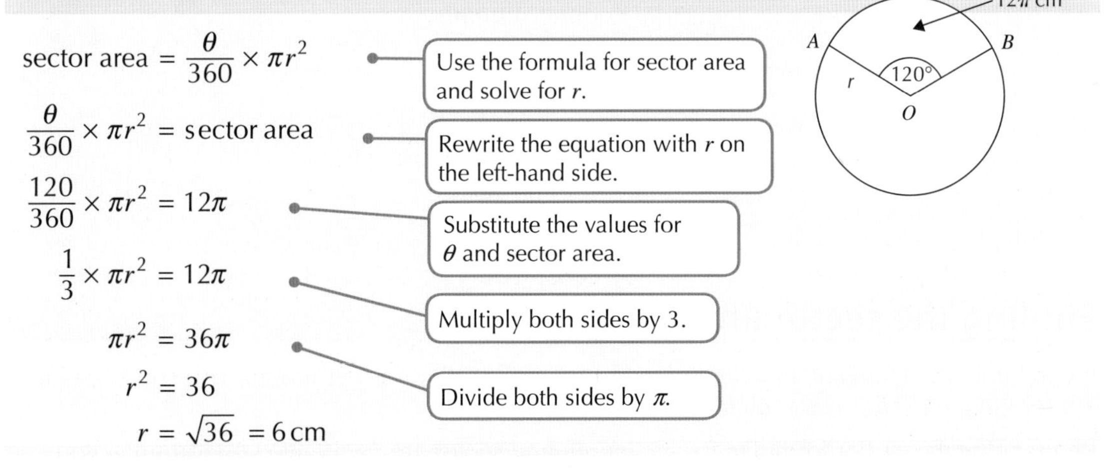

$\text{sector area} = \frac{\theta}{360} \times \pi r^2$
Use the formula for sector area and solve for r.
$\frac{\theta}{360} \times \pi r^2 = \text{sector area}$
Rewrite the equation with r on the left-hand side.
$\frac{120}{360} \times \pi r^2 = 12\pi$
Substitute the values for θ and sector area.
$\frac{1}{3} \times \pi r^2 = 12\pi$
Multiply both sides by 3.
$\pi r^2 = 36\pi$
Divide both sides by π.
$r^2 = 36$
$r = \sqrt{36} = 6 \text{ cm}$
$12\pi \text{ cm}^2$
A
B
r
120°
O

Exercise 9B

Do not use your calculator for this Exercise.

1 For each of the circles below, calculate:

i the length of the minor arc ii the area of the sector.

Leave your answers in terms of π.

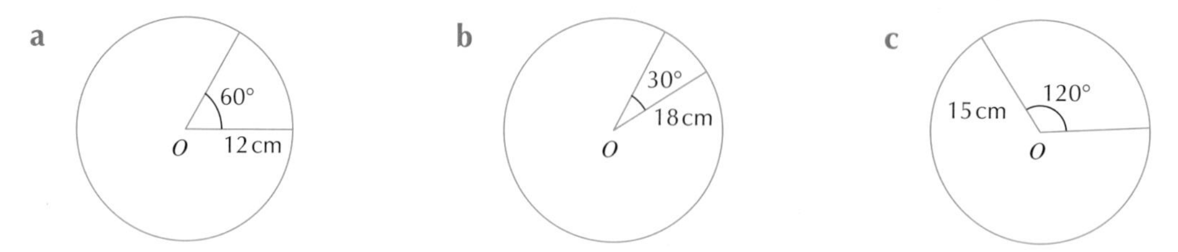

2 For each of the sectors below, calculate:

i the length of the curved side ii the area of the sector.

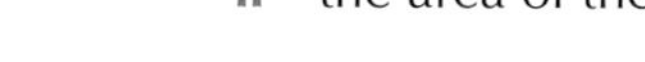

Leave your answers in terms of π.

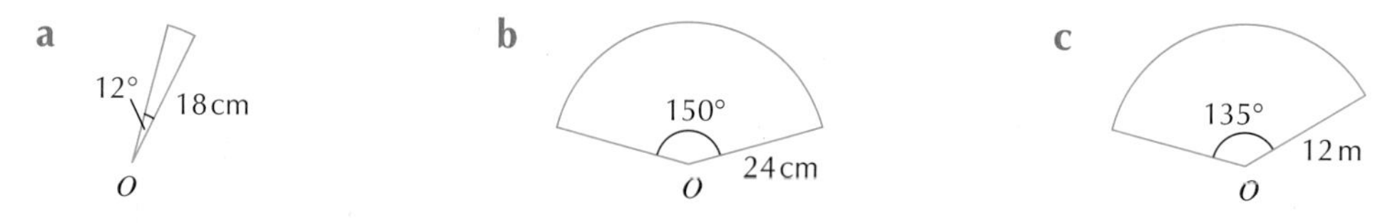

★ 3 a Find the length of the radius in a circle where the sector angle is 30° and the sector area is $3\pi\,\text{cm}^2$.

b Find the length of the radius in a circle where the sector angle is 72° and the sector area is $20\pi\,\text{cm}^2$.

c Find the length of the radius in a circle where the sector angle is 90° and the sector area is $\frac{\pi}{16}\,\text{cm}^2$.

4 This is a metal fan from an industrial cooling system.

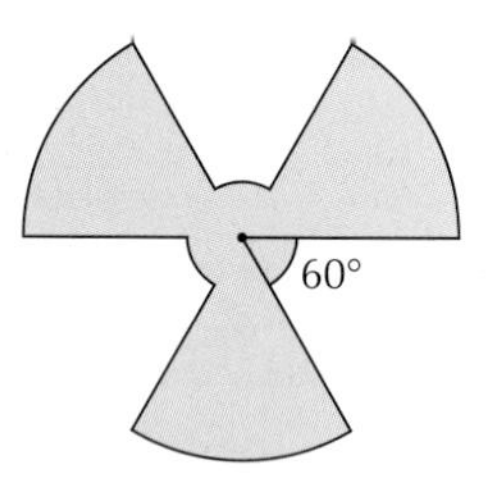

It is made from two concentric circles and has rotational symmetry of order 3.

The inner circle has a diameter of 4 cm and the outer circle has a diameter of 12 cm.

Calculate the area of metal required to make the fan. Leave your answer in terms of π.

Finding the sector angle

If you know the radius of a circle and either an arc length or a sector area, you can calculate the sector angle. The sector angle (θ) is a fraction of 360°.

Rearranging the earlier formulae for arc length and sector area will give you formulae you can use to calculate the sector angle.

Using arc length

The formula for arc length is:

$$\text{arc length} = \frac{\theta}{360} \times \pi d$$

The formula can be rearranged in the following way to find the angle sector.

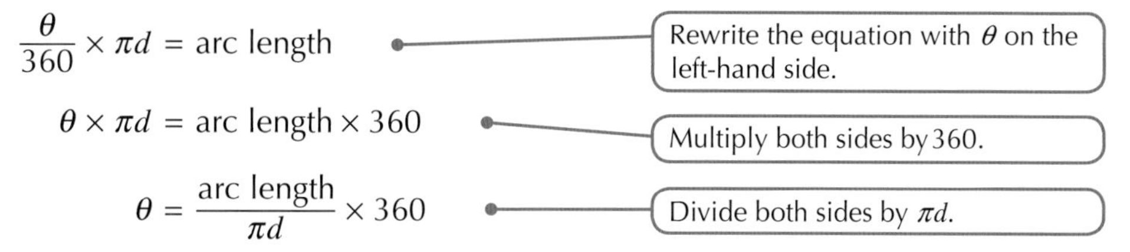

Rewrite the equation with θ on the left-hand side.
Multiply both sides by 360.
Divide both sides by πd.

$$\frac{\theta}{360} \times \pi d = \text{arc length}$$

$$\theta \times \pi d = \text{arc length} \times 360$$

$$\theta = \frac{\text{arc length}}{\pi d} \times 360$$

If you know the radius and an arc length AB then:

$$\textbf{sector angle } \theta = \frac{AB}{\pi d} \times 360°$$

Example 9.5

An arc AB of a circle with radius 8 cm has length 12 cm.
Find θ, the sector angle.

12 cm, A, B, 8 cm, θ, O

$$\theta = \frac{AB}{\pi d} \times 360°$$

$$= \frac{12}{16\pi} \times 360°$$

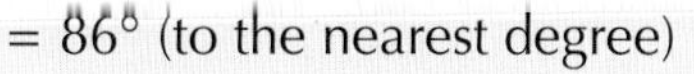

$= 86°$ (to the nearest degree)

Note on calculator use: Notice that in Example 9.5, although the denominator in the formula is πd, we write 16π when we substitute the numbers in. Most calculators understand that 16π is $16 \times \pi$, meaning you don't have to press the multiplication button. If you use $\pi \times 16$ you may have to use brackets to make sure the calculator does the calculation you want.

Using sector area

The formula for sector area can also be rearranged to find the sector angle.

$$\text{Sector area} = \frac{\theta}{360} \times \pi r^2$$

$$\frac{\theta}{360} \times \pi r^2 = \text{sector area}$$

$$\theta \times \pi r^2 = \text{sector area} \times 360$$

$$\theta = \frac{\text{sector area}}{\pi r^2} \times 360$$

If you know the radius and a sector area then:

$$\textbf{sector angle } \theta = \frac{\textbf{area } AOB}{\pi r^2} \times 360°$$

Example 9.6

A sector, AOB, of a circle with radius 1.4 m has an area of $0.8\,\text{m}^2$.

Find θ, the sector angle.

$$\theta = \frac{\text{area } AOB}{\pi r^2} \times 360°$$

$$= \frac{0.8}{1.4^2\pi} \times 360°$$

= 47° (to the nearest degree)

Writing the denominator with the π at the end will ensure your calculator does the correct calculation. On most calculators you won't need to press the × button.

Exercise 9C

You may use a calculator for Questions 1 and 4.

★ **1** For each of the circles below, calculate θ, the sector angle.

Round your answers to the nearest whole degree.

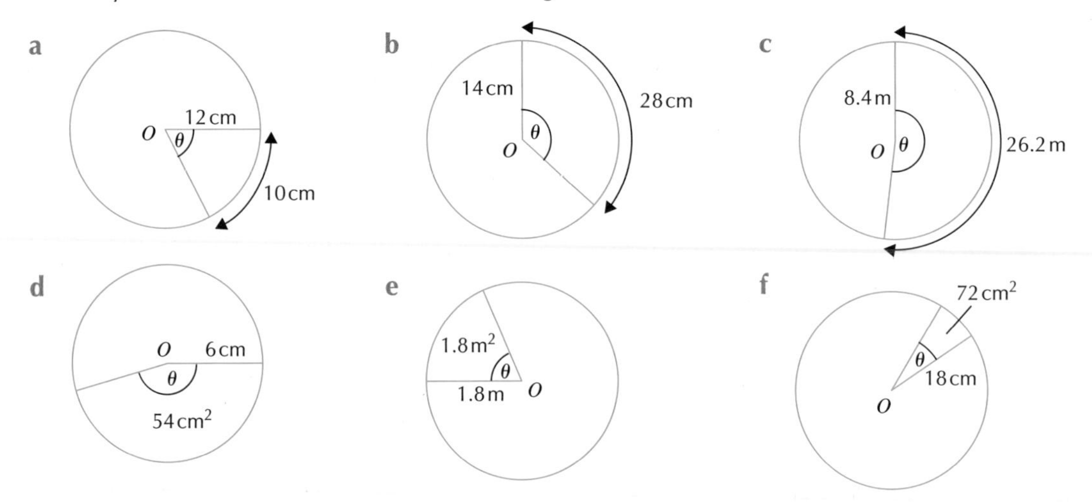

2 The length of the minor arc AB of a circle, centre O, with radius 10 cm is 4π cm. Without using a calculator, find:

a the acute angle AOB

b the area of sector AOB in terms of π.

3 The tip of the minute hand on this clock moves a distance of 2π cm in 5 minutes.

Without using a calculator, find:

a the angle the minute hand moves through in 5 minutes

b the length of the minute hand.

4 A gardener looks after a circular lawn with a radius of 12 m. She has been told to plant a flower bed in the shape of a sector. She has one bag of flower seed that will cover $50\,\text{m}^2$. What is the greatest possible sector angle of the flower bed if she uses the whole bag of seed?

Activity

Eratosthenes was a Greek mathematician who lived in the 3rd century BCE. He worked in the Great Library of Alexandria. Eratosthenes studied geometry and, having observed several lunar eclipses, believed the Earth to be shaped like a sphere. He came up with a clever method to calculate the circumference of the Earth. Work with a partner or in a group to research Eratosthenes' method, then prepare a poster for your classroom wall or give a presentation to your class describing how he did it.

- I can calculate the length of an arc or the area of a sector. ★ Exercise 9A Q2
- I can work with arc lengths and sector areas expressed in terms of π. ★ Exercise 9B Q3
- I know how to use arc length or sector area to calculate the sector angle. ★ Exercise 9C Q1
- I can solve problems involving arcs and sectors. ★ Exercise 9A Q4

For further assessment opportunities, see the Preparation for Assessment for Unit 1 on pages 89–92.

10 Calculating the volume of a standard solid

In this chapter you will learn how to:

- calculate the volume of a **sphere**
- calculate the volume of a **pyramid**
- calculate the volume of a **cone**
- calculate the volume of a **composite solid**
- solve problems involving the volume of a sphere, pyramid and cone.

You should already know

- how to calculate the volume of a triangular prism
- how to calculate the volume of a cylinder
- how to calculate the volume of other prisms given the area of the base.

Volume of a sphere

A **sphere** is a perfectly round geometrical object. It is the set of points which are equidistant from a given point in space. The distance is the radius, r, of the sphere and the given point is the centre of the sphere.

An interesting property of the sphere is that of *all* solids with volume V, the sphere will be the one with the smallest possible **surface area**, A. Similarly, of *all* solids with a given surface area A, the sphere will be the one with the greatest volume, V. These properties can be seen in soap bubbles. A soap bubble will hold a fixed volume and (due to surface tension), its surface area is the minimum for that volume. That is why a floating soap bubble takes the shape of a sphere.

The volume of a sphere with radius r can be calculated using the formula:

volume of a sphere $= V = \frac{4}{3}\pi r^3$

Example 10.1

Zorbing is an adventure sport that involves rolling downhill in a giant inflatable sphere.

Calculate the volume of a Zorb with a radius of 1.5 metres. Write your answer correct to 1 decimal place.

$V = \frac{4}{3}\pi r^3$ — Write down the formula.

$= \frac{4}{3}\pi \times 1.5^3$ — Substitute the known values.

$= 14.1\ \text{m}^3$ (to 1 d.p.)

Example 10.2

Calculate the radius of a Zorb with volume of $20\,\text{m}^3$.

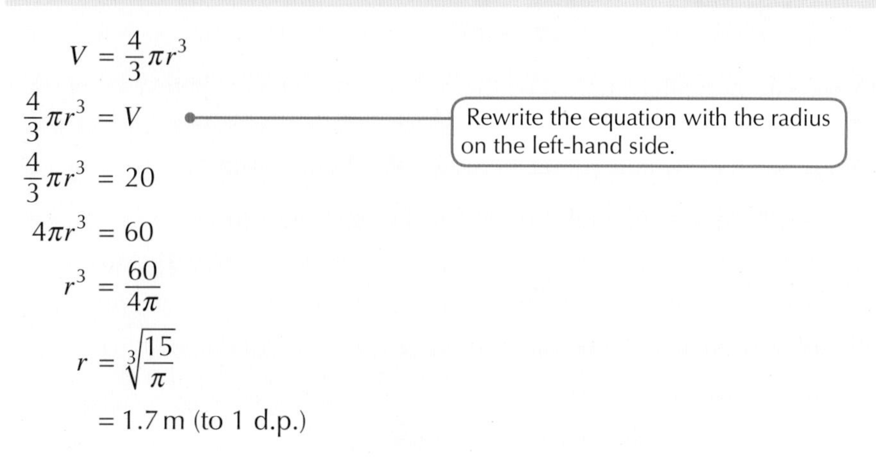

$$V = \frac{4}{3}\pi r^3$$

$$\frac{4}{3}\pi r^3 = V$$

Rewrite the equation with the radius on the left-hand side.

$$\frac{4}{3}\pi r^3 = 20$$

$$4\pi r^3 = 60$$

$$r^3 = \frac{60}{4\pi}$$

$$r = \sqrt[3]{\frac{15}{\pi}}$$

$$= 1.7\,\text{m (to 1 d.p.)}$$

Exercise 10A

You may use a calculator for this Exercise and you should round your answers to 1 decimal place unless otherwise stated.

★ **1** Calculate the volume of each of these spheres.

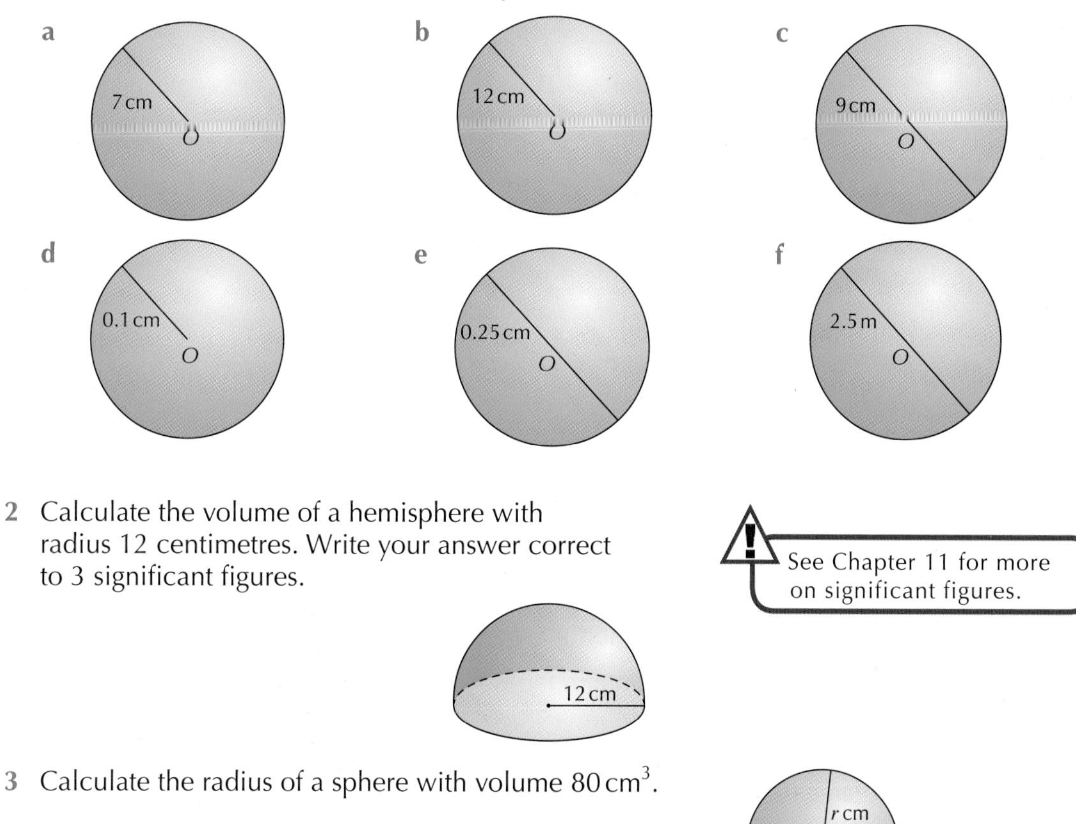

2 Calculate the volume of a hemisphere with radius 12 centimetres. Write your answer correct to 3 significant figures.

See Chapter 11 for more on significant figures.

3 Calculate the radius of a sphere with volume $80\,\text{cm}^3$.

r cm

$V = 80\,\text{cm}^3$

4 A solid metal sphere is melted down and recast into identical smaller spheres.

Rounding your answers appropriately, calculate the number of smaller spheres that can be made when:

a the original radius is 20 cm and the radius of the smaller balls is 4 cm

b the original radius is 5 cm and the radius of the smaller balls is $\frac{1}{2}$ cm

c the original radius is 1 m and the radius of the smaller balls is $\frac{1}{5}$ cm.

5 Metal spheres with radius 3 cm are packed into a rectangular box with dimensions 18 cm × 12 cm × 6 cm. When no more spheres will fit inside, the box is filled with water.

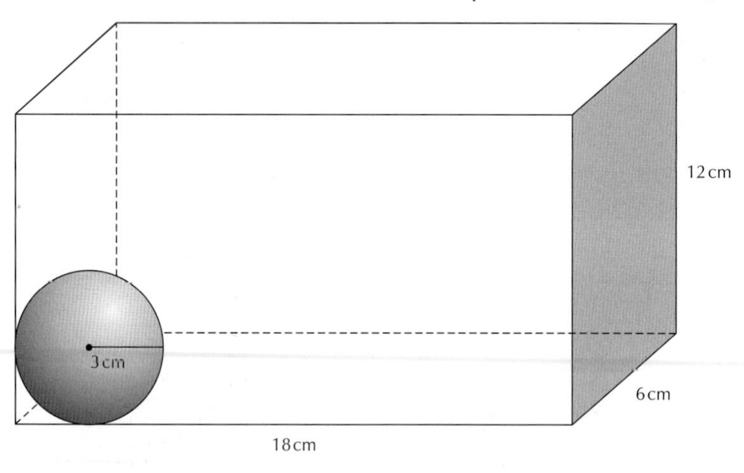

a Calculate how many spheres can fit in the box.

b Calculate the volume of water in the box.

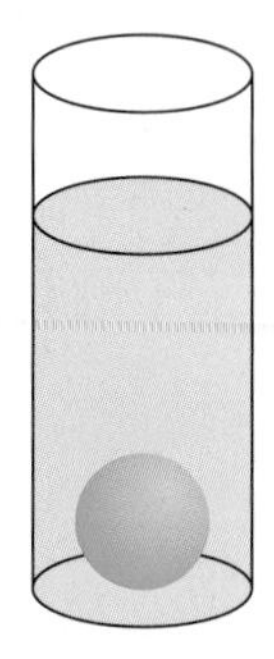

★ **6** A spherical ball is dropped into a cylinder of water.

Calculate the rise in water level if:

a the sphere has a radius of 4 cm and the cylinder has a radius of 7 cm

b the sphere has a radius of 3 cm and the cylinder has a radius of 4 cm.

7 A spherical ball is dropped into a cylinder of water.

Calculate the radius of the ball if:

a the cylinder has a radius of 8 cm and the water level rises 2 cm

b the cylinder has a radius of 20 cm and the water level rises 5 cm.

8 A sphere passes through the eight vertices of a cube with side 6 cm.

Calculate the volume of the sphere. Write your answer correct to 4 significant figures.

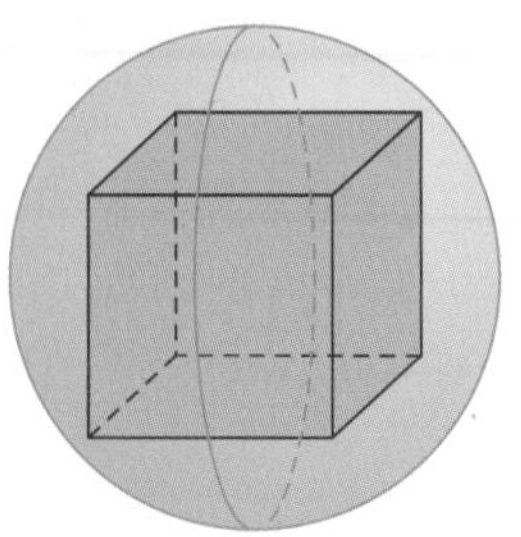

Volume of a pyramid

A **pyramid** has a base and triangular faces which rise to meet at the apex. The base can be *any* polygon. The diagram shows a square-based pyramid.

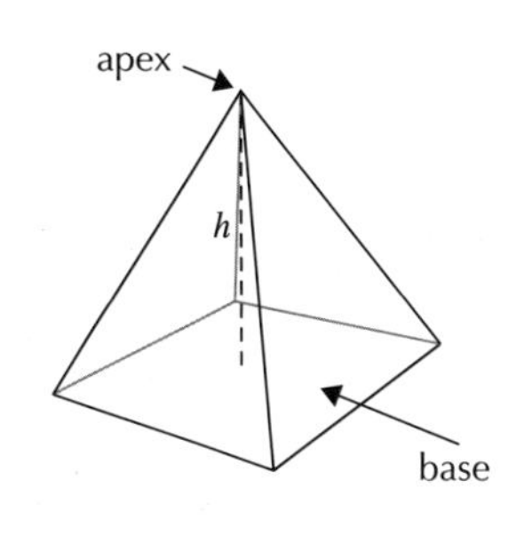

The volume of any pyramid can be calculated using the formula:

volume of a pyramid $= V = \frac{1}{3}Ah$

where A is the area of the base and h is the height of the pyramid. Note that the height, h, is measured vertically from the centre of the base to the apex. This is known as the **perpendicular height**.

Activity

The net of a square-based pyramid is shown. The base of the pyramid has sides $2x$ and the height of the pyramid is x.

- Make your own net and use it to make six identical square-based pyramids.
- Combine the pyramids together to form a cube.
- By considering the volume of the cube, can you derive the formula for the volume of one of the pyramids?

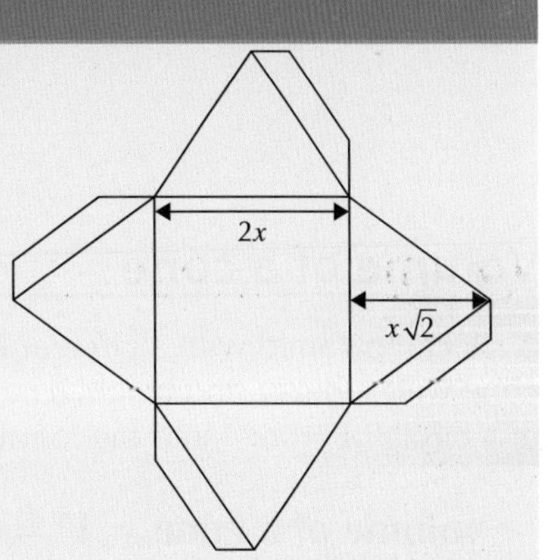

Example 10.3

Calculate the volume of a pyramid with a square base of side 7 cm and a height of 9 cm.

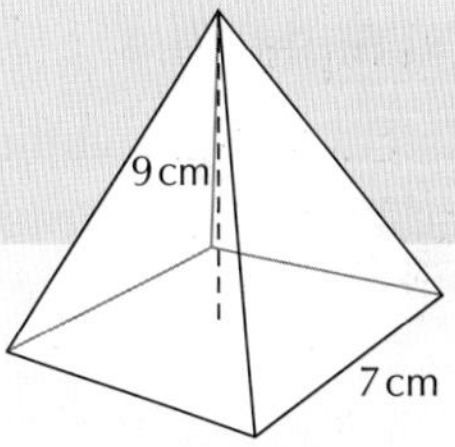

$V = \frac{1}{3}Ah$ — Write down the formula.

$= \frac{1}{3} \times 7^2 \times 9$

$= 147\text{ cm}^3$

Exercise 10B

You may use a calculator for this Exercise and you should round your answers to 1 decimal place unless otherwise stated.

> Be careful with finding the area of the triangular base in part **c**.

★ **1** Calculate the volume of each of these pyramids.

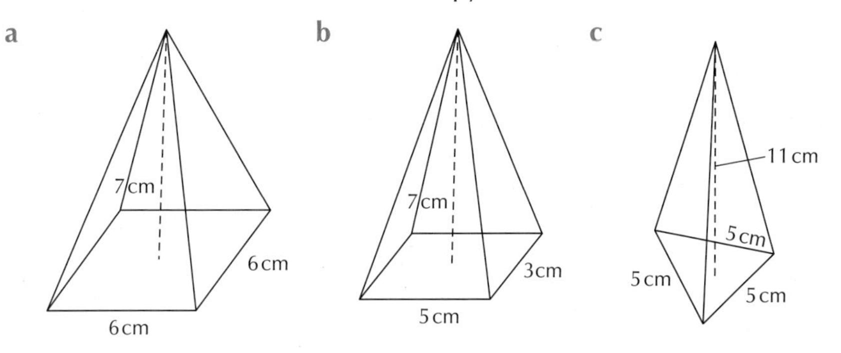

2 Calculate the height of a pyramid with volume $30\,\text{cm}^3$ and base area $8\,\text{cm}^2$. Write your answer correct to 3 significant figures.

★ 3 Six identical square-based pyramids fit together to form a cube of side 12 cm. Calculate the height of one of the pyramids.

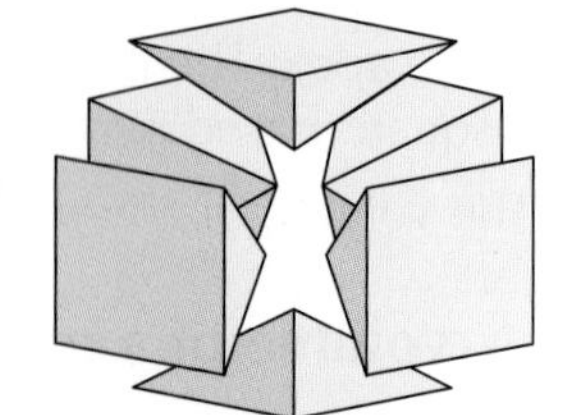

4 Attaching a square-based pyramid of height x onto each of the faces of a cube of side $2x$ forms a tetrahexahedron. Calculate the volume of the tetrahexahedron formed when:

a $x = 3\,\text{cm}$ **b** $x = 0.5\,\text{m}$

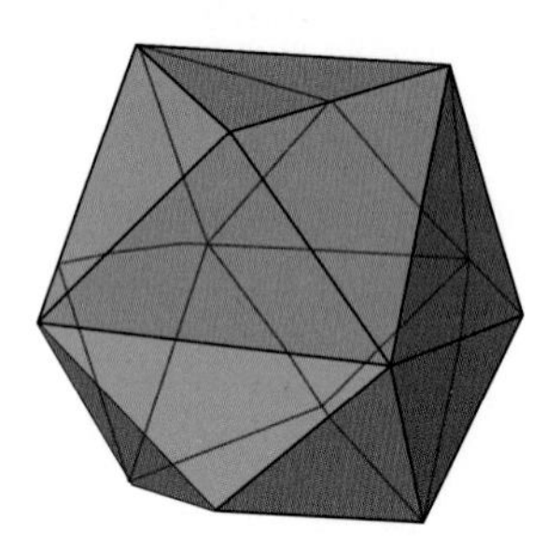

Volume of a cone

A special pyramid with a circular base is called a **cone**. The volume of a cone with radius r and height h can be calculated by substituting the area of a circle ($A = \pi r^2$) into the formula for the volume of a pyramid:

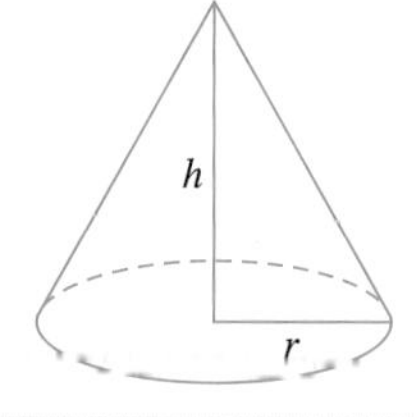

volume of a cone $= V = \frac{1}{3}\pi r^2 h$

Example 10.4

Calculate the volume of a cone with a height of 5 cm and a radius of 2 cm. Write your answer correct to 1 decimal place.

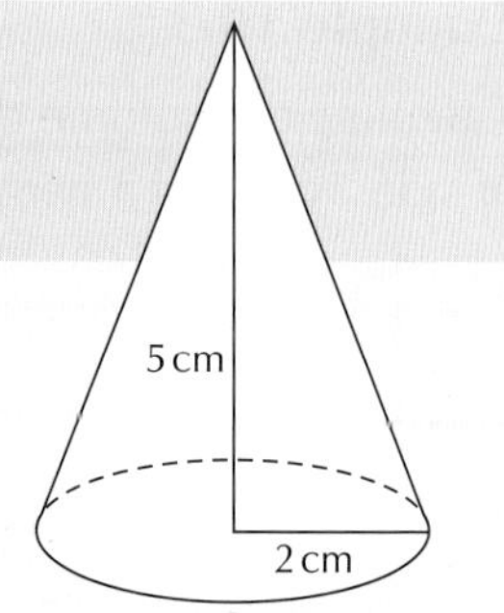

$$V = \frac{1}{3}\pi r^2 h$$

$$= \frac{1}{3}\pi \times 2^2 \times 5$$

$$= 20.9\ \text{cm}^3 \text{ (to 1 d.p.)}$$

Example 10.5

Calculate the height of a conical container with a capacity of 3 litres and a radius of 12 cm. Write your answer correct to 2 significant figures.

$$V = \frac{1}{3}\pi r^2 h$$

$$\frac{1}{3}\pi r^2 h = V$$

$$\frac{1}{3}\pi \times 12^2 \times h = 3000$$

3 litres = 3000 ml = $3000\,\text{cm}^3$.

$$48 \times \pi \times h = 3000$$

$$h = \frac{3000}{48 \times \pi}$$

$$= 20\,\text{cm (to 2 s.f.)}$$

Exercise 10C

You may use a calculator for this Exercise and you should round your answers to 1 decimal place unless otherwise stated.

★ **1** Calculate the volume of each of these cones.

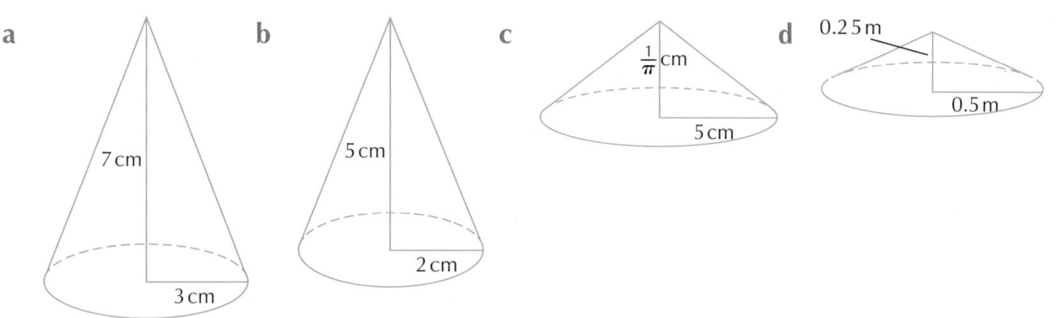

2 Calculate the height of a cone with volume $64\,\text{cm}^3$ and radius 3 cm. Write your answer correct to 3 significant figures.

3 This hat is in the shape of a cone. Calculate the diameter of the hat if it has volume $124\,\text{cm}^3$ and height 21 cm.

4 A frustum is a cone with the top sliced off.

A frustum is created from a cone with height 12 cm and base diameter 8 cm by making a horizontal cut 3 cm from the apex of the cone.

Calculate the volume of the frustum if the diameter of the top of the frustum is 3 cm.

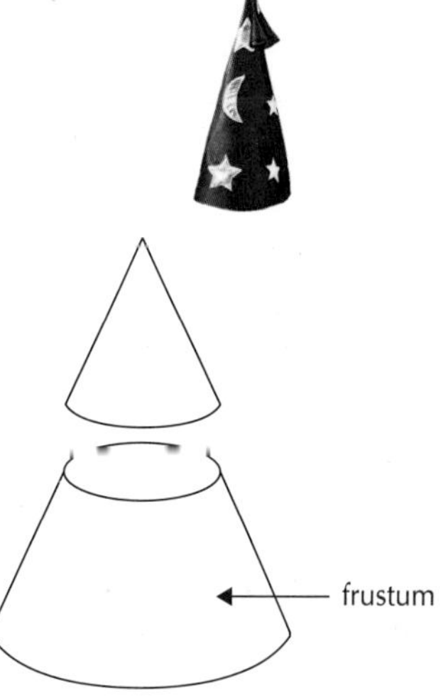

★ **5** The point at the tip of a javelin is formed by sharpening a cylinder of metal to produce a perfect cone at the end with no overall loss in length. If the diameter of the javelin is 3 cm, and the conical tip is of height 18 cm, calculate the volume of metal removed when creating the tip.

Volumes of composite solids

A **composite solid** is formed when two, or more, standard solids are combined to create a new solid. To find the volume of a composite solid we calculate the volumes of the standard solids it is made up of, then add (or subtract) these volumes as required.

Example 10.6

Calculate the volume of this solid made from a cone and a hemisphere. Write your answer correct to 1 decimal place.

$$V(\text{cone}) = \frac{1}{3}\pi r^2 h$$
$$= \frac{1}{3}\pi \times 4^2 \times 6$$
$$= \frac{96\pi}{3}$$
$$= 32\pi \approx 100.530\ldots\ \text{mm}^3$$

Leaving this answer in terms of π will avoid any loss of accuracy. If you use a number with a decimal fraction, avoid rounding errors by working with more decimal places or significant figures than you intend to round to at the end of the question.

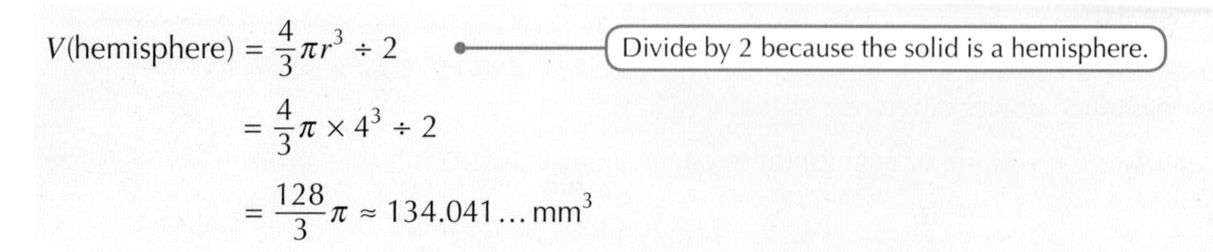

$V(\text{hemisphere}) = \frac{4}{3}\pi r^3 \div 2$ — Divide by 2 because the solid is a hemisphere.

$= \frac{4}{3}\pi \times 4^3 \div 2$

$= \frac{128}{3}\pi \approx 134.041\ldots \text{ mm}^3$

Total volume = V (cone) + V (hemisphere)

$= 32\pi + \frac{128}{3}\pi$

$= \frac{224}{3}\pi$

$= 234.6 \text{ mm}^3$

Exercise 10D

You may use a calculator for this Exercise and you should round your answers to 1 decimal place unless otherwise stated.

1 This toy is made from a cone of height 12 cm attached to a hemisphere with diameter 8 cm. Calculate the volume of the toy. Write your answer correct to 3 significant figures.

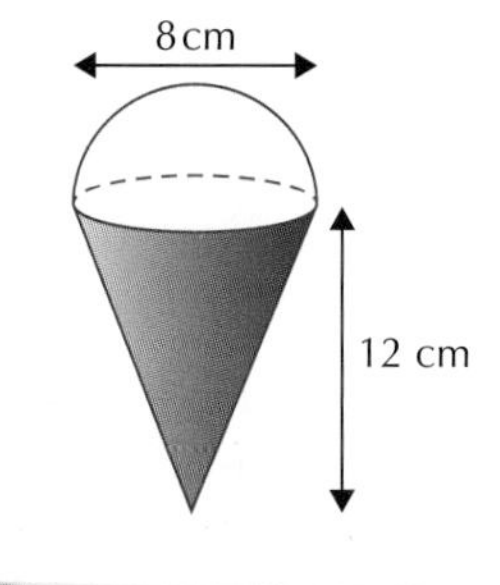

2 A commemorative British £2 coin was produced in 2009 to celebrate the 250th anniversary of the birth of Robert Burns. The coin is 28 mm in diameter and is 4 mm in height. If the diameter of the silver-coloured inner disk is 22 mm, calculate the volume of the gold-coloured outer ring.

3 This mould is made from a special type of plaster that can withstand the very high temperatures required in the manufacturing of silver jewellery. It is in the shape of a cuboid with a conical section removed. It produces silver cones with diameter 4 cm and height 6 cm. Calculate the volume of the mould.

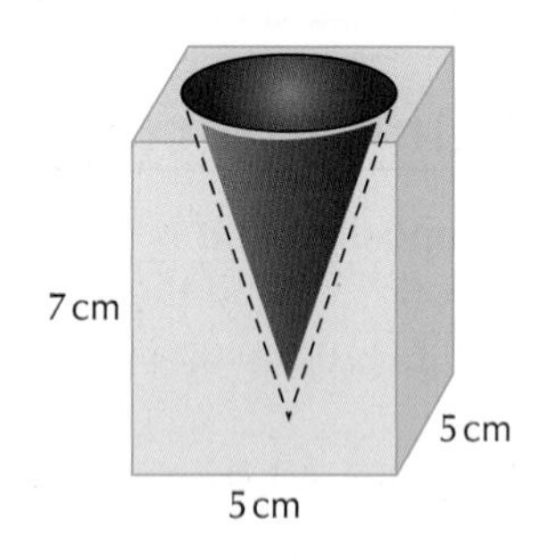

★ 4 A farmer's grain store is made from a cylinder and a hemisphere. Calculate the volume of the grain store if it has a diameter of 6 metres and a total height of 8 metres. Write your answer correct to 4 significant figures.

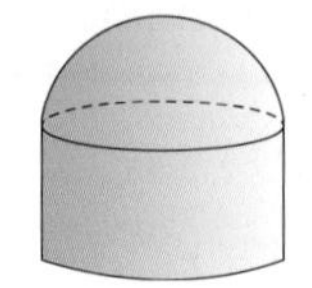

5 A pharmaceutical company is introducing a capsule to replace the standard pill version of their product. The original pill is in the shape of a cylinder with diameter 16 mm and height 3.375 mm. The new capsule is to be in the shape of a cylinder with two hemispherical ends. The diameter of the capsule cylinder is to be 6 mm. Calculate the total length of the capsule if the volume of the capsule and the pill are the same.

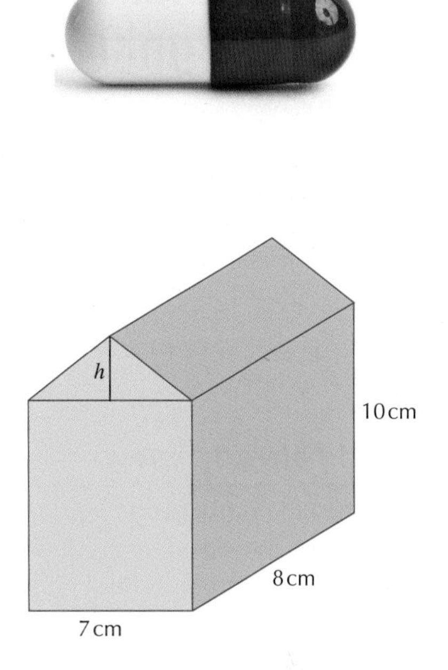

6 An engineer uses models for planning town spaces. The model he uses for houses is in the shape of a cuboid measuring 7 cm wide × 8 cm long × 10 cm high with a triangular prism on top.

a If x is 4 cm, calculate the volume of the model.

b If all dimensions of the model are doubled, work out the new volume.

Activity

Create a proof to verify the following two results established by Archimedes in his book *On the Sphere and Cylinder*. The circumscribed cylinder is the smallest cylinder that can contain the sphere. Note that the height of the cylinder will be twice the radius of the sphere.

Surface area of a sphere = $4\pi r^2$.

1 The volume of a sphere is $\frac{2}{3}$ that of the circumscribed cylinder.

2 The surface area of a sphere is $\frac{2}{3}$ of the total area of the circumscribed cylinder.

- I can use a formula to calculate the volume of a sphere. ★ Exercise 10A Q1
- I can use a formula to calculate the volume of a pyramid. ★ Exercise 10B Q1
- I can use a formula to calculate the volume of a cone. ★ Exercise 10C Q1
- I can use a combination of formulae to calculate the volume of a composite solid. ★ Exercise 10D Q3
- I can solve problems involving the volume of a standard solid. ★ Exercise 10A Q6 ★ Exercise 10B Q3 ★ Exercise 10C Q5

For further assessment opportunities, see the Preparation for Assessment for Unit 1 on pages 89–92.

11 Rounding to a given number of significant figures

In this chapter you will learn how to:

- understand what **significant figures** are and why they are used
- decide how many significant figures a number contains
- round to a given number of significant figures
- avoid losing accuracy in a calculation by rounding appropriately.

You should already know:

- how to round numbers and solutions to a given number of decimal places
- that different degrees of accuracy are acceptable in different situations
- that initial rounding has to be taken into account in practical situations
- how initial rounding can lead to inaccuracy in a calculation
- how to use measuring instruments with straightforward scales to measure length, weight, volume and temperature
- the concept of tolerance in measurements
- how to read scales to the nearest marked, unnumbered division with a functional degree of accuracy
- how to recognise numbers written using scientific notation.

What are significant figures and why are they used?

Science and technology are underpinned by mathematics. Using science we aim to understand how everything in (and beyond) the world we live in works. Scientific experimentation plays a large part in helping us in our quest and mathematics gives us the tools to analyse and describe the connections and patterns we discover. When we experiment we take measurements and these measurements always have a limited level of accuracy. These measurements are often used in calculations which give results which also have a limited level of accuracy. Depending on the type (or number) of calculations we do, we find our final answer can be far less accurate than the original measured value. **Significant figures** (sometimes called significant digits, and abbreviated to sig figs or s.f.) are used to keep track of the accuracy of measurements and allow us to be confident in the accuracy of calculations involving measured values.

Imagine you are measuring the radius of a circle and intend to calculate its area. Using a ruler, where the smallest marked division on the scale is one millimetre, you measure the radius to be 5.4 cm. Squaring this and multiplying the result by π gives an area of 91.608 841 778 678... cm^2.

Of course it is possible that the radius is *exactly* 5.4 cm but it is more likely that it is a little bit less or a little bit more, as shown below.

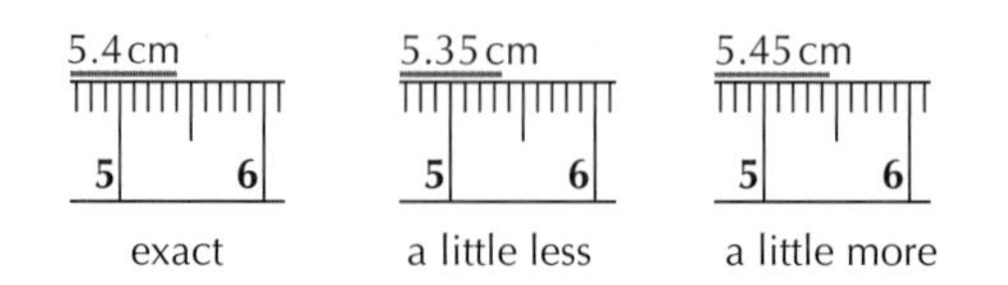

A standard **tolerance** for measured values is ± half of the smallest division on your measuring scale, meaning that the measured radius could be anything from 5.35 cm to 5.45 cm. The area of the circle could actually be as low as 89.92024… or as high as 93.31316…. So leaving our original answer with 16 digits of precision now seems a bit silly!

In this example our measured radius of 5.4 cm has 2 significant digits. This means that results from any calculation done using this value cannot meaningfully have more than 2 significant digits. Any digits after the first two are due to calculation rather than measurement and are meaningless. Therefore the area of the circle in the example above should be given as 92 cm^2 to 2 significant figures (2 s.f.).

Rules when using significant figures

Errors are a large part of experimental science and how we handle them in calculations is very interesting but can get quite complicated. At this level, you don't need to know how to do this but if you are interested you could ask your teacher (maths or science) to explain a little more. For now, all you need to know are the rules that are used to decide if a digit is significant or not. Follow these rules and you won't go wrong.

Rule 1: Non-zero digits are always significant

946.27 has 5 significant figures. Note that the '9' is the first (or most) significant figure as it has the largest place value and represents 900. The '7' is the least (or fifth) significant figure as it has the smallest place value and represents 7 hundredths.

Rule 2: One (or more) zeros with a non-zero digit on either side of it is significant

30 050.303 has 8 significant figures.

Rule 3: Leading zeros are never significant

The leading zeros in the number 0.000**045** are shaded. Only the 4 and the 5 are significant. This number has 2 significant figures.

Rule 4: Trailing zeros are not significant

670 000 has only 2 significant figures.

The only exception to Rule 4 is if we know **for certain** that the number is accurate. For example, usually the number 100 has 1 significant figure. However, in the context of a statement such as 'there are 100 pennies in one pound', then, since we know this to be exact, we say that (in this context), 100 has 3 significant figures.

Usually we wouldn't include the zeros when writing a number such as 65.300. Writing the zeros indicates that this is a measured value and that we are certain that the number is accurate. So we say that the number 65.300 has 5 significant digits. An example of this would be a set of electronic scales that are accurate to 0.001 of a gram. If you weigh an item on these scales and get a reading of 2.400 you can be sure that the last two digits have been measured accurately and the number has 4 significant figures.

Example 11.1

How many significant figures are in these numbers?

a 450 b 45.1 c 4005 d 4050.05 e 0.0405

a 2 b 3 c 4 d 6 e 3

Exercise 11A

1 Write down the number of significant figures in each of these numbers.

a 75284 b 4904 c 420 d 93 002

e 45 000 f 2090 g 1300 h 80

★ 2 Write down the number of significant figures in each of these numbers.

a 0.0045 b 0.003 c 0.0304 d 0.7008

e 4.173 f 24.8009 g 204.001 h 300.08

★ 3 Write down the number of significant figures in each of these numbers.

a 0.030 b 0.8000 c 0.01010 d 0.0050

e 4.50 f 28.090 g 302.3020 h 1200.1200

4 Write down the number of significant figures in each of these numbers.

a 4.52×10^5 b 4.003×10^7 c 6×10^{-4} d 6.0×10^{-9}

★ 5 Write down the number of significant figures in the numbers in the following statements.

a There are 360° in a full turn.

b There are 1000 millilitres in a litre.

c There are 25 000 000 litres of water in an Olympic-sized swimming pool.

d There are 1 000 000 cubic centimetres (cm^3) in one cubic metre (m^3).

e The capacity of Hampden Park is 52 000.

f 1.01 billion people use Facebook every month.

Rounding to a given number of significant figures

The rules for rounding with significant figures are the same as those for rounding to decimal places. If you are asked to round to n significant figures, examine the $(n + 1)$th significant figure. If it is less than 5, you round down, and if it is 5 or more, you round up.

Always check carefully whether a question asks you to round to significant figures or to decimal places because they often give different results.

Example 11.2

Round 0.05436 to 2 significant figures.

0.05436 — Use **Rule 3**: leading zeros are never significant.

→ 0.054 (to 2 s.f.) — Look at the 3rd significant figure; 3 is less than 5 so round down. Note that 0.05436 rounded to 2 decimal places gives 0.05.

Example 11.3

Round 2.3685 to 3 significant figures.

2.3685

→ 2.37 (to 3 s.f.)

Note that 2.3685 rounded to 3 decimal places gives 2.369.

Example 11.4

Round 5300.045 500 to 7 significant figures.

5300.045 500

→ 5300.046 (to 7 s.f.)

Use **Rule 2** (zeros with a non-zero digit on either side are significant), so the zeros in the tens and units columns are significant here. 5300.045 500 is accurate to 6 decimal places.

Example 11.5

Round 70.499 to 4 significant figures.

70.499

→ 70.50 (to 4 s.f.)

If asked for 4 significant figures you must give 4, even if it means including zeros.

Example 11.6

Round 5349 to 1 significant figure.

5349

→ 5000 (to 1 s.f.)

Include zeros to preserve place value.

Exercise 11B

1 Round each of the following numbers to 1 significant figure.

a 3421	b 87	c 9523	d 14080
e 0.0385	f 0.0049	g 0.349	h 0.955
i 232.45	j 8.09	k 200.5	l 9.885
m 5.71×10^8	n 3.018×10^5	o 8.0×10^{-5}	p 3.50×10^{-7}

★ 2 Round each of the following to the number of significant figures given in brackets.

a 6975 (2)	b 30055 (4)	c 449 (1)	d 35 230 (3)
e 0.845 (2)	f 0.0379 (2)	g 0.305 09 (3)	h 0.003 005 5 (4)
i 24.542 (3)	j 700.034 (2)	k 840.078 (4)	l 450.0045 (5)

The effects of rounding on overall accuracy

Rounding too early in a calculation or series of calculations can affect the accuracy. For example, rounding 100.1 to 100 before multiplying by 1000 means the difference between 100 100 and 100 000.

A more significant example is that of the American Patriot defence missile system, which is designed to track and intercept an incoming missile. On one occasion a malfunction was attributed to rounding errors created in the computer code that controlled the Patriot. Although the original error was as small as ± 0.000 000 095, this value was multiplied by the number of tenths of a second in 100 hours giving an error of ± 0.34 seconds. Since the incoming missile was travelling at about 1500 m/s, and would have covered over half a kilometre in that time, the originally tiny error became significant enough to cause the Patriot defence system to fail.

Example 11.7

An engineer is trying to calculate the height of this cuboid. He knows it has a volume of 2.2 cm^3 and that the base is a square of side 1.2 cm.

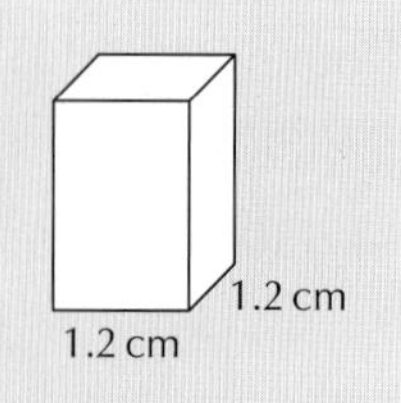

The engineer asks his assistant to calculate the area of the base which she does by squaring 1.2, i.e. $1.2^2 = 1.44$. Since the given dimensions have 2 significant figures, the assistant rounds her answer to 2 significant figures and tells the engineer that the base has an area 1.4 cm^2.

The engineer then calculates the height by dividing the volume of the cuboid by the area of its base, i.e. $2.2 \div 1.4 = 1.571...$ which he rounds to 2 significant figures, giving a height of 1.6 cm.

Later, the engineer checks the height is correct by calculating $2.2 \div 1.2^2 = 1.5277...$ which he rounds to 2 significant figures giving a height of 1.5 cm.

As this example shows, rounding before the very end of a calculation can affect overall accuracy.

Exercise 11C

★ **1** **a** Carry out each of these calculations rounding to 2 significant figures at **every** step.

i $2.3 \times 2.3 \times 1.2$ ii $3.6^2 \times \pi \div 2$ iii $\sqrt{(2.5)^2 + (1.5)^2}$

b Repeat the calculations in **a**, this time rounding to 2 significant figures only at the very end.

c List the calculations above that produced different answers.

> ⚠ For **ii** use the π key on your calculator.

Activity

There are more rules for significant figures when it comes to adding, subtracting, multiplying and dividing. Work with a partner, or in a group, to research these rules and then answer the following questions, giving your answer to the correct number of significant figures.

1 $8264 + 770 + 34.78$ **2** $5.772 - 2.48$

3 $(4.2 \times 10^3)(3.562 \times 10^4)$ **4** $8305 \div 0.020$

- I understand and can use the rules that determine if a digit is significant or not. ★ Exercise 11A Q2, Q3
- I understand why knowing whether a number is exact affects the appropriate number of significant figures. ★ Exercise 11A Q5
- I know how to round to a given number of significant figures. ★ Exercise 11B Q2
- I know that rounding throughout a calculation can have an effect on overall accuracy. ★ Exercise 11C Q1

For further assessment opportunities, see the Preparation for Assessment for Unit 1 on pages 89–92.

Preparation for Assessment: Unit 1

The questions in this section cover the minimum competence for the content of the course in Unit 1. They are a good preparation for your unit assessment. In an assessment you will get full credit only if your solution includes the appropriate steps, so make sure you show your thinking when writing your answers.

Remember that reasoning questions marked with the ⚙ symbol expect you to interpret the situation, identify a strategy to solve the problem and then clearly explain your solution. If a context is given you must relate your answer to this context.

The use of a calculator is allowed.

Applying numerical skills to simplify surds/expressions using the laws of indices (Chapters 1 and 2)

1 Simplify the following giving your answer as a surd in its simplest form.

a $\sqrt{20}$ b $\sqrt{18}$ c $\sqrt{55}$ d $\sqrt{630}$

2 Express $\sqrt{45} - \sqrt{5} + \sqrt{20}$ as a surd in its simplest form.

3 Two of the following have the same value:

$\sqrt{7} \times \sqrt{9}$ $2\sqrt{21}$ $3\sqrt{7}$

Which one has a different value? You must give a reason for your answer.

4 Find the mean of $\sqrt{8}$, $\sqrt{32}$ and $\sqrt{72}$.

5 Simplify the following.

a $6k^2 \div 3k^4$ b $3g^3 \times 5g^{-\frac{1}{3}}$ c $-4a^3 \times -5a^{-2}$

d $\dfrac{y^{\frac{5}{2}} \times y^{\frac{1}{2}}}{y^{-1}}$ e $4p^3 \div 3p^{-6}$ f $\dfrac{4a^5b^3}{8ab}$

6 A new phone has a resolution of 1136 × 740 pixels. Calculate the number of pixels in the screen and express your answer in scientific notation.

7 Light travels at a speed of 3.00×10^5 km/second. The greatest distance from the Sun to Mars is 2.38×10^8 km. How long does light from the Sun take to reach Mars?

8 The surface area of a sphere can be calculated using the formula $SA = 4\pi r^2$.

Consider the dwarf planet Pluto as a sphere with a diameter of 2360 km. Calculate the surface area of Pluto and express your answer in scientific notation to 3 significant figures.

Applying algebraic skills to manipulate expressions (Chapters 3, 4 and 5)

9 Expand and simplify as appropriate.

a $x(2x + y)$ b $t(t - 4) - 3(t + 4)$ c $3u(8 - 5u)$

d $(p - 3)(p - 6)$ e $(x + 3)(x - 4)$ f $(3s + 2)(2s - 1) - (4s + 3)$

10 Factorise.

a $6x + 24$ b $t^2 - 2t$ c $36 - a^2$

d $4t^2 - 49$ e $x^2 + 7x + 12$ f $x^2 + 3x - 10$

11 Express the following in the form $(x + p)^2 + q$.

a $x^2 + 6x + 15$ b $x^2 + 8x + 13$ c $x^2 - 10x - 5$ d $x^2 + 2x - 9$

Applying algebraic skills to manipulate expressions (Chapters 6 and 7)

12 Simplify the following fractions.

a $\dfrac{(x+1)(x+2)}{(x+2)^2}$ $x \neq -2$ b $\dfrac{(2x+2)^2}{(2x+2)(x-7)}$ $x \neq 7$ or -1

c $\dfrac{3c^2 + 15c}{2c + 10}$ $c \neq -5$ d $\dfrac{6z^3 - 2z^2}{6z - 2}$ $z \neq \frac{1}{3}$

13 Write each of the following as a single fraction.

a $\dfrac{1}{x} + \dfrac{x}{3y}$ $x, y \neq 0$ b $\dfrac{2}{b} - \dfrac{a}{4b}$ $a, b \neq 0$

c $\dfrac{r}{w} \times \dfrac{3}{w}$ $r, w \neq 0$ d $\dfrac{12}{3e} \div \dfrac{4}{6f}$ $e, f \neq 0$

14 Express each of the following as a single fraction.

a $s + \frac{2}{3} \times 3s - \frac{1}{4}$ b $\frac{c}{c} + 2 - \frac{2}{5} - c$

15 Write each of the following as a single fraction.

a $\dfrac{-x}{x^2 - 4} \div \dfrac{x}{(x+4)^2}$ b $\dfrac{x-2}{3-x} \times \dfrac{-x+4}{x-2}$

Applying geometric skills to manipulate expressions (Chapters 8, 9, 10 and 11)

16 Calculate the gradient of the line connecting each of the following pairs of points.

a A (2, 3) and B (4, 6) b C (–5, 1) and D (–2, –5)

17 a Find the gradient of the line connecting each of the following pairs of points.

i A (–2, –4) and B (2, 4)

ii C (–5, –5) and D (2, 4)

iii E (3, 7) and F (–1, –1)

b Which two of the three lines are parallel to each other?

18 The diagram shows a sector of a circle, centre O. The radius of the circle is 9.6 centimetres and angle AOB is 65°.

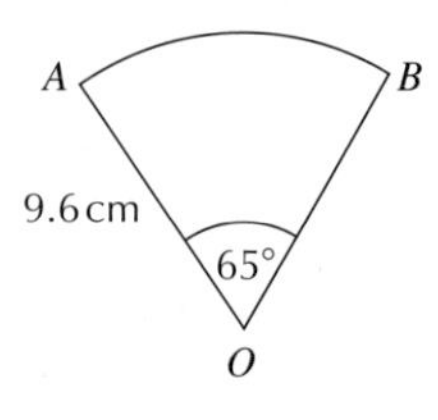

a Calculate the area of the sector.

b Find the length of the minor arc AB, giving your answer to 2 significant figures.

19 A manufacturer of playground equipment makes merry-go-rounds. Each merry-go-round is constructed of 12 identical sectors of radius 3.5 m.

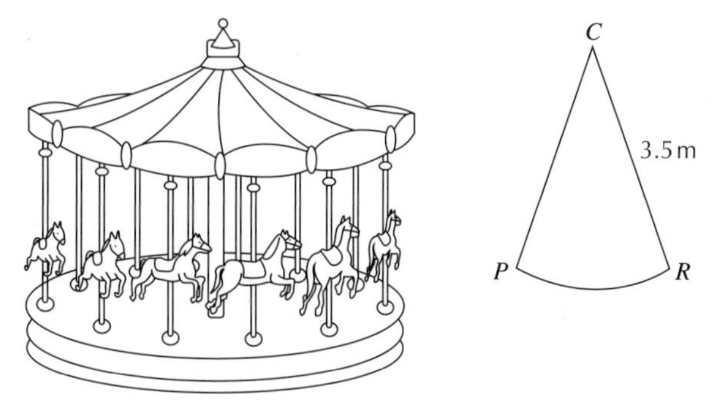

a Calculate the area of one sector of the merry-go-round.

b The manufacturer paints each merry-go-round using three colours. They have 20 pots of each colour in stock. Each pot contains enough paint to cover 3 m^2. How many merry-go-rounds can the manufacturer fully paint with its current stock of paint?

20 A dart board is divided into 20 equal sectors. The board has a diameter of 451 mm.

a Calculate the area of each sector. Give your answer to 4 significant figures.

b Each sector is bounded by wire. Calculate the length of wire around the outside of one sector.

21 Calculate the volume of a sphere with radius 68 mm. Give your answer to 3 significant figures.

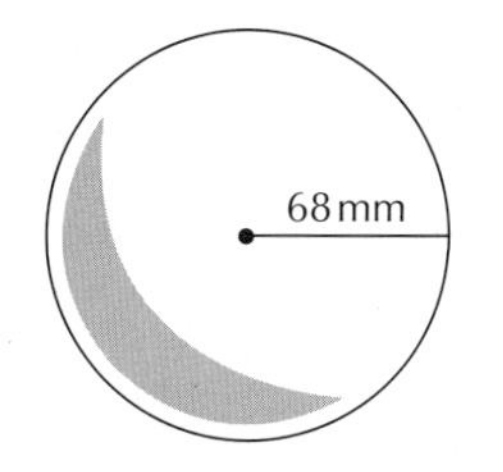

22 Calculate the volume of a cone with radius 23 cm and height 65 cm. Give your answer to 2 significant figures.

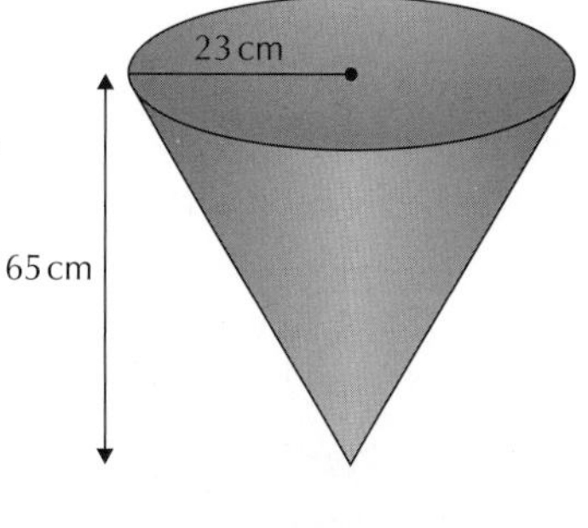

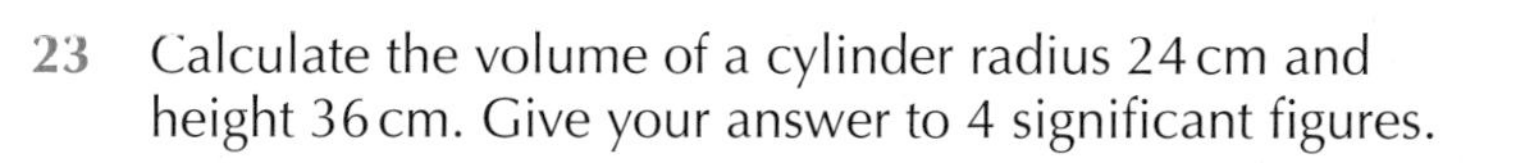

23 Calculate the volume of a cylinder radius 24 cm and height 36 cm. Give your answer to 4 significant figures.

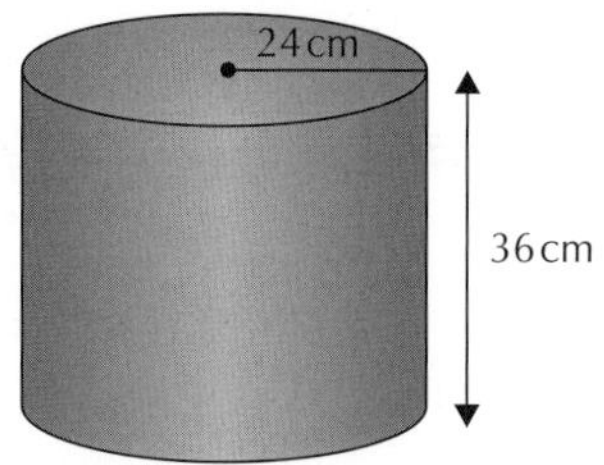

24 A toy manufacturer stores liquid resin in cylindrical containers with a radius of 35 cm and height of 50 cm.

a Calculate the volume of the cylinder.

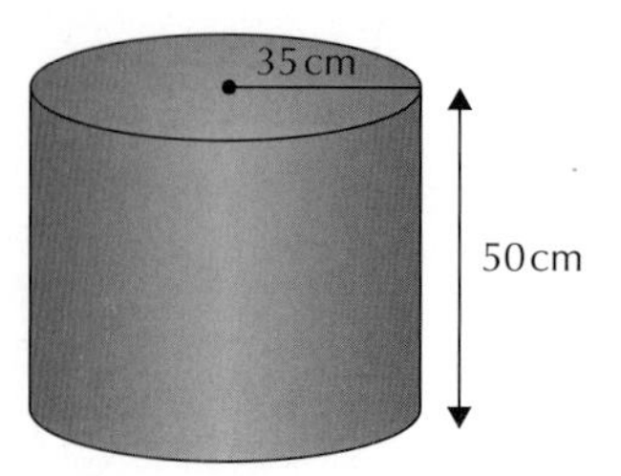

The manufacturer makes spherical balls from the resin.
Each ball has a radius of 5 cm.

b Calculate the volume of a single ball.

c How many balls can be made from the resin in one completely full cylinder?

25 The pen stylus for a tablet computer is constructed of a cylinder with a hemisphere attached to the top.

a The stylus has a radius of 6 mm and the cylinder has a height of 12 cm.
Calculate the volume of the stylus.

b The manufacturer creates each stylus using an injection moulding process. It buys its plastic in cubes of side length 50 cm.
How many pen styluses can be created from each cube of plastic?

26 Round the following to:

i 1 significant figure ii 2 significant figures iii 3 significant figures

a 50 894 b 0.256 c 58.647 d 0.004 587

12 Determining the equation of a straight line, given the gradient

In this chapter you will learn how to:

- find the **equation of a straight line** from a graph or table, expressing your answer in the form $y = mx + c$
- identify the **gradient** and ***y*-intercept** from a graph or equation
- use the formula $y - b = m(x - a)$, or equivalent, to find the equation of a straight line, given one point on the line and the gradient
- use the **general linear equation** $Ax + By + C = 0$
- identify the gradient and *y*-intercept from various forms of the equation of a straight line
- find the **midpoint** of a line segment
- use and apply function notation f(x)
- apply reasoning skills to solve related problems.

You should already know:

- when the gradient of a straight line is zero or undefined
- when a straight line has positive or negative gradient
- how to draw and recognise the graph of a linear equation
- how to find the gradient of a straight line using the formula $m = \frac{y_2 - y_1}{x_2 - x_1}$

Find the equation of a straight line in the form $y = mx + c$

Straight lines can be expressed in terms of the gradient m and the y-intercept using the equation $y = mx + c$. This equation can be used to draw the straight line. An extension of this is to be able to work out the equation of a line given suitable information such as pairs of points on the line.

Example 12.1

Find the equation of the line shown in:

a blue **b** red.

a y-intercept = 0 — Blue line goes through the origin (0, 0).

So: $c = 0$

There are two methods for working out the gradient of a line.

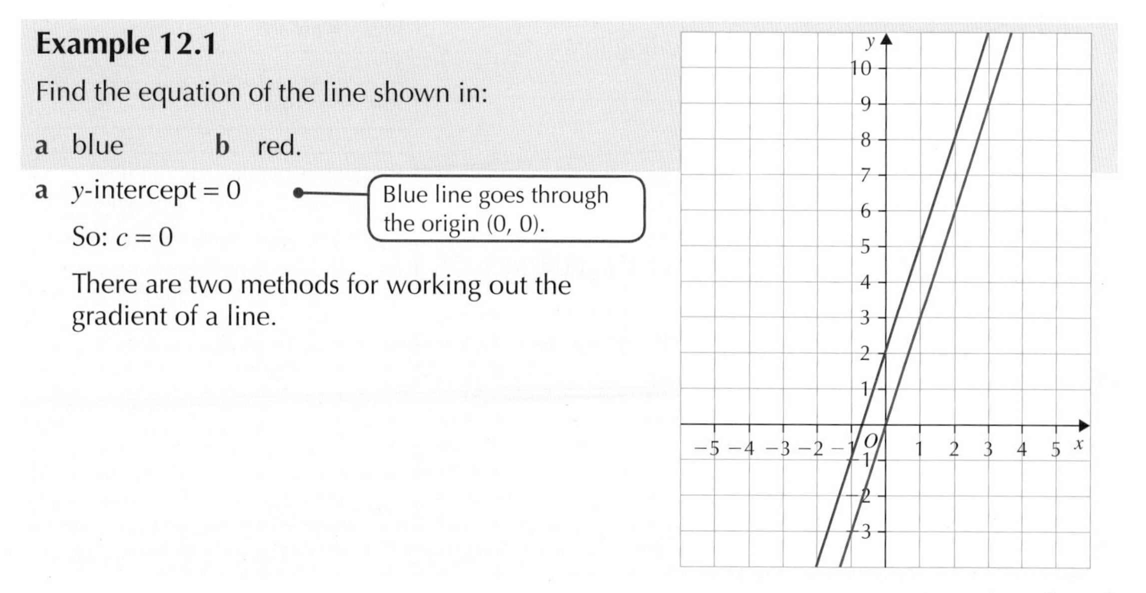

(continued)

Method 1

From the graph you can see that, for every increase of 1 in the x-direction, there is an increase of 3 in the y-direction, therefore the gradient is 3.

$$m = \frac{\text{vertical}}{\text{horizontal}} = \frac{3}{1} = 3$$

So: $m = 3$

Method 2

Choose and label two points on the graph.

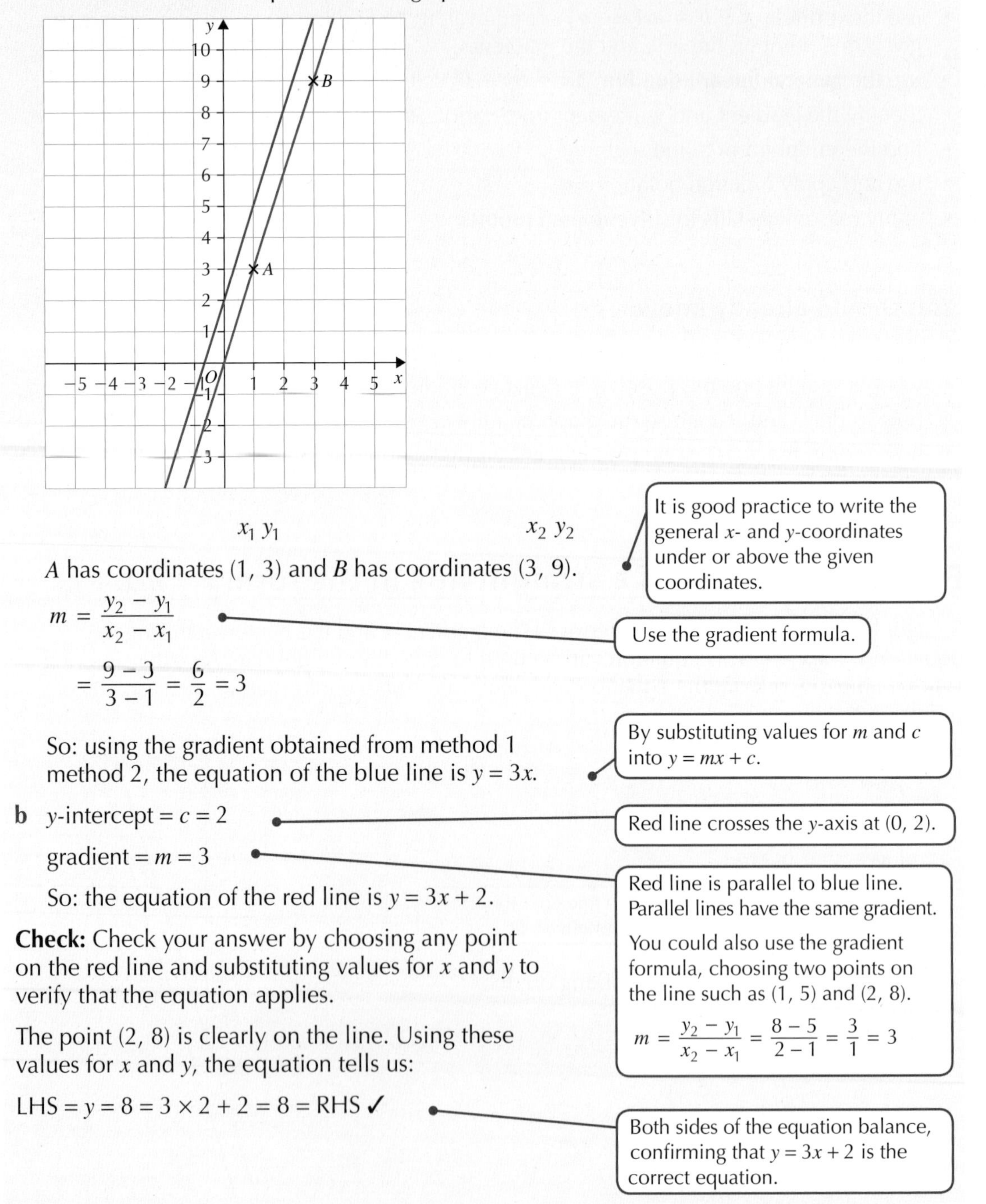

$\quad\quad\quad\quad x_1\ y_1 \quad\quad\quad\quad\quad\quad\quad x_2\ y_2$

A has coordinates (1, 3) and B has coordinates (3, 9).

It is good practice to write the general x- and y-coordinates under or above the given coordinates.

$$m = \frac{y_2 - y_1}{x_2 - x_1}$$

Use the gradient formula.

$$= \frac{9 - 3}{3 - 1} = \frac{6}{2} = 3$$

So: using the gradient obtained from method 1 method 2, the equation of the blue line is $y = 3x$.

By substituting values for m and c into $y = mx + c$.

b y-intercept $= c = 2$

Red line crosses the y-axis at (0, 2).

gradient $= m = 3$

Red line is parallel to blue line. Parallel lines have the same gradient.

You could also use the gradient formula, choosing two points on the line such as (1, 5) and (2, 8).

$$m = \frac{y_2 - y_1}{x_2 - x_1} = \frac{8 - 5}{2 - 1} = \frac{3}{1} = 3$$

So: the equation of the red line is $y = 3x + 2$.

Check: Check your answer by choosing any point on the red line and substituting values for x and y to verify that the equation applies.

The point (2, 8) is clearly on the line. Using these values for x and y, the equation tells us:

LHS $= y = 8 = 3 \times 2 + 2 = 8 =$ RHS ✓

Both sides of the equation balance, confirming that $y = 3x + 2$ is the correct equation.

Looking at the two lines in Example 12.1 you can see that for every value of x, the y value of the red line is 2 more than the y value of the blue line. A table of coordinate values also shows this:

x	−2	−1	0	1	2	3
$y = 3x$	−6	−3	0	3	6	9
$y = 3x + 2$	−4	−1	2	5	8	11

Example 12.2

Find the equation of the straight line.

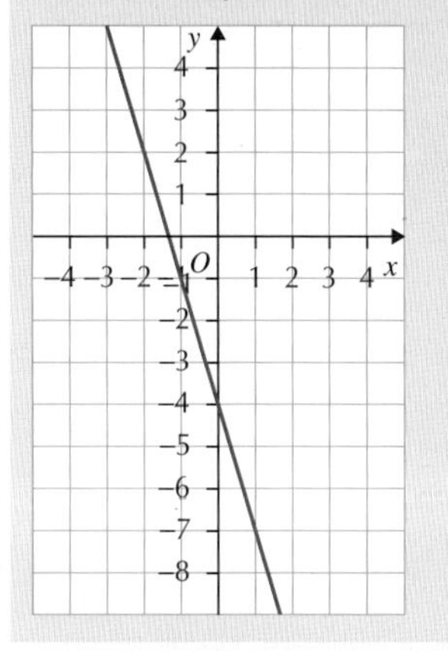

y-intercept $= c = -4$

So: $y = mx - 4$

Find the value of the y-intercept, which is the point at which $x = 0$. The line intercepts the y-axis at (0, −4).

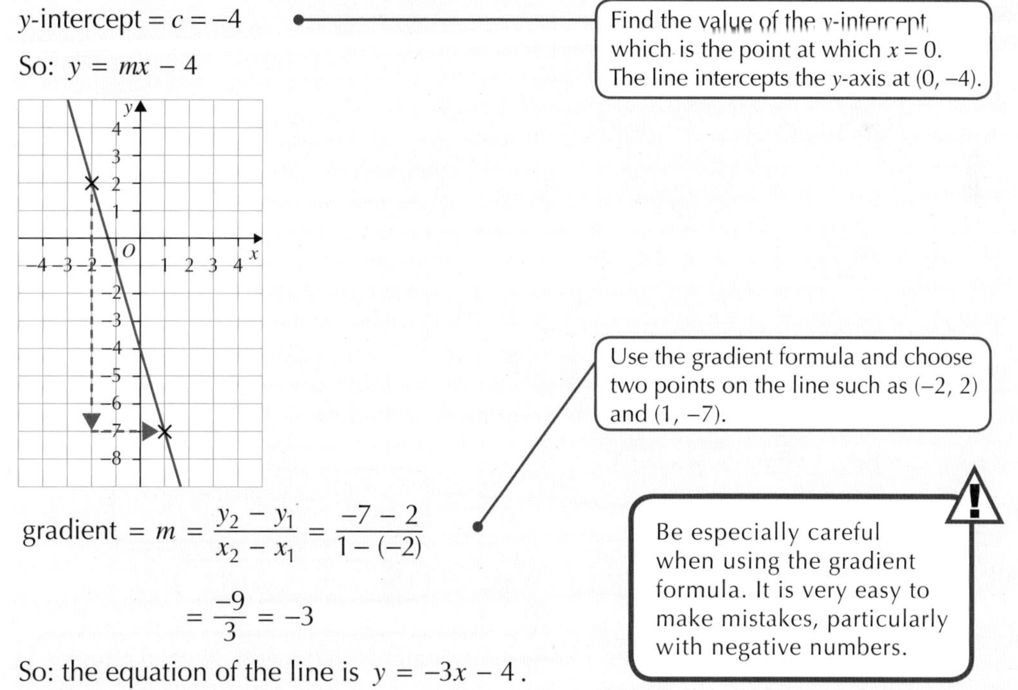

Use the gradient formula and choose two points on the line such as (−2, 2) and (1, −7).

$$\text{gradient} = m = \frac{y_2 - y_1}{x_2 - x_1} = \frac{-7 - 2}{1 - (-2)}$$

$$= \frac{-9}{3} = -3$$

So: the equation of the line is $y = -3x - 4$.

Be especially careful when using the gradient formula. It is very easy to make mistakes, particularly with negative numbers.

Example 12.3

Find the equation of the straight line.

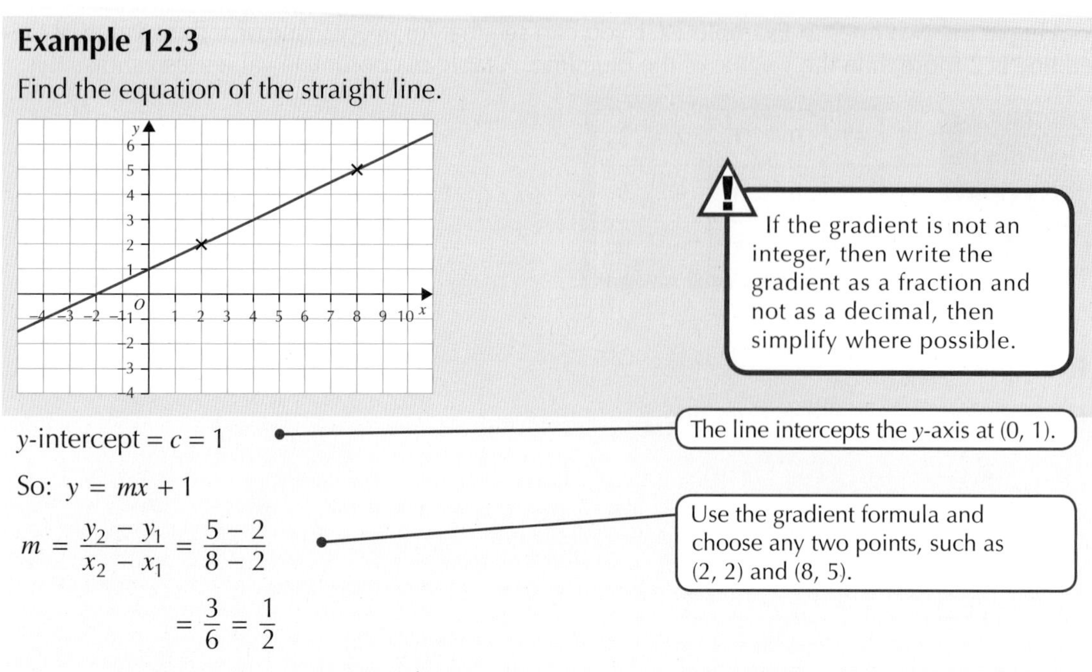

If the gradient is not an integer, then write the gradient as a fraction and not as a decimal, then simplify where possible.

y-intercept $= c = 1$

The line intercepts the y-axis at (0, 1).

So: $y = mx + 1$

$$m = \frac{y_2 - y_1}{x_2 - x_1} = \frac{5-2}{8-2}$$

Use the gradient formula and choose any two points, such as (2, 2) and (8, 5).

$$= \frac{3}{6} = \frac{1}{2}$$

So: the equation of the line is $y = \frac{1}{2}x + 1$.

Example 12.4

For the given line, express p in terms of t.

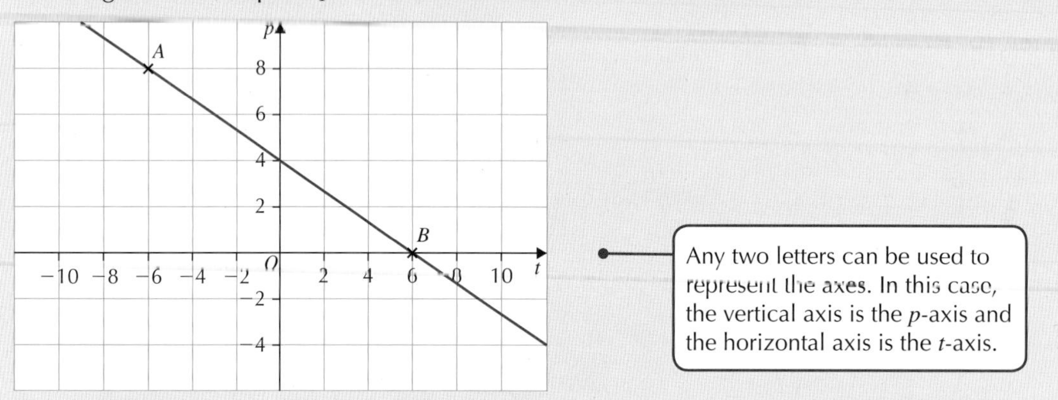

Any two letters can be used to represent the axes. In this case, the vertical axis is the p-axis and the horizontal axis is the t-axis.

'Express p in terms of t' is simply asking for the equation of the line.

So $y = mx + c$ can be rewritten as $p = mt + c$.

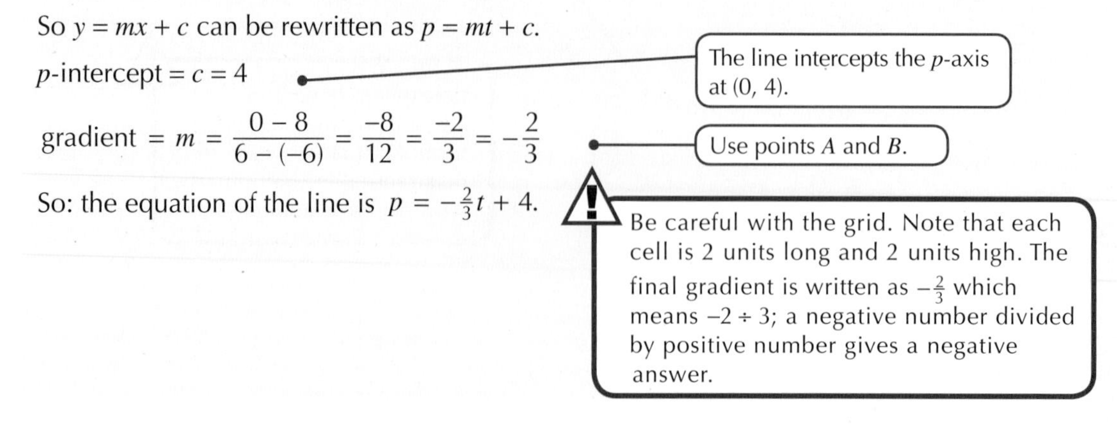

p-intercept $= c = 4$

The line intercepts the p-axis at (0, 4).

$$\text{gradient} = m = \frac{0-8}{6-(-6)} = \frac{-8}{12} = \frac{-2}{3} = -\frac{2}{3}$$

Use points A and B.

So: the equation of the line is $p = -\frac{2}{3}t + 4$.

Be careful with the grid. Note that each cell is 2 units long and 2 units high. The final gradient is written as $-\frac{2}{3}$ which means $-2 \div 3$; a negative number divided by positive number gives a negative answer.

Example 12.5

A straight line passes through points A (0, 5) and B (4, –3) as shown. Find the equation of the line.

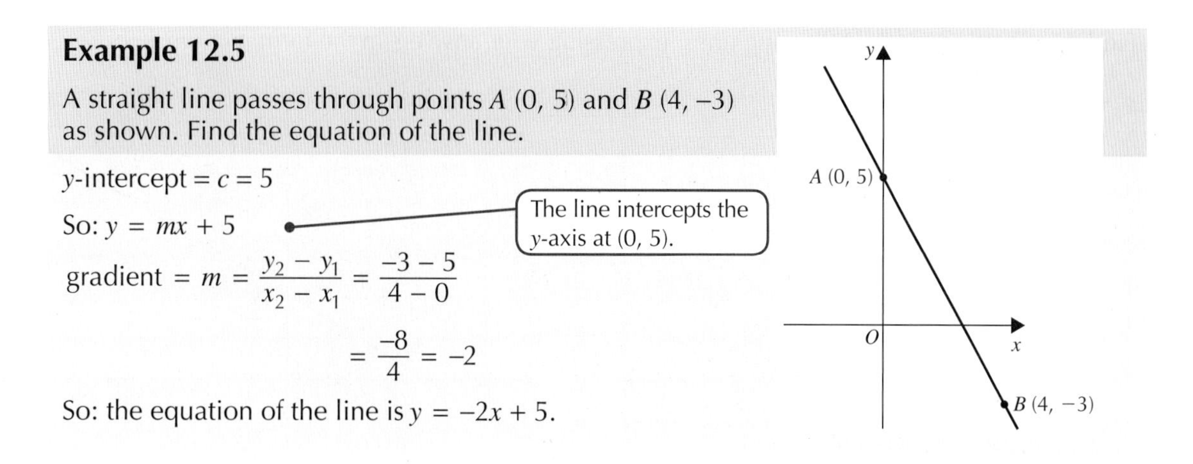

y-intercept = c = 5

So: $y = mx + 5$

$$\text{gradient} = m = \frac{y_2 - y_1}{x_2 - x_1} = \frac{-3-5}{4-0}$$

$$= \frac{-8}{4} = -2$$

So: the equation of the line is $y = -2x + 5$.

Exercise 12A

1 Find the equations of the following lines, expressing your answers in the form $y = mx + c$.

a

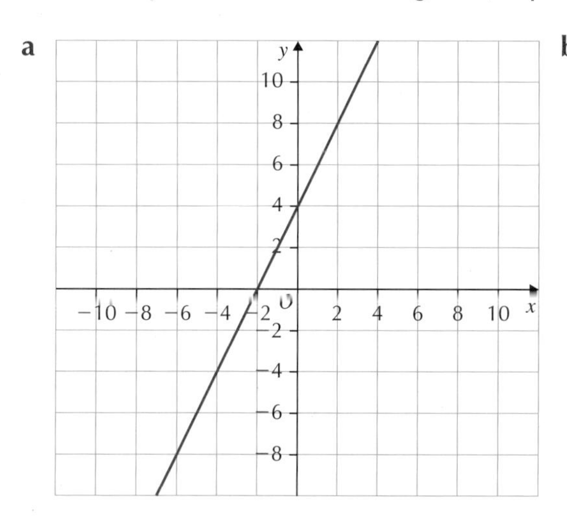

b

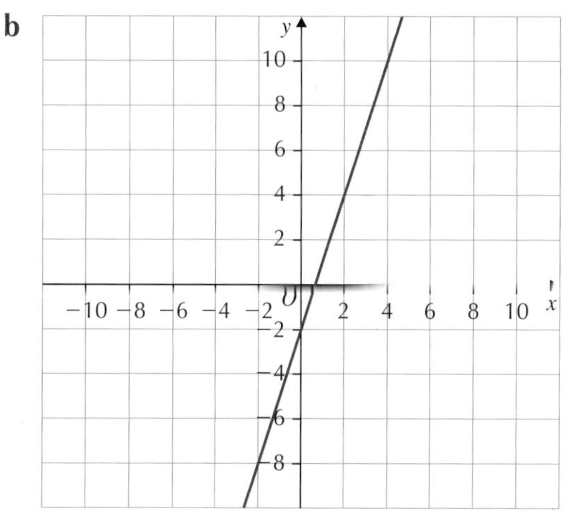

c

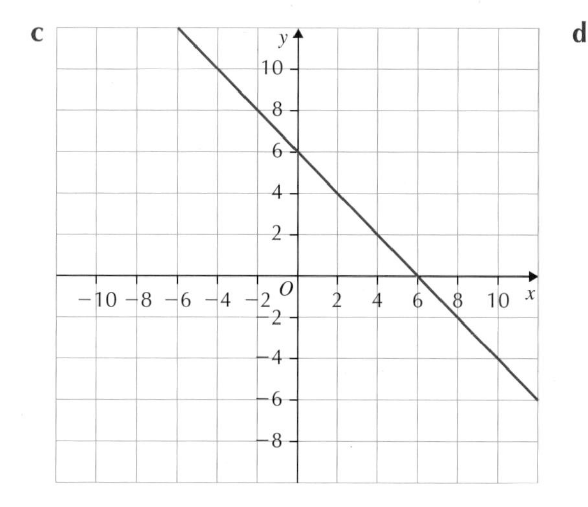

d

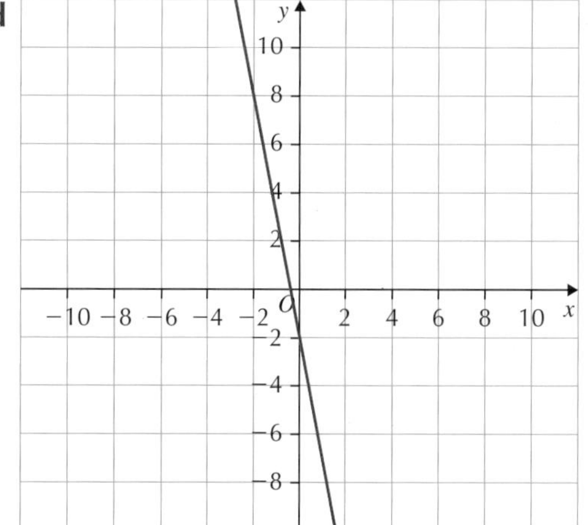

★ **2** Find the equations of the following lines, expressing your answers in the form $y = mx + c$.

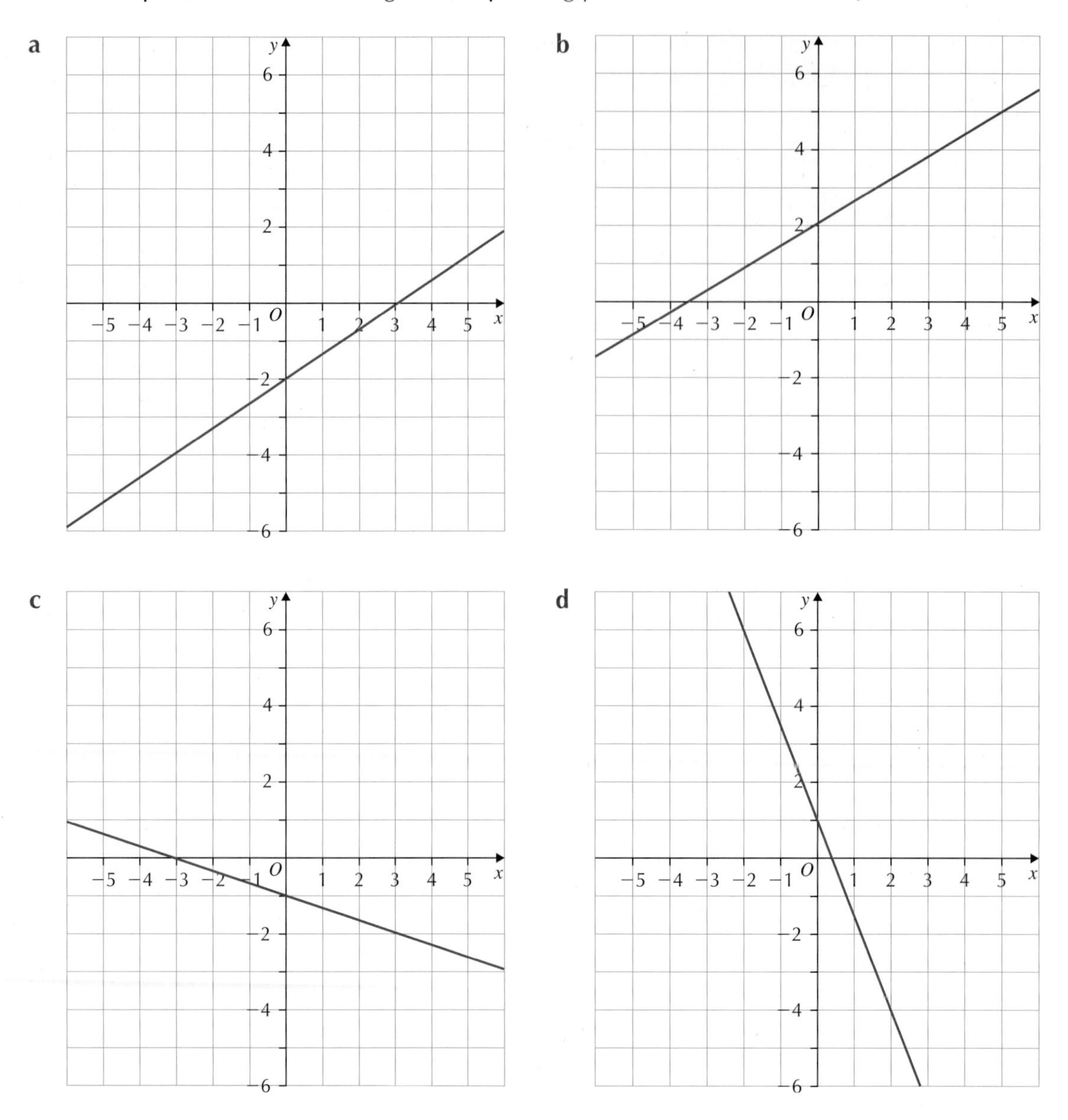

3 Find the equations of the following lines.

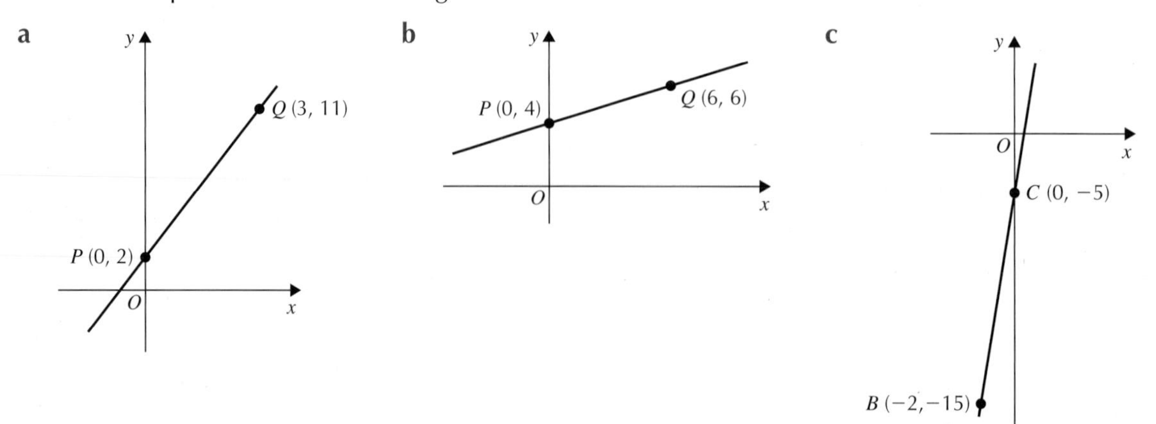

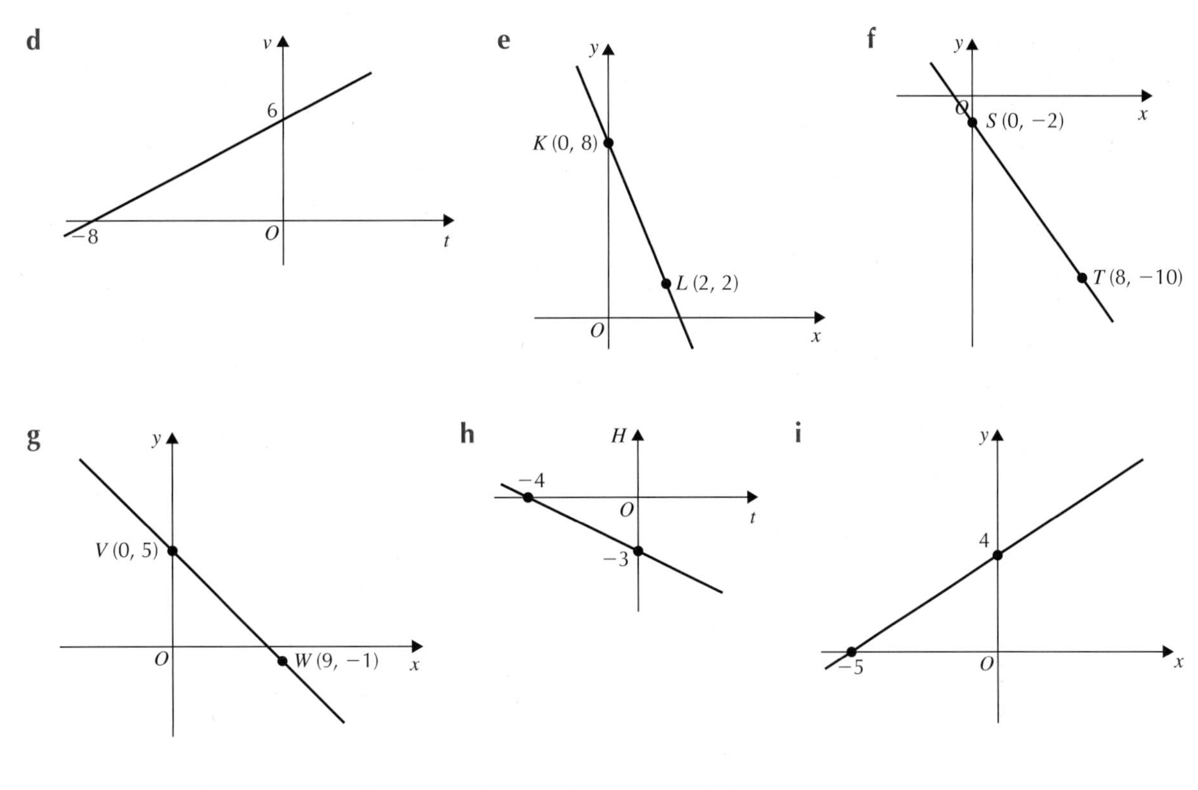

4 The graph shows the taxi fare, £C, charged by a council for a journey of m miles.

a Express C in terms of m

b After a night out in Inverness, Shona and her four friends need a taxi to take them home to Wick, a distance of 104 miles. They share the fare equally. How much will each of them have to pay?

c John's journey cost £43.90. How far did he travel?

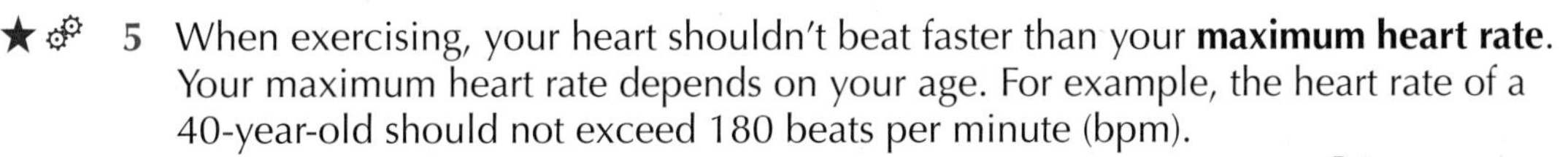

★ **5** When exercising, your heart shouldn't beat faster than your **maximum heart rate**. Your maximum heart rate depends on your age. For example, the heart rate of a 40-year-old should not exceed 180 beats per minute (bpm).

a The relationship between the maximum heart rate, R, and age, t, can be expressed in the form $R = a - bt$. Using the graph, find the values of a and b.

b Lorna is 55 years old. When running on the treadmill her heart rate is 172 bpm. Should Lorna slow down? Justify your answer.

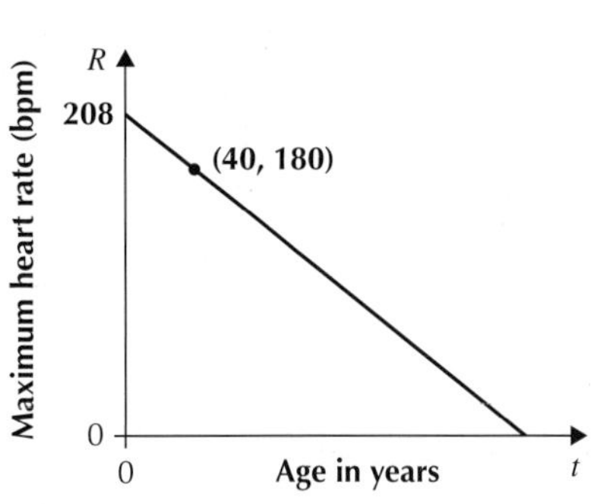

6 To measure temperature, two scales are commonly used: Celsius (or Centigrade) and Fahrenheit. European countries tend to use the Celsius scale, whereas the United States uses the Fahrenheit scale.

The table below shows some temperatures in degrees Celsius, along with their equivalents in degrees Fahrenheit.

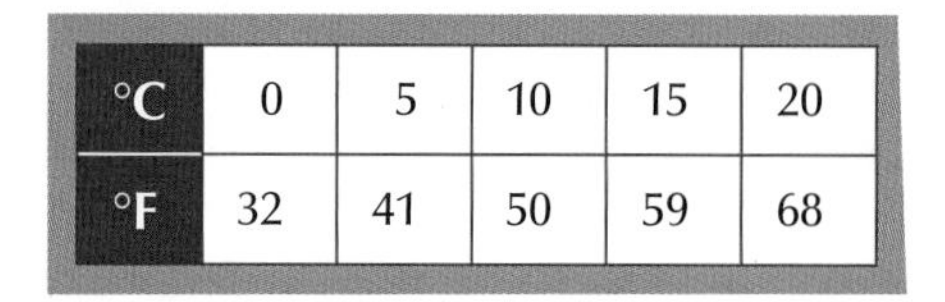

°C	0	5	10	15	20
°F	32	41	50	59	68

a The equation connecting degrees Celsius, C, and degrees Fahrenheit, F, is:

$$F = pC + q$$

Find the values of p and q.

b On a particularly hot summer's day in Scotland the temperature is 30 °C. An American tourist asks if you know how hot it is. How would you reply?

c The lowest ever recorded temperature on the surface of the Earth was −128.6°F. What would this be in degrees Celsius?

Use the formula $y - b = m(x - a)$ to find the equation of a straight line

The equation of the line $y = mx + c$ can be rewritten as $y - b = m(x - a)$. This can be used when you know the gradient of the line and one point on the line.

Example 12.6

Write down the equation of the straight line shown. It passes through points P (2, 5) and E and F.

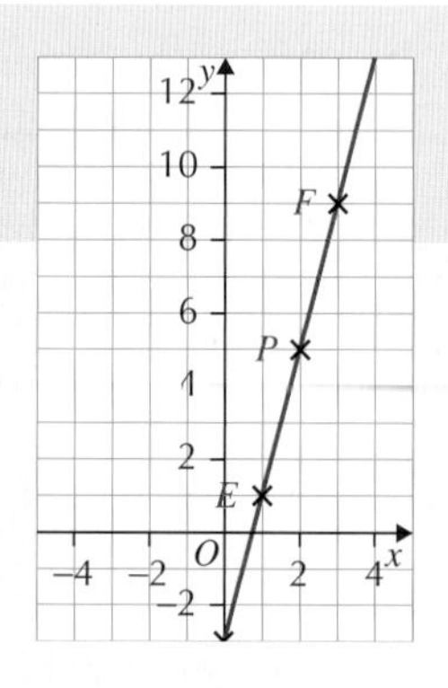

$$m_{EP} = \frac{1-5}{1-2} = \frac{-4}{-1} = 4$$

Find the gradient of the line using P and either E or F.

$$m_{FP} = \frac{9-5}{3-2} = \frac{4}{1} = 4$$

gradient = m = 4

So: $y = 4x + c$

If you know the gradient of the line you can choose any x and use it to find y. That is, the equation allows you to find any point $Q(x, y)$ on the line:

$$m_{PQ} = \frac{y-5}{x-2} = 4$$

Use P (2, 5).

$$y - 5 = 4(x - 2)$$

Rearrange the formula.

So: the equation of the line is $y - 5 = 4(x - 2)$.

You can see from Example 12.6 that the coordinates of P (2, 5) and the gradient 4 appear in the equation of the line. This method can be applied to **any** straight line passing through a known point (a, b) and having gradient m. The **general equation** for any such line is:

$$\boldsymbol{y - b = m(x - a)}$$

Example 12.7

A straight line has gradient 2 and passes through the point (6, 3). Write down the equation of the line.

$a = 6, b = 3, m = 2$ — Identify values of a, b and m.

$y - 3 = 2(x - 6)$ — Substitute values into the general equation $y - b = m(x - a)$.

Note: It is possible to write this equation in the form $y = mx + c$ by expanding the brackets and simplifying:

$y - 3 = 2(x - 6)$

$y - 3 = 2x - 12$ — Expand the brackets.

$y = 2x - 9$ — Add 3 to both sides.

So: $y - 3 = 2(x - 6)$ and $y = 2x - 9$ are different forms of the equation for the same line.

Example 12.8

A straight line has gradient $\frac{3}{4}$ and passes through the point (8, −5). Find the equation of the line, expressing your answer in the form $y = mx + c$.

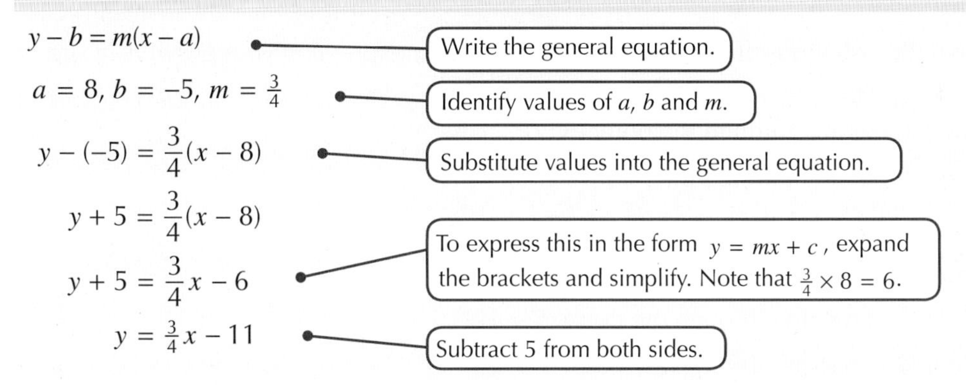

y − b = m(x − a)
Write the general equation.
a = 8, b = −5, m = 3/4
Identify values of a, b and m.
y − (−5) = 3/4(x − 8)
Substitute values into the general equation.
y + 5 = 3/4(x − 8)
y + 5 = 3/4x − 6
To express this in the form y = mx + c, expand the brackets and simplify. Note that 3/4 × 8 = 6.
y = 3/4x − 11
Subtract 5 from both sides.

Example 12.9

A straight line passes through the points (3, −5) and (−1, 7).

Find the equation of the line and express it in the form $y - b = m(x - a)$.

$$\text{gradient} = m = \frac{y_2 - y_1}{x_2 - x_1}$$ — Use the gradient formula.

$$= \frac{7 - (-5)}{-1 - 3}$$

$$= \frac{12}{-4} = -3$$

(continued)

$a = 3$, $b = -5$, $m = -3$

Identify values for a, b and m; choose (3, –5) as the known point.

$y - (-5) = -3(x - 3)$

Substitute values into the general equation.

So: $y + 5 = -3(x - 3)$

Note: Notice that choosing (–1, 7) as the known point gives the equation:

$y - 7 = -3(x + 1)$

Although the two equations appear to be different, they are equivalent, once they are expanded and simplified.

$y + 5 = -3(x - 3)$	$y - 7 = -3(x + 1)$
$y + 5 = -3x + 9$	$y - 7 = -3x - 3$
$y = -3x + 4$	$y = -3x + 4$

As shown in Example 12.9, if you know more than one point on a line you can choose which point to use in the formula.

Exercise 12B

1 Write down the equations of the following straight lines in the form $y - b = m(x - a)$.

a gradient 2, passing through the point (6, 1)

b gradient 5, passing through the point (3, –8)

c gradient –4, passing through the point (–1, 5)

d gradient $\frac{1}{3}$, passing through the point (–2, –9)

e gradient t, passing through the point (c, d)

2 Write down the equations of the following straight lines, expressing your answers in the form $y = mx + c$.

a gradient 3, passing through the point (5, 2)

b gradient 2, passing through the point (–4, 6)

c gradient –8, passing through the point (2, –3)

d gradient $\frac{1}{2}$, passing through the point (–6, –8)

e gradient $-\frac{2}{3}$, passing through the point (5, –12)

f gradient $-\frac{3}{5}$, passing through the point (–4, 10)

g gradient t, passing through the point (c, d)

★ **3** Find the equations of the straight lines passing through the following pairs of points, expressing your answers in the form $y - b = m(x - a)$.

a (2, 7) and (5, 10) **b** (–1, 2) and (4, 11) **c** (–5, 1) and (1, 13)

d (3, –4) and (–1, –12) **e** (9, –7) and (–1, –2) **f** (–8, 8) and (–3, 11)

4 Find the equations of the straight lines passing through the following pairs of points, expressing your answers in the form $y = mx + c$.

a $\left(\frac{1}{2}, \frac{3}{4}\right)$ and $\left(\frac{2}{3}, \frac{5}{6}\right)$ **b** (1.6, 2.4) and (2, 4)

c (s, t) and (t, s) **d** $\left(\sqrt{32}, \sqrt{18}\right)$ and $\left(\sqrt{128}, \sqrt{50}\right)$

Remember to simplify the gradient as much as possible. Refer to Chapter 1 for help with surds.

★ 5 A linear equation can be used to describe the relationship between a person's ideal body weight, w kilograms, and their height, h inches.

One formula predicts the ideal body weight of a man who is 70 inches (5 feet 10 inches) tall to be 71 kilograms. The same formula predicts the ideal weight of a man who is 80 inches (6 feet 8 inches) tall to be 90 kilograms.

a Using the information above, express w in terms of h.

b This formula is only used for men who are over 50 inches (4 feet 2 inches) tall. Explain why the formula can't be used for men who are particularly short.

6 Here are some facts about an isosceles triangle ABC:

- $AB = AC$
- the gradient of BC is 0
- A has coordinates (3, 1)
- the gradient of AB is 5

What is the equation of AC?

7 A straight line passes through the points (c, c^2) and (d, d^2). Show that the equation of the line is $y = (c + d)x - cd$.

Use and apply function notation f(x)

A **function** takes numbers in and applies a rule to them to give a new set of numbers. A simple example is $f(x) = x^2$. In this case, the function squares the number, so for $x = 2$, $f(x) = 4$, for $x = 5$, $f(x) = 25$, and so on.

Much of modern mathematics uses functions with greater or lesser degrees of complexity and can often require lengthy calculations.

Functions are expressed in terms of their **domain** and their **range**. The domain is the set of **input** numbers, and the range is the set of **output** numbers after the function has been applied to the input numbers.

Consider the line with equation $y = 2x + 3$. The table shows corresponding y values for given x values.

x	−3	−2	−1	0	1	2	3
y	−3	−1	1	3	5	7	9

Think of the equation as a function which connects the x numbers and the y numbers: the x numbers go in and we use our rule to get the y numbers out.

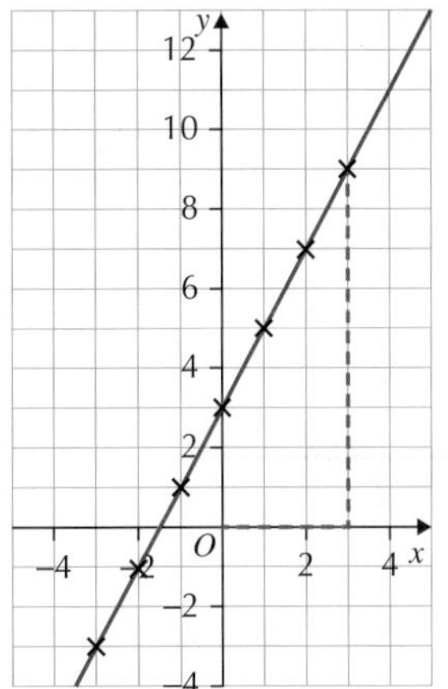

The graph illustrates this function, where the rule is 'multiply by 2 and add 3'.

So:

$f(2) = 7 \quad f(0) = 3 \quad f(-1) = 1$

More generally we can write $f(x) = 2x + 3$.

Every point on the graph has an x-coordinate and a y-coordinate. The x-coordinate represents a number in the domain, and the y-coordinate represents a number in the range.

For the function $f(x) = 2x + 3$ the following statements are equivalent:

- the point $(3, 9)$ is on the graph of the function $y = 2x + 3$
- $f(3) = 9$

Note also that y and $f(x)$ are the same, so we can write $y = f(x)$.

Example 12.10

For the function $f(x) = 3x - 4$ find:

a $f(2)$ **b** $f(-3)$.

a $f(x) = 3x - 4$

$f(2) = 3(2) - 4$ — Substitute x with 2.

$= 6 - 4 = 2$ — Calculate f(2).

b $f(x) = 3x - 4$

$f(-3) = 3(-3) - 4$ — Substitute x with −3.

$= -9 - 4 = -13$ — Calculate f(−3).

Example 12.11

The function g is given by $g(x) = x^2 - 4x$. Find the value of $g(-5)$.

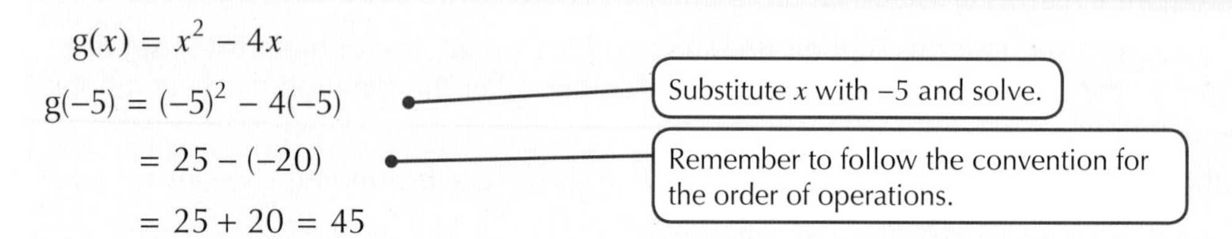

$g(x) = x^2 - 4x$
$g(-5) = (-5)^2 - 4(-5)$
$= 25 - (-20)$
$= 25 + 20 = 45$
Substitute x with −5 and solve.
Remember to follow the convention for the order of operations.

Example 12.12

The function h is given by $h(x) = 7 - 4x$. Find the value of x for which $h(x) = 13$.

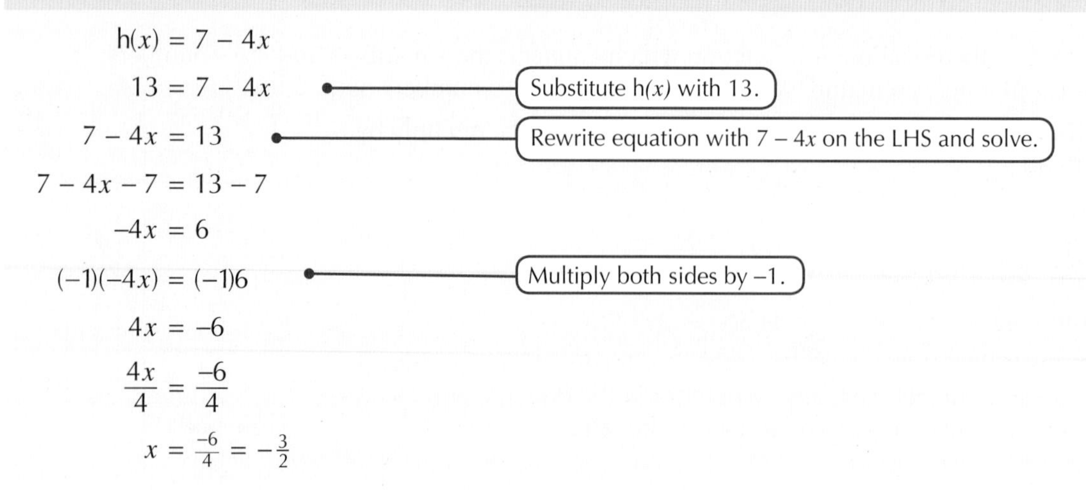

$h(x) = 7 - 4x$
$13 = 7 - 4x$
$7 - 4x = 13$
$7 - 4x - 7 = 13 - 7$
$-4x = 6$
$(-1)(-4x) = (-1)6$
$4x = -6$
$\frac{4x}{4} = \frac{-6}{4}$
$x = \frac{-6}{4} = -\frac{3}{2}$
Substitute h(x) with 13.
Rewrite equation with 7 − 4x on the LHS and solve.
Multiply both sides by −1.

Exercise 12C

1 For the function $f(x) = 5x - 2$:

a find f(4), f(0) and f(−2)

b work out the value of b if the point $(1, b)$ is on the graph of $y = f(x)$

c find the value of x for which $f(x) = 13$.

2 For the function $g(x) = 1 - 2x$:

a find g(5), g(−3) and $g\left(\frac{2}{3}\right)$

b find the value of p for which $g(p) = 11$.

3 For the function $f(x) = \frac{2}{3}x - 5$:

a find f(−6) and $f\left(\frac{3}{4}\right)$

b work out the value of p if the point $(p, 1)$ is on the graph of $y = f(x)$.

4 For the function $f(x) = x^2 - 4x - 5$:

a find f(3), f(0) and f(−2)

b find the values of x for which $f(x) = 0$

c find the value of x for which $f(x) = -9$.

d What feature of the graph of $y = f(x)$ does your answer to part **c** confirm?

5 For the function $h(t) = 100t - 5t^2$:

a find h(2) and h(−3)

b find the values of t for which $h(t) = 0$.

6 For the function $f(x) = x^3 - x^2 - 2$:

a find f(3) and f(−1)

b find $f\left(\frac{1}{2}\right)$.

★ **7** Functions f and g are given by $f(x) = 5x + 1$ and $g(x) = -x + 9$.

Find the value of x for which $f(x) = g(x)$.

8 Functions f and g are given by $f(x) = 3x + 1$ and $g(x) = -2x - 1$.

Solve the equation $4f(x) - 3g(x) = 1$.

9 For the functions $p(x) = \dfrac{3x - 2}{4}$ and $q(x) = \dfrac{x - 1}{3}$:

a express $p(x) - q(x)$ as a single function

b solve the equation $p(x) = q(x)$.

10 For the function $f(x) = 4^x$:

a evaluate f(3), f(−2) and $f\left(\frac{3}{2}\right)$

b solve the equation $(f(x))^3 = \frac{1}{4}f(x)$.

The general linear equation $Ax + By + C = 0$

The following are three different **forms** of equation used to describe a straight line:

- $x = a$ and $y = b$, which represent the equations of **vertical** lines (gradient undefined) and **horizontal** lines (gradient 0)
- $y = mx + c$, which represents a straight line having gradient m and intercepting the y-axis at the point $(0, c)$
- $y - b = m(x - a)$, which represents a straight line having gradient m and which passes through the point (a, b).

See Chapter 8 for the gradients of vertical and horizontal lines.

There is a fourth form for writing down the equation of a straight line. The **general linear equation** is of the form:

$$Ax + By + C = 0$$

where A, B and C are constants.

For example, the equation $3x + 2y - 5 = 0$ is a linear equation written in general form, where $A = 3$, $B = 2$ and $C = -5$.

It is important in National 5 and also in Higher that you can carry out the following processes:

- rearrange a linear equation from the form $Ax + By + C = 0$ into the form $y = mx + c$ and hence determine the gradient of the line
- given the equation of a line in any form, be able to determine where the line intercepts both the x- and y-axes.

Example 12.13

Find the gradient of the straight line having equation $2x + 3y + 6 = 0$.

To identify the gradient of a straight line it is easiest to have its equation in the form $y = mx + c$.

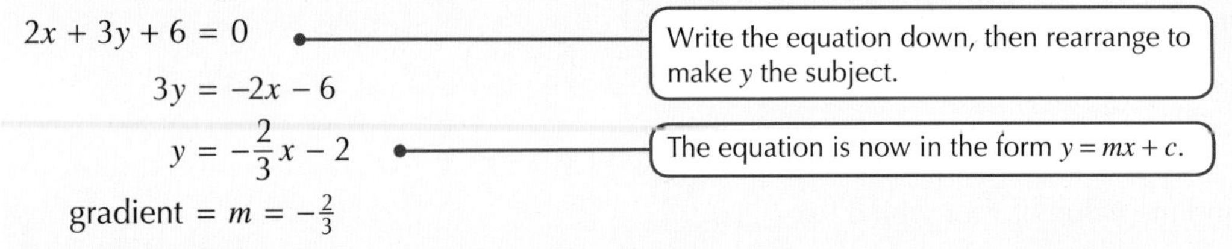

Write the equation down, then rearrange to make y the subject.
The equation is now in the form $y = mx + c$.

$2x + 3y + 6 = 0$

$3y = -2x - 6$

$y = -\frac{2}{3}x - 2$

gradient $= m = -\frac{2}{3}$

Example 12.14

A straight line has equation $4y - 3x + 9 = 0$.

a Find the gradient of the line.

b Find the coordinates of the points where the line intercepts:

i the y-axis

ii the x-axis

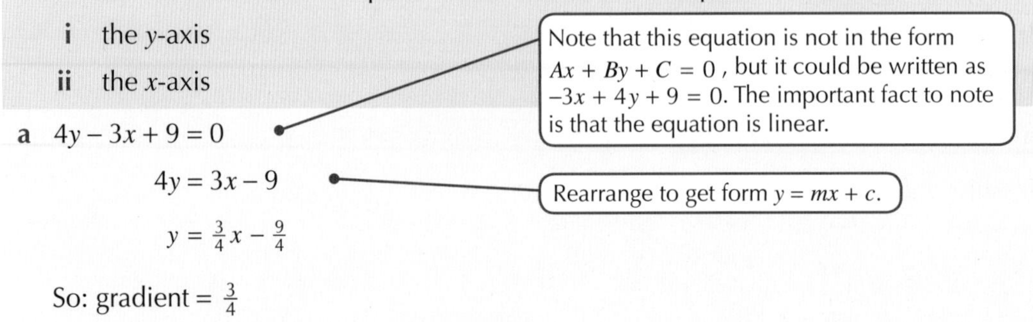

a $4y - 3x + 9 = 0$

$4y = 3x - 9$

$y = \frac{3}{4}x - \frac{9}{4}$

So: gradient $= \frac{3}{4}$

(continued)

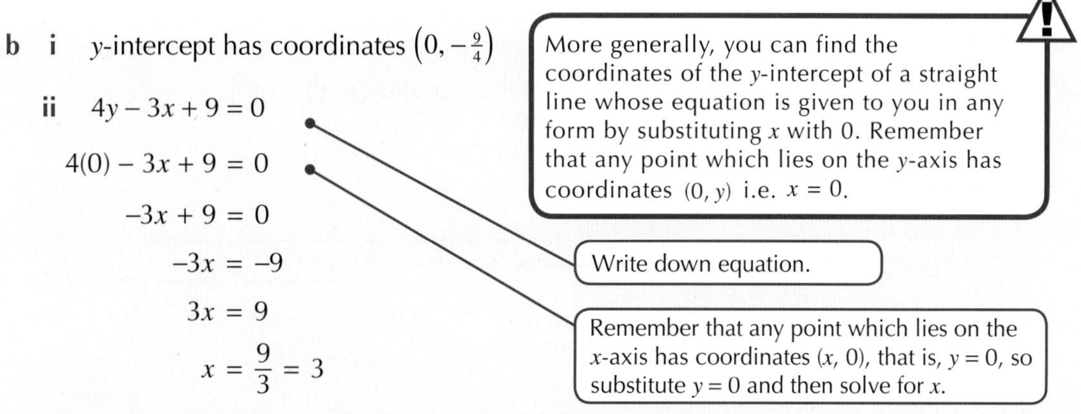
More generally, you can find the coordinates of the y-intercept of a straight line whose equation is given to you in any form by substituting x with 0. Remember that any point which lies on the y-axis has coordinates $(0, y)$ i.e. $x = 0$.
Write down equation.
Remember that any point which lies on the x-axis has coordinates $(x, 0)$, that is, $y = 0$, so substitute $y = 0$ and then solve for x.

b **i** y-intercept has coordinates $\left(0, -\frac{9}{4}\right)$

ii $4y - 3x + 9 = 0$

$4(0) - 3x + 9 = 0$

$-3x + 9 = 0$

$-3x = -9$

$3x = 9$

$x = \frac{9}{3} = 3$

When $y = 0$, $x = 3$, so the line intercepts the x-axis at $(3, 0)$.

Notice that in Example 12.14 part **b i**, for the given line, you could also rearrange the equation into the form $y = mx + c$ but this is not necessary:

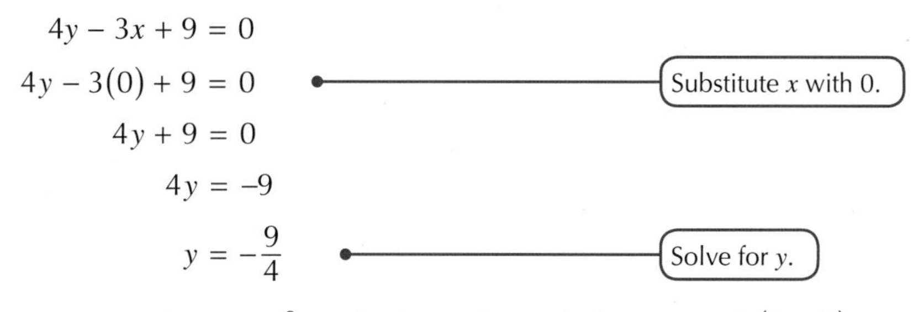
Substitute x with 0.
Solve for y.

$4y - 3x + 9 = 0$

$4y - 3(0) + 9 = 0$

$4y + 9 = 0$

$4y = -9$

$y = -\frac{9}{4}$

When $x = 0$, $y = -\frac{9}{4}$, so the line intercepts the y-axis at $\left(0, -\frac{9}{4}\right)$.

Example 12.15

A straight line has equation $y = -\frac{3}{4}x + 6$. Find the coordinates of the point where this line intercepts the x-axis.

$0 = -\frac{3}{4}x + 6$ — Substitute y with 0 and solve for x.

$-\frac{3}{4}x + 6 = 0$

$-\frac{3}{4}x = -6$

$-3x = -24$

$3x = 24$

$x = \frac{24}{3} = 8$

When $y = 0$, $x = 8$, so the line intercepts the x-axis at $(8, 0)$.

Example 12.16

A straight line has equation $y = 3x - 2$. Express this equation in the form $Ax + By + C = 0$.

$y = 3x - 2$

$y - 3x + 2 = 0$

$-3x + y + 2 = 0$ — Rewrite in the form $Ax + By + C = 0$.

So: the equation of the line is $-3x + y + 2 = 0$.

Exercise 12D

1 Rearrange the following equations into the form $y = mx + c$.

a $y + 4x = -2$ b $3x + 2y = 5$ c $5x - y + 2 = 0$

d $-2x - 7y + 4 = 0$ e $\frac{1}{2}y + 3x = -6$ f $\frac{1}{5}y - \frac{1}{4}x = 1$

★ 2 For each line in Question 1, determine the gradient of the line and the point where the line cuts the y-axis.

★ 3 For the following straight lines, find the coordinates of the points where each intercepts:

i the y-axis ii the x-axis

a $y = 2x - 8$ b $y = 5x + 10$ c $y = -3x + 5$

d $y = \frac{1}{2}x - 3$ e $y = \frac{3}{5}x + 9$ f $y = -\frac{4}{3}x - 16$

★ 4 For the following straight lines, find the coordinates of the points where each intercepts:

i the y-axis ii the x-axis

a $x + y = 6$ b $y - 2 = 4(x - 1)$ c $x - 2y = 4$

d $y = \frac{3}{8}(x + 12)$ e $\frac{2}{3}x - 4 = y$ f $y = -\frac{4}{3}x - 16$

g $y = 15 - 3x$ h $12 - 3x - 4y = 0$ i $\frac{3}{4}y = \frac{1}{2}x + 18$

j $\frac{5}{6}y + x = 10$ k $\frac{x}{3} - \frac{y}{4} = 2$ l $y - \frac{1}{2} = \frac{2}{3}\left(x - \frac{3}{4}\right)$

5 Find the equation of the line passing through $(3, -2)$ which is parallel to the line with equation $x - 3y = 2$.

6 Find the equation of the line passing through $(-2, 5)$ which is perpendicular to the line with equation $4y + 3x - 5 = 0$.

See Chapter 8.

7 The straight line with equation $8y - 3x + 24 = 0$ intercepts the x- and y-axes at the points P and Q respectively.

Determine the area, in square units, of the triangle POQ where O is the origin.

8 Three of the vertices, P, Q and R, of a parallelogram $PQRS$, have coordinates (5, 2), (9, 10) and (15, 14) respectively.

a Find the coordinates of the fourth vertex S.

b **i** Determine the equation of the diagonal QS.

ii Find the coordinates of the point where QS intercepts the y-axis.

c The diagonal PR intersects the x-axis at the point M. Determine the coordinates of M.

9 The following information is known about a rectangle $ABCD$:

- A and C have coordinates (−2, 4) and (10, 13), respectively
- AB is horizontal
- the diagonals AC and BD intercept the y-axis at the points P and Q
- the diagonals AC and BD meet at the point M.

Determine the area of triangle MPQ.

Finding the midpoint of a line segment

An important skill, frequently used in Higher maths, is finding the midpoint of a line segment.

GO! Activity

a On a coordinate diagram, plot the following pairs of points and draw the line segment which connects each pair of points.

A (2, 5) and B (4, 9)	C (7, 1) and D (9, 5)	E (−3, −4) and F (1, 2)
G (5, −3) and H (−5, 5)	J (−1, −6) and K (8, −2)	L (0, −8) and M (4, 0)

b Write down the midpoint of each line segment.

c Can you suggest a quick way of finding the midpoint of the line segment connecting any two points? Explain your answer.

The midpoint, M, of the line segment connecting $P(x_1, y_1)$ and $Q(x_2, y_2)$ has coordinates:

$$\left(\frac{x_1 + x_2}{2}, \frac{y_1 + y_2}{2}\right)$$

To find the midpoint of a line segment, you simply need to find the average of the two x-coordinates and the average of the two y-coordinates.

Example 12.17

P and Q have coordinates (5, −3) and (1, −5). Find the midpoint of PQ.

$$M = \left(\frac{x_1 + x_2}{2}, \frac{y_1 + y_2}{2}\right)$$

$$= \left(\frac{5+1}{2}, \frac{-3+(-5)}{2}\right) = \left(\frac{6}{2}, \frac{-8}{2}\right) = (3, -4)$$

So: midpoint of PQ is (3, −4).

Example 12.18

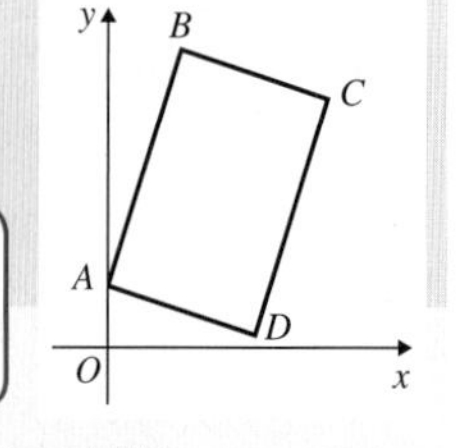

The diagram shows a rectangle $ABCD$. A is on the y-axis and the equation of AB is $y = 3x + 4$. C has coordinates (10, 10).

Find the midpoint of BD.

A has coordinates (0, 4).

Since A is on the y-axis (where $x = 0$), the y-intercept must be 4 because the equation of AB is $y = 3x + 4$.

The midpoint of BD will be the same as the midpoint of AC.

The diagonals of a rectangle bisect each other.

The midpoint of AC is $\left(\frac{0+10}{2}, \frac{4+10}{2}\right)$

$= \left(\frac{10}{2}, \frac{14}{2}\right) = (5, 7)$

So: the midpoint of BD is (5, 7).

Exercise 12E

★ **1** Find the midpoint of the line segment which connects the following pairs of points.

a P (5, 2) and Q (3, 4)

b A (−1, 4) and B (7, 10)

c C (−8, −1) and D (−6, −5)

d R (10, 3) and S (1, 2)

e V (u, v) and U (v, u)

f G ($2g$, h) and H (g, $2h$)

2 The vertices of a parallelogram have coordinates A (−4, −5), B (0, 3), C (15, 8) and D (11, 0). Show that the diagonals bisect each other.

3 PR is a diagonal of a rhombus $PQRS$. The coordinates of P and R are (−4, 6) and (20, 6).

a Find the equation of the diagonal QS.

b Given that the area of the rhombus is 216 cm^2, determine the coordinates of S, given that S lies below the x-axis.

4 P and Q have coordinates (a, a^2) and (b, b^2). M is the midpoint of PQ.

The line L is perpendicular to PQ and passes through M.

Show that the equation of L can be written as $2(a + b)y + 2x = (a + b)(1 + a^2 + b^2)$.

- I can determine the equation of a line in the form $y = mx + c$ or $y - b = m(x - a)$. ★ Exercise 12A Q2 ★ Exercise 12B Q3
- I can solve problems using the equation of a straight line. ★ Exercise12A Q5 ★ Exercise 12B Q5
- I can use function notation to solve problems. ★ Exercise 12C Q7
- I can rearrange a linear equation to find the gradient of a line. ★ Exercise 12D Q2
- I can use algebra to find where the graph of a straight line intercepts the x- and y-axes. ★ Exercise 12D Q3, Q4
- I can find the midpoint of a line segment. ★ Exercise 12E Q1

For further assessment opportunities, see the Preparation for Assessment for Unit 2 on pages 288–291.

13 Working with linear equations and inequations

In this chapter you will learn how to:

- solve more complex linear equations involving brackets
- solve linear equations containing fractions
- solve more complex linear inequations
- apply knowledge to solve related problems.

You should already know

- how to solve simple linear equations and inequations
- how to evaluate a formula or expression
- how to multiply out brackets and simplify an algebraic expression
- how to perform arithmetic operations with numerical and algebraic fractions.

Solving linear equations containing brackets

In this section, you will learn to solve linear equations which require fluency in algebra. Expanding brackets and simplifying are crucial skills, especially in Higher mathematics.

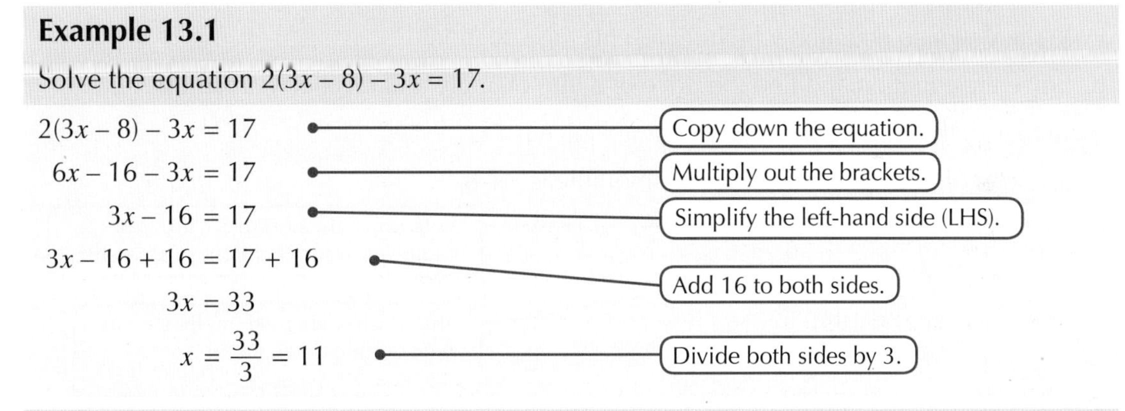

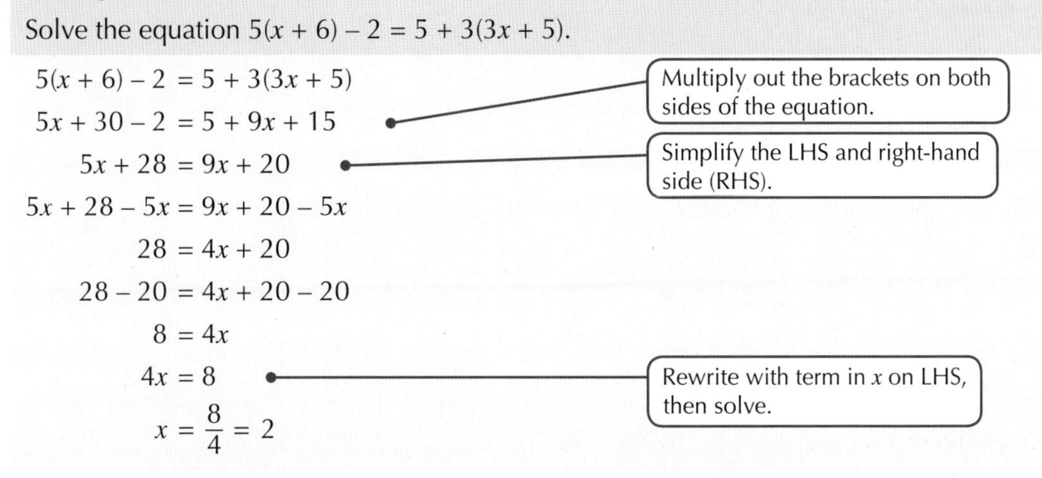

2(3x – 8) – 3x = 17

6x – 16 – 3x = 17

3x – 16 + 16 = 17 + 16
Add 16 to both sides.

Example 13.2
Solve the equation 5(x + 6) – 2 = 5 + 3(3x + 5).
5(x + 6) – 2 = 5 + 3(3x + 5)
5x + 30 – 2 = 5 + 9x + 15
Multiply out the brackets on both sides of the equation.
5x + 28 = 9x + 20
Simplify the LHS and right-hand side (RHS).
5x + 28 – 5x = 9x + 20 – 5x
28 = 4x + 20
28 – 20 = 4x + 20 – 20
8 = 4x
4x = 8
Rewrite with term in x on LHS, then solve.
x = 8/4 = 2

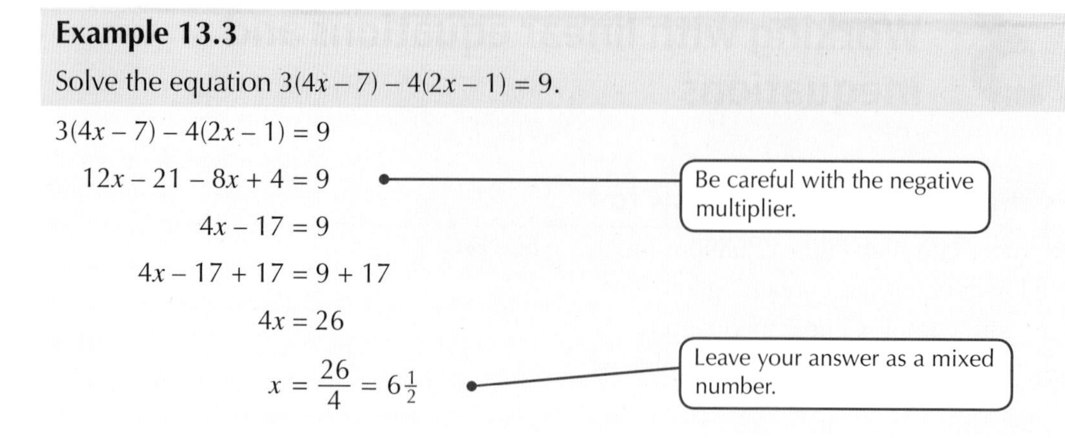

Example 13.3

Solve the equation $3(4x - 7) - 4(2x - 1) = 9$.

$$3(4x - 7) - 4(2x - 1) = 9$$

$$12x - 21 - 8x + 4 = 9$$

Be careful with the negative multiplier.

$$4x - 17 = 9$$

$$4x - 17 + 17 = 9 + 17$$

$$4x = 26$$

$$x = \frac{26}{4} = 6\tfrac{1}{2}$$

Leave your answer as a mixed number.

Example 13.4

A litre carton of orange juice costs x pence.

a Write down the cost of 5 cartons of orange juice.

The cost of a litre carton of apple and raspberry smoothie is 82 pence more than the cost of a litre carton of orange juice.

b Write down the cost of a carton of apple and raspberry smoothie.

Jenny buys 5 litre cartons of orange juice and 3 litre cartons of apple and raspberry smoothie. Altogether she pays £19.90.

c Write down an equation which represents this situation.

d Solve the equation and find the price of a litre of apple and raspberry smoothie.

a In pence, the cost of 5 cartons of orange juice is $5x$.

b The cost of an apple and raspberry smoothie is $x + 82$.

c $5x + 3(x + 82)$

Write down the cost of 5 cartons of orange juice and 3 cartons of smoothie.

$5x + 3(x + 82) = 1990$

Write down the equation. Be careful when writing down the RHS: £19.90 = 1990 pence.

d

$$5x + 3(x + 82) = 1990$$

$$5x + 3x + 246 = 1990$$

$$8x + 246 = 1990$$

$$8x + 246 - 246 = 1990 - 246$$

$$8x = 1744$$

$$x = \frac{1744}{8} = 218$$

A carton of orange juice costs 218 pence.

So: a one litre carton of smoothie costs 218 + 82 = 300 pence = £3.00.

Exercise 13A

1 Solve the following, giving your answer as a fraction or mixed number.

a $3x - 2 = x + 5$ b $8x + 9 = 5x - 1$ c $2x + 7 = 4x + 10$

d $5x + 2 = 5 - 3x$ e $7x - 8 = 8x + 9$ f $x + 6 = 7x - 2$

2 Solve the following.

a $3(2x + 1) = 21$ b $2(4x - 5) = 22$ c $4(3 - x) = 24$

d $27 = 3(2x - 5)$ e $4(2x - 5) - 3x = 30$ f $8(3x + 2) - 15x = 43$

3 Solve the following.

a $5(3x - 2) - 8 = 6 + 7x$ b $2(3 - 5x) + 4 = 7(5 - 2x) - 1$

c $9(4x + 3) = 5(2 - x) + x - 2$ d $4(x + 1) - 3(2x - 5) = 11$

e $20 - (1 - x) - 2(2 - 3x) = 1$ f $8 - 3(5x + 1) = 2(x - 8) + 4$

g $14 - 5(x + 1) = 3(2 - 3x) + 6x - 5$ h $8(3 - 2x) = 7(13 - x)$

★ 4 Solve the following, giving your answer as a fraction or mixed number where required.

a $6(3x - 2) + 5(1 - 2x) = 5$ b $3(5 - x) - 2(4 - 3x) = 1$

c $6x + 1 - (x - 7) = 6 - 2(5 - 4x)$ d $10 - (2 - x) = 3(2x + 7)$

e $8x - 3 = 1 - 6(2 - 3x)$ f $2(7x - 3) - 5(4x - 1) = 2$

g $15 - (8 - x) = 4(2x - 3) + 2$ h $9x - 5 - (4 - x) = 14 + 3(2x + 1)$

★ 5 Two rectangles have the dimensions shown.

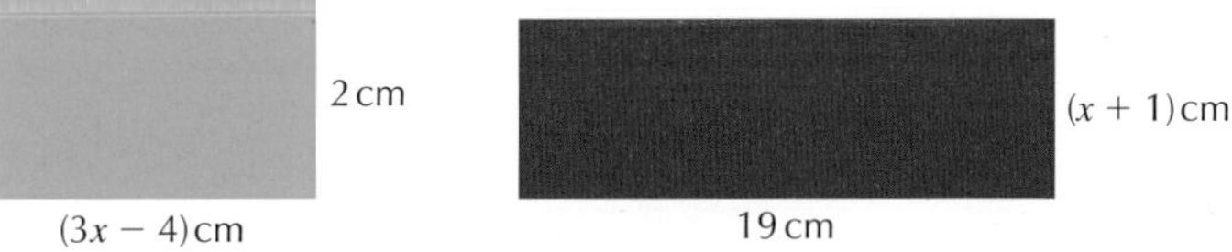

The area of the red rectangle is greater than the area of the blue rectangle. The difference in their areas is 131 cm^2. Determine the area of the red rectangle.

6 Seonaid and Kyle's teacher gives them both the same number.

Kyle multiplies the number by 6 and then subtracts 4.

Seonaid adds 3 to the number and then multiplies by 4.

Kyle and Seonaid end up with same answer.

a Form an equation and use it to find which number Kyle and Seonaid were given by their teacher.

b What number did Kyle and Seonaid end up with?

7 Deirdre practised her mental maths skills by playing a game three times.

In game 2 she scored 6 less than 5 times her score in game 1.

In game 3 she scored 5 times the average of her scores from games 1 and 2.

Deirdre scored 567 points in total.

Use the information given to set up an equation and use it to find how much she scored in each of the three games.

Activity

Solving equations is a vital application of mathematics. As an example, consider an athlete running in the inside lane in the Olympic 400 metres sprint final.

Here is some information about an Olympic running track:

- the inside perimeter of the track consists of two semi-circular arcs and two straight sections
- a semi-circular arc has a diameter of 73 m.
- a straight section has a length of 84.39 m.

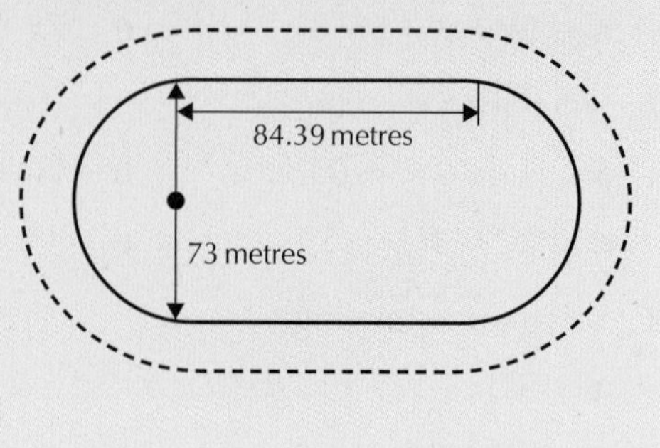

At what (constant) distance from the edge of the track should an athlete in the inside lane run so that they run exactly 400 metres?

Solving linear equations containing fractions

In this section, you will learn how to solve linear equations containing fractions. When solving an equation which contains fractions you should follow these steps:

1 multiply both sides by the lowest common multiple (LCM) to clear fractions

2 multiply out brackets and simplify the LHS and RHS

3 solve as before.

Example 13.5

Solve the equation $\frac{x}{5} = 6$.

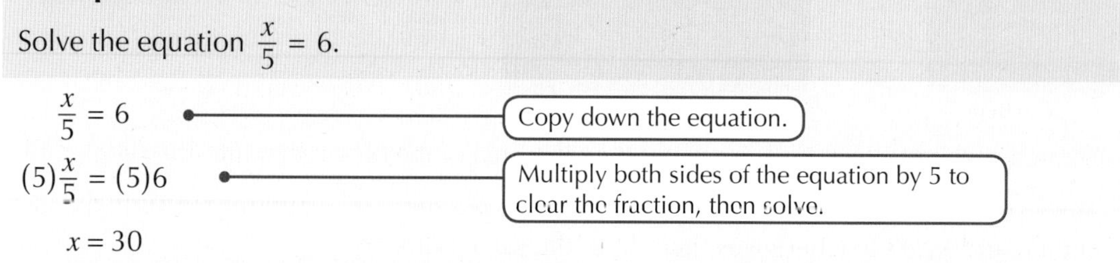

$\frac{x}{5} = 6$
Copy down the equation.
$(5)\frac{x}{5} = (5)6$
Multiply both sides of the equation by 5 to clear the fraction, then solve.
$x = 30$

Example 13.6

Solve the equation $\frac{3}{4}(2x - 1) = 9$.

$$\frac{3}{4}(2x - 1) = 9$$

$$(4)\frac{3}{4}(2x - 1) = (4)9$$

Multiply both sides of the equation by 4 to clear the fraction.

$$3(2x - 1) = 36$$

$$6x - 3 = 36$$

Multiply out the brackets, then solve.

$$6x - 3 + 3 = 36 + 3$$

$$6x = 39$$

$$x = \frac{39}{6} = 6\frac{1}{2}$$

Example 13.7

Solve the equation $\frac{5x + 2}{4} - \frac{2x - 3}{7} = -1$.

$$\frac{5x + 2}{4} - \frac{2x - 3}{7} = -1$$

$$(28)\frac{5x + 2}{4} - (28)\frac{2x - 3}{7} = (28)(-1)$$

Multiply both sides of the equation by the LCM of 4 and 7 (28), and simplify where possible.

$$(7)(5x + 2) - (4)(2x - 3) = -28$$

$$35x + 14 - 8x + 12 = -28$$

$$27x + 26 = -28$$

Simplify the LHS, then solve.

$$27x + 26 - 26 = -28 - 26$$

$$27x = -54$$

$$x = \frac{-54}{27} = -2$$

Example 13.8

A square has length x cm. A rectangle has length $3x$ cm and breadth $(x - 4)$ cm.

It is known that one third of the perimeter of the square is equal to one fifth of the perimeter of the rectangle. Determine the breadth of the rectangle.

Perimeter of the square = $4x$

Perimeter of the rectangle = $3x + (x - 4) + 3x + (x - 4)$
$= 8x - 8$

In a word problem you should always try to connect the information given. The problem mentions perimeter so it is sensible to write down expressions for the perimeter of the square and the rectangle.

$\frac{1}{3}$ of $4x = \frac{1}{5}$ of $(8x - 8)$

Use information given about perimeter to write another equation.

$$\frac{4x}{3} = \frac{8x - 8}{5}$$

Rewrite the equation so that it's ready to solve.

$$(15)\frac{4x}{3} = (15)\frac{8x - 8}{5}$$

Multiply both sides by the LCM of 3 and 5 (15), then solve.

$$20x = 3(8x - 8)$$

$$20x = 24x - 24$$

$$-4x = -24$$

$$4x = 24$$

$$x = 6$$

The breadth of the rectangle is $(x - 4) = (6 - 4) = 2$ cm.

Write your answer in context.

Exercise 13B

1 Solve the following.

a $\frac{x}{6} = 5$ **b** $\frac{x}{9} = -4$ **c** $\frac{x+7}{3} = 5$

d $\frac{x-4}{5} = 2$ **e** $\frac{1}{5}x = 8$ **f** $\frac{1}{9}x = 6$

g $\frac{3}{4}x = 18$ **h** $\frac{3x+9}{5} = 6$ **i** $\frac{2}{3}(9 - 4x) = 14$

2 Solve the following, giving your answer as a fraction or mixed number where necessary.

a $\frac{x}{2} + \frac{x}{4} = 5$ **b** $\frac{x}{2} - \frac{x}{3} = 4$ **c** $\frac{5x}{3} + \frac{x}{6} = 1$

d $\frac{2x}{5} - \frac{x}{2} = 3$ **e** $\frac{1}{4}x + \frac{1}{5}x = 20$ **f** $\frac{x}{8} = 2 + \frac{3x}{4}$

★ **3** Solve the following, giving your answer as a fraction or mixed number where necessary.

a $\frac{x-3}{2} + \frac{4x}{3} = 15$ **b** $\frac{x+1}{5} - \frac{x-1}{6} = 2$ **c** $\frac{2x-1}{3} + \frac{3x+1}{4} = 1$

d $\frac{4x+1}{3} + \frac{x+2}{5} = -2$ **e** $2x - \frac{(3x-1)}{4} = 4$ **f** $\frac{3(x+1)}{4} - \frac{4(x-2)}{3} = -1$

g $\frac{1}{5}(2x-3) - \frac{2}{3}(4-x) = -4$ **h** $\frac{x+3}{2} - \frac{5}{6}(1-2x) = 1$

4 There are x biscuits in a family-sized tin. At a party, 45 of them are eaten, and three eighths of the biscuits remain. Set up an equation and solve it to find how many biscuits were in the tin to start with.

5 I think of a number. I multiply this number by 3, add 5 and divide the result by 8. My answer is four ninths of the original number. Form an equation and solve it to find the original number.

Solving more complex linear inequations

You have previously met inequalities and solved simple linear inequations. In this section, you will learn to solve more complex linear inequations.

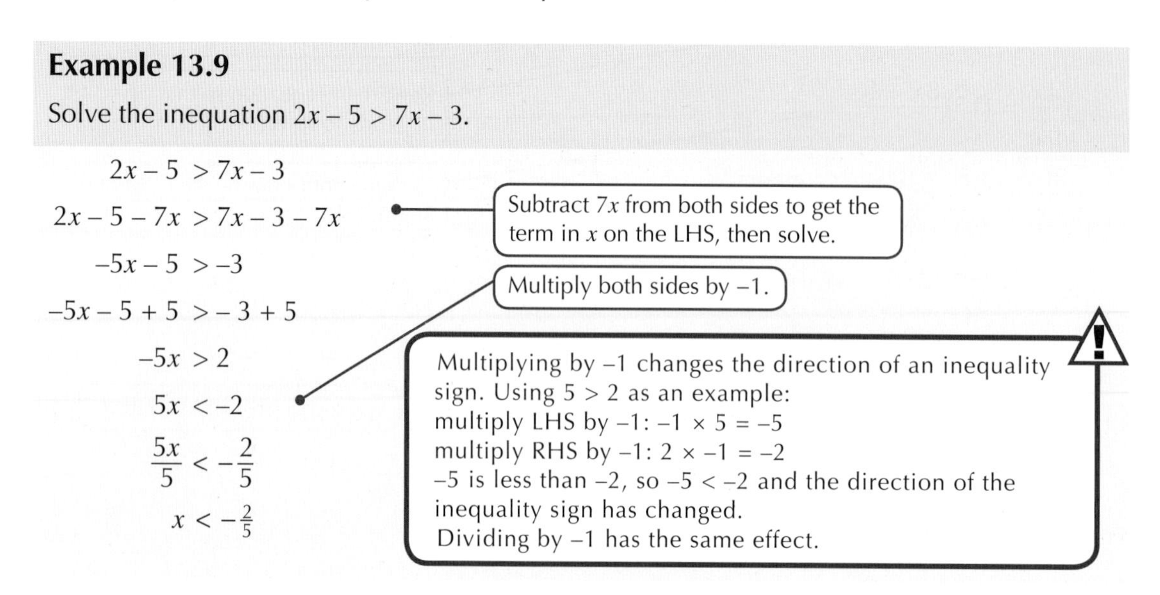

Example 13.9

Solve the inequation $2x - 5 > 7x - 3$.

$2x - 5 > 7x - 3$

$2x - 5 - 7x > 7x - 3 - 7x$

$-5x - 5 > -3$

$-5x - 5 + 5 > -3 + 5$

$-5x > 2$

$5x < -2$

$\frac{5x}{5} < -\frac{2}{5}$

$x < -\frac{2}{5}$

Subtract $7x$ from both sides to get the term in x on the LHS, then solve.
Multiply both sides by –1.
Multiplying by –1 changes the direction of an inequality sign. Using 5 > 2 as an example:
multiply LHS by –1: –1 × 5 = –5
multiply RHS by –1: 2 × –1 = –2
–5 is less than –2, so –5 < –2 and the direction of the inequality sign has changed.
Dividing by –1 has the same effect.

Example 13.10

Solve the inequation $20 - 2(3x + 8) \geqslant 8 - 5x$.

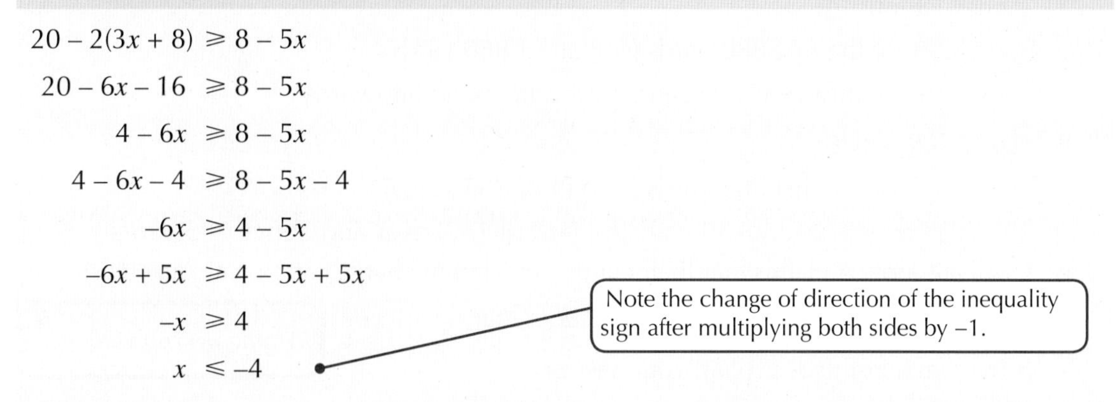

$$\begin{aligned} 20 - 2(3x + 8) &\geqslant 8 - 5x \\ 20 - 6x - 16 &\geqslant 8 - 5x \\ 4 - 6x &\geqslant 8 - 5x \\ 4 - 6x - 4 &\geqslant 8 - 5x - 4 \\ -6x &\geqslant 4 - 5x \\ -6x + 5x &\geqslant 4 - 5x + 5x \\ -x &\geqslant 4 \\ x &\leqslant -4 \end{aligned}$$

Note the change of direction of the inequality sign after multiplying both sides by –1.

Exercise 13C

★ **1** Solve the following.

a $4x + 2 > x + 11$ **b** $7x - 5 < 2x + 30$ **c** $6x + 8 \geqslant 2x - 12$

d $3x + 7 < 15 - x$ **e** $12 - 5x > 3x - 4$ **f** $3x + 6 \leqslant 12 - 3x$

g $7x + 5 > 4x - 10$ **h** $1 - 5x > -2 + 4x$ **i** $2x - 9 < 3 - x$

★ **2** Solve the following.

a $5(x - 2) - 3x > 2 - 6x$ **b** $15 - 2(1 - 3x) > x + 6$

c $2 - (2 - x) \geqslant 2(4x - 5) - 5x$ **d** $4(3x - 1) < 8 - 3(2x + 1)$

e $5(2 - x) - (8 - x) > 7$ **f** $2(3x + 7) - 3(1 - 4x) \leqslant 1 - 2x$

g $3(8 - 2x) \geqslant 4 - 2(6 - x)$ **h** $3x - 2(5x + 1) < 4(1 - x)$

i $20 > 3(1 + 2x) - 4(1 + 3x)$ **j** $-18 < 9x + 4 - (2 - x)$

k $7x - (4 - 5x) > 3(5x - 8) + 2$ **l** $3(1 - 5x) - 8(1 - 2x) > 5x - 3$

m $8x - (4 - 5x) < 3(5x + 2)$ **n** $8x - 2(6x - 1) \geqslant 2 - 4(5 + 2x)$

3

Pukka Plumbing
Call-out charge £60
Labour charge £25 per hour
24-hour emergency service

Perfect Plumbers
Call-out charge £25
Labour charge £35 per hour
24-hour emergency service

a For Pukka Plumbing write down a formula to find the cost, £C, of a job lasting h hours.

b Repeat for Perfect Plumbers.

It's 2.30 in the morning and Kyle has discovered a burst pipe in the kitchen. The stopcock won't work and the ground floor is in serious danger of flooding.

c Should Kyle call Pukka Plumbing or Perfect Plumbers? Explain your answer.

4 Here is some information about the maximum taxi fare which can be charged in two local authorities:

- Local authority A:
  - » £3.00 for the first 880 yards (or part of 880 yards)
  - » 10 pence for every 110 yards over and above 880 yards
- Local authority B:
  - » £2.50 for the first 785 yards (or part of 785 yards)
  - » 10 pence for every 130 yards (or part of 130) over and above 785 yards

a Local authority A claims that their general maximum charge can be calculated as:

£3 for the first mile and £1.60 per mile after that

Is this claim justified? Explain your answer.

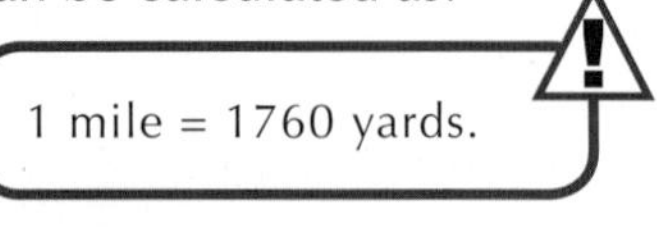

b How might local authority B describe their general maximum charge?

c Andy travels frequently on business and often has to take taxis in both local authorities. He claims that the taxi fare in local authority A is more than that in local authority B. How would you respond to this claim?

- I can solve more complicated linear equations containing brackets. ★ Exercise 13A Q4
- I can solve more complicated linear equations containing fractions. ★ Exercise 13B Q3
- I can set up and solve a linear equation given suitable information. ★ Exercise 13A Q5
- I can solve more complicated linear inequalities. ★ Exercise 13C Q1, Q2

For further assessment opportunities, see the Preparation for Assessment for Unit 2 on pages 288–291.

14 Working with simultaneous equations

In this chapter you will learn how to:

- solve a pair of simultaneous equations graphically
- solve a pair of simultaneous equations by substitution
- solve a pair of simultaneous equations using the elimination method
- from given text, create and solve a pair of simultaneous equations.

You should already know:

- how to solve linear equations
- how to evaluate a formula or expression
- how to find the equation of a straight line given suitable information.

Solving a pair of simultaneous equations graphically

In this section, you will revise how to draw straight lines and use your drawing to find the point where two straight lines intersect.

To draw a straight line accurately, follow these steps:

1 find where the line intersects the y-axis (by choosing $x = 0$)

2 find where the line intersects the x-axis (by choosing $y = 0$)

3 find any other point on the line

4 draw the line (remembering to extend it beyond the three plotted points).

Example 14.1

Two straight lines have equations $x + y = 6$ and $x + 2y = 8$.

a Draw the two lines on a single coordinate grid.

b Write down the coordinates of the point where the lines intersect.

a For the first line: $x + y = 6$

$0 + y = 6$

$y = 6$

Copy down the equation.
Substitute x with 0.

So: y-intercept has coordinates (0, 6).

$x + y = 6$

$x + 0 = 6$

$x = 6$

Substitute y with 0.

So: x-intercept has coordinates (6, 0).

(continued)

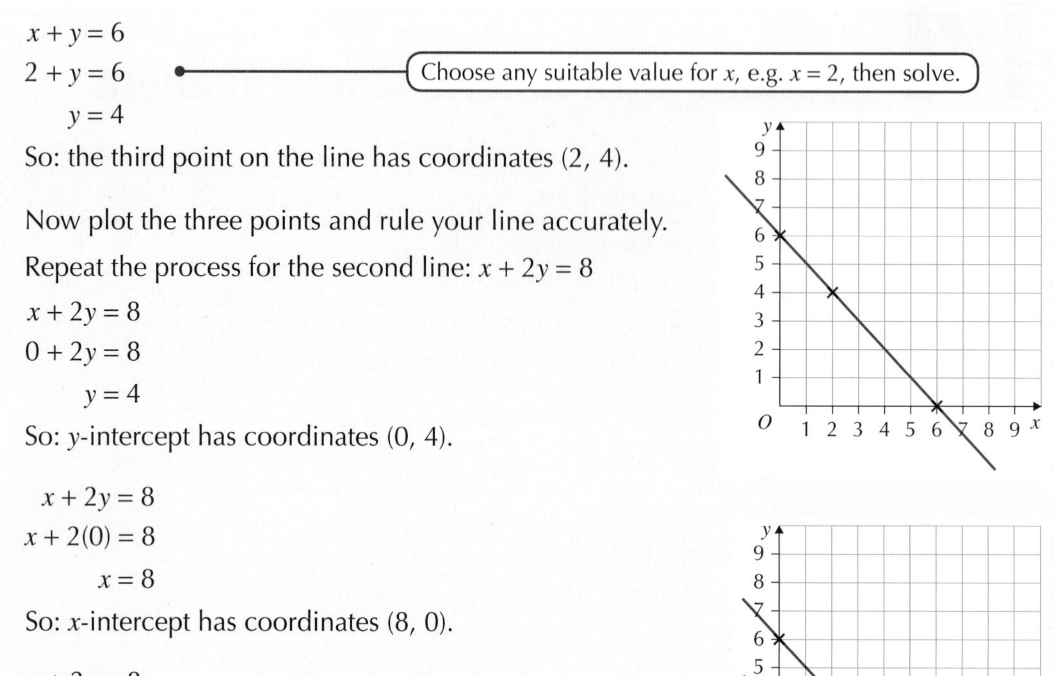
Choose any suitable value for x, e.g. x = 2, then solve.
y
9
8
7
6
5
4
3
2
1
O
1 2 3 4 5 6 7 8 9 x

$x + y = 6$

$2 + y = 6$

$y = 4$

So: the third point on the line has coordinates (2, 4).

Now plot the three points and rule your line accurately.

Repeat the process for the second line: $x + 2y = 8$

$x + 2y = 8$

$0 + 2y = 8$

$y = 4$

So: y-intercept has coordinates (0, 4).

$x + 2y = 8$

$x + 2(0) = 8$

$x = 8$

So: x-intercept has coordinates (8, 0).

$x + 2y = 8$

$6 + 2y = 8$

$y = 1$

So: the third point on the line has coordinates (6, 1).

Plot the three points and neatly rule an accurate line.

b From the graph it is clear that the lines intersect at the point (4, 2). The point (4, 2) is the **only** point which is on **both** lines.

Check: You can check that you have the correct answer by substituting for x and y into one of the given equations.

Using equation 2:

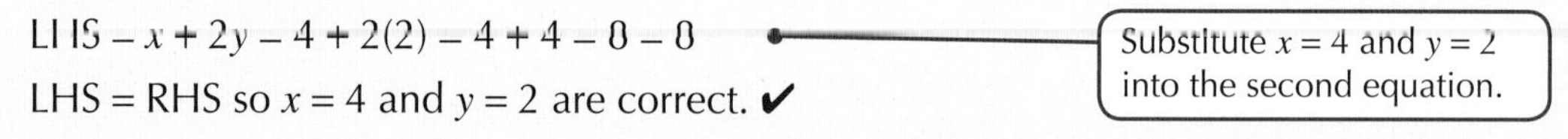
Substitute x = 4 and y = 2
into the second equation.

LHS = $x + 2y = 4 + 2(2) = 4 + 4 = 8 = 8$

LHS = RHS so $x = 4$ and $y = 2$ are correct. ✔

Example 14.1 asked for the coordinates of the point where the two lines intersect, so the answer is given as (4, 2). If, however, a question asks you to solve a system of equations, then write your answer as $x = 4$, $y = 2$.

A **system of equations** is a group or set of two or more equations with the same variables. In most of the examples in this chapter, the variables are x and y.

Exercise 14A

1 Find graphically the point of intersection of the following pairs of straight lines.

a $x + y = 5$ $x - y = 3$	**b** $x - y = 7$ $x + y = -3$	**c** $x + y = -6$ $x - y = -2$	**d** $x + 2y = 10$ $x - 2y = -2$
e $3y + 2x = 12$ $x - y = 9$	**f** $y = 2x + 6$ $y = x + 1$	**g** $y = 4x - 8$ $y = 2x + 4$	**h** $y = -3x + 9$ $y = 2x - 1$

★ 2 Find the solution to each of the following systems of equations using a graphical method.

a $x + y = 3$
$x - y = -5$

b $y = 2x - 6$
$y = -x + 3$

c $y - x = 5$
$-2x + 3y = 13$

d $y = 8 - x$
$y = 3x$

e $y = 5 - x$
$y = 3x + 1$

f $y = 2 - x$
$y = 6 - 2x$

GO! Activity

1 **a** Sketch the lines $x + y = 6$ and $x + y = 10$.
b What is the gradient of each line?
c Explain why the system of equations $x + y = 6$ and $x + y = 10$ has no solutions.

2 **a** Sketch the lines $y - 4x = 12$ and $-4x + y = 4$.
b What is the gradient of each line?
c Explain why the system of equations $y - 4x = 12$ and $-4x + y = 4$ has no solutions.

3 For each system below, decide whether or not it has a solution. Justify your answers.

a $3x + 2y = 5$
$3x + 2y = 9$

b $y - x = 5$
$2y - 2x = 8$

c $x - 3y = 4$
$3x - y = 9$

d $4y - x = 6$
$4y + x = 6$

Solving a pair of simultaneous equations by substitution

In an examination you will often be asked to solve a system of equations **algebraically**. This means that you are not allowed to find the solution graphically. There are two methods for solving a system of equations algebraically. In this section you will learn the first of these, the **substitution method**.

When solving a system using this method, you are finding the coordinates of the point of intersection of two straight lines. At the point of intersection, the x-value and the y-value will be the same for both lines. The substitution method simply uses this fact.

Example 14.2

Solve the system of equations

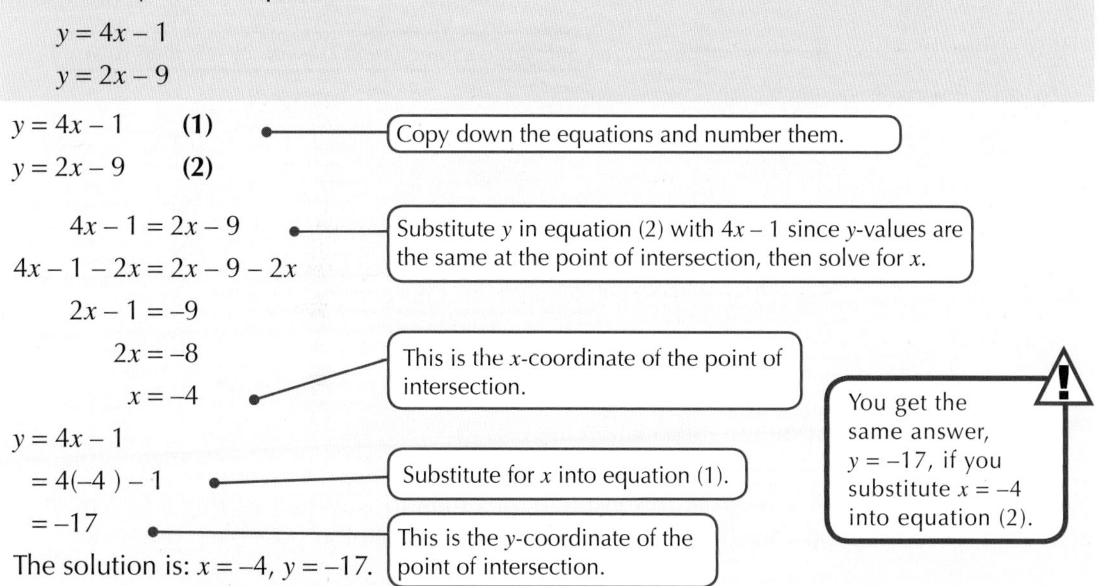

y = 4x – 1
y = 2x – 9
y = 4x – 1 (1)
y = 2x – 9 (2)
Copy down the equations and number them.
4x – 1 = 2x – 9
4x – 1 – 2x = 2x – 9 – 2x
2x – 1 = –9
2x = –8
x = –4
Substitute y in equation (2) with 4x – 1 since y-values are the same at the point of intersection, then solve for x.
This is the x-coordinate of the point of intersection.
y = 4x – 1
= 4(–4) – 1
= –17
Substitute for x into equation (1).
This is the y-coordinate of the point of intersection.
The solution is: x = –4, y = –17.
You get the same answer, y = –17, if you substitute x = –4 into equation (2).

Example 14.3

Solve the system of equations

$y = 2x - 8$

$x + 3y = -3$

$y = 2x - 8$ **(1)**

$x + 3y = -3$ **(2)**

$x + 3(2x - 8) = -3$ — Substitute y in equation (2) with $2x - 8$, then solve.

$x + 6x - 24 = -3$

$7x - 24 = -3$

$7x = 21$

$x = 3$

$x + 3y = -3$

$3 + 3y = -3$ — Substitute for x in equation (2), and solve.

$3y = -6$

$y = -2$

The solution is: $x = 3$, $y = -2$.

Example 14.4

Find the coordinates of the point of intersection of the straight lines

$3x + 4y = 5$

$x = 2y - 5$

$3x + 4y = 5$ **(1)**

$x = 2y - 5$ **(2)**

$3(2y - 5) + 4y = 5$ — Substitute x in equation (1) with $2y - 5$, then solve.

$6y - 15 + 4y = 5$

$10y - 15 = 5$

$10y = 20$

$y = 2$

$x = 2y - 5$ — Substitute for y in equation 2.

$= 2(2) - 5$

$= -1$

The coordinates of the point of intersection are (–1, 2).

Note: In Example 14.4 it was easier to substitute for x than for y. When solving a system of equations by substitution you should always try to be as efficient as possible.

Exercise 14B

1 Solve the following pairs of equations using the substitution method.

a $y = 4x + 5$, $y = x - 1$ **b** $y = 3x - 8$, $y = x + 2$ **c** $y = 2x + 6$, $y = 5x - 9$ **d** $y = 6x + 1$, $y = 2x - 3$

e $y = 3x - 5$, $y = -x + 3$ **f** $y = 8x + 10$, $y = -2x + 5$ **g** $y = 3 - x$, $y = 7 - 3x$ **h** $y = 9 - 4x$, $y = 2x - 3$

★ **2** Find the point of intersection for each of the following pairs of straight lines.

a $x + 2y = 7$, $y = 2x + 1$ **b** $3y - x = 2$, $y = 2x - 6$ **c** $2x - 5y = 1$, $y = x + 1$ **d** $y = 3x + 9$, $x - 3y = -11$

e $x = 5 - 2y$, $3y - 2x = -17$ **f** $3x + y = 7$, $x = -6 - 2y$ **g** $y = 2 - x$, $x = 4 + y$ **h** $y = 16 - 3x$, $x = 4y + 1$

3 Solve the following pairs of equations using the substitution method.

a $y = \frac{1}{2}x - 1$, $y = \frac{2}{3}x - 2$ **b** $y = -\frac{2}{3}x - 6$, $y = \frac{3}{5}x + 13$ **c** $y = -\frac{3}{2}x + 1$, $y = \frac{1}{4}x + 8$

d $y = \frac{3}{4}x - 1$, $y = -\frac{7}{6}x - 24$ **e** $y = -\frac{1}{6}x + 2$, $y = \frac{1}{3}x + \frac{7}{2}$ **f** $y = \frac{4}{5}x - 3$, $y = \frac{1}{2}x - \frac{3}{2}$

Solving a pair of simultaneous equations by elimination

The second method for solving a system of linear equations algebraically is called the **elimination** method. The given pair of equations can be added or subtracted to eliminate either x or y. You then solve the system as before.

There are two types of system:

- systems ready for elimination
- systems needing preparation for elimination.

There are three steps involved in the elimination method:

1. add or subtract the equations to eliminate one of the variables and obtain a value for x or y
2. substitute the value obtained into one of the equations and solve for the second variable
3. check your answer by substituting your values into the equation not used in previous step.

Systems which are ready for elimination

This section focuses on systems which are ready for elimination.

Whether you add or subtract depends on the signs of the terms. The following examples demonstrate when you add and when you subtract.

Example 14.5

Solve the system of equations

$2x + y = 11$

$x + y = 7$

$2x + y = 11$ **(1)**

$x + y = 7$ **(2)**

$$\begin{array}{r} 2x + y = 11 \\ -\quad x + y = 7 \\ \hline x \quad\;\; = 4 \end{array}$$

Eliminate y by **subtracting** equation (2) from equation (1).

So: $x = 4$

$2x + y = 11$

$2(4) + y = 11$

Substitute for x into equation (1), then solve.

$8 + y = 11$

$y = 11 - 8 = 3$

Use equation (2) to check that your values for x and y are correct.

So: $y = 3$

Check:

LHS $= x + y = 4 + 3 = 7 =$ RHS ✔

So: $x = 4$, $y = 3$ is the correct solution.

It is important to check your values because this will show you whether or not you have made a mistake in the earlier parts of the question.

Example 14.6

Solve the pair of equations

$5x - 2y = 16$

$x - 2y = 8$

$5x - 2y = 16$ **(1)**

$x - 2y = 8$ **(2)**

$$\begin{array}{r} 5x - 2y = 16 \\ -\quad x - 2y = 8 \\ \hline 4x \qquad = 8 \\ x \qquad = 2 \end{array}$$

Eliminate y by **subtracting** equation (2) from equation (1), then solve for x.

So: $x = 2$

$5x - 2y = 16$

Substitute for x into equation (1).

$5(2) - 2y = 16$

$10 - 2y = 16$

$-2y = 6$

$2y = -6$

$y = \frac{-6}{2} = -3$

(continued)

Check:

LHS = $x - 2y = 2 - 2(-3) = 2 - (-6) = 2 + 6 = 8$ = RHS ✔

LHS = RHS so $x = 2$, $y = -3$ is the correct solution.

Use equation (2) to check that your values for x and y are correct.

Example 14.7

Solve the pair of equations

$-3x + 8y = -14$

$3x + 5y = 1$

$-3x + 8y = -14$ **(1)**

$3x + 5y = 1$ **(2)**

$$\begin{array}{rr} & -3x + 8y = -14 \\ + & 3x + 5y = 1 \\ \hline & 13y = -13 \end{array}$$

Eliminate x by **adding** equations (1) and (2). Then find the value of y.

$$y = \frac{-13}{13}$$

$y = -1$

$-3x + 8y = -14$

Write down equation (1).

$-3x + 8(-1) = -14$

Substitute for y into equation (1) and solve for x.

$-3x - 8 = -14$

$-3x = -6$

$3x = 6$

$x = 2$

Check:

LHS = $3x + 5y = 3(2) + 5(-1) = 6 - 5 = 1$ = RHS ✔

So: $x = 2$, $y = -1$ is the correct solution.

Note: In Examples 14.5 and 14.6 the y terms had the same sign (both positive in Example 14.5 and both negative in Example 14.6) so we **subtracted**. In Example 14.7 the x terms were opposite in sign so we **added**.

Exercise 14C

★ **1** Solve each of the following pairs of equations by either adding or subtracting.

a $x + y = 10$
$2x - y = 8$

b $2x - y = 10$
$4x + y = 14$

c $3x + y = -1$
$3x - 2y = -7$

d $x + 5y = 11$
$2x + 5y = 1$

e $x - 2y = 6$
$3x - 2y = 2$

f $2x + 5y = 3$
$-2x - y = -7$

g $-x - 3y = -9$
$x - 2y = 1$

h $4x - y = 20$
$3x - y = 17$

i $6x + 5y = 9$
$x - 5y = 19$

Systems which require preparation for elimination

This section shows different ways to prepare systems for elimination.

Example 14.8

Solve the system of equations

$$3x + 2y = -12$$

$$x - y = 1$$

If you add or subtract the equations as they are, then you will still have an equation containing x and y terms.

In a situation like this, you must **choose** a **variable** to eliminate and then **multiply** one or both equations to make elimination possible.

Suppose you decide to eliminate x.

$3x + 2y = -12$ **(1)**

$x - y = 1$ **(2)**

Copy down the original system, labelling the equations (1) and (2).

$3x + 2y = -12$ **(1)**

$3x - 3y = 3$ **(3)**

Multiply equation (2) by 3. Label this new equation (3).

The system can now be solved in the usual way.

$$\begin{aligned} 3x + 2y &= -12 \\ -\ 3x - 3y &= 3 \\ \hline 5y &= -15 \\ y &= \frac{-15}{5} \\ &= -3 \end{aligned}$$

Eliminate x by subtracting equation (3) from equation (1), then solve for y.

$$\begin{aligned} 3x + 2y &= -12 \\ 3x + 2(-3) &= -12 \\ 3x - 6 &= -12 \\ 3x &= -6 \\ x &= \frac{-6}{3} \\ &= -2 \end{aligned}$$

Substitute for y into equation (1).

Check:

LHS $= x - y = -2 - (-3) = -2 + 3 = 1 =$ RHS ✔

Use equation (2) to check your answer.

So: $x = -2$, $y = -3$ is the correct solution.

Example 14.9

Solve the system of equations

$2x + 3y = 2$

$3x - 4y = 20$

$2x + 3y = 2$ **(1)**

$3x - 4y = 20$ **(2)**

Suppose you choose to eliminate y:

$8x + 12y = 8$ **(3)**

$9x - 12y = 60$ **(4)**

Multiply equation (1) by 4 and equation (2) by 3. Label the new equations (3) and (4), then eliminate y and solve.

$$\begin{aligned} 8x + 12y &= 8 \\ + \; 9x - 12y &= 60 \\ \hline 17x \quad &= 68 \\ x \quad &= \frac{68}{17} \\ &= 4 \end{aligned}$$

$$\begin{aligned} 2x + 3y &= 2 \\ 2(4) + 3y &= 2 \\ 8 + 3y &= 2 \\ 3y &= -6 \\ y &= \frac{-6}{3} = -2 \end{aligned}$$

Substitute for x into equation (1) as it is usually easier to use the original equations.

Check:

LHS = $3x - 4y = 3(4) - 4(-2) = 12 - (-8) = 12 + 8 = 20$ = RHS ✔

Use equation (2) to check.

So: $x = 4$, $y = -2$ is the correct solution.

Exercise 14D

★ **1** Solve each of the following pairs of equations by elimination.

a $x - 2y = 1$, $2x + y = 7$

b $4x + 3y = 11$, $x - y = 8$

c $x - 5y = 13$, $3x - y = -9$

d $2x + y = 10$, $3x - 4y = 26$

e $a - 3b = 5$, $5a + 2b = -9$

f $4p - q = 17$, $3p - 2q = 19$

g $-2s + 5t = 2$, $4s - 3t = -22$

h $3c - d = -3$, $2c + 4d = 5$

2 Solve each of the following pairs of equations by elimination.

a $3x - 5y = 11$, $2x + 3y = 1$

b $4x + 5y = -7$, $3x + 2y = -7$

c $7x - 3y = -8$, $2x + 4y = -12$

d $2x + 5y = 0$, $3x - 8y = 31$

e $2x - 3y = -27$, $3x + 2y = -8$

f $3x + 4y = -4$, $7x + 6y = -11$

g $5p - 4q = 22$, $3p + 5q = -9$

h $2f - 3g = 6$, $5f - 4g = 1$

3 Sally solved the system of equations $3x - 5y = 17$, $x + 3y = 1$ and obtained the solution $x = -5$, $y = 2$.

Explain why Sally's answer is wrong.

Create and solve a pair of simultaneous equations from text

Chapter 13 covers setting up and solving a linear equation given suitable information. At National 5 you are also expected to be able to set up and solve a system of equations.

Example 14.10

Tickets for a golf club ceilidh cost £9 for members and £15 for non-members.

a The total amount collected by selling tickets was £735. m tickets were sold to members and n tickets to non-members. Write down an equation containing m and n which represents this information.

b 65 tickets were sold altogether. Write down a second equation containing m and n.

c Find how many tickets were sold to:

i members **ii** non-members.

a Amount of money from members (in £) = $9m$

Amount of money from non-members (in £) = $15n$

Total money made = £735

So equation is: $9m + 15n = 735$

b $m + n = 65$

c Solve simultaneously the pair of equations obtained in parts **a** and **b**.

$9m + 15n = 735$ **(1)**
$m + n = 65$ **(2)**

Copy down the original system, labelling the equations (1) and (2).

$9m + 15n = 735$ **(1)**
$9m + 9n = 585$ **(3)**

Multiply equation (2) by 9. Label this new equation (3), then solve for n.

$$\begin{array}{r} 9m + 15n = 735 \\ -\ 9m + 9n = 585 \\ \hline 6n = 150 \end{array}$$

Eliminate m by subtracting equation (3) from equation (1).

$$n = \frac{150}{6} = 25$$

$9m + 15n = 735$

Substitute for n into equation (1).

$$9m + 15(25) = 735$$
$$9m + 375 = 735$$
$$9m = 360$$
$$m = \frac{360}{9} = 40$$

(continued)

Check:

LHS = $m + n = 40 + 25 = 65$ = RHS ✔ — Use equation (2) to check your answer.

So: $m = 40$, $n = 25$ is the correct solution.

So, **i** 40 tickets were sold to members and **ii** 25 were sold to non-members.

Exercise 14E

1 **a** 5 apples and 3 bananas cost £1.78. Write down an equation to represent this information.

b 2 apples and 1 banana cost £0.64. Write down an equation to represent this information.

c Find the cost of:

i one apple **ii** one banana.

★ **2** Brass is made by mixing together copper and zinc. Two types of brass are made by mixing different quantities of copper and zinc.

The first type uses 6 cubic centimetres of copper and 5 cubic centimetres of zinc. This mixture weighs 89 grams.

a Let c be the mass of 1 cubic centimetre of copper and z be the mass of 1 cubic centimetre of lead. Write down an equation containing c and z.

The second type of brass uses 8 cubic centimetres of copper and 3 cubic centimetres of zinc. This mixture weighs 93 grams.

b Write down a second equation containing c and z.

c Find the mass of:

i 1 cubic centimetre of copper **ii** 1 cubic centimetre of zinc.

3 The cost, £C, of a taxi fare for a journey of m miles, is given by this formula:

$$C = pm + q$$

The fare for a journey of 66 miles is £108.

a Write down an equation containing p and q.

The fare for a journey of 45 miles is £74.40.

b Write down a second equation containing p and q.

c In an emergency, John needs a taxi to travel 120 miles. He has £195. Will this be enough to cover the cost? Justify your answer.

4 The function f is defined by $f(x) = ax^2 + bx$.

Given that $f(3) = 9$ and $f(-2) = 14$, determine the value of $f(5)$.

5 At a supermarket, 5 tomatoes and 3 onions cost £2.35, whereas 2 tomatoes and 7 onions cost £2.39.

Alex goes to the checkout with 9 tomatoes and 4 onions. He has £6 of cash.

Does Alex have enough cash on him to buy this produce? You must explain your answer.

6 Two mobile phone providers have these different rates for calls and picture messages:

- 'A Goner' charges 10 pence per minute for any call made to a landline and 37 pence for each picture message
- 'Lob Me It' charges 12 pence per minute for landline calls and 35 pence per minute for each picture message.

Lindsay is a customer with 'A Goner'. She received a bill of £33.50 for landline calls and picture messages.

Brian is a customer with 'Lob Me It'. He received a bill for £35.50 for landline calls and picture messages.

Lindsay and Brian spent the same number of minutes making landline calls and also sent the same number of picture messages.

How many picture messages did each of them send?

- I can find the solution to a system of equations graphically. ★ Exercise 14A Q2
- I can use an appropriate method to solve a system of equations. ★ Exercise 14B Q2 ★ Exercise 14C Q1 ★ Exercise 14D Q1
- I can solve word problems by setting up and solving simultaneous equations. ★ Exercise 14E Q2

For further assessment opportunities, see the Preparation for Assessment for Unit 2 on pages 288–291.

15 Changing the subject of a formula

In this chapter you will learn how to:

- change the subject of a simple formula
- change the subject of a formula containing a fraction
- change the subject of a formula containing brackets, roots or powers
- apply reasoning skills to solve related problems.

You should already know:

- how to apply the conventions for the order of operations
- how to perform arithmetic operations on numerical and algebraic fractions
- how to evaluate a formula by substitution
- how to solve linear equations
- how to perform calculations involving powers and roots.

Changing the subject of a simple formula

In this first section, you will learn how to change the subject of a simple formula, which doesn't contain any fractions, brackets, powers or roots. The phrase 'make x the subject of the formula' means solve the equation and find x.

Example 15.1

Make x the subject of the formula $x + b = c$.

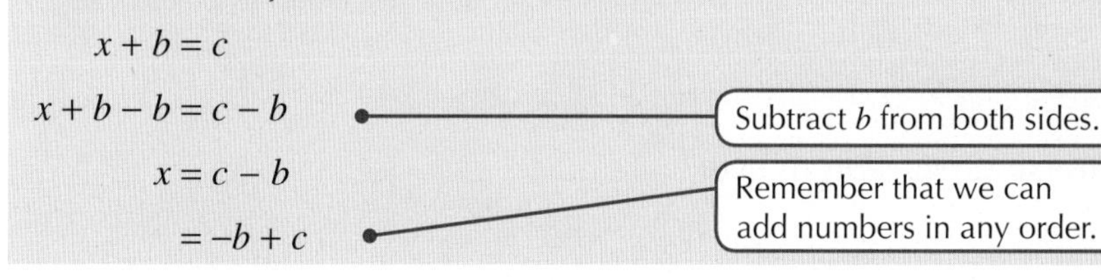

$x + b = c$
$x + b - b = c - b$
Subtract b from both sides.
$x = c - b$
$= -b + c$
Remember that we can add numbers in any order.

Example 15.2

Make x the subject of the formula $y = 5x$.

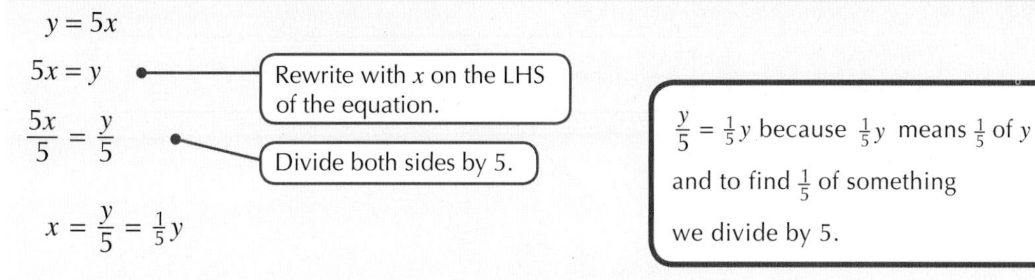

$y = 5x$
$5x = y$
Rewrite with x on the LHS of the equation.
$\frac{5x}{5} = \frac{y}{5}$
Divide both sides by 5.
$x = \frac{y}{5} = \frac{1}{5}y$
$\frac{y}{5} = \frac{1}{5}y$ because $\frac{1}{5}y$ means $\frac{1}{5}$ of y
and to find $\frac{1}{5}$ of something
we divide by 5.

Example 15.3

The formula $C = 2\pi r$ is used to find the circumference, C, of a circle when we know the radius r. Make r the subject of the formula.

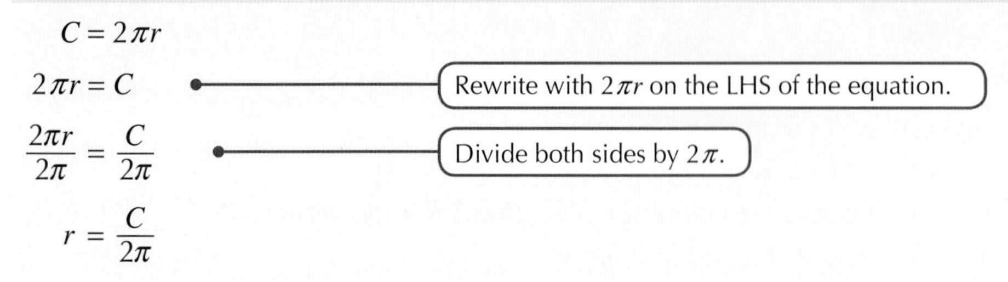

Rewrite with $2\pi r$ on the LHS of the equation.
Divide both sides by 2π.

$$C = 2\pi r$$
$$2\pi r = C$$
$$\frac{2\pi r}{2\pi} = \frac{C}{2\pi}$$
$$r = \frac{C}{2\pi}$$

Example 15.4

A straight line has equation $y = 3x - 5$. Make x the subject of the formula.

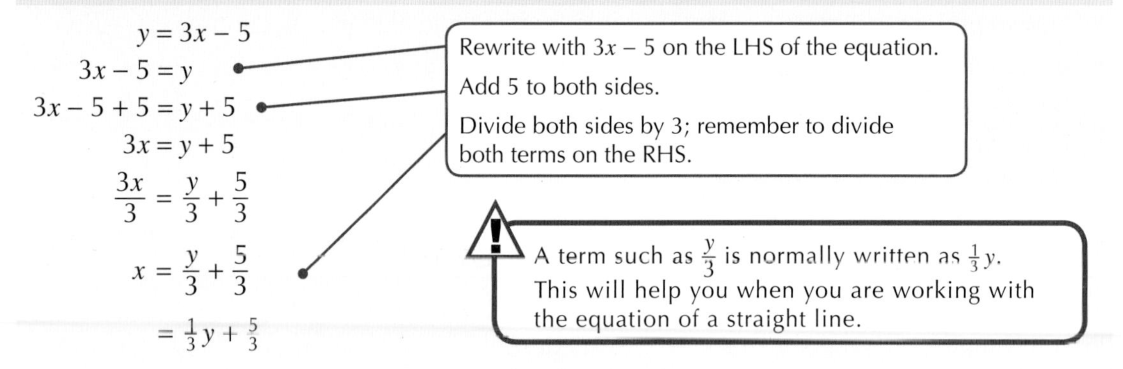

Rewrite with $3x - 5$ on the LHS of the equation.
Add 5 to both sides.
Divide both sides by 3; remember to divide both terms on the RHS.
A term such as $\frac{y}{3}$ is normally written as $\frac{1}{3}y$. This will help you when you are working with the equation of a straight line.

$$y = 3x - 5$$
$$3x - 5 = y$$
$$3x - 5 + 5 = y + 5$$
$$3x = y + 5$$
$$\frac{3x}{3} = \frac{y}{3} + \frac{5}{3}$$
$$x = \frac{y}{3} + \frac{5}{3}$$
$$= \tfrac{1}{3}y + \tfrac{5}{3}$$

Example 15.5

Make x the subject of the formula $y = 2 - 6x$.

$$y = 2 - 6x$$
$$2 - 6x = y$$
$$2 - 6x - 2 = y - 2$$
$$-6x = y - 2$$

There are two different ways to proceed from here.

Method 1

$$-6x = y - 2$$
$$\frac{-6x}{-6} = \frac{y}{-6} - \frac{2}{-6}$$

Divide both sides by -6.

$$x = \frac{y}{-6} - \frac{2}{-6}$$
$$= -\tfrac{1}{6}y + \tfrac{1}{3}$$

Remember that $\frac{y}{-6} = -\frac{y}{6} = -\frac{1}{6}y$ and $\frac{2}{-6} = \frac{-2}{6} = -\frac{1}{3}$.

(continued)

Method 2

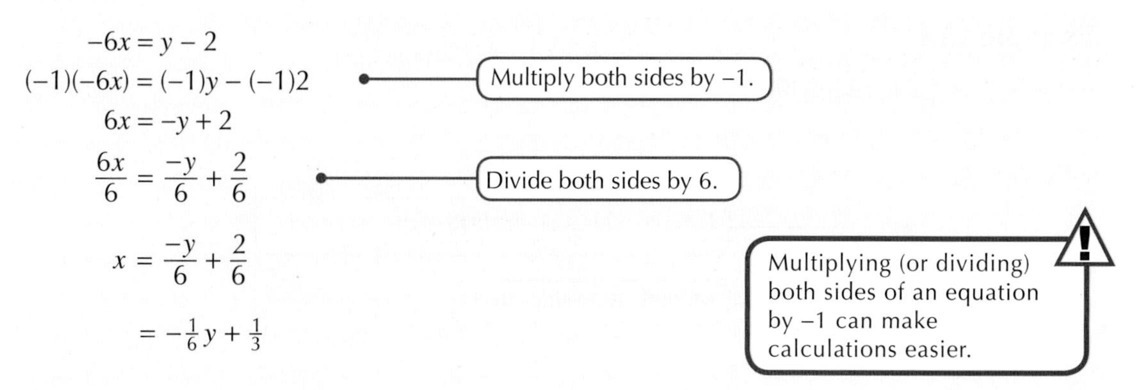

$-6x = y - 2$
$(-1)(-6x) = (-1)y - (-1)2$
Multiply both sides by −1.
$6x = -y + 2$
$\frac{6x}{6} = \frac{-y}{6} + \frac{2}{6}$
Divide both sides by 6.
$x = \frac{-y}{6} + \frac{2}{6}$
$= -\frac{1}{6}y + \frac{1}{3}$
Multiplying (or dividing) both sides of an equation by −1 can make calculations easier.

Exercise 15A

1 Make x the subject of the following formulae.

a $x + 2 = p$ **b** $x + t = w$ **c** $x - 5 = q$

d $10 + x = a$ **e** $f - x = 4$ **f** $mn - x = k$

★ **2** Make x the subject of the following formulae.

a $y = 2x$ **b** $y = 4x$ **c** $y = 3x + 2$

d $y + 5x = 4$ **e** $y - 7x = 3$ **f** $y = -5x + 1$

g $a + bx = c$ **h** $1 - pqrx = m$ **i** $m + 3nx = 7p$

3 For each of the following maths and science formulae, change the subject of the formula to the letter in brackets.

a $d = vt$ (t) **b** $W = mg$ (m) **c** $V = IR$ (R)

d $V = lbh$ (b) **e** $F = ma$ (a) **f** $E = mgh$ (m)

g $Q = It$ (I) **h** $l = Nhf$ (h) **i** $v = u + at$ (t)

In science it is very important to be able to change the subject of a formula.

Changing the subject of a formula containing fractions

Formulae can become complicated very quickly, especially when they contain fractions.

There are two types of formulae containing fractions you need to be able to rearrange:

- those where the formula contains a fraction where the intended subject, for example x, appears within the numerator, for example:

  $y = \frac{3x}{4} - 5$ or $y - \frac{x+1}{2} = 4$

- those where the formula contains a fraction where the intended subject, for example x, appears within the denominator, for example:

  $y = \frac{5}{x}$ or $q = \frac{1}{x-3} + t$

Formulae with the intended subject in the numerator

Example 15.6

Make x the subject of the formula $y = \frac{x}{3}$.

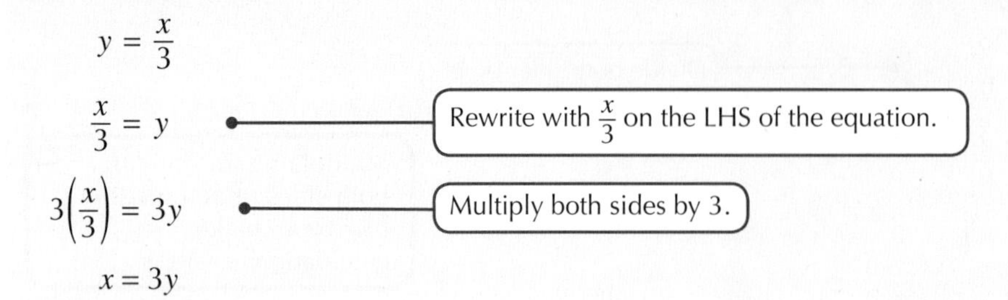

$$y = \frac{x}{3}$$

$$\frac{x}{3} = y$$ Rewrite with $\frac{x}{3}$ on the LHS of the equation.

$$3\left(\frac{x}{3}\right) = 3y$$ Multiply both sides by 3.

$$x = 3y$$

Example 15.7

Make x the subject of the formula $y = \frac{2}{5}x$.

$$y = \frac{2}{5}x$$

$$\frac{2}{5}x = y$$

$$5\left(\frac{2}{5}x\right) = 5y$$

$$2x = 5y$$

$$\frac{2x}{2} = \frac{5y}{2}$$ Divide both sides by 2.

$$x = \frac{5y}{2}$$

$$= \frac{5}{2}y$$

Example 15.8

Make p the subject of the formula $f = \frac{4 - p}{g}$.

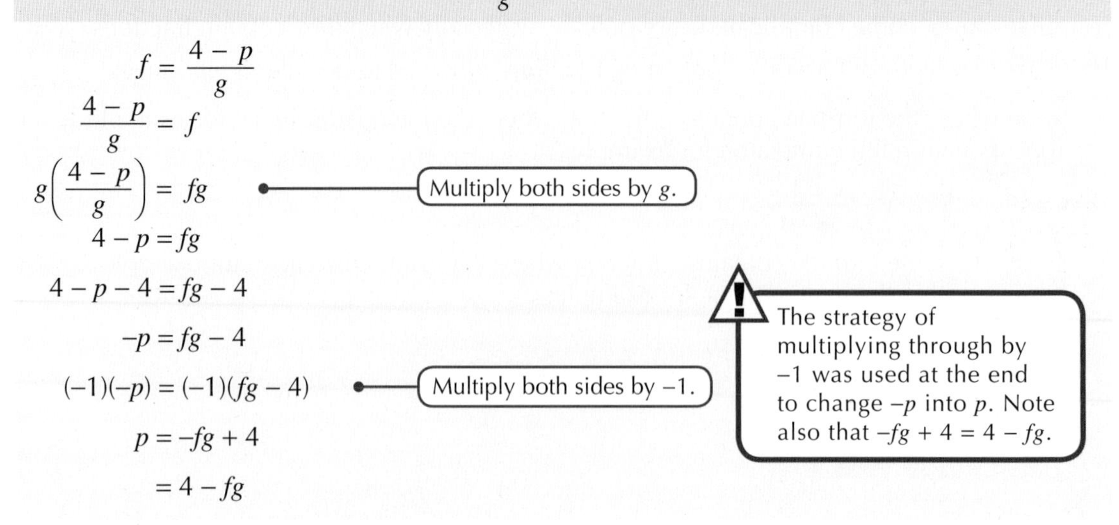

$$f = \frac{4 - p}{g}$$

$$\frac{4 - p}{g} = f$$

$$g\left(\frac{4 - p}{g}\right) = fg$$ Multiply both sides by g.

$$4 - p = fg$$

$$4 - p - 4 = fg - 4$$

$$-p = fg - 4$$

$$(-1)(-p) = (-1)(fg - 4)$$ Multiply both sides by −1.

$$p = -fg + 4$$

$$= 4 - fg$$

The strategy of multiplying through by −1 was used at the end to change $-p$ into p. Note also that $-fg + 4 = 4 - fg$.

Example 15.9

Make k the subject of the formula $q = \frac{k}{2} - \frac{p}{3}$.

The RHS of the equation contains two terms, both of which are fractions. There are two methods you can use to remove the fractions:

- method 1: multiply by the lowest common multiple (LCM) of the denominators
- method 2: express fractions as a single fraction and multiply by LCM.

Method 1 Multiply by the LCM

$$q = \frac{k}{2} - \frac{p}{3}$$

$$6q = 6\left(\frac{k}{2}\right) - 6\left(\frac{p}{3}\right)$$ — Multiply both sides by LCM of 2 and 3, which is 6.

$$6q = 3k - 2p$$ — Simplify fractions on the RHS.

Method 2 Express the two fractions as a single fraction and then multiply by the LCM

$$q = \frac{k}{2} - \frac{p}{3}$$

$$q = \frac{3k}{6} - \frac{2p}{6}$$ — Make denominators the same, then express RHS as a single fraction.

$$= \frac{3k - 2p}{6}$$

$$6q = 6\left(\frac{3k - 2p}{6}\right)$$ — Multiply both sides by 6.

$$6q = 3k - 2p$$ — Simplify RHS.

Both methods have eliminated the fractions. Now proceed to rearrange as before.

$$6q = 3k - 2p$$

$$3k - 2p = 6q$$ — Rewrite equation with $3k - 2p$ on LHS.

$$3k - 2p + 2p = 6q + 2p$$ — Add $2p$ to both sides.

$$3k = 6q + 2p$$

$$\frac{3k}{3} = \frac{6q}{3} + \frac{2p}{3}$$ — Divide both sides by 3.

$$k = \frac{6q}{3} + \frac{2p}{3}$$

$$= 2q + \tfrac{2}{3}p$$

Exercise 15B

1 Make x the subject of the following formulae.

a $y = \frac{x}{4}$ **b** $y = \frac{x}{5}$ **c** $y = \frac{3x}{7}$ **d** $y = \frac{2x}{3}$

e $y = \frac{1}{6}x$ **f** $y = \frac{4}{5}x$ **g** $y = \frac{x + 3}{8}$ **h** $y = \frac{x + 1}{4}$

i $y = \frac{x - 3}{2}$ **j** $y = \frac{3x + 5}{4}$ **k** $y = \frac{5x - 1}{3}$ **l** $y = \frac{2 - 3x}{5}$

★ **2** Make x the subject of the following formulae.

a $y = \frac{x}{2} + 6$ **b** $y = \frac{x}{3} + 1$ **c** $y = \frac{x}{7} - 2$ **d** $\frac{1}{6}x - y = 4$

e $y + \frac{4}{5}x = 0$ **f** $y = \frac{x+5}{3} - 1$ **g** $y = \frac{2x+1}{2} + 5$ **h** $y = 3 - \frac{4x-3}{2}$

★ **3** Change the subject of the following maths and science formulae to the letter in brackets.

a $g = \frac{w}{m}$ (w) **b** $I = \frac{V}{R}$ (V) **c** $\frac{F}{a} = m$ (F)

d $h = \frac{E}{mg}$ (E) **e** $t = \frac{v-u}{a}$ (v) **f** $a = \frac{v-u}{t}$ (u)

★ **4** To exercise safely, your heart should not beat more than a certain number of times per minute. This is called your maximum heart rate, H, and can be found using the formula

$$H = 208 - \tfrac{7}{10}A$$

where A is your age in years.

a Calculate the maximum heart rate for a 30-year old.

b Make A the subject of the formula.

c Anna is cycling on an exercise bike, and her heart monitor indicates that Anna's heart rate is 176 beats per minute. Anna is 46 years old. Is she putting herself at risk? You should explain your answer.

5 The volume of blood, V litres, in an adult woman's body can be calculated using the formula

$$V = \frac{53.7W + 984}{1000}$$

where W is the woman's weight in kilograms.

a The average woman weighs 70 kilograms and her body should contain approximately 8 pints of blood. Does the formula predict this? Justify your answer.

b Make W the subject of the formula.

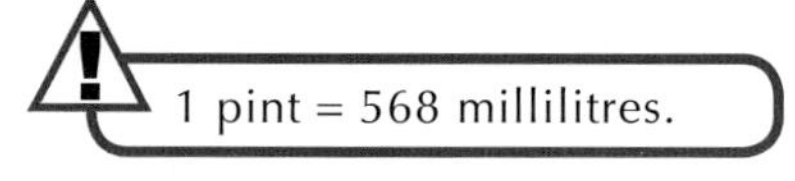

c The following statement is a used to estimate a woman's weight if her volume of blood is known.
'Subtract 1 litre from the volume of blood and multiply the answer by 20'.

i Check that this works for a woman who has 4.5 litres of blood.

ii Show why the statement provides a good approximation for any given volume.

Formulae with the intended subject in the denominator

This section shows how to rearrange a formula when the intended subject is in the denominator of a fraction in the formula, such as when you want to make x the subject of $y = 5 - \frac{3}{x}$.

If you want to make x the subject, provided there are no brackets, powers or roots in the original formula, then follow these steps:

1. make sure the LHS of the equation contains only the term containing x
2. ensure that that the RHS is written as a single fraction
3. turn the LHS and RHS 'upside down'
4. clear the fraction on the LHS in the usual way.

Example 15.10

Make x the subject of the formula $y = \frac{5}{x}$.

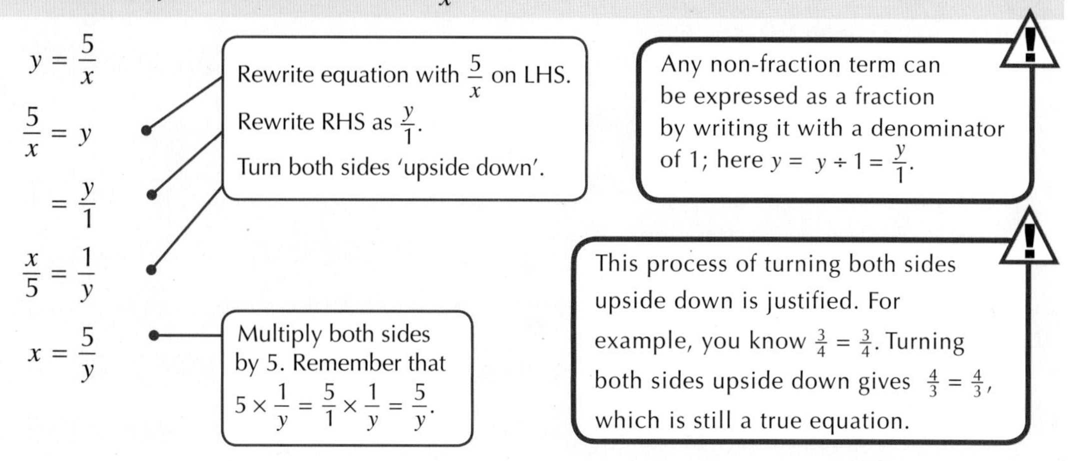

$y = \frac{5}{x}$
$\frac{5}{x} = y$
$= \frac{y}{1}$
$\frac{x}{5} = \frac{1}{y}$
$x = \frac{5}{y}$
Rewrite equation with $\frac{5}{x}$ on LHS.
Rewrite RHS as $\frac{y}{1}$.
Turn both sides 'upside down'.
Multiply both sides by 5. Remember that $5 \times \frac{1}{y} = \frac{5}{1} \times \frac{1}{y} = \frac{5}{y}$.
Any non-fraction term can be expressed as a fraction by writing it with a denominator of 1; here $y = y \div 1 = \frac{y}{1}$.
This process of turning both sides upside down is justified. For example, you know $\frac{3}{4} = \frac{3}{4}$. Turning both sides upside down gives $\frac{4}{3} = \frac{4}{3}$, which is still a true equation.

Example 15.11

Make x the subject of the formula $y = 4 + \frac{3}{x}$.

$$y = 4 + \frac{3}{x}$$

$$\frac{3}{x} = y - 4$$

$$= \frac{y-4}{1}$$

$$\frac{x}{3} = \frac{1}{y-4}$$

$$x = \frac{3}{y-4}$$

Example 15.12

Make g the subject of the formula $T = 5 - \frac{l}{g}$.

$$T = 5 - \frac{l}{g}$$

$$5 - \frac{l}{g} = T$$

$$-\frac{l}{g} = T - 5$$

$$\frac{l}{g} = 5 - T$$

$$= \frac{5-T}{1}$$

$$\frac{g}{l} = \frac{1}{5-T}$$

$$g = \frac{l}{5-T}$$

Example 15.13

Make t the subject of the formula $G = \frac{m}{t-4}$.

$$G = \frac{m}{t-4}$$

$$\frac{m}{t-4} = G$$

$$= \frac{G}{1}$$

$$\frac{t-4}{m} = \frac{1}{G}$$

$$t - 4 = \frac{m}{G}$$

$$t = \frac{m}{G} + 4$$

Example 15.14

Make p the subject of the formula $H = \frac{\pi}{k - pr}$.

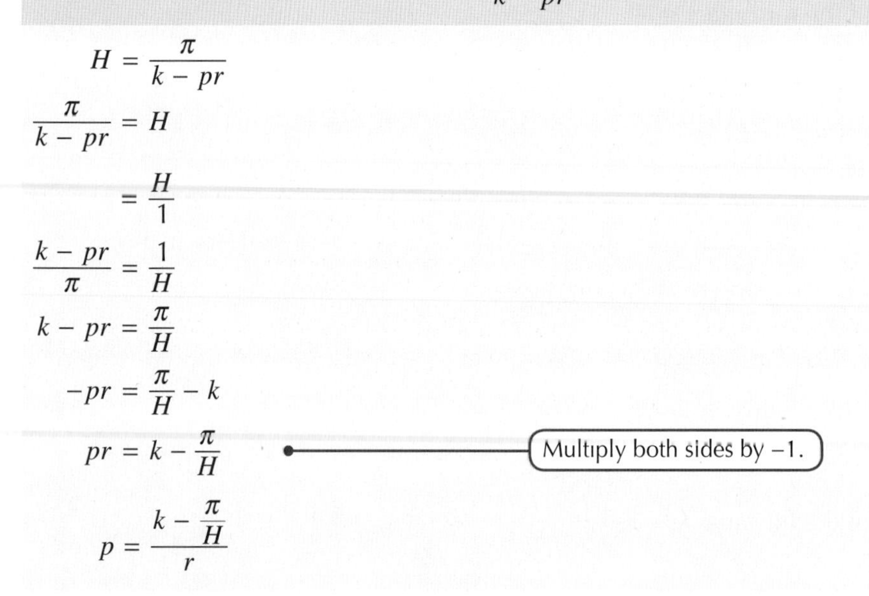

$$H = \frac{\pi}{k - pr}$$

$$\frac{\pi}{k - pr} = H$$

$$= \frac{H}{1}$$

$$\frac{k - pr}{\pi} = \frac{1}{H}$$

$$k - pr = \frac{\pi}{H}$$

$$-pr = \frac{\pi}{H} - k$$

$$pr = k - \frac{\pi}{H}$$

Multiply both sides by –1.

$$p = \frac{k - \frac{\pi}{H}}{r}$$

Example 15.15

In optical physics, the formula $\frac{1}{f} = \frac{1}{u} + \frac{1}{v}$ is used. Make u the subject of the formula.

$$\frac{1}{f} = \frac{1}{u} + \frac{1}{v}$$

$$\frac{1}{u} = \frac{1}{f} - \frac{1}{v}$$

$$= \frac{v}{fv} - \frac{f}{fv} = \frac{v - f}{fv}$$

Express the RHS as a single fraction.

$$\frac{u}{1} = \frac{fv}{v - f}$$

$$u = \frac{fv}{v - f}$$

Exercise 15C

★ **1** Make x the subject of the following formulae.

a $y = \frac{1}{x}$ **b** $y = \frac{8}{x}$ **c** $y = \frac{3}{x + 5}$

d $y = \frac{2}{x - 1}$ **e** $y = \frac{3}{2x + 5}$ **f** $y = \frac{7}{1 - 6x}$

g $y = \frac{1}{2x} + 6$ **h** $y = \frac{3}{4x} - 8$ **i** $y = \frac{4}{x + 3} + 5$

j $y = \frac{p}{q - x} - 1$ **k** $y = f - \frac{g}{hx}$ **l** $y = \frac{\pi}{mx + n} - p$

★ **2** In physics, when two resistors are connected in parallel, the total resistance, R, is given by the formula

$$\frac{1}{R} = \frac{1}{A} + \frac{1}{B}$$

where A and B are the two individual resistances.

a Make B the subject of the formula.

When three resistors are connected in parallel, the total resistance is given by the formula

$$\frac{1}{R} = \frac{1}{A} + \frac{1}{B} + \frac{1}{C}$$

where A, B and C are the individual resistances.

b Make A the subject of the formula.

Changing the subject of a formula containing brackets, roots or powers

You follow a slightly different process if the formula you are working with contains brackets, powers or roots. When making x the subject of a formula, you should follow these steps:

1. if x is inside a pair of brackets, multiply out the brackets first
2. isolate any powers or roots (of x) on the LHS of the equation
3. remove any powers or roots
4. complete the solution if required.

Example 15.16

The area, A, of a circle is given by $A = \pi r^2$, where r is the radius. Make r the subject of the formula.

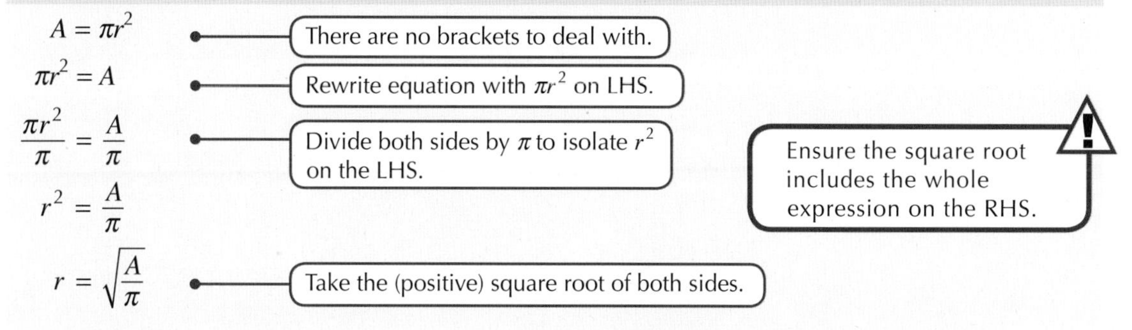

$A = \pi r^2$
There are no brackets to deal with.
$\pi r^2 = A$
Rewrite equation with πr^2 on LHS.
$\frac{\pi r^2}{\pi} = \frac{A}{\pi}$
Divide both sides by π to isolate r^2 on the LHS.
$r^2 = \frac{A}{\pi}$
$r = \sqrt{\frac{A}{\pi}}$
Take the (positive) square root of both sides.
Ensure the square root includes the whole expression on the RHS.

Example 15.17

In medicine, the Mosteller Formula is used to calculate the approximate surface area of the human body, S, in square metres. The formula is

$$S = \frac{\sqrt{WH}}{60}$$

where W and H are the weight (kg) and height (cm), respectively.

Make H the subject of the formula.

$$S = \frac{\sqrt{WH}}{60}$$

$$\frac{\sqrt{WH}}{60} = S$$

$$\sqrt{WH} = 60S$$ Multiply both sides by 60 to isolate the square root.

$$WH = 3600S^2$$ Square both sides.

$$H = \frac{3600S^2}{W}$$ Divide both sides by W.

Example 15.18

The volume, V, of a sphere with radius, r, can be calculated using the formula

$$V = \tfrac{4}{3}\pi r^3$$

Make r the subject of the formula.

$$V = \frac{4}{3}\pi r^3$$

$$\frac{4}{3}\pi r^3 = V$$

$$4\pi r^3 = 3V$$

$$r^3 = \frac{3V}{4\pi}$$

$$r = \sqrt[3]{\frac{3V}{4\pi}}$$ Take the cube root of both sides.

Example 15.19

In physics, the distance, s, travelled by an object can be found using the formula

$$s = \frac{t(u+v)}{2}$$

Make v the subject of the formula.

$$s = \frac{t(u+v)}{2}$$

$$\frac{t(u+v)}{2} = s$$

$$t(u+v) = 2s$$ Multiply both sides by 2.

$$\frac{t(u+v)}{t} = \frac{2s}{t}$$ Divide both sides by t.

$$u+v = \frac{2s}{t}$$

$$v = \frac{2s}{t} - u$$

Example 15.20

Make d the subject of the formula $T = k\sqrt{d} + e$

$$T = k\sqrt{d} + e$$

$$k\sqrt{d} + e = T$$

$$k\sqrt{d} = T - e$$

$$\sqrt{d} = \frac{T-e}{k}$$

$$d = \left(\frac{T-e}{k}\right)^2 = \frac{(T-e)^2}{k^2}$$ Square both sides of the equation.

Example 15.21

The base, b, of a square-based pyramid can be found using the formula

$$b = \sqrt{\frac{3V}{h}}$$

where V and h are the volume and height, respectively.

Make h the subject of the formula.

$$b = \sqrt{\frac{3V}{h}}$$

$$\sqrt{\frac{3V}{h}} = b$$

$$\frac{3V}{h} = b^2$$

$$= \frac{b^2}{1}$$

$$\frac{h}{3V} = \frac{1}{b^2}$$ Multiply both sides by $3V$.

$$h = \frac{3V}{b^2}$$

Exercise 15D

1 Make x the subject of the following formulae.

a $y = x^2$ **b** $y = 4x^2$ **c** $y = \frac{2}{3}x^2$ **d** $y = \frac{3}{5}x^2$

e $y = \sqrt{x}$ **f** $y = 3\sqrt{x}$ **g** $y = \frac{1}{2}\sqrt{x}$ **h** $y = \sqrt{x+5}$

i $y = 5x^2 + 2$ **j** $y = 6 - x^2$ **k** $y = \frac{3}{5}x^2 + 4$ **l** $y = 2\sqrt{x-3}$

★ **2** Make x the subject of the following formulae.

a $y = x^3$ **b** $y = 8x^3$ **c** $y = \frac{3\sqrt{x+1}}{4}$ **d** $y = (x+2)^3$

e $y = \frac{2\sqrt{x}}{3} + 5$ **f** $y = \frac{5\sqrt{x}}{2} - 1$ **g** $y = 6\sqrt{x-3} + 1$ **h** $y = \frac{1}{\sqrt{x}}$

i $y = \frac{4}{\sqrt{x}}$ **j** $y = \frac{3}{\sqrt{x+4}}$ **k** $y = \frac{5}{2\sqrt{x}}$ **l** $y = \frac{7}{4\sqrt{x-1}}$

★ **3 a** Make r the subject of the formula $M + 5 = 4r^2$.

b Make q the subject of the formula $I = 5q^3 - f$.

c Make k the subject of the formula $W = \frac{3}{k^2} - g$.

d Make d the subject of the formula $P = v - \frac{1}{4d^2}$.

4 a Make h the subject of the formula $P = 3\sqrt{h} - 10$.

b Make t the subject of the formula $\frac{g}{\sqrt{t}} = Q - v$.

c Make r the subject of the formula $L = 2 - \pi\sqrt{\frac{r}{g}}$.

d Make b the subject of the formula $V - 4s = 3\sqrt{\frac{1}{bh}}$.

5 a The formula for finding the volume of a cylinder is $V = \pi r^2 h$, where r is the radius and h is the height.

i Make h the subject of the formula.

ii Make r the subject of the formula.

b The formula for finding the volume of a cone is $V = \frac{1}{3}\pi r^2 h$, where r is the radius and h is the height.

i Make h the subject of the formula.

ii Make r the subject of the formula.

c The formula for finding the energy of a body of mass m moving with velocity v is $E = \frac{1}{2}mv^2$. Make v the subject of the formula.

d The formula for finding the distance travelled is given by $s = ut + \frac{1}{2}at^2$, where u is the initial speed, a is acceleration and t is time. Make a the subject of the formula.

★ **6** The gravitational force of attraction, F, between two bodies is given by $F = \frac{GMm}{d^2}$, where G is the universal gravitational constant, M and m are the masses of the two bodies and d is the distance between them.

a Make d the subject of the formula.

b If the distance between the two bodies is doubled, what is the effect on the force of attraction?

7 For gas molecules of given mass m at temperature T, their average velocity, v, can be found using the formula $v = \sqrt{\frac{3kT}{m}}$, where k is the Boltzmann constant.

a Make T the subject of the formula.

b Make m the subject of the formula.

c What happens to the velocity of the gas molecules if the temperature is multiplied by 4?

8 The speed of a car, v (miles per hour), just before braking, can be calculated from its braking distance, d (feet), using the formula $v = 2\sqrt{5d + 25} - 10$.

> Braking distance is the distance the car travels before coming to a standstill. It assumes 'ideal' driving conditions, that is, the weather is dry, it is daytime and the visibility is good.

a Calculate the speed (just before braking) of a car whose braking distance is 240 feet.

b Change the subject of the formula to d and show that $d = \frac{v^2}{20} + v$.

c The maximum permitted speed (for cars) on British motorways is 70 miles per hour. How much further would it take for a car travelling at 80 miles per hour to completely stop compared to one travelling at 70 miles per hour? You may assume ideal driving conditions.

d Show that the extra distance, A, needed to stop when driving at p miles per hour compared to q miles per hour (where $p > q$) is given by the formula

$$A = \frac{(p-q)(p+q+20)}{20}$$

9 In athletics, the decathlon consists of 10 track and field events. Formulae are used to calculate how many points an athlete is awarded in a particular event.
For the 100 metre sprint, the number of points awarded is calculated using the formula

$$P = 25.4347(18 - t)^{1.81}$$

where t is the time in seconds.

a In March 2013 the world record for the men's 100 metre sprint is 9.58 seconds, held by Jamaica's Usain Bolt. Calculate how many points this time would receive in the decathlon.

In a decathlon athletes need to be good in a range of events and not just one or two. Decathletes aim to score around 900 points in each event.

b Change the subject of the above formula to t and calculate the time which would be required in the 100 metre sprint to score 900 points. Give your answer to the nearest hundredth of a second.

10 The distance from Achy to Breakey is d miles. A train travels from Achy to Breakey at a constant speed of a miles per hour. On the way back, from Breakey to Achy, the train travels at a constant speed of b miles per hour.

a Write down expressions for the time taken for the train to travel from:

i Achy to Breakey **ii** Breakey to Achy

b Show that the average speed of the train, v miles per hour, for the whole journey is given by the formula $v = \frac{2ab}{a + b}$.

c Change the subject of the formula to b.

GO! Activity

Investigate the formulae used for other events in the decathlon and work out what score would be achieved by an athlete who equalled the current world record in all 10 events. Investigate also the times and distances which would be needed to score 900 points in every event.

- I can change the subject of a simple formula.
★ Exercise 15A Q2
- I can change the subject of a formula when the formula contains brackets, fractions, powers or roots.
★ Exercise 15B Q2, Q3, Q4 ★ Exercise 15C Q1
★ Exercise 15D Q2, Q3
- I can apply my skills to solve related problems.
★ Exercise 15C Q2 ★ Exercise 15D Q6

For further assessment opportunities, see the Preparation for Assessment for Unit 2 on pages 288–291.

16 Recognise and determine the equation of a quadratic function from its graph

In this chapter you will learn how to:

- recognise and determine the equation of a quadratic of the form $y = kx^2$
- recognise and determine the equation of a quadratic of the form $y = (x + p)^2 + q$
- apply knowledge to solve related problems.

You should already know:

- how to determine the equation of a straight line in the form $y - b = m(x - a)$
- how to use function notation $f(x)$.

Recognise and determine the equation of a quadratic of the form $y = kx^2$

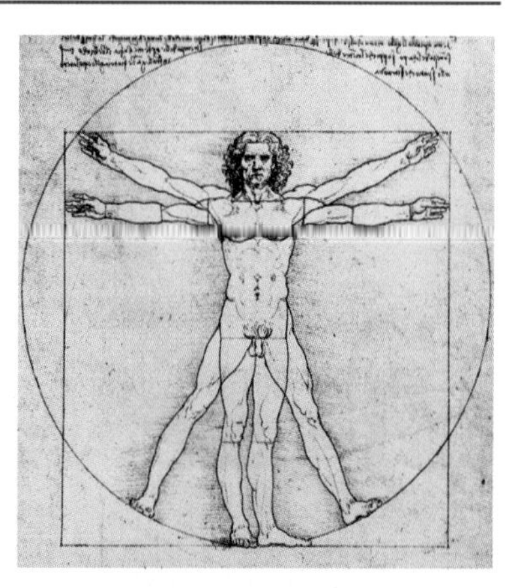

Kicking a penalty in rugby. The path of a golf ball. The orbits of planets around the Sun. The braking distance of a car. Leonardo da Vinci, his *Vitruvian Man* and dynamic symmetry in art. Simple supply and demand models in business. Load-bearing arched bridges. The reflector of a car headlight. Forecasting the weather. The basic principles of aviation. Mobile phones. What do these all have in common? The answer is simple: quadratic functions. The applications described are only the tip of the iceberg. The quadratic function is among the most important in mathematics and its uses are virtually endless.

The graph of a quadratic function is called a **parabola**. The world's longest suspension bridge is the *Akashi Kaikyo* in Japan, which is almost 2 kilometres long. Engineers use mathematics to analyse the forces involved and, in order to provide stability, the shape of a bridge cable must be a parabola.

This chapter looks at finding the equation of a quadratic function from its graph, while Chapters 17 and 18 look at sketching quadratic functions and identifying the features of quadratic functions. Chapter 19 covers quadratic equations in greater depth.

Activity

You may find it useful in this activity to check your answers using a graphing calculator.

The simplest quadratic function is $f(x) = x^2$.

1 a Copy and complete the table for the function $f(x) = x^2$.

x	−3	−2	−1	0	1	2	3
x^2	9	4					

b Plot these points on a coordinate diagram and connect them with smooth curve.

Note the following key features of the graph of $y = x^2$:

- the y-axis is an axis of symmetry; the equation of the axis of symmetry is $x = 0$
- the graph has a **minimum turning point** at (0, 0)
- the graph never goes below the x-axis.

c Explain why the graph displays the three features described above.

Note that the **minimum value** of the function is 0 and this occurs when $x = 0$.

2 a Copy and complete the table for the function $f(x) = 2x^2$.

x	−3	−2	−1	0	1	2	3
x^2	18	8					

b Plot these points on the same coordinate diagram you used in Question 1.

c Explain why the graph of the function $y = 2x^2$ has the same key features as $y = x^2$.

d Compare the shape of the graph $y = 2x^2$ with the shape of $y = x^2$. Describe any differences between them.

3 Repeat Question 2 for these functions:

a $f(x) = 3x^2$ b $f(x) = \frac{1}{2}x^2$.

4 a Copy and complete the table for the function $f(x) = -x^2$.

x	−3	−2	−1	0	1	2	3
$-x^2$	−9						

$-x^2$ means $-1 \times x^2$.

b Plot these points on a coordinate diagram and connect them with smooth curve.

Note the following key features of the graph of $y = -x^2$:

- the y-axis is an axis of symmetry; the equation of the axis of symmetry is $x = 0$
- the graph has a **maximum turning point** at (0, 0)
- the graph never goes above the x-axis.

c Explain why the graph displays the three features described above.

Note that the **maximum value** of the function is 0 and this occurs when $x = 0$.

5 Repeat Question 4 for these functions:

a $f(x) = -2x^2$ b $f(x) = -3x^2$ c $f(x) = -\frac{1}{2}x^2$

Key features of the graph of $y = kx^2$

From the activity you will have established the following key features of the graph of $y = kx^2$:

- if $k > 0$ then the graph of the function $y = kx^2$ has a minimum turning point
- if $k > 0$ the minimum value of the function $f(x) = kx^2$ is equal to 0 and this occurs when $x = 0$
- if $k < 0$ then the graph of the function $y = kx^2$ has a maximum turning point
- if $k < 0$ the maximum value of the function $f(x) = kx^2$ is equal to 0 and this occurs when $x = 0$
- the y-axis is the axis of symmetry of the graph having equation $y = kx^2$
- the equation of the axis of symmetry of the graph $y = kx^2$ is $x = 0$.

You now know the key features of the graphs of simple quadratic functions of the form $y = kx^2$. You can use this to work out the equation of the function if you are given the graph, that is, work out k. Remember that the **graph** and the **equation** give the **same information**. To find the equation, there are three steps you should follow:

1 choose any point on the graph
2 substitute the values of the coordinate point into the equation $y = kx^2$
3 solve the equation to find k.

Example 16.1

The diagram shows the parabola with equation $y = kx^2$.

What is the value of k?

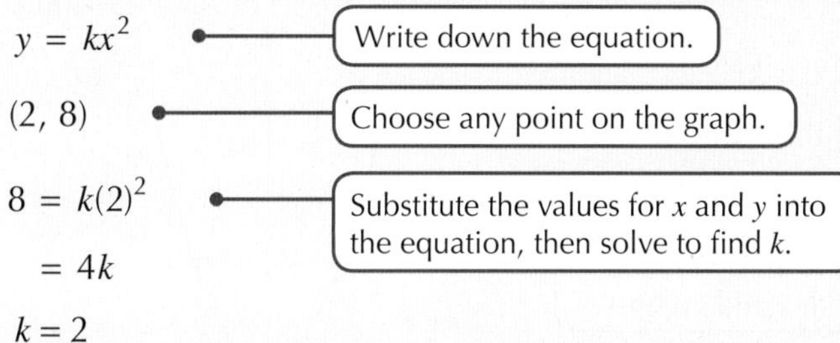

y = kx²
Write down the equation.
(2, 8)
Choose any point on the graph.
8 = k(2)²
= 4k
Substitute the values for x and y into the equation, then solve to find k.

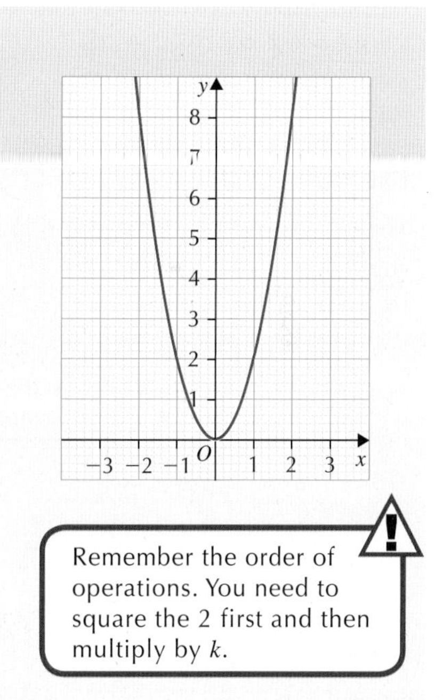

Remember the order of operations. You need to square the 2 first and then multiply by k.

$k = 2$

So: the equation of the parabola is $y = 2x^2$.

Check: To confirm that the value for k is correct do the following:

- look at the graph to check that k has the correct sign (is positive or negative)
- substitute another point from the graph to confirm your value of k.

The graph has a minimum turning point so k must be positive. ✓

Choose another point on the graph, such as (1, 2) and substitute the values for x and y into the general question $y = kx^2$:

LHS = 2

RHS = $k(1)^2$ — Put LHS = RHS and solve the equation to find k.

$2 = k(1)^2$

$= k(1)$

$k = 2$ ✓

Example 16.2

The diagram shows the parabola with equation $y = kx^2$.

What is the value of k?

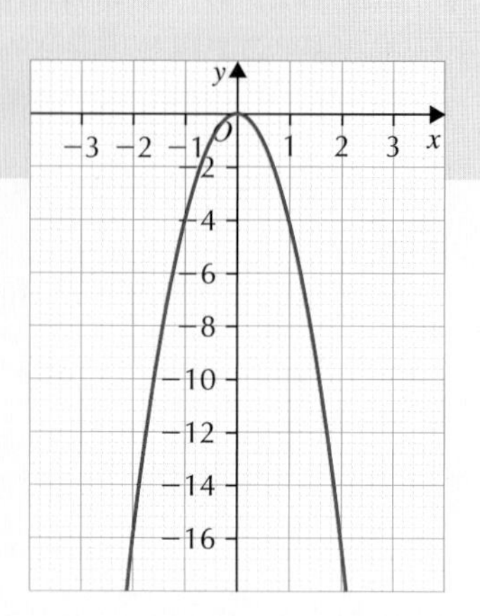

$y = kx^2$

$(2, -16)$

$-16 = k(2)^2$

$= 4k$

$k = -4$

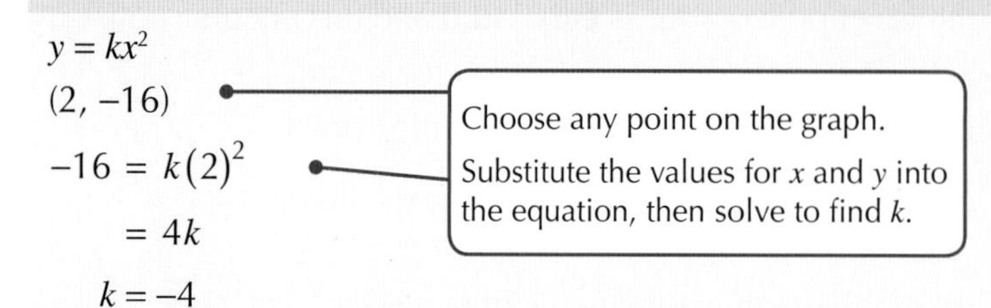

So: the equation of the parabola is $y = -4x^2$.

Check: The graph has a maximum turning point so k must be negative. ✓

Choose another point on the graph, such as (1, –4) and substitute the values for x and y into the general equation $y = kx^2$:

LHS = –4

RHS = $k(1)^2$

$-4 = k(1)^2$

$k = -4$ ✓

Example 16.3

The diagram shows part of the graph $f(x) = kx^2$.

The graph passes through (2, 12). Find the value of k.

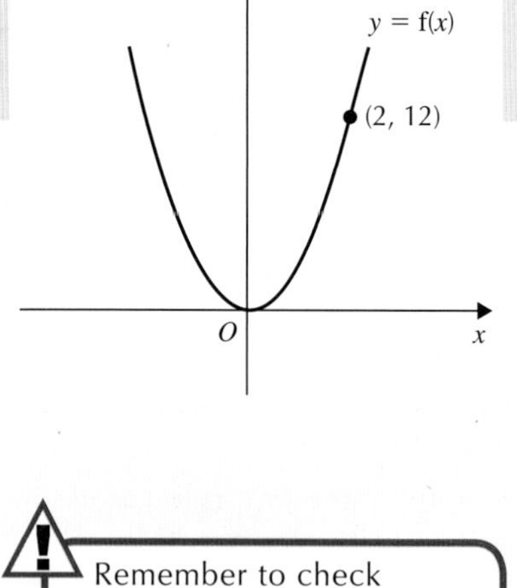

$y = kx^2$

$12 = k(2)^2$

$= 4k$

$k = 3$

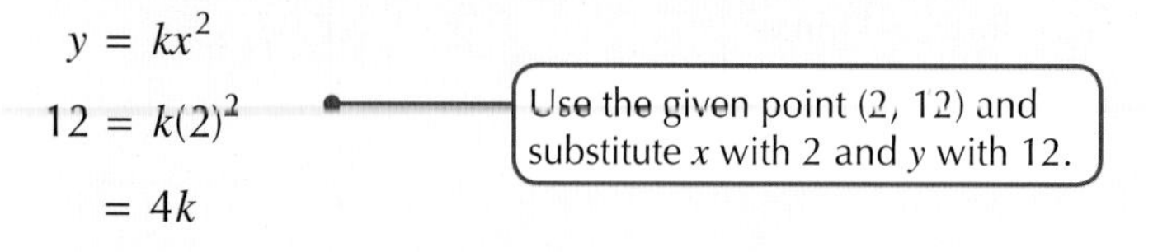

Use the given point (2, 12) and substitute x with 2 and y with 12.

So: the equation of the parabola is $y = 3x^2$.

Remember to check your answer.

Exercise 16A

★ **1** The equation of each of the following graphs is $y = kx^2$. For each one, identify the value of k. Write the equation of the function.

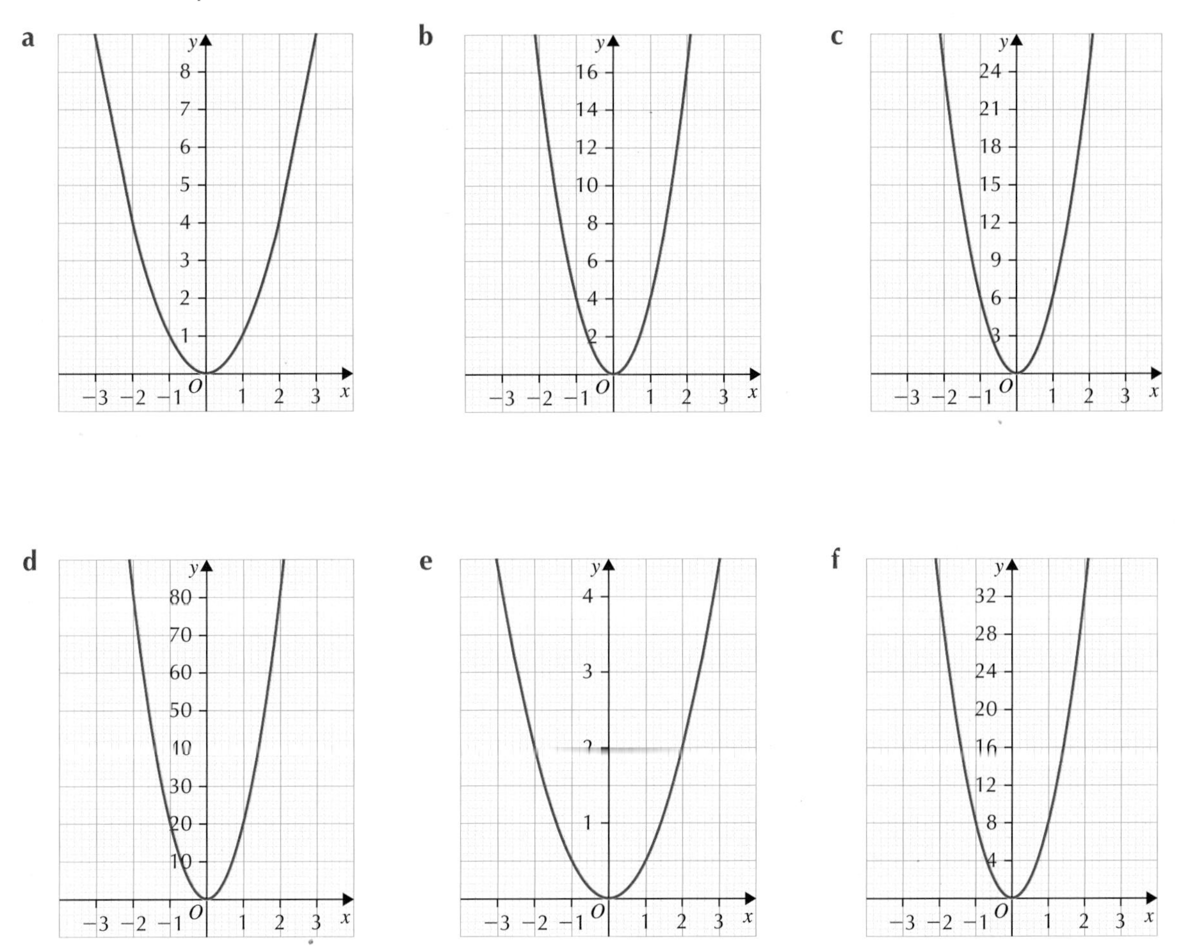

2 The equation of each of the following graphs is $y = kx^2$. For each one, identify the value of k. Write the equation of the function.

a

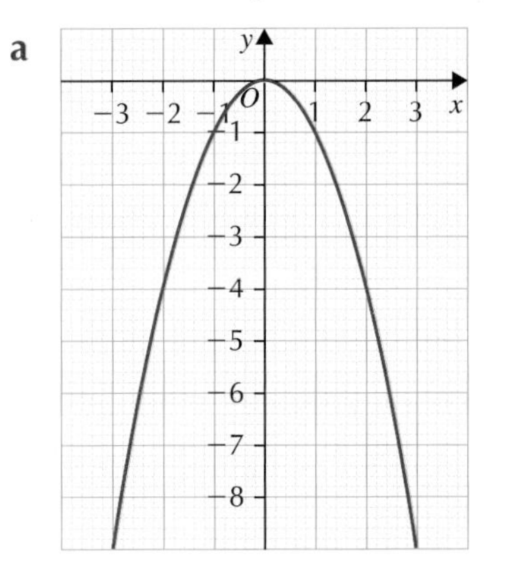

b

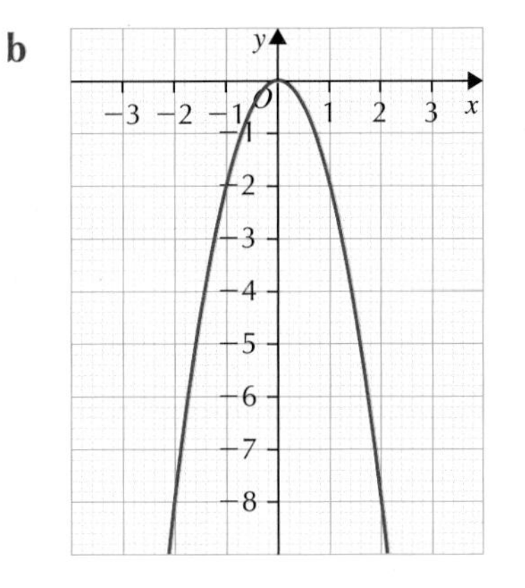

c

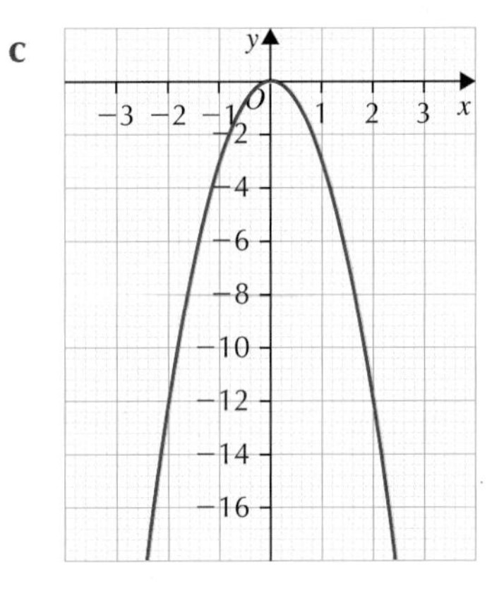

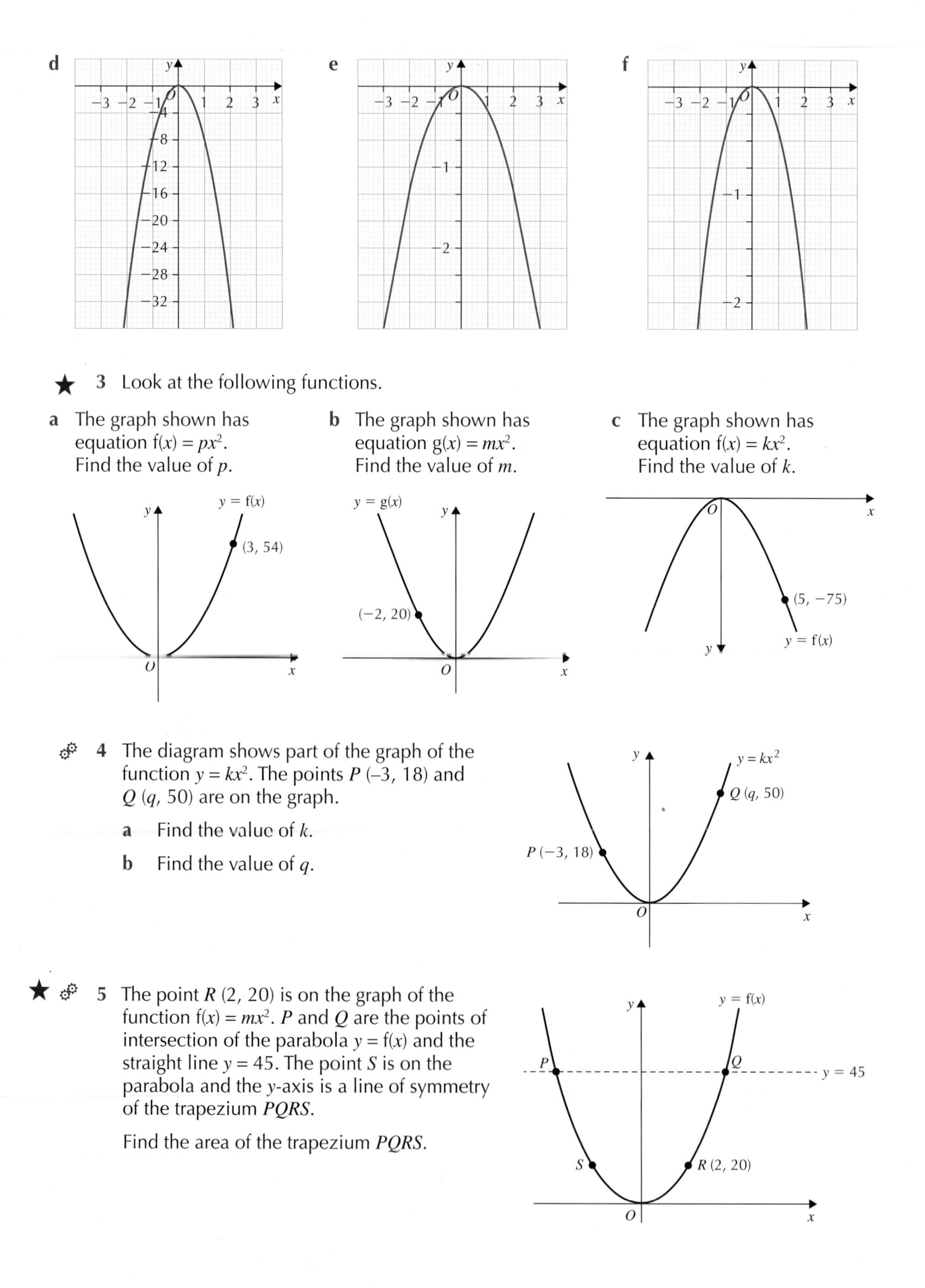

★ **3** Look at the following functions.

a The graph shown has equation $f(x) = px^2$. Find the value of p.

b The graph shown has equation $g(x) = mx^2$. Find the value of m.

c The graph shown has equation $f(x) = kx^2$. Find the value of k.

4 The diagram shows part of the graph of the function $y = kx^2$. The points P (–3, 18) and Q (q, 50) are on the graph.

a Find the value of k.

b Find the value of q.

★ **5** The point R (2, 20) is on the graph of the function $f(x) = mx^2$. P and Q are the points of intersection of the parabola $y = f(x)$ and the straight line $y = 45$. The point S is on the parabola and the y-axis is a line of symmetry of the trapezium $PQRS$.

Find the area of the trapezium $PQRS$.

6 The diagram shows the graph of the parabola with equation $f(x) = -\frac{1}{6}x^2$. B has coordinates (0, –6).

A and C are points on the graph of $y = f(x)$. The kite $OABC$ is symmetrical about the y-axis and has area 12 square units. Determine the coordinates of A and C.

7 The diagram shows the graph of the parabola with equation $y = kx^2$. The points R (–2, 2) and T (t, 8) are on the parabola.

a Find the value of k.

b Find the equation of the straight line RT.

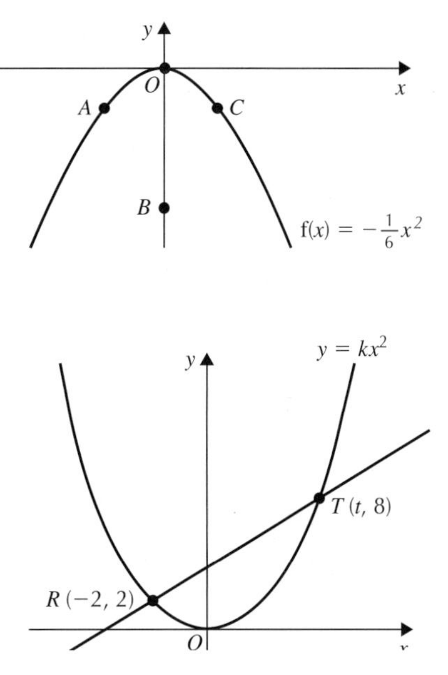

8 A quadratic function has equation $f(x) = kx^2$. P and Q are points on the parabola with x-coordinates p and q, respectively. A straight line, parallel to PQ, is drawn from the origin, O, to the point R. The x-coordinate of R is r.

Show that $r = k(p + q)$.

Recognise and determine the equation of a quadratic of the form $y = (x + p)^2 + q$

This section looks at quadratic functions of the form $y = (x + p)^2 + q$. This form is called the **completed square** form of the quadratic function. The algebraic process of completing the square is covered in Chapter 5.

Remember these two key points relating to the graph of $y = x^2$:

- for all values of x, $x^2 \geqslant 0$
- the smallest or **minimum** value of x^2 is 0 and this occurs when $x = 0$.

Activity

You can draw these graphs by hand or use a graphing calculator.

Consider the function $f(x) = x^2 + 5$. When $x = 0$, the minimum possible value for x^2 is 0, so the minimum value of $x^2 + 5$ must be 5.

1 a Copy and complete the table for the function $f(x) = x^2 + 4$.

x	–2	–1	0	1	2
$x^2 + 4$	8				

b Plot the points and draw the graph of $y = f(x)$.

c Write down the coordinates of the minimum turning point.

d What is the minimum value of the function f?

e Write down the equation of the axis of symmetry of the graph $y = f(x)$.

(continued)

2 Repeat Question 2 for these functions:

i $f(x) = x^2 + 7$ **ii** $g(x) = x^2 + 5$ **iii** $h(x) = x^2 - 4$ **iv** $m(x) = x^2 - 8$

3 For each of the following functions write down:

i the minimum value **ii** the value of x when the minimum occurs.

a $f(x) = x^2 + 1$ **b** $g(x) = x^2 + 6$ **c** $h(x) = x^2 - 3$ **d** $p(x) = x^2 - 8$

4 a The table shows points which are on the graph of the function $f(x) = x^2$.

x	–2	–1	0	1	2
x^2	4	1	0	1	4

Draw the graph of $y = f(x)$.

b Copy and complete the table for the function $f(x) = (x - 1)^2$.

x	–1	0	1	2	3
$x - 1$	–2				
$(x - 1)^2$	4				

c Draw the graph of the function $y = (x - 1)^2$ on the same coordinate grid as part **a**.

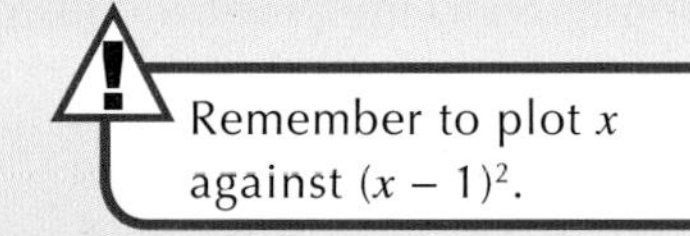
Remember to plot x against $(x - 1)^2$.

d **i** What is the minimum value of the function $f(x) = (x - 1)^2$?

ii For which value of x does the minimum occur?

iii Write down the coordinates of the minimum turning point of the graph of $y = f(x)$.

e Write down the equation of the axis of symmetry of the graph $y = f(x)$.

5 a Copy and complete the table for the function $f(x) = (x - 3)^2$.

x	1	2	3	4	5
$x - 3$	–2	–1	0	1	2
$(x - 3)^2$	4				

b Draw the graph of $f(x) = (x - 3)^2$.

c **i** What is the minimum value of the function $f(x) = (x - 3)^2$?

ii For which value of x does the minimum occur?

iii Write down the coordinates of the minimum turning point of the graph of $y = f(x)$.

d Write down the equation of the axis of symmetry of the graph $y = f(x)$.

(continued)

6 **a** Copy and complete the table for the function $f(x) = (x + 2)^2$.

x	–4	–3	–2	–1	0
$x + 2$	–2	–1	0	1	2
$(x + 2)^2$	4				

b Draw the graph of $f(x) = (x + 2)^2$.

c **i** What is the minimum value of the function $f(x) = (x + 2)^2$?

ii For which value of x does the minimum occur?

iii Write down the coordinates of the minimum turning point of the graph of $y = f(x)$.

d Write down the equation of the axis of symmetry of the graph $y = f(x)$.

7 **a** Copy and complete the table for the function $f(x) = (x + 4)^2$.

x	–6	–5	–4	–3	–2
$x + 4$	–2	–1	0	1	2
$(x + 4)^2$	4				

b Draw the graph of $f(x) = (x + 4)^2$.

c **i** What is the minimum value of the function $f(x) = (x + 4)^2$?

ii For which value of x does the minimum occur?

iii Write down the coordinates of the minimum turning point of the graph of $y = f(x)$.

d Write down the equation of the axis of symmetry of the graph $y = f(x)$.

8 For each of the following functions write down:

i the minimum value

ii the value of x when the minimum occurs

iii the coordinates of the minimum turning point of the graph of $y = f(x)$

iv the equation of the axis of symmetry of the graph $y = f(x)$.

a $f(x) = (x - 2)^2$ **b** $f(x) = (x - 5)^2$ **c** $f(x) = (x + 1)^2$

d $f(x) = (x + 6)^2$ **e** $f(x) = (x + a)^2$ **f** $f(x) = (x - a)^2$

9 **a** Copy and complete the table for the functions $f(x) = (x - 2)^2$ and $g(x) = (x - 2)^2 + 3$.

x	0	1	2	3	4
$(x - 2)^2$	4	1	0	1	4
$(x - 2)^2 + 3$	7				

b Draw the graphs of $f(x) = (x - 2)^2$ and $g(x) = (x - 2)^2 + 3$ on the same set of axes.

c **i** What is the minimum value of the function $g(x) = (x - 2)^2 + 3$?

ii For which value of x does the minimum value occur?

iii Write down the coordinates of the minimum turning point of the graph of $g(x) = (x - 2)^2 + 3$.

(continued)

10 a Copy and complete the table for the functions $f(x) = (x + 1)^2$ and $g(x) = (x + 1)^2 - 5$.

x	−3	−2	−1	0	1
$(x + 1)^2$	4	1	0	1	4
$(x + 1)^2 - 5$	−1				

b Draw the graphs of $f(x) = (x + 1)^2$ and $g(x) = (x + 1)^2 - 5$ on the same set of axes.

c i What is the minimum value of the function $g(x) = (x + 1)^2 - 5$?

ii For which value of x does the minimum value occur?

iii Write down the coordinates of the minimum turning point of the graph of $g(x) = (x + 1)^2 - 5$.

11 For each of the following functions write down:

i the minimum value

ii the value of x when the minimum occurs

iii the coordinates of the minimum turning point of the graph of $y = f(x)$.

a $f(x) = (x - 4)^2 + 3$ b $f(x) = (x - 8)^2 - 1$

c $f(x) = (x + 3)^2 + 10$ d $f(x) = (x + 6)^2 - 4$

12 For each of the following functions write down:

i the maximum value

ii the value of x when the maximum occurs

iii the coordinates of the maximum turning point of the graph of $y = f(x)$.

This question is asking about maximum values, not minimum.

a $f(x) = -x^2 + 1$ b $f(x) = -x^2 + 6$

c $f(x) = -x^2 - 3$ d $f(x) = -x^2 - 8$

Key features for the graph of $y = (x + p)^2 + q$

From the activity you will have established the following points.

- The quadratic function $f(x) = x^2 + q$ has a minimum value of q when $x = 0$. (See Q3.)
- The quadratic function $f(x) = -x^2 + q$ has a maximum value of q when $x = 0$. (See Q12.)
- The following two statements are equivalent:
  - » the quadratic function $f(x) = x^2 + q$ has a minimum value q whenever $x = 0$
  - » the graph of the quadratic function $y = x^2 + q$ has a minimum turning point at $(0, q)$. (See Q3.)
- The function $f(x) = (x + p)^2$ has a minimum value of 0 when $x = -p$. (See Q8.)
- The graph of the function $f(x) = (x + p)^2 + q$ has a minimum turning point at $(-p, q)$. (See Q11.)
- The equation of the axis of symmetry of the graph $f(x) = (x + p)^2 + q$ is $x = -p$. (See Q8.)

These are the **key points** from the activity.

- When $x = -p$ the function $f(x) = (x + p)^2 + q$ takes its minimum value of q.
- The graph of the quadratic function $f(x) = (x + p)^2 + q$ has a minimum turning point at $(-p, q)$.
- The equation of the axis of symmetry of the graph $f(x) = (x + p)^2 + q$ is $x = -p$.

Example 16.4

The equation of the quadratic function in the diagram is of the form $y = (x + p)^2 + q$, where p and q are integers. Write down the values of p and q.

Coordinates of the minimum turning point are $(2, 4)$.

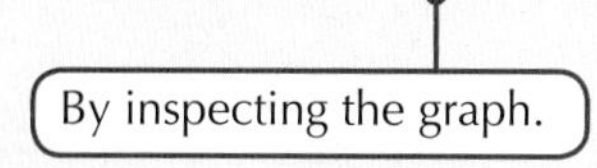

This means that the minimum value of the function is 4 and this occurs when $x = 2$.

The equation of the graph must be $y = (x - 2)^2 + 4$.

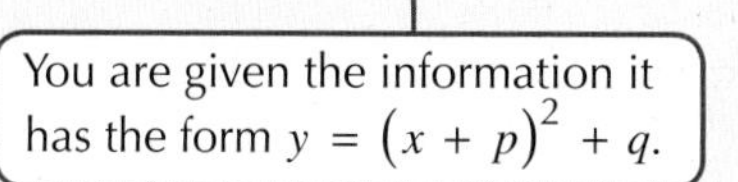

You are given the information it
has the form $y = (x + p)^2 + q$.

So: $p = -2$ and $q = 4$.

Example 16.4 shows how the given graph could have been obtained by **translating** the graph of $y = x^2$ **2 units horizontally to the right** followed by **4 units vertically upwards**.

Example 16.5

The equation of the quadratic function in the diagram is of the form $y = -(x + p)^2 + q$, where p and q are integers. Write down the values of p and q.

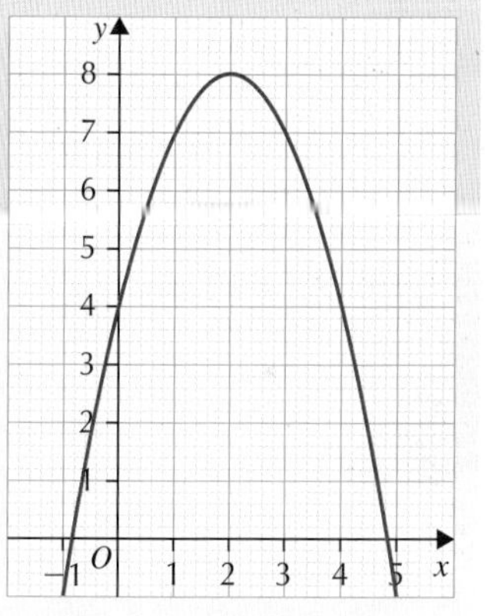

Coordinates of the maximum turning point are $(2, 8)$. By inspecting the graph.

This means that the maximum value of the function is 8 and this occurs when $x = 2$.

The equation of the graph must be $y = -(x - 2)^2 + 8$.

So: $p = -2$ and $q = 8$.

Example 16.6

The diagram shows the graph of the quadratic function $f(x) = (x - p)^2 + q$. Write down the values of p and q.

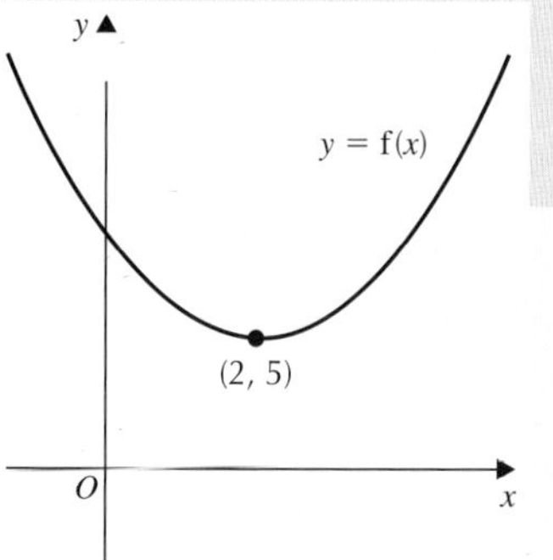

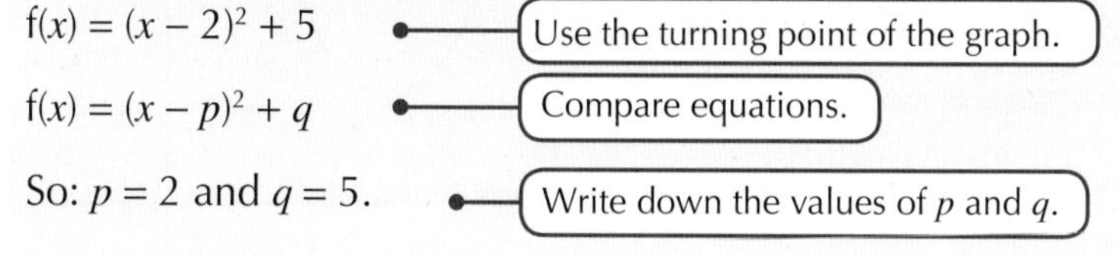

Use the turning point of the graph.
Compare equations.
Write down the values of p and q.

$f(x) = (x - 2)^2 + 5$

$f(x) = (x - p)^2 + q$

So: $p = 2$ and $q = 5$.

Exercise 16B

★ **1** The equation of each of the parabolas shown is $f(x) = (x + p)^2 + q$. For each, identify the values of p and q and write the equation of the function.

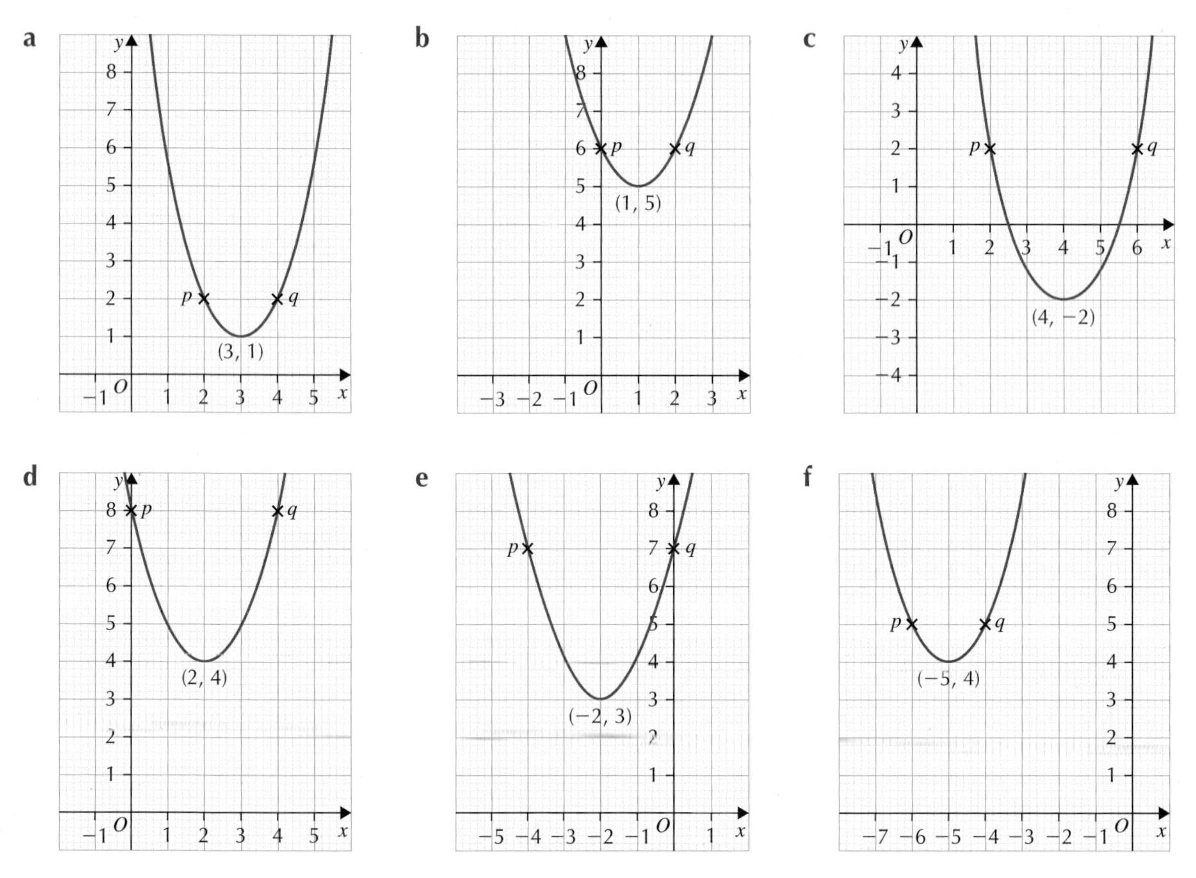

★ **2** The equation of each of the parabolas shown is $g(x) = (x + p)^2 + q$ or $g(x) = -(x + p)^2 + q$. For each, identify the values of p and q.

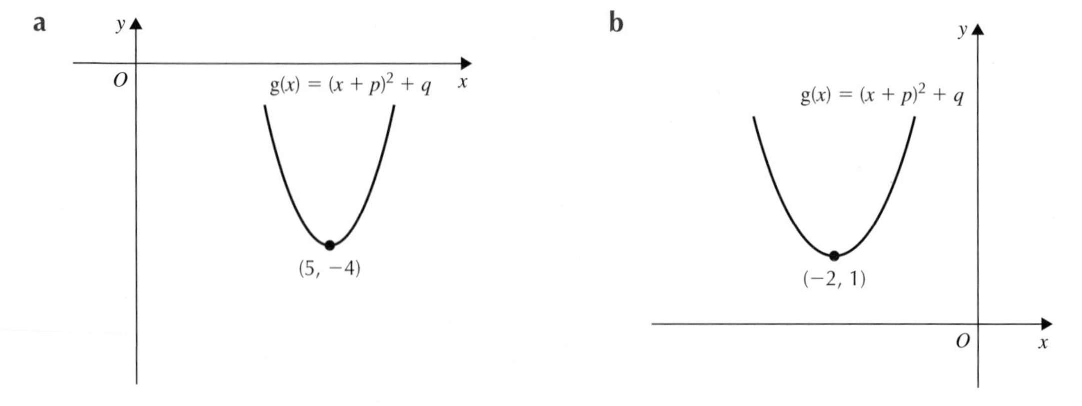

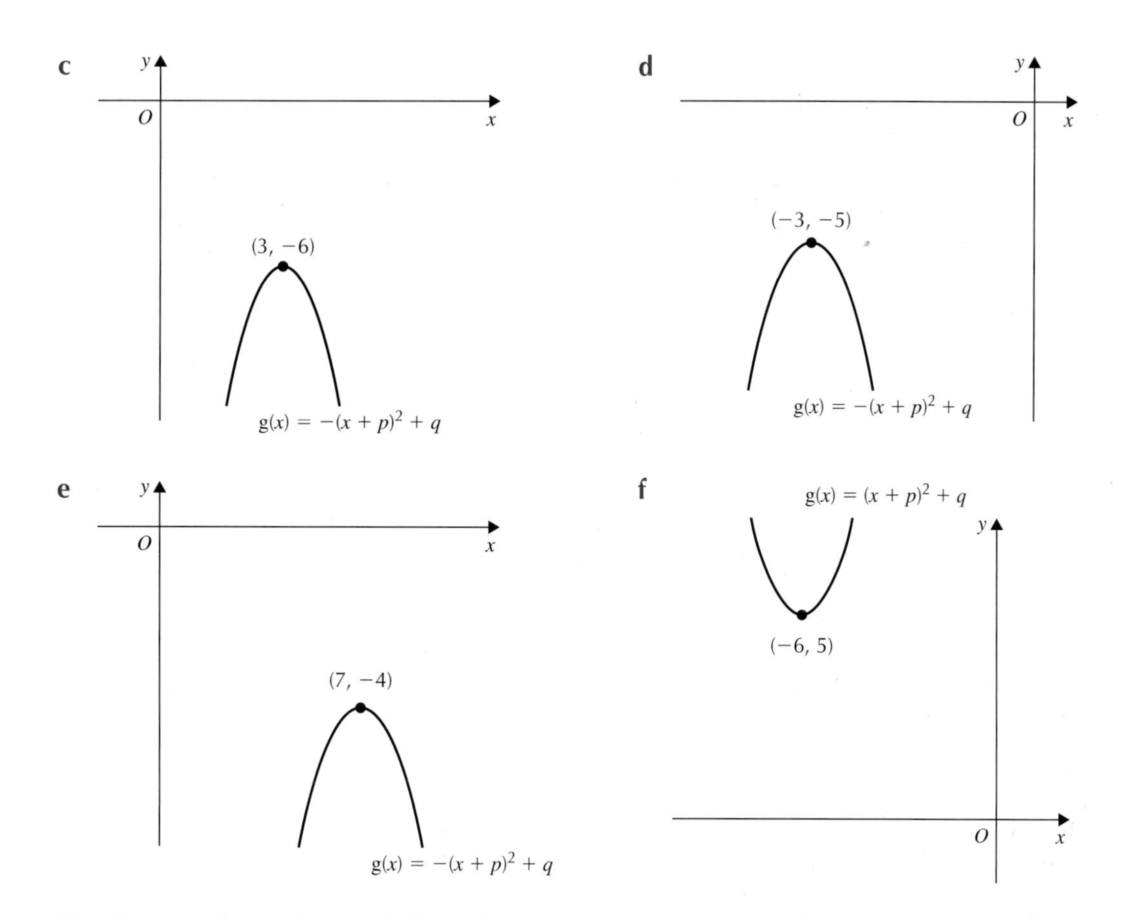

9 The diagram shows the parabola with equation $y = (x - 4)^2 + 2$. The vertical line $x = 1$ intersects the parabola at the point B.

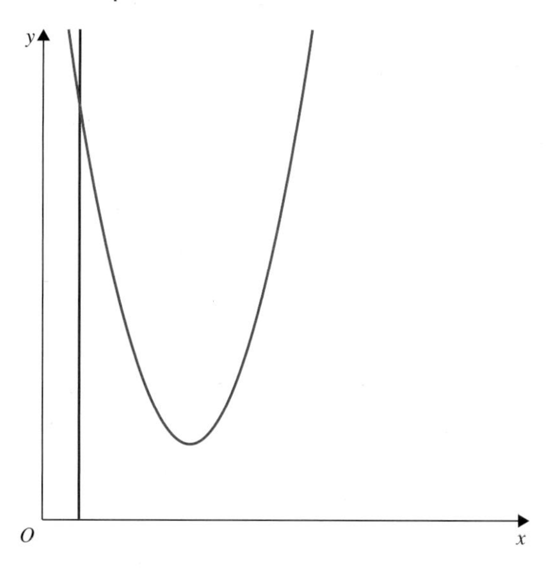

a Write down the equation of the axis of symmetry of the parabola.

b Find the coordinates of B.

4 The quadratic function g has equation $g(x) = (x - p)^2 + q$. The graph of $y = g(x)$ passes through the points (−1, 12) and (7, 12). Find the values of p and q.

> Note that the function contains the term $(x - p)^2$.

★ 5 The function h is given by $h(x) = (x + p)^2 + q$. The graph of $y = h(x)$ has a minimum turning point at A and intersects the y-axis at B (0, 21). Triangle OAB has area 21 square units. Find the values of p and q.

- I can determine the equation of a parabola of the form $y = kx^2$ or $y = (x + p)^2 + q$. ★ Exercise 16A Q1, Q3 ★ Exercise 16B Q1, Q2
- I can apply my knowledge of the parabola to solve problems. ★ Exercise 16A Q5 ★ Exercise 16B Q5

For further assessment opportunities, see the Preparation for Assessment for Unit 2 on pages 288–291.

17 Sketching a quadratic function

This chapter will show you how to:

- sketch the graph of a quadratic function of the form $y = (x - m)(x - n)$
- sketch the graph of a quadratic function of the form $y = (x + p)^2 + q$ or $y = -(x + p)^2 + q$
- sketch the graph of the function $y = ax^2 + bx + c$.

You should already know:

- how to recognise and determine the equation of a quadratic of the form $y = kx^2$
- how to recognise and determine the equation of a quadratic of the form $y = (x + p)^2 + q$

Sketch the graph of a quadratic function of the form $y = (x - m)(x - n)$

This section looks at quadratic functions of the form $y = (x - m)(x - n)$. This form is called the **root form** of the quadratic function. The **roots** of the function are the x-coordinates of the points where the graph intersects the x-axis. Finding the roots of a quadratic equation is covered in Chapter 19. In this first section we look at how to sketch the graph of $y = (x - m)(x - n)$.

Activity

In this activity, a graphing calculator may also be useful.

1 a Copy and complete the table for the function $y = (x - 1)(x - 7)$.

x	0	1	3	5	7	9
$(x - 1)$	–1	0				
$(x - 7)$	–7	–6				
$(x - 1)(x - 7)$	7	0				

b Using the table draw the graph of the function $y = (x - 1)(x - 7)$.

c Write down the roots of the function $y = (x - 1)(x - 7)$.

d Write down the coordinates of the minimum turning point of the graph of $y = f(x)$.

2 a Copy and complete the table for the function $y = (x + 2)(x - 4)$.

x	–4	–2	0	2	4	6
$(x + 2)$	–2	0				
$(x - 4)$	–8	–6				
$(x + 2)(x - 4)$	16	0				

(continued)

b Using the table draw the graph of the function $y = (x + 2)(x - 4)$.

c Write down the roots of the function $y = (x + 2)(x - 4)$.

d Write down the coordinates of the minimum turning point of the graph of $y = f(x)$.

3 a Copy and complete the table for the function $y = x(x - 6)$.

x	−4	−2	0	2	4	6	8
$(x - 6)$	−10	−8					
$x(x - 6)$	40	16					

b Using the table draw the graph of the function $y = x(x - 6)$.

c Write down the roots of the function $y = x(x - 6)$.

d Write down the coordinates of the minimum turning point of the graph of $y = f(x)$.

4 For each of the following functions write down:

i the roots

ii the x-coordinate of the turning point

iii the minimum value of the function.

a $f(x) = (x - 2)(x - 6)$ b $f(x) = (x - 3)(x - 5)$ c $f(x) = (x + 1)(x - 3)$

d $f(x) = x(x + 4)$ e $f(x) = (x + 1)(x + 7)$ f $f(x) = (x + 5)(x - 5)$

Key features of the graph of the quadratic function $f(x) = (x - m)(x - n)$

From the previous activity you should have noticed these key features of the graph of the quadratic function $f(x) = (x - m)(x - n)$:

- the roots of the function are m and n
- the graph of $y = f(x)$ intersects the x-axis at the points $(m, 0)$ and $(n, 0)$
- the average of the two roots gives the x-coordinate of the turning point. The y-coordinate of the turning point can be found by substituting this value of x into the equation.

You can use these key features in order to sketch the graph. To make a sketch of a quadratic function given in root form, follow these steps:

1 find the roots by setting $y = 0$

2 find the y-intercept by setting $x = 0$

3 find the turning point

4 make a neat sketch.

Note:

Sketching a graph is different from drawing a graph. You should be able to sketch the graph of a quadratic function on plain paper, that is, without a grid. To do this, you should:

- rule axes
- mark and label key points: x-intercepts (roots), y-intercept and turning points
- join points with a smooth curve.

Example 17.1

Sketch the graph $y = (x - 2)(x - 4)$. Mark clearly where the graph crosses the axes and state the coordinates of the turning point.

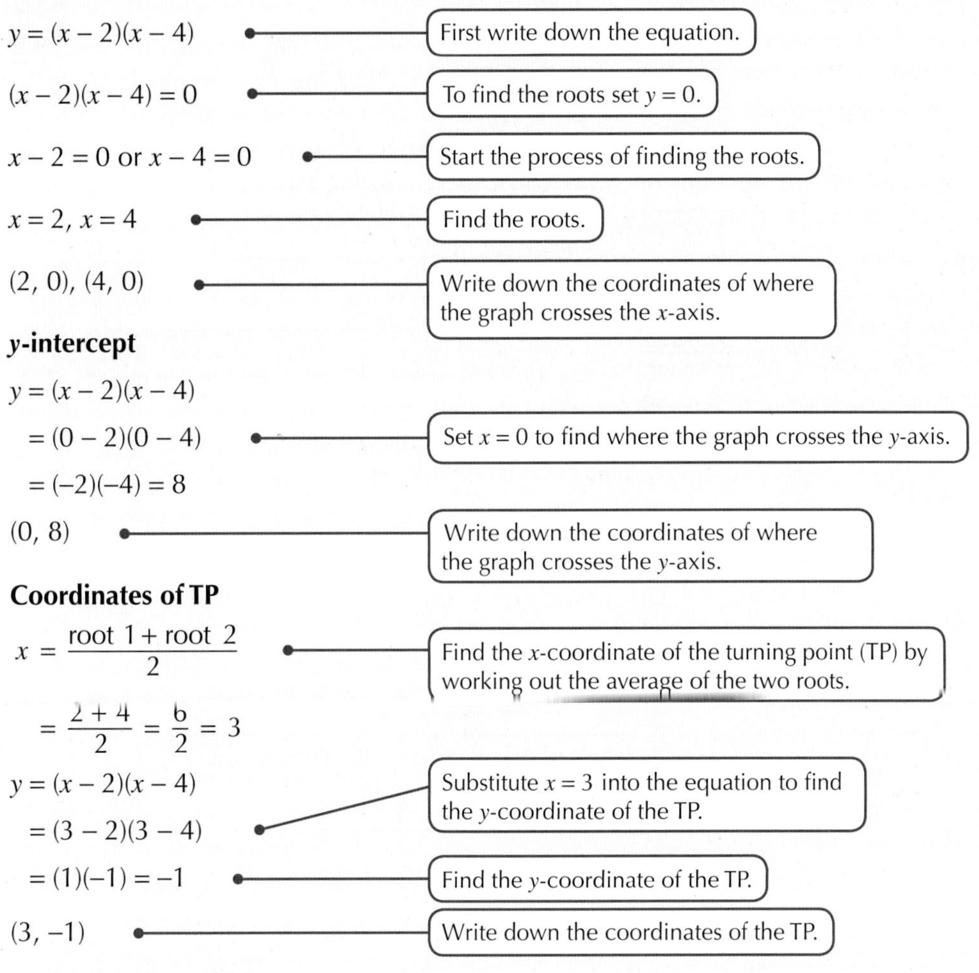

Roots

$y = (x - 2)(x - 4)$ — First write down the equation.

$(x - 2)(x - 4) = 0$ — To find the roots set $y = 0$.

$x - 2 = 0$ or $x - 4 = 0$ — Start the process of finding the roots.

$x = 2, x = 4$ — Find the roots.

$(2, 0), (4, 0)$ — Write down the coordinates of where the graph crosses the x-axis.

y-intercept

$y = (x - 2)(x - 4)$

$= (0 - 2)(0 - 4)$ — Set $x = 0$ to find where the graph crosses the y-axis.

$= (-2)(-4) = 8$

$(0, 8)$ — Write down the coordinates of where the graph crosses the y-axis.

Coordinates of TP

$x = \frac{\text{root 1} + \text{root 2}}{2}$ — Find the x-coordinate of the turning point (TP) by working out the average of the two roots.

$= \frac{2 + 4}{2} = \frac{6}{2} = 3$

$y = (x - 2)(x - 4)$

$= (3 - 2)(3 - 4)$ — Substitute $x = 3$ into the equation to find the y-coordinate of the TP.

$= (1)(-1) = -1$ — Find the y-coordinate of the TP.

$(3, -1)$ — Write down the coordinates of the TP.

Sketching the graph

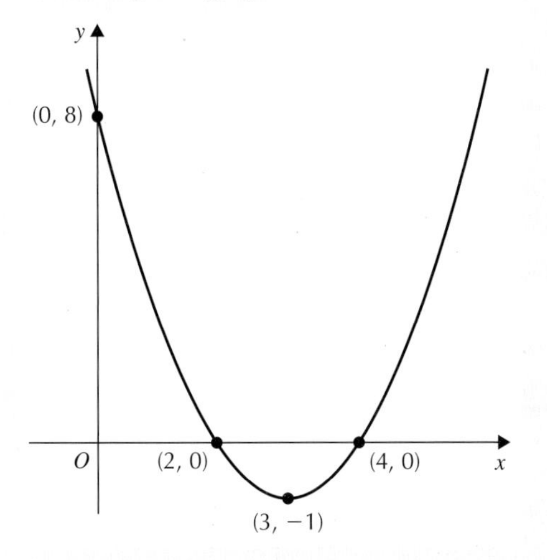

Example 17.2

Sketch the graph $y = (x + 1)(x - 2)$. Mark clearly where the graph crosses the axes and state the coordinates of the turning point.

Roots

$y = (x + 1)(x - 2)$

$(x + 1)(x - 2) = 0$ — To find the roots set $y = 0$.

$x + 1 = 0$ or $x - 2 = 0$

$x = -1, x = 2$

$(-1, 0), (2, 0)$ — Write down the coordinates of where the graph crosses the x-axis.

***y*-intercept**

$y = (x + 1)(x - 2)$

$= (0 + 1)(0 - 2)$ — Set $x = 0$ to find where the graph crosses the y-axis.

$= (1)(-2) = -2$

$(0, -2)$ — Write down the coordinates of where the graph crosses the y-axis.

Coordinates of TP

$x = \frac{\text{root 1} + \text{root 2}}{2}$ — Find the x-coordinate of the TP.

$= \frac{-1 + 2}{2} = \frac{1}{2}$

$y = (x + 1)(x - 2)$

$= \left(\frac{1}{2} + 1\right)\left(\frac{1}{2} - 2\right)$ — Find the y-coordinate of the TP by substituting $x = \frac{1}{2}$ into the equation.

$= \left(\frac{3}{2}\right)\left(-\frac{3}{2}\right)$

$= -\frac{9}{4}$

$\left(\frac{1}{2}, -\frac{9}{4}\right)$ — Write down the coordinates of the TP.

Sketching the graph

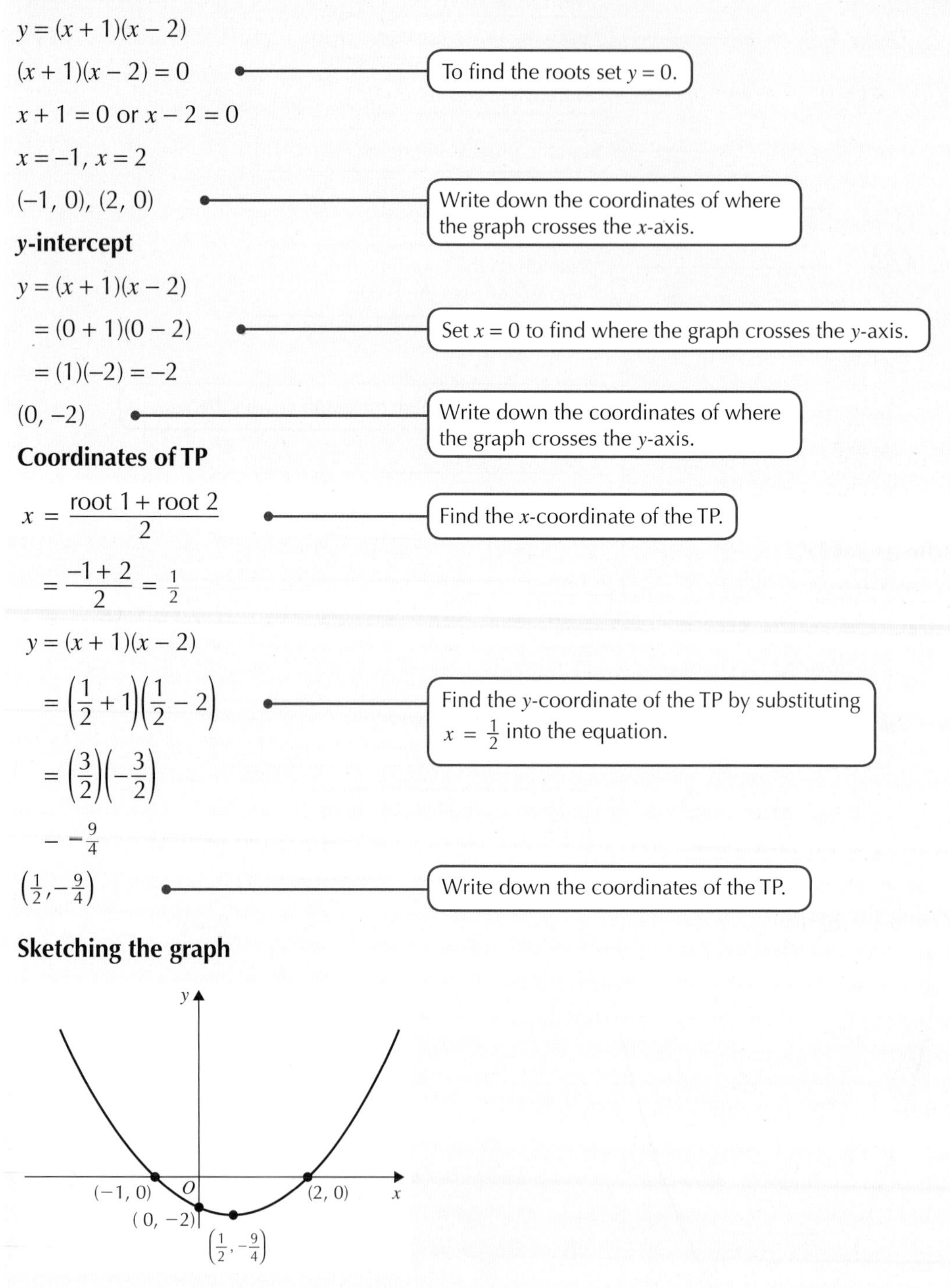

Exercise 17A

★ 1 Make neat sketches of the following quadratic functions.

a $y = (x-1)(x-5)$ b $y = (x-2)(x-8)$ c $y = (x+1)(x-3)$

d $y = (x+5)(x-1)$ e $f(x) = (x-4)(x+6)$ f $g(x) = (x+5)(x+3)$

g $h(x) = (x+1)(x-7)$ h $p(x) = (x-3)(x-7)$ i $m(x) = (x-3)(x+4)$

2 Make neat sketches of the following quadratic functions.

a $y = (x-3)(x-2)$ b $y = (x+1)(x+2)$ c $y = (x-4)(x-1)$

d $y = x(x-3)$ e $f(x) = (x+4)(x+3)$ f $f(x) = (x-6)(x+2)$

3 Each of the following graphs has an equation of the form $y = (x-m)(x-n)$. For each, identify the values of m and n.

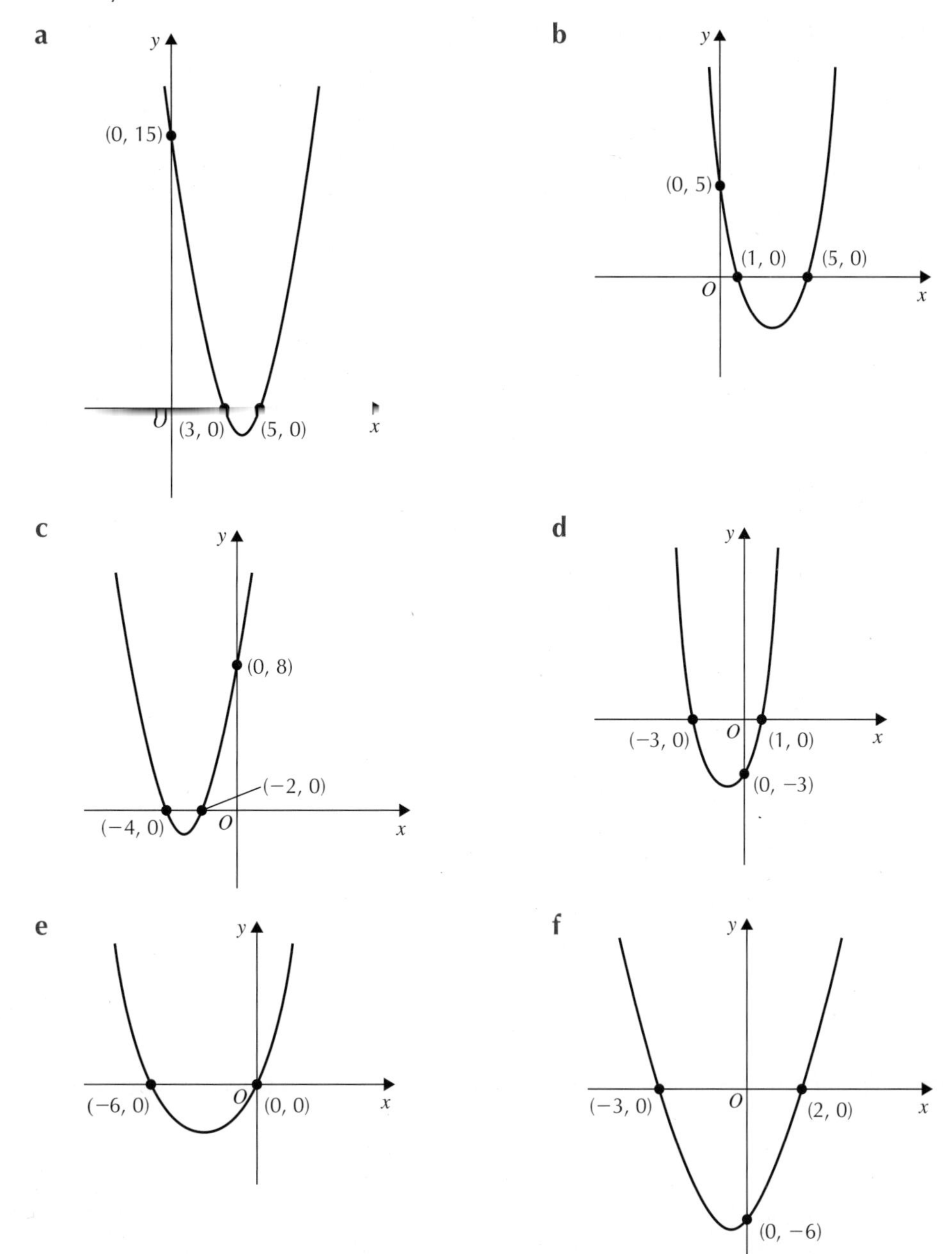

Sketch the graph of a quadratic function of the form $y = (x + p)^2 + q$ or $y = -(x + p)^2 + q$

This section looks at how to sketch quadratic functions of the form $y = (x + p)^2 + q$ and $y = -(x + p)^2 + q$. This form is called the **completed square** form of the quadratic function. The algebraic process of completing the square is covered in Chapter 5.

You can use the key features of quadratic functions in order to sketch the graph. To make a sketch of a quadratic function given in completed square form, follow these three steps:

1. find the y-intercept by setting $x = 0$
2. find the turning point
3. make a neat sketch, plotting and labelling the y-intercept and turning point.

Quadratic functions of the form $y = (x + p)^2 + q$

Example 17.3

Sketch the graph of the function $f(x) = (x - 1)^2 + 5$.

***y*-intercept**

$f(x) = (x - 1)^2 + 5$ — First write down the equation.

$y = f(0) = (0 - 1)^2 + 5$ — Find the y-intercept by making $x = 0$.

$= (-1)^2 + 5$

$= 6$

$(0, 6)$ — Write down the coordinates of the y-intercept.

Coordinates of TP

$(1, 5)$ — Identify the coordinates of the TP.

Sketching the graph

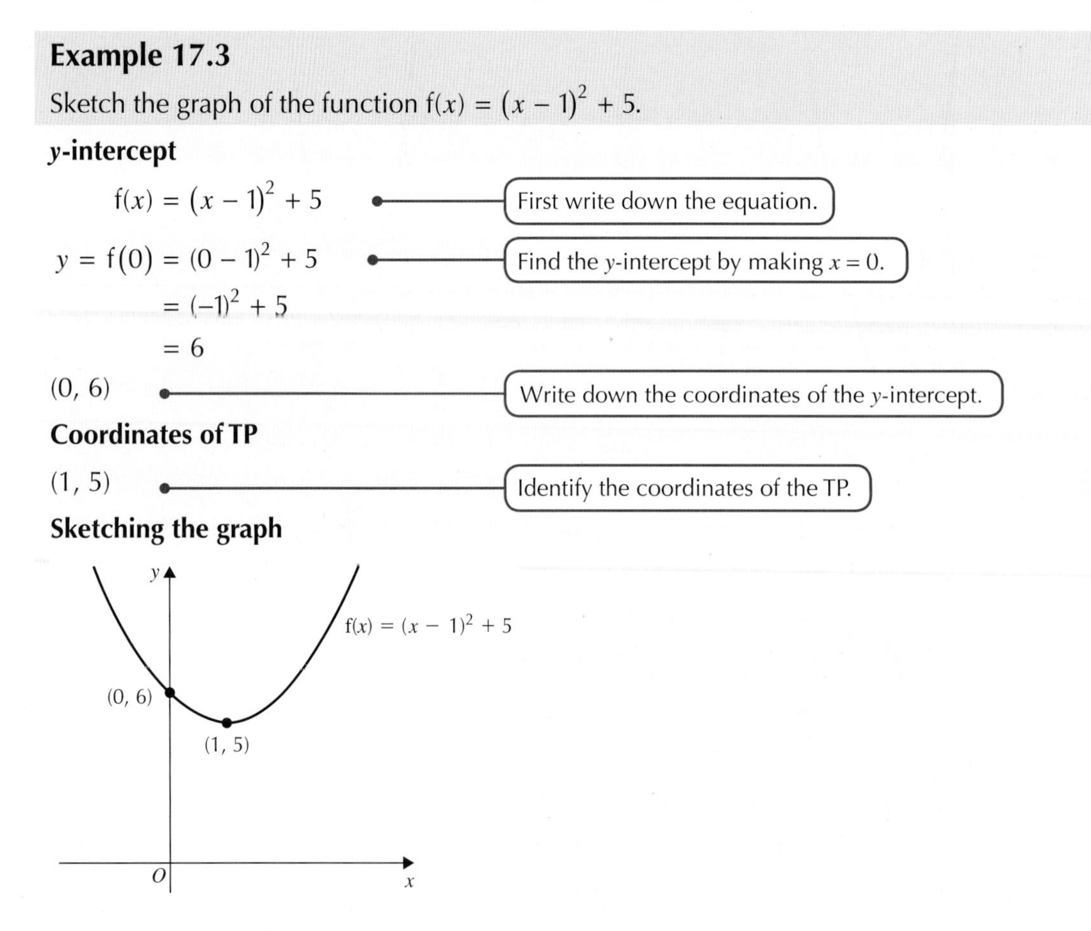

Example 17.4

Sketch the graph of the function $g(x) = (x + 3)^2 - 4$.

y-intercept

$$g(x) = (x + 3)^2 - 4$$
$$y = g(0) = (0 + 3)^2 - 4$$
$$= (3)^2 - 4$$
$$= 9 - 4 = 5$$

(0, 5) — Write down the coordinates of the y-intercept.

Coordinates of TP

(−3, −4) — Identify the coordinates of the TP.

Sketching the graph

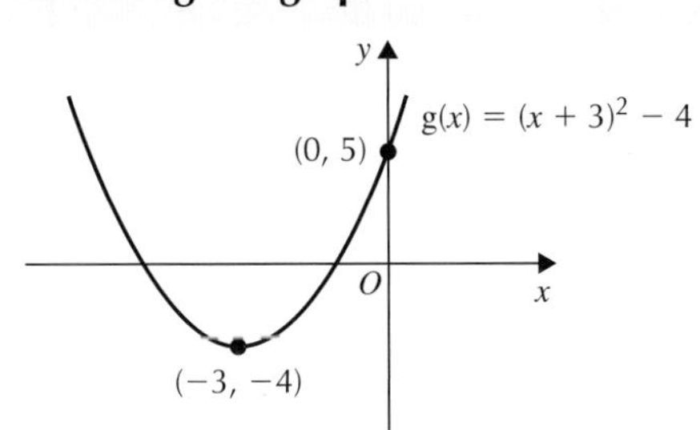

Exercise 17D

★ 1 Make neat sketches of the following quadratic functions.

a $f(x) = (x - 1)^2 + 4$ b $g(x) = (x + 2)^2 - 1$ c $h(x) = (x - 4)^2 - 8$

d $f(x) = (x + 5)^2 - 5$ e $g(x) = (x - 5)^2 + 5$ f $h(x) = 10 + (x - 3)^2$

2 a Explain why the function $y = x^2$ has a minimum value of 0.

b Explain why the function $y = -x^2$ has a maximum value of 0.

3 a The maximum value of the function $y = -x^2$ is 0. Explain why the maximum value of the function $y = -x^2 + 10$ must be 10.

b What is the maximum value of the function $y = 10 - x^2$? Explain why.

4 a Explain why the function $y = (x - 3)^2$ has a minimum value of 0.

b Explain why the function $y = -(x - 3)^2$ must have a maximum value of 0.

5 a What is the minimum value of the function $y = (x + 2)^2 + 8$?

b Explain why the maximum value of the function $y = -(x + 2)^2 + 8$ must be 8.

c Kayleigh claims that the functions $y = -(x + 2)^2 + 8$ and $y = 8 - (x + 2)^2$ are the same. Is she right? Justify your answer.

6 The function $y = (x + p)^2 + q$ has a minimum value of q when $x = -p$. Explain what happens when the function is changed to $y = -(x + p)^2 + q$.

Quadratic functions of the form $y = -(x + p)^2 + q$

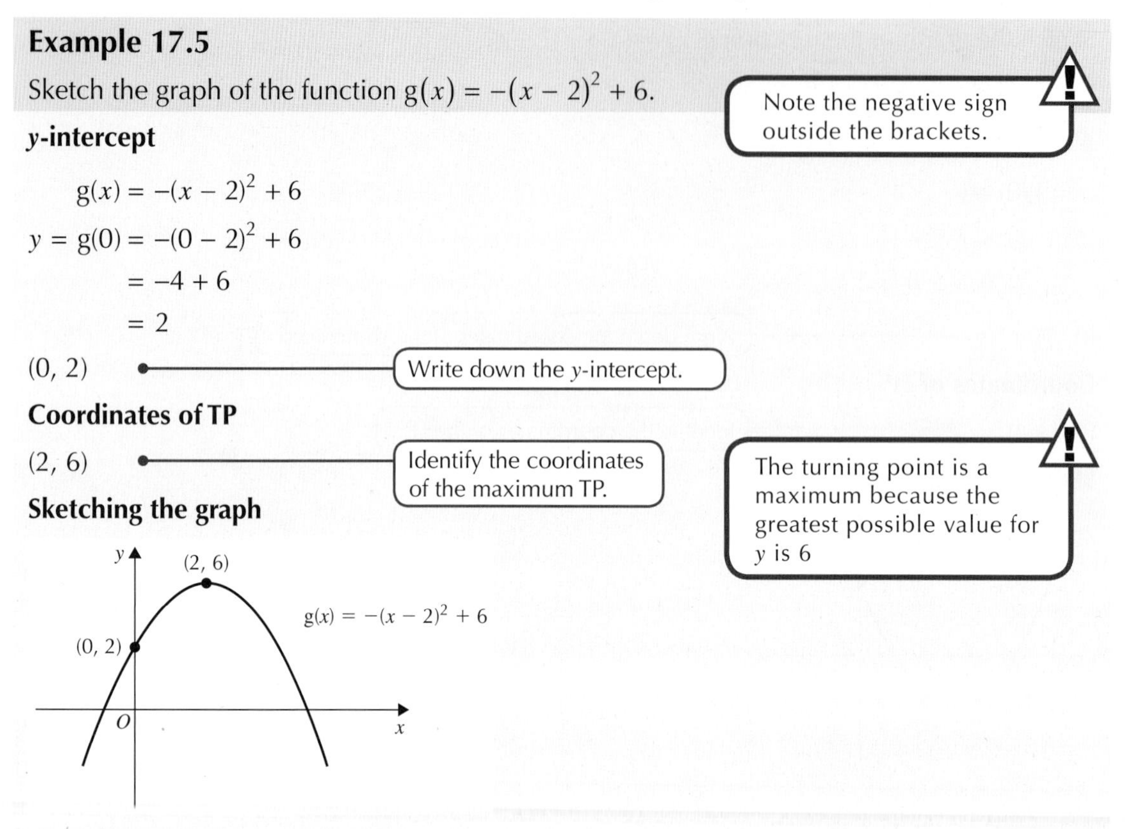

Example 17.5
Sketch the graph of the function g(x) = −(x − 2)² + 6.
Note the negative sign outside the brackets.
y-intercept
g(x) = −(x − 2)² + 6
y = g(0) = −(0 − 2)² + 6
= −4 + 6
= 2
(0, 2)
Write down the y-intercept.
Coordinates of TP
(2, 6)
Identify the coordinates of the maximum TP.
The turning point is a maximum because the greatest possible value for y is 6
Sketching the graph
y
(2, 6)
g(x) = −(x − 2)² + 6
(0, 2)
O
x

Exercise 17C

★ **1** Make neat sketches of the following quadratic functions.

a $f(x) = -(x - 2)^2 + 3$ **b** $g(x) = -(x + 1)^2 + 9$ **c** $h(x) = -(x - 2)^2 - 4$

d $f(x) = -(x - 4)^2 - 1$ **e** $h(x) = 8 - (x - 1)^2$ **f** $f(x) = -4 - (x + 3)^2$

2 The function $y = x^2$ has a minimum value of 0.

a Explain why the function $y = kx^2$ has a minimum value of 0, $k > 0$.

b What happens when $k < 0$?

3 Consider the function $y = (x - 2)^2$.

a What is the minimum value of this function?

b What is the minimum value of the function $y = 3(x - 2)^2$?

c If $k > 0$ what is the minimum value of the function $y = k(x - 2)^2$?

d Compare the turning points of the graphs $y = (x - 2)^2$ and $y = k(x - 2)^2$. What do you notice?

4 **a** $f(x) = (x + 2)^2 - 5$. Write down the coordinates of the minimum turning point of the graph of $y = f(x)$.

b Explain why the graph of the function $g(x) = 4(x + 2)^2 - 5$ has the same turning point as the graph of $y = f(x)$.

c Would the graph of $y = k(x + 2)^2 - 5$ $(k > 0)$, always have the same turning point as $y = (x + 2)^2 - 5$? Explain your answer.

5 **a** Write down the maximum value of the function $y = -(x + 3)^2 + 15$.

b What would be the maximum value of the functions:

i $y = -2(x + 3)^2 + 15$ **ii** $y = -3(x + 3)^2 + 15$

c What can you say about the turning points of the graphs of the functions $y = -(x + 3)^2 + 15$ and $y = -k(x + 3)^2 + 15$?

6 **a** Write down the coordinates of the maximum turning point of the graph of $y = -(x + p)^2 + q$.

b Write down the coordinates of the maximum turning point of the graph of $y = -k(x + p)^2 + q$.

Sketch the graph of a quadratic function of the form $y = ax^2 + bx + c$

This section looks at how to sketch quadratic functions of the form $y = ax^2 + bx + c$. This equation is called the **general form** of the quadratic function. You need skills developed in Chapter 19 to be able to quickly sketch the graph of a quadratic function in this form.

When sketching the graph of a quadratic function expressed in general form there are four key points to consider:

- the shape of the parabola
- where the graph crosses the y-axis
- the coordinates of the maximum/minimum turning point
- where the graph cuts the x-axis, if at all; that is, find the roots if they exist.

Shape of the parabola

The value of a determines the shape of the parabola.

- When x becomes large positive, x^2 is large positive.
- When x becomes large negative, x^2 is still large positive.
- If $a > 0$ then ax^2 is positive, giving this shape of graph:

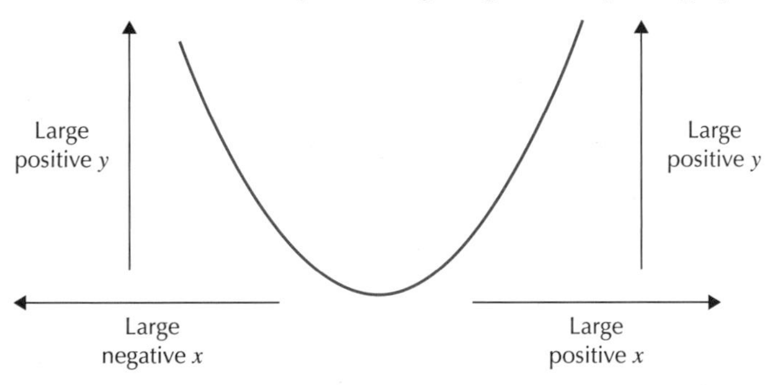

- If a is negative ($a < 0$) then ax^2 is negative, giving this shape of graph:

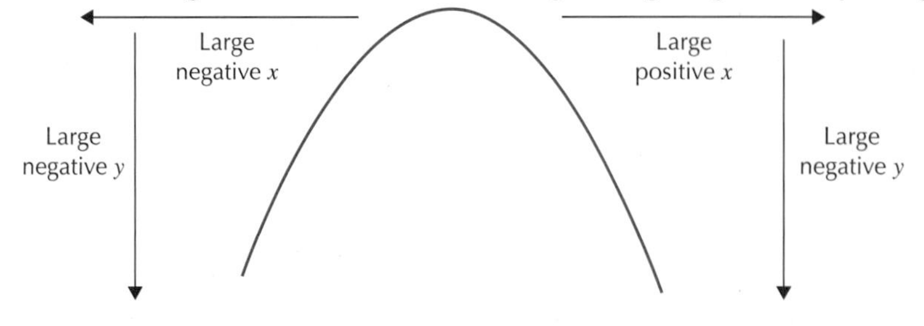

y-intercept

To find where the graph intercepts the y-axis, make $x = 0$.

Coordinates of turning point

The graph will have a **minimum** or **maximum** turning point.

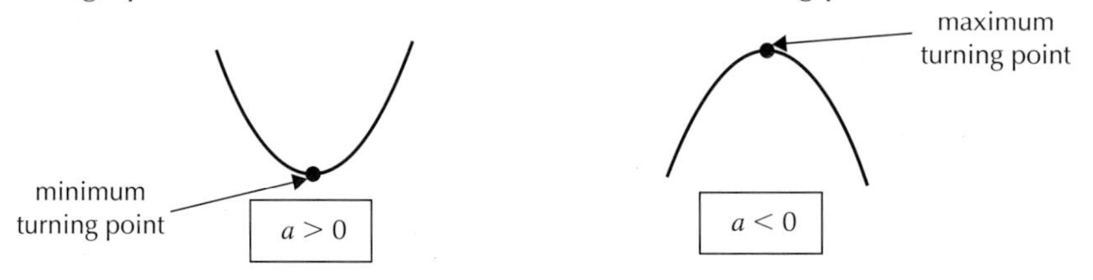

- Calculate $-\frac{b}{2a}$ to give the x-coordinate of the turning point. Substitute this value of x into the equation to find the y-coordinate.
- Note that $ax^2 + bx + c$ (general form) can be written as $a\left(x + \frac{b}{2a}\right)^2 + \frac{4ac - b^2}{4a}$ (completed square form). This confirms $x = -\frac{b}{2a}$ giving a minimum or maximum value.

x-intercepts (roots)

The parabola $y = ax^2 + bx + c$ intersects the x-axis when $y = 0$. When $y = 0$ you have to solve the equation $ax^2 + bx + c = 0$.

It is useful to know whether or not the equation has roots before attempting to find them.

First, calculate the **discriminant**, $b^2 - 4ac$. The table shows what you should do next. (The discriminant is covered in Chapter 19.)

Value of $b^2 - 4ac$	Nature of roots	Graph	How to proceed
$b^2 - 4ac > 0$ and a square number	Real, rational and distinct	Intersects x-axis at two distinct points (2 distinct roots)	Solve $ax^2 + bx + c = 0$ by factorising or by using the quadratic formula
$b^2 - 4ac > 0$ and not a square number	Real, irrational and distinct	Intersects x-axis at two distinct points (2 distinct roots)	Solve $ax^2 + bx + c = 0$ by using the quadratic formula
$b^2 - 4ac = 0$	Real and equal	Graph touches the x-axis at one point (that is, the x-axis is a tangent to the graph)	Solve $ax^2 + bx + c = 0$ by factorising or by using the quadratic formula
$b^2 - 4ac < 0$	No real roots	Graph does not intersect the x-axis	Sketch graph using y-intercept and turning point

Example 17.6

Sketch the graph of the function $y = x^2 + 2x - 8$.

$a = 1, b = 2, c = -8$ — Compare $y = x^2 + 2x - 8$ with $y = ax^2 + bx + c$ and identify values of a, b and c.

Shape

Since $a > 0$.

***y*-intercept**

$y = x^2 + 2x - 8$ — Write down the equation, then substitute $x = 0$ to find the y-intercept.

$y = (0)^2 + 2(0) - 8 = -8$

$(0, -8)$ — Write down the coordinates of the y-intercept.

Coordinates of TP

$x = -\dfrac{b}{2a}$ — Write down the equation to find the value of the x-coordinate of the TP, then substitute values for a and b.

$= -\dfrac{2}{2(1)} = -\dfrac{2}{2} = -1$

$y = (-1)^2 + 2(-1) - 8$ — Substitute the value of the x-coordinate of the TP into the equation to find the y-coordinate, then complete the calculation.

$= 1 - 2 - 8$

$= -9$

$(-1, 9)$ — Write down the coordinates of the TP.

***x*-intercepts (roots)**

$b^2 - 4ac$ — Write down the formula for the discriminant and substitute values for a, b and c, then complete the calculation.

$= (2)^2 - 4(1)(-8)$

$= 4 - 4 \times -8$

$= 4 + 32$

$= 36$

$b^2 - 4ac > 0$, so the graph intersects the x-axis at two distinct points (there are 2 distinct roots).

Solve $x^2 + 2x - 8 = 0$ to find the roots:

$x^2 + 2x - 8 = 0$

$(x + 4)(x - 2) = 0$ — Factorise.

$x + 4 = 0$ or $x - 2 = 0$

$x = -4, x = 2$ — Identify the roots.

$(-4, 0)$ and $(2, 0)$ — Identify the coordinates of the x-intercepts.

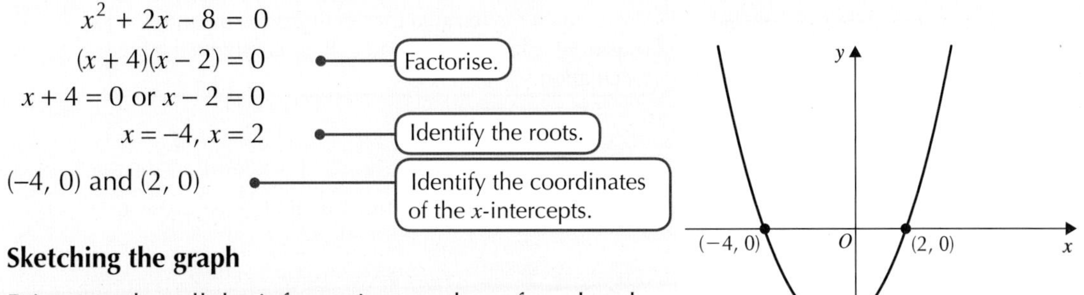

Sketching the graph

Bring together all the information you have found and sketch the graph, labelling key features.

Example 17.7

Sketch the graph of the function $y = 2x^2 - 5x - 3$.

$a = 2, b = -5, c = -3$ — Identify values of a, b and c.

Shape

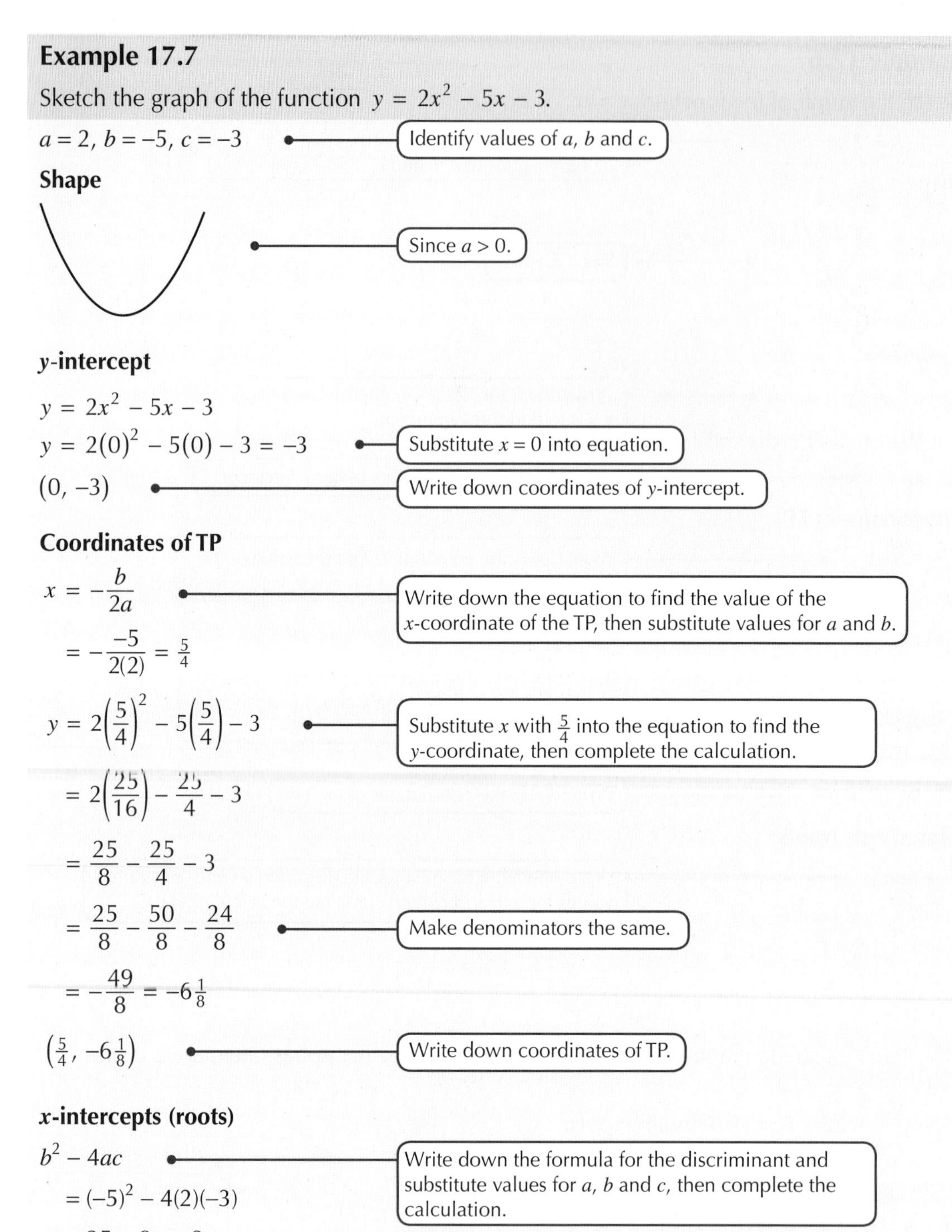

Since $a > 0$.

***y*-intercept**

$y = 2x^2 - 5x - 3$

$y = 2(0)^2 - 5(0) - 3 = -3$ — Substitute $x = 0$ into equation.

$(0, -3)$ — Write down coordinates of y-intercept.

Coordinates of TP

$$x = -\frac{b}{2a}$$

$$= -\frac{-5}{2(2)} = \frac{5}{4}$$

Write down the equation to find the value of the x-coordinate of the TP, then substitute values for a and b.

$$y = 2\left(\frac{5}{4}\right)^2 - 5\left(\frac{5}{4}\right) - 3$$

Substitute x with $\frac{5}{4}$ into the equation to find the y-coordinate, then complete the calculation.

$$= 2\left(\frac{25}{16}\right) - \frac{25}{4} - 3$$

$$= \frac{25}{8} - \frac{25}{4} - 3$$

$$= \frac{25}{8} - \frac{50}{8} - \frac{24}{8}$$

Make denominators the same.

$$= -\frac{49}{8} = -6\tfrac{1}{8}$$

$\left(\frac{5}{4}, -6\frac{1}{8}\right)$ — Write down coordinates of TP.

***x*-intercepts (roots)**

$b^2 - 4ac$

$= (-5)^2 - 4(2)(-3)$

$= 25 - 8 \times -3$

$= 25 + 24 = 49$

Write down the formula for the discriminant and substitute values for a, b and c, then complete the calculation.

$b^2 - 4ac > 0$, so the graph intersects the x-axis at two distinct points (there are 2 distinct roots).

(continued)

Solve $2x^2 - 5x - 3 = 0$ to find the roots:

$2x^2 - 5x - 3 = 0$

$(2x + 1)(x - 3) = 0$ — Factorise.

$2x + 1 = 0$ or $x - 3 = 0$

$x = -\frac{1}{2}, x = 3$ — Identify the roots.

$(-\frac{1}{2}, 0)$ and $(3, 0)$ — Identify the coordinates of the x-intercepts.

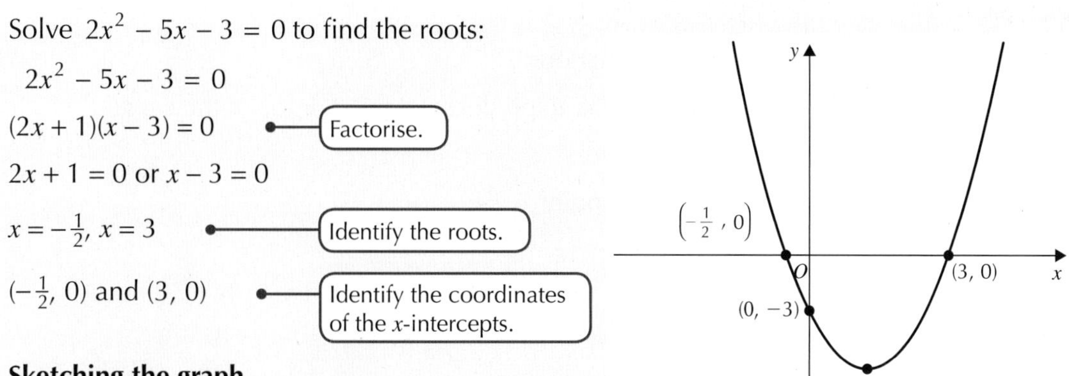

Sketching the graph

Bring together all the information you have found and sketch the graph, labelling key features.

Example 17.8

Sketch the graph of the function $g(x) = -5 - 4x - x^2$.

$a = -1, b = -4, c = -5$

Shape

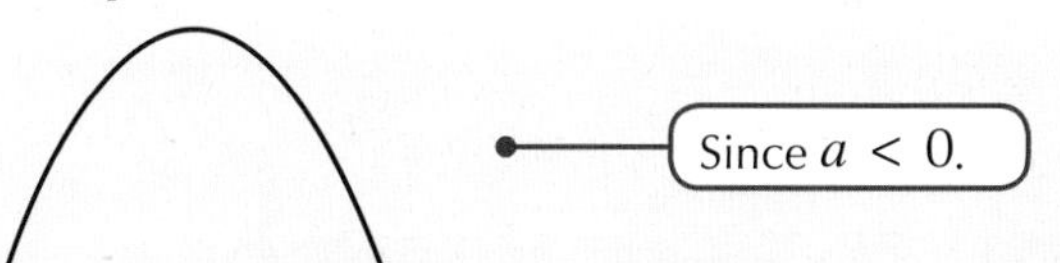

***y*-intercept**

$y = -(0)^2 - 4(0) - 5 = -5$

$(0, -5)$

Coordinates of TP

$$x = -\frac{b}{2a}$$

$$= -\frac{(-4)}{2(-1)}$$

$$= -\frac{(-4)}{-2} = -2$$

$y = -(-2)^2 - 4(-2) - 5$ — Be careful working this out.

$= -4 + 8 - 5 = -1$

$(-2, -1)$ — Write down coordinates of TP.

***x*-intercepts (roots)**

$b^2 - 4ac$

$= (-4)^2 - 4(-1)(-5)$

$= 16 + 4 \times -5$

$= 16 - 20 = -4$

$b^2 - 4ac < 0$, so the graph **does not** intersect the x-axis.

Sketching the graph

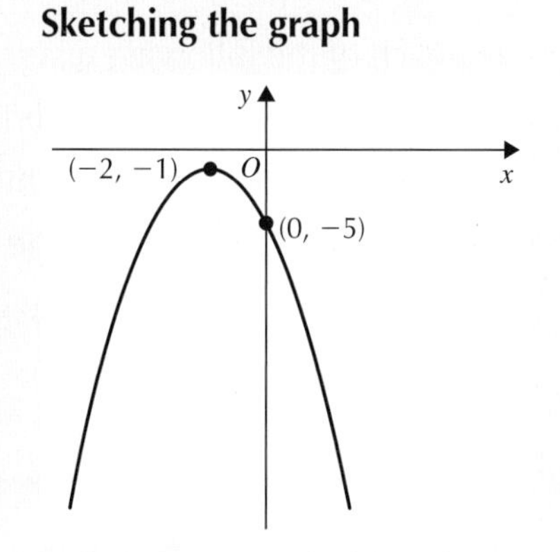

Exercise 17D

1 For each of the following quadratic functions:

i calculate $b^2 - 4ac$ and verify that the graph does not meet or touch the x-axis

ii find the y-intercept

iii find the coordinates of the turning point

iv make a neat sketch of the graph.

a $y = x^2 - 2x + 4$ **b** $y = x^2 + 6x + 10$ **c** $y = x^2 - 8x + 30$

d $y = -x^2 + 2x - 5$ **e** $y = -x^2 - 2x - 3$ **f** $y = -x^2 + 4x - 8$

2 a For the functions in Question 1 parts **a** to **c**:

i express them in the form $(x + p)^2 + q$

ii write down the minimum value of each function.

See Chapter 5 for more on completing the square.

b Explain why your answers to part **a** show that none of these parabolas intersect the x-axis.

★ **3** For each of the following quadratic functions:

i calculate $b^2 - 4ac$ and verify that the graph meets the x-axis at two distinct points

ii find the roots and the y-intercept

iii find the coordinates of the minimum turning point

iv make a neat sketch of the graph.

a $y = x^2 - 4x + 3$ **b** $y = x^2 + 4x - 12$ **c** $y = x^2 - 2x$

d $y = x^2 - 2x - 24$ **e** $y = x^2 + x - 30$ **f** $y = x^2 - 2x - 3$

g $y = x^2 - 8x + 15$ **h** $y = x^2 - 9$ **i** $y = x^2 + 4x$

j $y = x^2 - 10x + 21$ **k** $y = x^2 - x - 2$ **l** $y = x^2 - 5x + 4$

4 For each of the following quadratic functions:

i calculate the discriminant, $b^2 - 4ac$

ii describe the roots of each function

iii explain why the x-axis is a tangent to the graph of each function.

a $f(x) = x^2 + 8x + 16$ **b** $y = x^2 - 6x + 9$ **c** $g(x) = -x^2 + 2x - 1$

5 For each of the following quadratic functions:

i calculate $b^2 - 4ac$ and verify that the graph meets the x-axis at two distinct points

ii find the roots and the y-intercept

iii find the coordinates of the maximum turning point

iv make a neat sketch of the graph.

a $y = -x^2 + 6x - 5$ **b** $y = -x^2 + 2x + 15$ **c** $y = -x^2 - 2x + 8$

d $y = 16 - x^2$ **e** $y = 100 - x^2$ **f** $y = 3 + 2x - x^2$

g $y = 15 - 2x - x^2$ **h** $y = 32 + 4x - x^2$ **i** $y = 12 + x - x^2$

6 Make neat sketches of the quadratic functions.

a $y = 2x^2 - 3x + 1$ **b** $y = 2x^2 - 3x - 2$

c $y = 3x^2 + 6x - 9$ **d** $y = 4x^2 + 4x - 3$

Activity

You may find this activity challenging.

a By completing the square show that $x^2 + \frac{b}{a}x + \frac{c}{a}$ can be written as $\left(x + \frac{b}{2a}\right)^2 + \frac{c}{a} - \frac{b^2}{4a^2}$

The expression $\left(x + \frac{b}{2a}\right)^2 + \frac{c}{a} - \frac{b^2}{4a^2}$ has a minimum value when $x = -\frac{b}{2a}$.

b **i** Given $f(x) = x^2 + \frac{b}{a}x + \frac{c}{a}$ find $f\left(-\frac{b}{2a} + t\right)$.

ii Show that $f\left(-\frac{b}{2a} + t\right) = f\left(-\frac{b}{2a} - t\right)$.

iii What feature of the parabola with equation $y = f(x)$ does your answer to part **ii** confirm?

c Repeat part **b** for the function $f(x) = ax^2 + bx + c$.

d Show that $a(x - m)(x - n)$ can be written as $ax^2 - (m + n)x + mn$.

e Write down the roots of the equation $a(x - m)(x - n) = 0$.

f Show that the x-coordinate of the turning point of the parabola $y = ax^2 + bx + c$ is halfway between the roots.

Linking together the different forms of a quadratic function

This section looks at how the different forms of a quadratic function are related and how it is possible to convert between one form and another using your algebra skills.

Consider the function $f(x) = (x - 2)(x - 6)$. This function has roots 2 and 6. The graph of $y = f(x)$ intercepts the x-axis at (2, 0) and (6, 0). The graph has a minimum turning point at (4, –4).

By expanding brackets it can be shown that $f(x) = x^2 - 8x + 12$.

By completing the square it can be shown that $f(x) = (x - 4)^2 - 4$.

Notice that $f(x)$ can be written in three different forms but it is still the same function:

- $f(x) = x^2 - 8x + 12$ general form
- $f(x) = (x - 2)(x - 6)$ root form
- $f(x) = (x - 4)^2 - 4$ completed square form

The diagram shows how you can convert between the three different forms of a quadratic function.

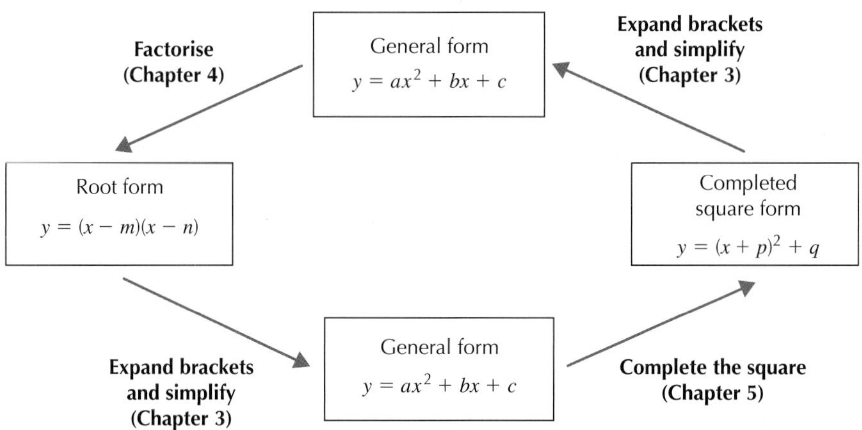

Exercise 17E

1 The table below shows a number of quadratic functions. Each function has been written in a particular form. Copy and compete the table, expressing each function in the remaining two forms.

For example, Function 1 is given in root form. You need to express it in general form and completed square form.

Function	General form	Completed square form	Root form
a Function 1			$(x-1)(x-3)$
b Function 2	x^2+4x+3		
c Function 3		$(x-1)^2-16$	
d Function 4	$x^2-4x-12$		
e Function 5			$(x+4)(x+6)$
f Function 6		$(x+3)^2-4$	

2 A function is given by $f(x) = (x-3)(x-5)$. The function can be expressed in the form $f(x) = (x-4)^2 + q$. Determine the value of q.

3 A parabola has equation $y = x^2 - 2x + 5$. Use two different methods to show that the graph of the parabola does not cross or touch the x-axis.

4 A parabola has equation $y = x^2 + bx + c$. The coordinates of the minimum turning point are $(-2, -9)$. Determine the values of b and c.

★ 5 A function has equation $f(x) = (x+p)^2 + q$. The graph of $y = f(x)$ intersects the x-axis at $(-4, 0)$ and $(-8, 0)$. Determine the value of q.

6 A parabola has equation $y = (x-a)(x-b)$. It has a minimum value when $x = 4$ and passes through the point $(2, -5)$. Determine the values of a and b.

7 A quadratic function is given by $f(x) = (x-3)(x+t)$. The minimum value of f is -16. Find the possible values of t.

- I can sketch the graph of a quadratic function given in any suitable form. ★ Exercise 17A Q1 ★ Exercise 17B Q1 ★ Exercise 17C Q1 ★ Exercise 17D Q3
- I can solve problems which require me to link the different forms of a quadratic function. ★ Exercise 17E Q5

For further assessment opportunities, see the Preparation for Assessment for Unit 2 on pages 288–291.

18 Identifying features of a quadratic function

In this chapter you will learn how to:

- identify the nature and coordinates of the turning point and the equation of the axis of symmetry of the function $y = \pm(x + p)^2 + q$
- apply knowledge of the different forms of a quadratic function to solve related problems.

You should already know:

- how to sketch the graph of a quadratic function of the form $y = (x - m)(x - n)$
- how to sketch the graph of a quadratic function of the form $y = (x + p)^2 + q$ or $y = -(x + p)^2 + q$
- how to solve quadratic equations of the form $ax^2 + bx + c = 0$.

Identify the nature and coordinates of the turning point and the equation of the axis of symmetry of the function $y = \pm(x + p)^2 + q$

Chapter 16 explores the graphs of the functions $y = kx^2$ and $y = (x + p)^2 + q$.
Chapter 17 explores the graphs of the functions $y = (x + p)^2 + q$ and $y = -(x + p)^2 + q$.
This section looks at the features of the graphs of these functions and uses them to solve simple problems.

Example 18.1

A parabola has equation $y = (x - 3)^2 + 4$.

a Write down the equation of its axis of symmetry.

b Write down the coordinates of the turning point of the parabola and state whether it is a maximum or minimum.

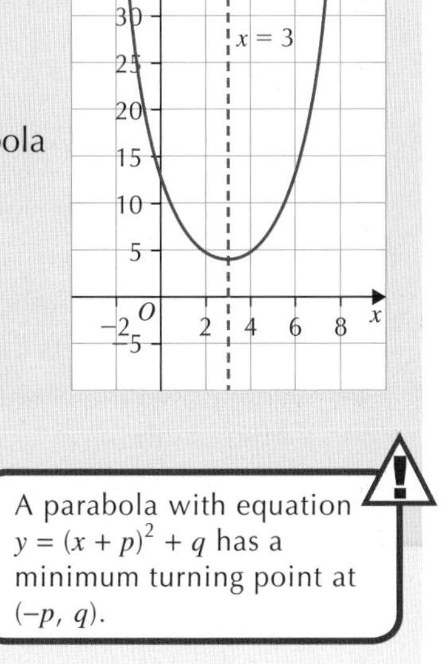

A parabola with equation $y = (x + p)^2 + q$ has a minimum turning point at $(-p, q)$.

a The equation of the axis of symmetry is $x = 3$.

b The parabola has a **minimum** turning point at (3, 4).

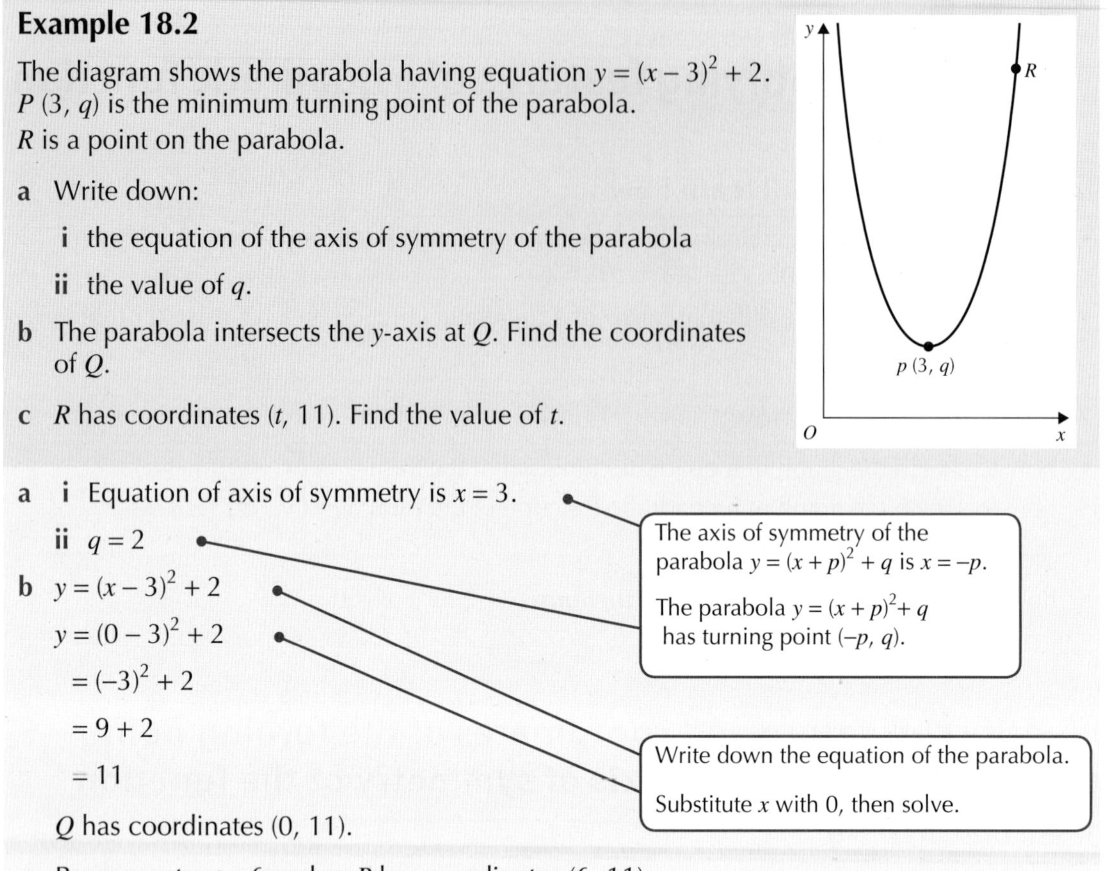

Example 18.2

The diagram shows the parabola having equation $y = (x - 3)^2 + 2$.
$P\ (3, q)$ is the minimum turning point of the parabola.
R is a point on the parabola.

a Write down:

i the equation of the axis of symmetry of the parabola

ii the value of q.

b The parabola intersects the y-axis at Q. Find the coordinates of Q.

c R has coordinates $(t, 11)$. Find the value of t.

a **i** Equation of axis of symmetry is $x = 3$.

ii $q = 2$

The axis of symmetry of the parabola $y = (x + p)^2 + q$ is $x = -p$.

The parabola $y = (x + p)^2 + q$ has turning point $(-p, q)$.

b $y = (x - 3)^2 + 2$

$y = (0 - 3)^2 + 2$

$= (-3)^2 + 2$

$= 9 + 2$

$= 11$

Write down the equation of the parabola.

Substitute x with 0, then solve.

Q has coordinates $(0, 11)$.

c By symmetry $t = 6$ and so R has coordinates $(6, 11)$.

Exercise 18A

★ **1** For each equation of a parabola given below write down:

i the equation of its axis of symmetry

ii the coordinates and nature of the turning point.

a $y = (x - 2)^2 + 5$	**b** $f(x) = (x - 4)^2 + 1$	**c** $y = (x - 8)^2 - 3$
d $g(x) = -(x - 6)^2 - 2$	**e** $y = \left(x - \frac{1}{2}\right)^2 + \frac{3}{4}$	**f** $y = (x + 1)^2 + 9$
g $h(x) = -(x + 7)^2 + 2$	**h** $y = (x + 3)^2 - 5$	**i** $f(x) = -(x + 10)^2 - 4$

2 The graph shows a parabola with equation $y = (x + a)^2 + b$.
The parabola intersects the x-axis at $P\ (2,0)$ and $Q\ (10,0)$.

a Write down the equation of the axis of symmetry of the parabola.

b Write down the value of a.

c Find the value of b.

d The parabola intersects the y-axis at R.
Find the coordinates of R.

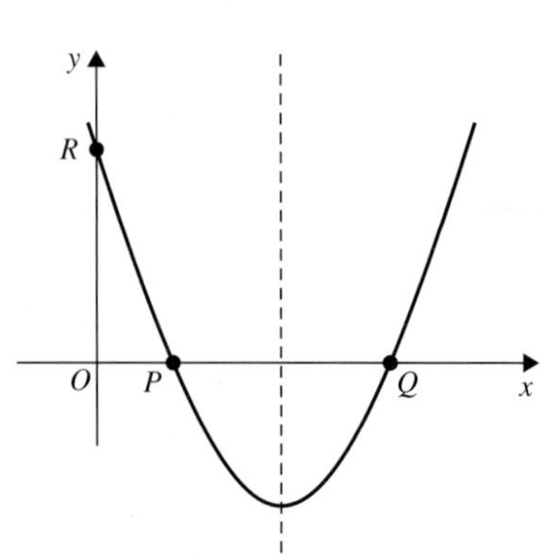

3 The diagram shows the parabola with equation $y = -(x - a)^2 + b$. The parabola intersects the x-axis at the points S and T. The length of ST is 10 units. The equation of the axis of symmetry is $x = 8$.

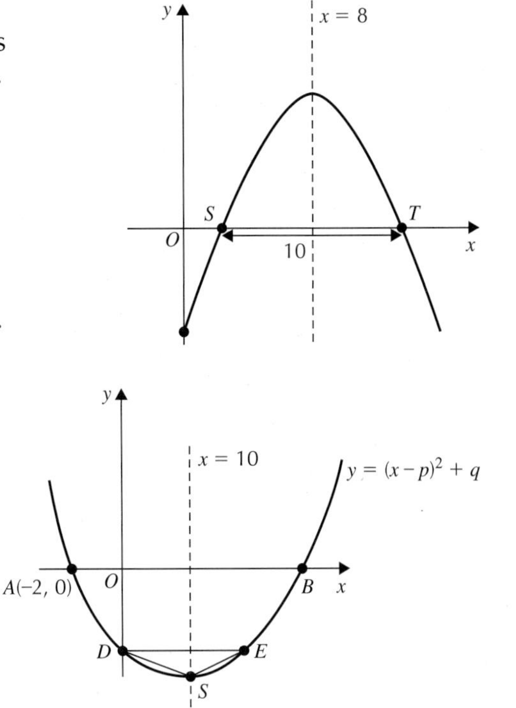

a Write down the value of a.

b Find the coordinates of S and T.

c Find the coordinates of the maximum turning point of the parabola.

d Find where the parabola intersects the y-axis.

★ 4 The diagram shows the parabola with equation $y = (x - p)^2 + q$. The parabola intersects the x-axis at A and B. A has coordinates $(-2, 0)$. S is the minimum turning point of the parabola. The line $x = 10$ is the axis of symmetry of the parabola.

a Write down the coordinates of B.

b Find the coordinates of S.

c The parabola intersects the y-axis at D. The line DE is horizontal. Determine the area of the triangle DES.

5 The diagram shows two parabolas having equations $f(x) = (x - 2)^2 + 5$ and $g(x) = -(x - p)^2 + q$.

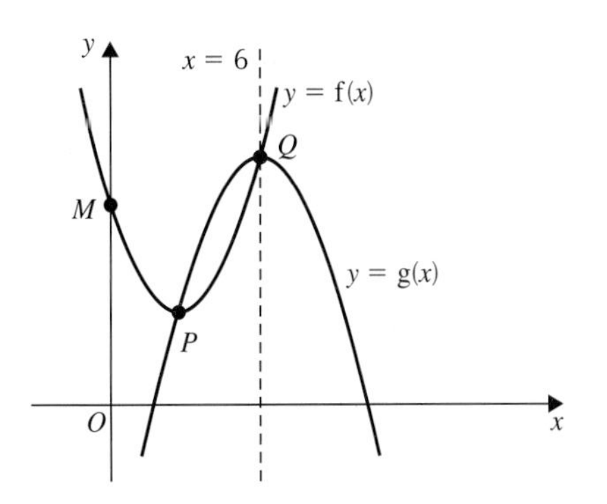

The following information is known about the parabolas:

- P and Q are on both parabolas
- P is the minimum turning point of one parabola
- the vertical line passing through Q has equation $x = 6$.

a Determine the values of p and q.

The graph of $y = f(x)$ intersects the y-axis at M.

b i Find the coordinates of M.

ii The length of MQ is $a\sqrt{b}$ units. Determine the values of a and b.

6 A parabola has equation $f(x) = a - (x - b)^2$. The line $x = 6$ is the equation of the axis of symmetry of the parabola. The parabola intersects the x-axis at the points A and B. A has coordinates $(1, 0)$. The parabola intersects the y-axis at P. Q is the point on the parabola symmetrical to P.

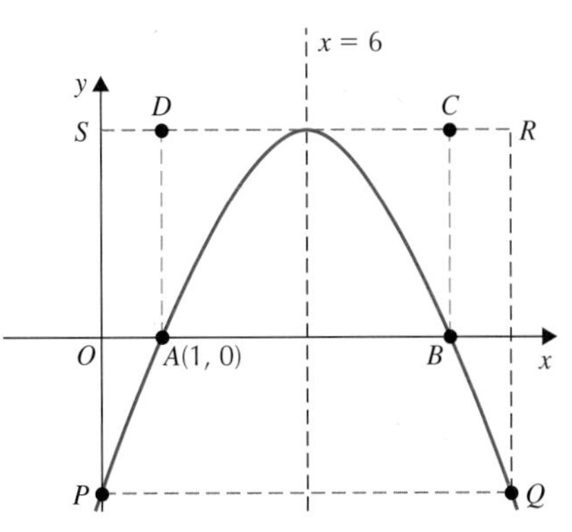

The ratio of the areas of rectangles $PQRS$ and $ABCD$ can be expressed in the form $\left(\frac{m}{n}\right)^3$, where m and n are integers. Determine the values of m and n.

Apply knowledge of the different forms of a quadratic function to solve related problems

This section links together your knowledge of the graphs of quadratic functions to solve problems.

In problem solving questions you usually need to find the turning point and/or the roots of a parabola. You may also need to work out other information. The table summarises how to find the key features of a parabola depending on the form of the quadratic function.

Form	Nature of TP	How to find coordinates of TP	How to find roots
$y = ax^2 + bx + c$ (general form)	$a > 0$ minimum TP $a < 0$ maximum TP	x-coordinate $= -\frac{b}{2a}$ Substitute value for x into equation to find y-coordinate (See Chapter 17)	Solve $ax^2 + bx + c = 0$ by factorising/quadratic formula (remember also that you can work out $b^2 - 4ac$ to check the nature of the roots) (See Chapter 19)
$y = (x + p)^2 + q$ (completed square form)	minimum TP	Coordinates are $(-p, q)$ (See Chapter 16)	If they exist, solve $(x + p)^2 + q = 0$ (multiply out brackets, simplify and then either factorise or use quadratic formula) (See Chapter 19)
$y = -(x + p)^2 + q$ (completed square form)	maximum TP	Coordinates are $(-p, q)$ (See Chapter 16)	If they exist, solve $-(x + p)^2 + q = 0$ (multiply out brackets, simplify and then either factorise or use quadratic formula) (See Chapter 19)
$y = (x - m)(x - n)$ (root form)	minimum TP	x-coordinate of TP is half-way between the roots, so $x = \frac{m + n}{2}$ Substitute value for x into equation to find y-coordinate (See Chapter 17)	The roots are m and n
$y = -(x - m)(x - n)$ (root form)	maximum TP	x-coordinate of TP is half-way between the roots, so $x = \frac{m + n}{2}$ Substitute value for x into equation to find y-coordinate (See Chapter 17)	The roots are m and n

Example 18.3

The rate at which cars leave a car park after 6 pm is given by the quadratic function

$$R = -\tfrac{1}{30}t^2 + 2t + 10$$

where R is the rate in cars per minute and t is the time in minutes after 6 pm.

a At what rate were cars leaving the car park at 6.20 pm?

b At what times did cars leave the car park at the rate of 10 cars per minute?

c At what time did cars leave the car park at a maximum rate? Determine this rate (in cars per minute).

a $y = -\frac{1}{30}x^2 + 2x + 10$ — Rewrite the function in a more familiar form.

$= -\frac{1}{30}(20)^2 + 2(20) + 10$ — At 6.20 pm, $x = 20$ so substitute $x = 20$ into the equation, then solve.

$= 37$ — Evaluate y using a calculator and round appropriately.

So: at 6.20 pm cars were leaving the car park at the rate of 37 cars per minute.

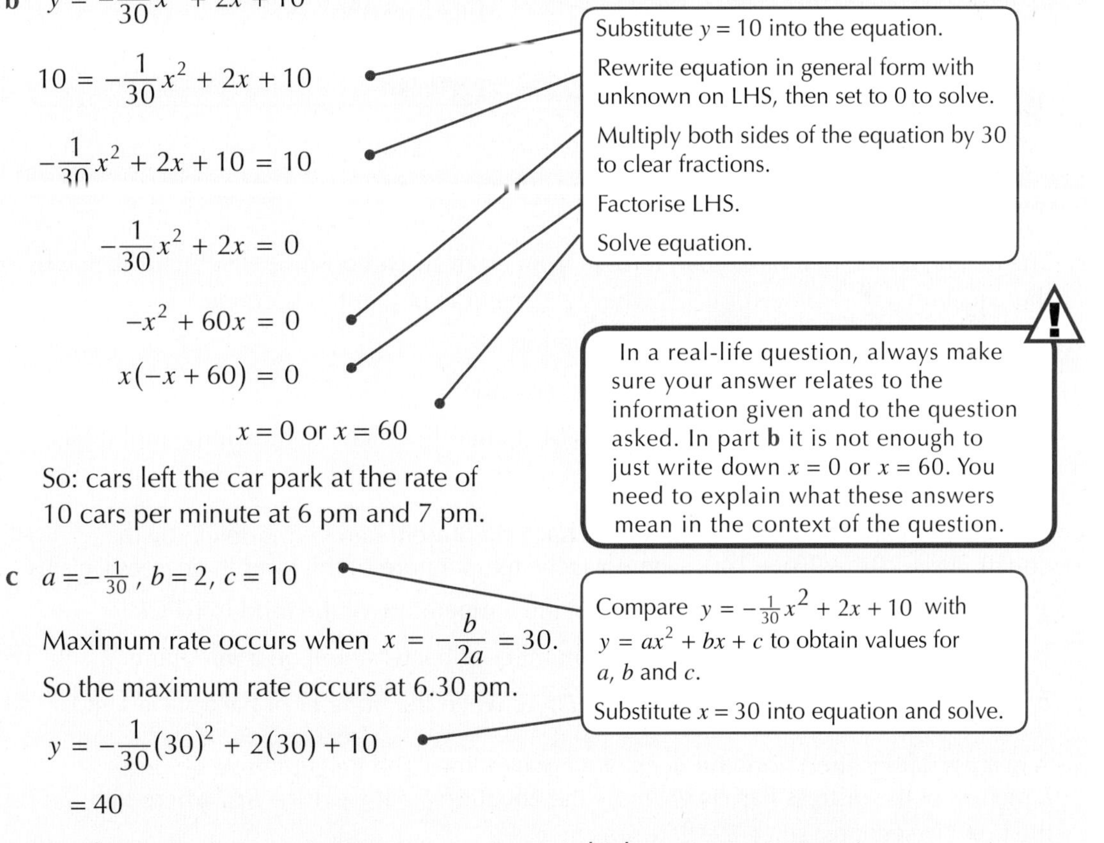

b $y = -\frac{1}{30}x^2 + 2x + 10$

$10 = -\frac{1}{30}x^2 + 2x + 10$ — Substitute $y = 10$ into the equation.

$-\frac{1}{30}x^2 + 2x + 10 = 10$ — Rewrite equation in general form with unknown on LHS, then set to 0 to solve.

$-\frac{1}{30}x^2 + 2x = 0$

$-x^2 + 60x = 0$ — Multiply both sides of the equation by 30 to clear fractions.

$x(-x + 60) = 0$ — Factorise LHS.

$x = 0$ or $x = 60$ — Solve equation.

So: cars left the car park at the rate of 10 cars per minute at 6 pm and 7 pm.

In a real-life question, always make sure your answer relates to the information given and to the question asked. In part **b** it is not enough to just write down $x = 0$ or $x = 60$. You need to explain what these answers mean in the context of the question.

c $a = -\frac{1}{30}$, $b = 2$, $c = 10$ — Compare $y = -\frac{1}{30}x^2 + 2x + 10$ with $y = ax^2 + bx + c$ to obtain values for a, b and c.

Maximum rate occurs when $x = -\frac{b}{2a} = 30$.

So the maximum rate occurs at 6.30 pm.

$y = -\frac{1}{30}(30)^2 + 2(30) + 10$ — Substitute $x = 30$ into equation and solve.

$= 40$

So: the maximum rate is 40 cars per minute (which occurs at 6.30 pm).

Example 18.4

The width of a rectangle is w centimetres. The length of the rectangle is 5 centimetres more than its width.

a Given that the area of the rectangle is 24 square centimetres, show that $w^2 + 5w - 24 = 0$.

b Find the length and width of the rectangle.

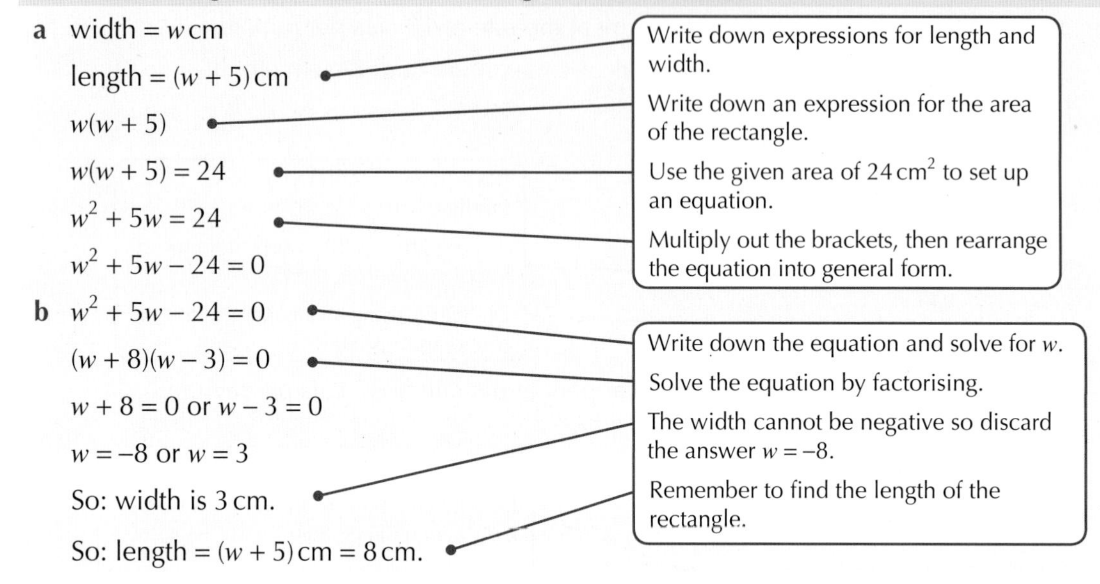

a width = w cm
length = (w + 5) cm
w(w + 5)
w(w + 5) = 24
w² + 5w = 24
w² + 5w – 24 = 0
b w² + 5w – 24 = 0
(w + 8)(w – 3) = 0
w + 8 = 0 or w – 3 = 0
w = –8 or w = 3
So: width is 3 cm.
So: length = (w + 5) cm = 8 cm.
Write down expressions for length and width.
Write down an expression for the area of the rectangle.
Use the given area of 24 cm² to set up an equation.
Multiply out the brackets, then rearrange the equation into general form.
Write down the equation and solve for w.
Solve the equation by factorising.
The width cannot be negative so discard the answer w = –8.
Remember to find the length of the rectangle.

Exercise 18B

1 The height, h feet, of a basketball thrown from an overhead position can be modelled by the equation $h = -16t^2 + 40t + 11$, where t is the time of flight in seconds.

a From what height was the basketball thrown?

b Calculate the maximum height of the basketball.

c Assuming the basketball misses the basket entirely, calculate how long until it hits the floor.

2 In some refrigerated food, the number of bacteria present can be modelled by the equation $y = 20x^2 - 20x + 140$, where x is the temperature of the food in degrees Celsius.

a How many bacteria are present when the temperature of the food is 10°C?

b Calculate the temperature when the number of bacteria will be a minimum.

c Calculate the temperature, T, where $T > 0$, when the number of bacteria is 1580.

★ **3** A distress flare is fired from the deck of an ocean liner. The height above sea level, h metres, of the distress flare is given by the equation $h = 45 + 40t - 5t^2$, where t is the time (in seconds) since the flare was fired.

a From what height above sea level was the flare fired?

b Calculate the maximum height of the flare.

c How long does it take until the flare enters the sea?

4 The corner of a swimming pool is to be cordoned off with rope so that young children can be given lessons in an enclosed area. The area to be cordoned off is rectangular and the pool attendants have 20 metres of rope.

a The length of the enclosure is x metres. Show that its area, $A\,\text{m}^2$, is given by the equation $A = 20x - x^2$.

b Find the maximum area which can be cordoned off.

5 A concert venue can hold 15 000 people. A promoter calculates that the profit, £P, made when the selling price of a ticket is £x, is given by the equation $P = 15000x - 100x^2$.

a Find the price that should be charged for a ticket to ensure a maximum profit.

b Calculate the maximum profit.

c If the price of a ticket exceeds £T, the promoter will make a loss. Find the value of T.

6 In a scientific study, the percentage, P, of people indoors who felt comfortable at temperature T (° Fahrenheit) is given by the equation $P = -18807 + 527.3x - 3.678x^2$.

a Calculate the temperature at which the highest percentage of people felt comfortable and calculate this percentage.

b A formula used to convert between degrees Celsius, C, and degrees Fahrenheit, F, is

$$C = \frac{5}{9}(F - 32)$$

Determine the percentage of people who felt comfortable at 25 °C.

c Between what range of temperatures in degrees Fahrenheit would more than 80% of people feel comfortable?

7 If you begin an exercise by walking and keep increasing your speed, then you will soon break into a run. The speed at which walking spontaneously becomes running is called the 'preferred transition speed'. For humans, the preferred transition speed is around 2 metres per second.

As you move from walking to running, the rate at which you consume energy also changes. Using suitable units, the rate at which you consume energy, y, is given by the equation $y = 278.7x^2 - 317.6x + 171.3$, where x is your speed as a percentage of the preferred transition speed.

For example, if you are walking at 10% of the preferred transition speed, then $x = 0.1$. When $x = 0.1$, $y = 278.7(0.1)^2 - 317.6(0.1) + 171.3 = 142$.

a At what percentage of the preferred transition speed will your energy consumption be at a minimum?

b What is this minimum rate of consumption?

Activity

A firm manufactures water tanks from rectangular sheets of metal 3 metres long by 2 metres wide. Squares of side x metres are cut out of each corner as shown in the diagram.

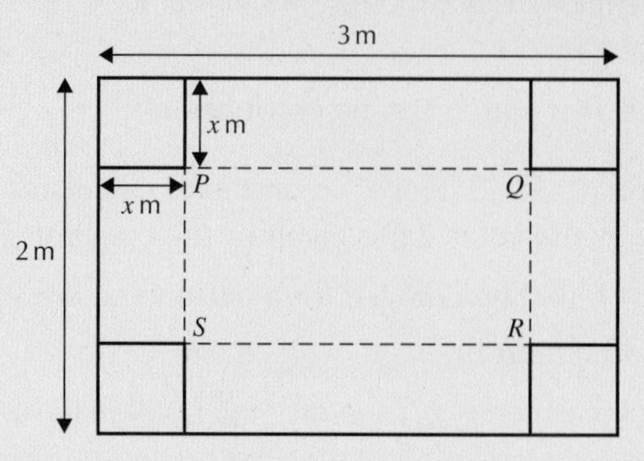

The rectangle *PQRS* forms the base of the tank and the remaining parts are folded upwards and sealed to form the open tank.

a **i** Find and simplify an expression for the volume of water that the tank can hold.
ii Find and simplify an expression for the internal surface area of the water tank.

Each sheet of metal purchased by the manufacturer costs £30.

The internal surface area is to be coated with an anti-corrosive agent which costs £22.50 per square metre.

When selling a tank, the manufacturer charges a standing charge of £165 plus £350 for every cubic metre of capacity.

b Using your answers to part **a** and neglecting all other costs, find and simplify an expression for the profit made on each tank.

c What is the range of possible values for x?

The value of x which gives the maximum profit can be found by solving the equation $210x^2 - 341x + 105 = 0$.

d Find the value of x which gives the maximum profit and hence determine the maximum profit.

- I can apply my knowledge of the equation $y = (x + p)^2 + q$ to solve problems. ★ Exercise 18A Q1, Q4
- I can apply my knowledge and skills of quadratic functions and graphs to solve real-life problems. ★ Exercise 18B Q3

For further assessment opportunities, see the Preparation for Assessment for Unit 2 on pages 288–291.

19 Working with quadratic equations

In this chapter you will learn how to:

- solve a quadratic equation **algebraically**
- solve a quadratic equation using the **quadratic formula**
- use the relationship between the **roots** of a quadratic equation and its corresponding graph
- find the points of intersection of a parabola and a straight line
- solve problems using quadratic equations
- use the **discriminant** of a quadratic equation in a question involving the nature of its roots.

You should already know:

- how to factorise a quadratic expression
- how to solve a linear equation
- how to solve a pair of simultaneous equations by substitution
- how to use formulae for the area and volume of common 2D shapes and 3D objects.

Solving a quadratic equation algebraically

A quadratic equation is one which can be written in the general or standard form:

$ax^2 + bx + c = 0, a \neq 0$

To solve a quadratic equation, given in standard form, algebraically, the quadratic expression must first be factorised and made equal to zero. For example:

$x^2 + 3x - 4 = 0$

$(x + 4)(x - 1) = 0$

If a product of terms is equal to zero then at least one of these terms is itself equal to zero, so either:

$(x + 4) = 0$ or $(x - 1) = 0$

which gives:

$x = -4$ or $x = 1$

The values −4 and 1 are the **roots** of this particular equation. You can check by substitution that each root does satisfy the equation:

$x = -4$: LHS $= (-4)^2 + 3 \times (-4) - 4 = 0 =$ RHS ✓

$x = 1$: LHS $= 1^2 + 3 \times 1 - 4 = 0 =$ RHS ✓

Solving a quadratic equation in standard form algebraically

Solving a quadratic equation in standard form algebraically involves:

- **factorising** the quadratic expression to create a product of terms
- setting each of these terms equal to **zero**
- **solving** for each term.

Example 19.1

Solve the following quadratic equations algebraically.

a $x(x - 3) = 0$ **b** $4x^2 - 9 = 0$ **c** $x^2 + 4x + 3 = 0$ **d** $2x^2 + 3x - 5 = 0$

Note that the first of these is already in factorised form.

a $x(x - 3) = 0$

$x = 0$ or $x - 3 = 0$

$x = 0$ or $x = 3$

b $4x^2 - 9 = 0$

$(2x - 3)(2x + 3) = 0$ — Factorise to create a product of terms.

$2x - 3 = 0$ or $2x + 3 = 0$ — Solve for each term.

$2x = 3$ or $2x = -3$

$x = \frac{3}{2}$ or $x = -\frac{3}{2}$

c $x^2 + 4x + 3 = 0$

$(x + 3)(x + 1) = 0$

$x + 3 = 0$ or $x + 1 = 0$

$x = -3$ or $x = -1$

d $2x^2 + 3x - 5 = 0$

$(2x + 5)(x - 1) = 0$

$2x + 5 = 0$ or $x - 1 = 0$

$2x = -5$ or $x = 1$

$x = -\frac{5}{2}$ or $x - 1$

Exercise 19A

1 Solve the following equations algebraically.

a $(x - 4)(x - 2) = 0$ **b** $3x(x + 4) = 0$ **c** $(2x - 3)(x + 2) = 0$

d $(2x + 3)(2x + 3) = 0$ **e** $x(2x + 7) = 0$ **f** $(2x - 1)(x + 7) = 0$

g $4x(5 - 3x) = 0$ **h** $(x + 1)(3x - 2) = 0$ **i** $12x(4 + 3x) = 0$

2 Find the roots of the following equations.

a $4x^2 - x = 0$ **b** $6x^2 + 9x = 0$ **c** $15x - 25x^2 = 0$ **d** $4x^2 - 10x = 0$

e $5x^2 - 5x = 0$ **f** $16x - 4x^2 = 0$ **g** $11x + x^2 = 0$ **h** $4x - 6x^2 = 0$

3 Solve the following quadratic equations algebraically.

a $4x^2 - 9 = 0$ **b** $25p^2 - 16 = 0$ **c** $4 - m^2 = 0$ **d** $x^2 - 81 = 0$

e $x^2 - 49 = 0$ **f** $9x^2 - 100 = 0$ **g** $121 - 81q^2 = 0$ **h** $64 - 4t^2 = 0$

4 Find the solutions to the following equations.

a $x^2 + 8x + 15 = 0$ **b** $t^2 - 4t + 3 = 0$ **c** $x^2 - 3x - 10 = 0$

d $x^2 - 5x + 6 = 0$ **e** $x^2 - 8x - 20 = 0$ **f** $z^2 + 14z + 45 = 0$

g $y^2 + 4y - 12 = 0$ **h** $w^2 + w - 6 = 0$ **i** $r^2 + 5r - 14 = 0$

5 Solve the following equations algebraically.

a $2r^2 + 3r + 1 = 0$ **b** $-t^2 + 7t - 12 = 0$ **c** $3s^2 - 4s - 4 = 0$

d $-2p^2 - 7p - 3 = 0$ **e** $3w^2 + 5w - 12 = 0$ **f** $-6x^2 + 31x - 5 = 0$

g $-12x^2 + 24x - 12 = 0$ **h** $2m^2 + 7m - 15 = 0$ **i** $5p^2 + 13p - 18 = 0$

★ **6** Find the roots of the following quadratic equations.

a $p^2 + 4p = 0$ **b** $x^2 + 14x + 49 = 0$ **c** $2x^2 - 3x - 5 = 0$

d $36 - p^2 = 0$ **e** $18m + 12m^2 = 0$ **f** $-5x^2 - 38x - 21 = 0$

g $8x^2 - 50 = 0$ **h** $-6x^2 + 22x + 40 = 0$ **i** $98 - 32m^2 = 0$

j $6a^2 - 33a + 15 = 0$ **k** $12x^2 - 75 = 0$ **l** $5x^2 + 35x + 60 = 0$

Dealing with quadratic equations which are not in standard form

Exercise 19A requires two conditions to be met in order to solve a quadratic equation:

- it is given in the standard form $ax^2 + bx + c = 0$, $a \neq 0$
- it is possible to factorise $ax^2 + bx + c$.

These two conditions are not always met.

For equations which are not in standard form, some manipulation of the given equation is needed before they can be solved.

Example 19.2

Solve the following equations algebraically.

a $x^2 = 3x + 10$ **b** $x(x - 4) = 2x - 5$ **c** $(x + 3)^2 = 9$

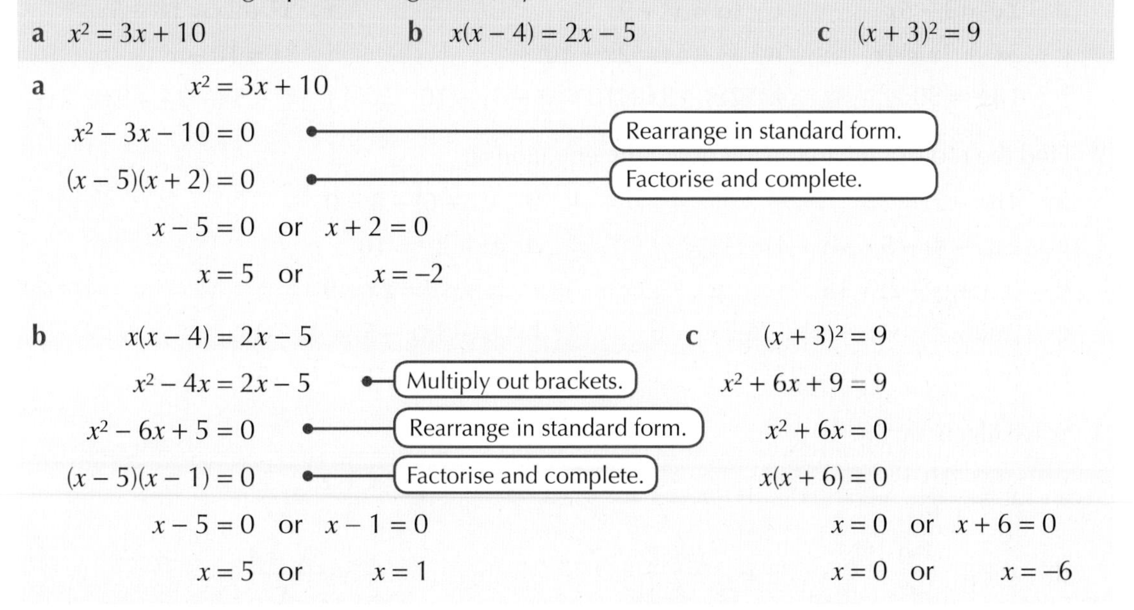

Example 19.3

Find the solutions of the following equations.

a $x + 5 = \frac{14}{x}$ **b** $x = \frac{15}{x + 2}$

a

$$x + 5 = \frac{14}{x}$$

$x(x + 5) = 14$ — Multiply each side by x to remove the fraction.

$x^2 + 5x = 14$ — Multiply out brackets.

$x^2 + 5x - 14 = 0$ — Rearrange in standard form.

$(x - 2)(x + 7) = 0$ — Factorise and complete.

$x - 2 = 0$ or $x + 7 = 0$

$x = 2$ or $x = -7$

b

$$x = \frac{15}{x + 2}$$

$x(x + 2) = 15$ — Multiply each side by $(x + 2)$ to remove the fraction.

$x^2 + 2x = 15$

$x^2 + 2x - 15 = 0$

$(x + 5)(x - 3) = 0$

$x + 5 = 0$ or $x - 3 = 0$

$x = -5$ or $x = 3$

Exercise 19B

1 Solve the following equations algebraically.

a $4x^2 = 8x$	**b** $4x^2 = 9$	**c** $x^2 = 3x - 2$
d $2x^2 = 3 - 5x$	**e** $6x = x^2 + 9$	**f** $x^2 + 2x = 18 - 5x$
g $3x^2 + 5x = 4x$	**h** $x^2 + 5x = 2x + 10$	**i** $3x = 10 - x^2$
j $18x^2 = 50$	**k** $2x^2 + 5x + 10 = x^2 - 4x - 10$	**l** $x^2 - 5x - 24 = 9x - 2x^2$

★ **2** Find the roots of the following quadratic equations.

a $4x(x - 2) = 2x$	**b** $x(x - 6) + 8 = 0$
c $2x(x + 4) - 5 = x(x + 4)$	**d** $(x - 2)^2 = 16$
e $(x + 4)(x + 2) = 15$	**f** $(x + 3)^2 = 2x + 9$
g $2x(x + 2) = 4x - 2x^2 + 25$	**h** $(x - 3)(x - 5) = 4x - 17$
i $(x + 2)^2 + (x + 1)^2 = 25$	**j** $169 - (x - 2)^2 = (x + 5)^2$

★ **3** Solve algebraically.

a $x = \frac{10}{x + 3}$	**b** $x - 3 = \frac{28}{x}$	**c** $\frac{x + 2}{3} = \frac{5}{x}$
d $x + 8 - \frac{20}{x} = 0$	**e** $\frac{x + 2}{x} = x$	**f** $\frac{x}{2} = \frac{3x - 4}{x}$

Solving quadratic equations using the quadratic formula

Quadratic equations which are not initially written in standard form can sometimes be solved by first dealing with fractions, multiplying out brackets and rearranging as necessary, as shown in Exercise 19B. However, there are quadratic equations which cannot be factorised. These equations cannot be solved algebraically but can be solved using the **quadratic formula**.

The quadratic formula states that for any quadratic equation, $ax^2 + bx + c = 0$, $a \neq 0$, the roots are given by:

$$x = \frac{-b \pm \sqrt{b^2 - 4ac}}{2a}$$

Example 19.4

Solve the equation $2x^2 + 4x + 1 = 0$ giving your answer to 2 decimal places.

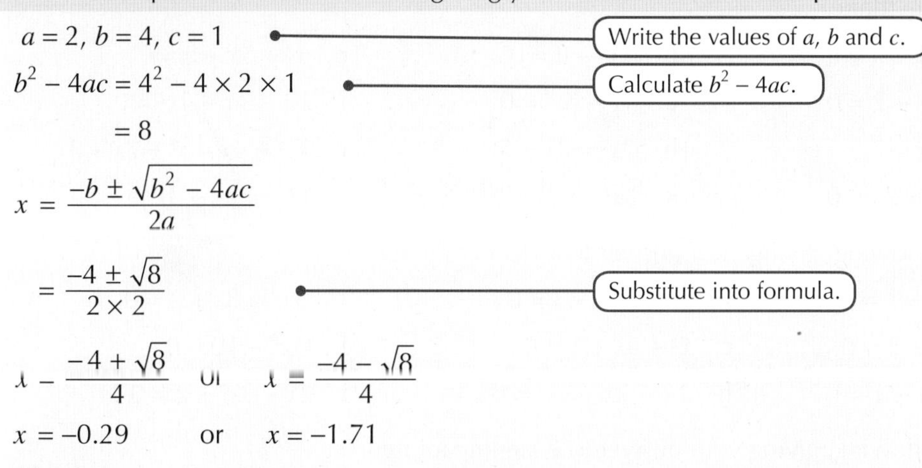

$a = 2, b = 4, c = 1$
Write the values of a, b and c.
$b^2 - 4ac = 4^2 - 4 \times 2 \times 1$
Calculate $b^2 - 4ac$.
$= 8$
$x = \frac{-b \pm \sqrt{b^2 - 4ac}}{2a}$
$= \frac{-4 \pm \sqrt{8}}{2 \times 2}$
Substitute into formula.
$x = \frac{-4 + \sqrt{8}}{4}$ or $x = \frac{-4 - \sqrt{8}}{4}$
$x = -0.29$ or $x = -1.71$

Check

These solutions can also be checked by substitution as follows, but, remember rounding has taken place when the square roots are taken.

$x = -0.29$: LHS $= 2 \times (-0.29)^2 + 4 \times (-0.29) + 1 = 0.0082 \approx 0 =$ RHS ✓

$x = -1.71$: LHS $= 2 \times (-1.71)^2 + 4 \times (-1.71) + 1 = 0.0082 \approx 0 =$ RHS ✓

The match is good enough.

Example 19.5

Solve the equation $3x^2 - 2x = 7$ giving your answer to 2 significant figures.

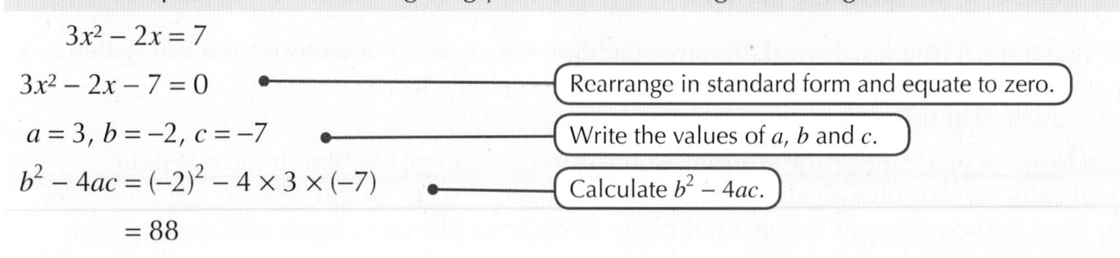

(continued)

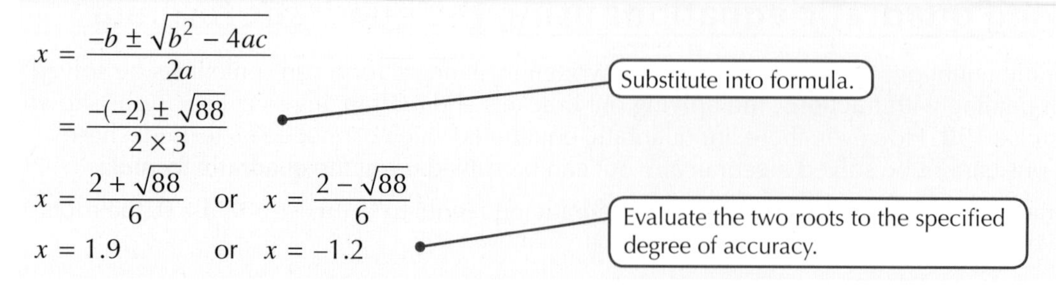

$x = \frac{-b \pm \sqrt{b^2 - 4ac}}{2a}$
$= \frac{-(-2) \pm \sqrt{88}}{2 \times 3}$
Substitute into formula.
$x = \frac{2 + \sqrt{88}}{6}$ or $x = \frac{2 - \sqrt{88}}{6}$
$x = 1.9$ or $x = -1.2$
Evaluate the two roots to the specified degree of accuracy.

Note that the values of a, b and c can be substituted directly into the formula, but by finding $b^2 - 4ac$ separately you break the work up into more manageable steps and minimise the chances of making a mistake.

Exercise 19C

1 Write down the values of a, b and c for the following quadratic equations. (You do not need to solve them.)

a $3x^2 + 2x - 4 = 0$ **b** $4x^2 - 8 = 0$ **c** $x^2 + 5x - 2 = 0$

d $4x - 3x^2 + 2 = 0$ **e** $4x - 7x^2 = 0$ **f** $12 - 3x - 4x^2 = 0$

g $3x^2 + 2x = 7$ **h** $2x^2 = 3x - 5$ **i** $3x^2 + 4x = x^2 - 2x + 3$

j $(x - 3)^2 + 2x = 0$ **k** $5x(x - 2) = 4$ **l** $7x^2 = (2x + 3)^2$

★ **2** Solve the following, giving your answer to 1 decimal place.

a $x^2 + 3x - 1 = 0$ **b** $2x^2 + 4x - 3 = 0$ **c** $3x^2 + 8x + 2 = 0$

d $x^2 - 7x + 2 = 0$ **e** $x^2 + 4x + 1 = 0$ **f** $3x^2 - 10 = 0$

g $2x^2 + 3x - 1 = 0$ **h** $12 - 2x - 3x^2 = 0$ **i** $2x - 3x^2 + 2 = 0$

3 Solve the following, giving your answer to 2 significant figures.

a $3x^2 - 5x + 1 = 0$ **b** $x^2 - 8x + 7 = 0$ **c** $4x(x - 3) + 2 = 0$

d $(x + 5)^2 = 7$ **e** $(2x - 1)(x - 3) - 4 = 0$ **f** $x = \frac{3x + 2}{2x}$

g $x - 7 = \frac{3}{x}$ **h** $(x - 2)^2 + (x - 3)^2 = 18$

Factorise or formula?

You now have two methods for solving a quadratic equation:

- by factorising
- by applying the quadratic formula.

The quadratic formula will work for any quadratic equation but if factorisation is possible it is usually simpler to solve algebraically. It is useful to be able to work out quickly which is the better method to use.

In an exam or assessment, the wording of the question can give a hint. If the question specifically says to solve algebraically, you will be required to factorise. If the question asks you to round your answer to a given number of decimal places or significant figures, you should use the quadratic formula.

There may be no clue in the question, in which case, you need a general test for factorising using the value of $b^2 - 4ac$ in the quadratic formula.

- If $b^2 - 4ac$ is a perfect square, then the quadratic will factorise and you can use either method (although in an exam it not advisable to use the formula unless you are having difficulty factorising).
- If $b^2 - 4ac$ is not a perfect square, then the quadratic will not factorise and only the formula will work.

For example, with $a = 1$, $b = 8$, $c = 9$:

$$\begin{aligned} b^2 - 4ac &= 64 - 36 \\ &= 28 \end{aligned}$$

In this case, $b^2 - 4ac$ is not a perfect square so the quadratic will not factorise – only the formula will work.

With $a = 3$, $b = -5$, $c = -2$:

$$\begin{aligned} b^2 - 4ac &= 5^2 - 4 \times 3 \times (-2) \\ &= 25 - (-24) \\ &= 49 \end{aligned}$$

In this case, $b^2 - 4ac$ is a perfect square so the quadratic will factorise – the equation can be solved algebraically or by the formula.

Exercise 19D

★ **1** Which of the following will require the use of the quadratic formula in order to solve them? (You do not need to solve them.)

a $4x^2 + 3x - 2 = 0$ **b** $x^2 - 6x + 5 = 0$ **c** $3x^2 - 2x - 4 = 0$

d $5x^2 - 20x = 0$ **e** $5 + 3x - 2x^2 = 0$ **f** $x^2 - 4x + 2 = 0$

g $2x^2 = 4x + 3$ **h** $4x(x - 3) - 2 = 0$ **i** $x^2 + 3(2x + 1) = 0$

Link between the roots of a quadratic equation and the graph of the corresponding quadratic function

There is a relationship between the roots of a quadratic equation and the points at which the graph of the corresponding quadratic function intersects the x-axis. The roots are defined as the values for which the quadratic equation is zero, so we can show graphically that the roots of the equation correspond to the points where the graph cuts the x-axis.

Consider the graph of $y = x^2 + 3x - 4$.

What do you know about the points A and B where the parabola intersects the x-axis?

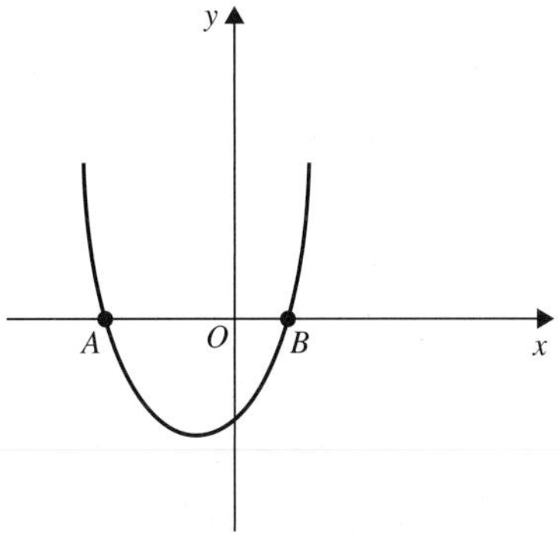

You know that:

$y = x^2 + 3x - 4$ because the points lie on the curve.

$y = 0$ because the points lie on the x-axis.

As each of these statements is true it follows that it must also be true that:

$x^2 + 3x - 4 = 0$

If we find the roots of this equation, we find where the graph cuts the x-axis. Conversely, if we know where the graph cuts the x-axis, we can read its roots straight from the graph.

Example 19.6

Use the graph of $y = x^2 - 3x - 10$ as shown here to solve the equation $x^2 - 3x - 10 = 0$.

The graph cuts the x-axis at $x = -2$ and $x = 5$.

The roots of the equation $x^2 - 3x - 10 = 0$ are therefore $x = -2$ and $x = 5$.

Example 19.7

Find where the following graphs of quadratic functions cut the x-axis.

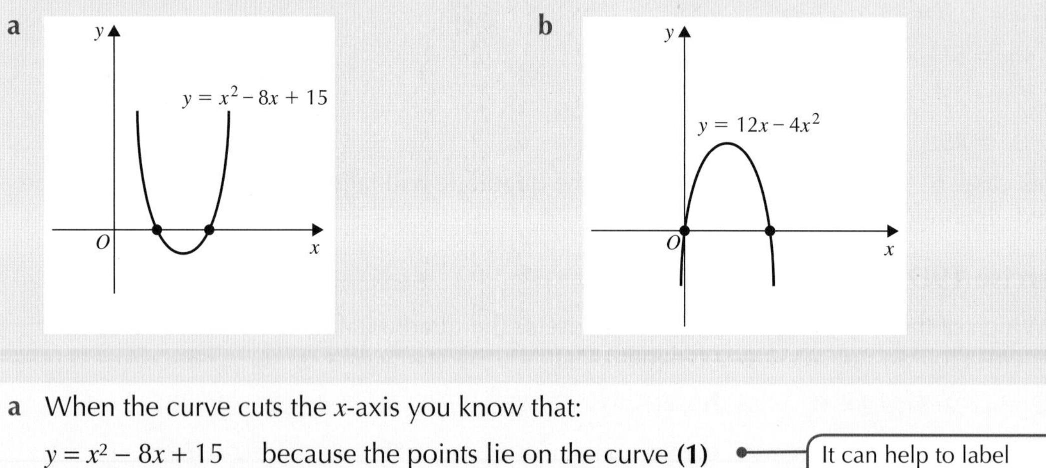

a When the curve cuts the x-axis you know that:

$y = x^2 - 8x + 15$ because the points lie on the curve **(1)**

$y = 0$ because the points lie on the x-axis **(2)**

It can help to label these two equations for reference later in the solution.

It follows that:

$x^2 - 8x + 15 - 0$

$(x - 3)(x - 5) = 0$

$x - 3 = 0$ or $x - 5 = 0$

$x = 3$ or $x = 5$

The graph cuts the x-axis at (3, 0) and (5, 0).

b When the curve cuts the x-axis you know that:

$y = 12x - 4x^2$ because the points lie on the curve **(1)**

$y = 0$ because the points lie on the x-axis **(2)**

It follows that:

$12x - 4x^2 = 0$

$4x(3 - x) = 0$

$4x = 0$ or $3 - x = 0$

$x = 0$ or $x = 3$

The graph cuts the x-axis at (0, 0) and (3, 0).

Exercise 19E

★ **1** Find where the following curves cut the x-axis.

a $y = x^2 - 10x + 24$ **b** $y = x^2 - 3x - 10$ **c** $y = 25 - x^2$

d $y = x^2 + 4x - 12$ **e** $y = 2x^2 - 7x - 15$ **f** $y = 3x^2 + 12x$

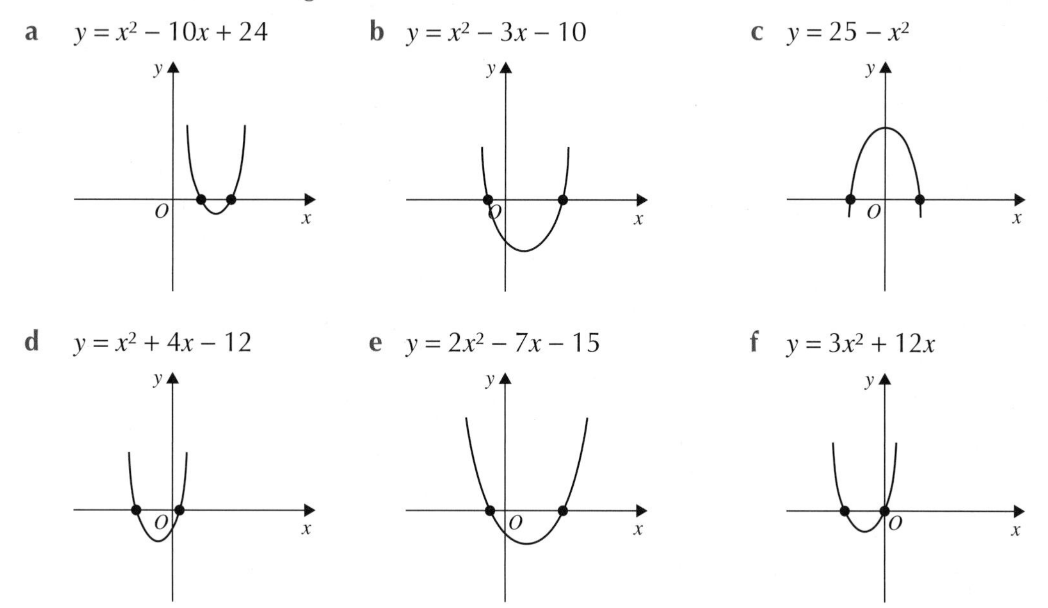

2 Find where the following curves cut the x-axis, correct to 1 decimal place.

a $y = x^2 - 10x + 1$ **b** $y = 2x^2 - 3x - 10$ **c** $y = 12 - 5x^2$

d $y = 3x^2 + 5x + 1$ **e** $y = 2x^2 - 7x + 4$ **f** $y = 1 + 4x - 2x^2$

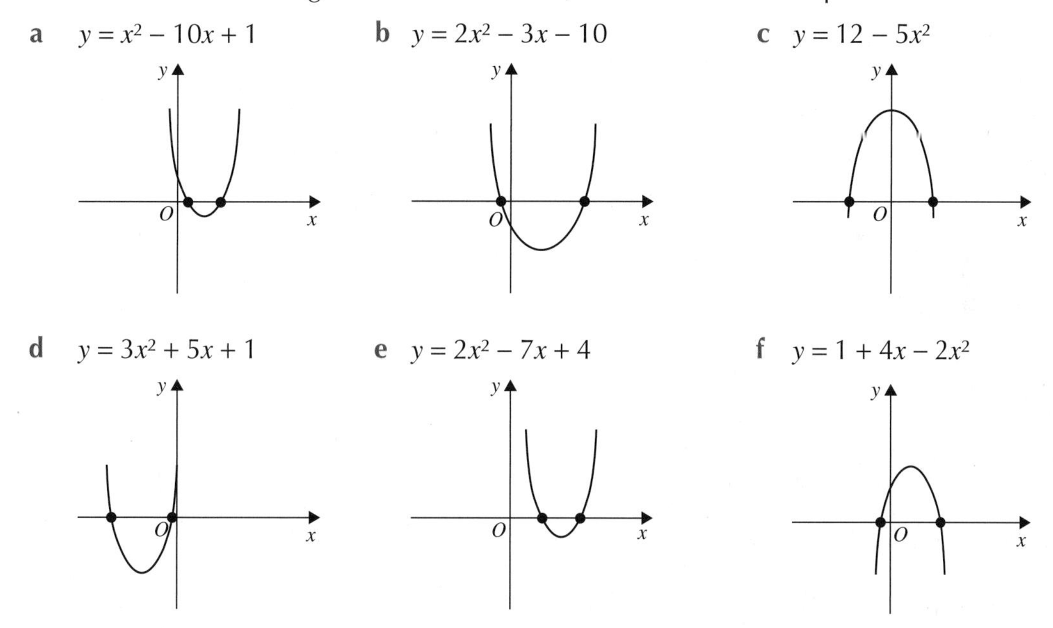

Finding points of intersection of a parabola and a straight line

The previous section showed how to find where a parabola intersects the x-axis. A similar method can be used to find the points of intersection of a parabola and any straight line as shown in Examples 19.8 to 19.10.

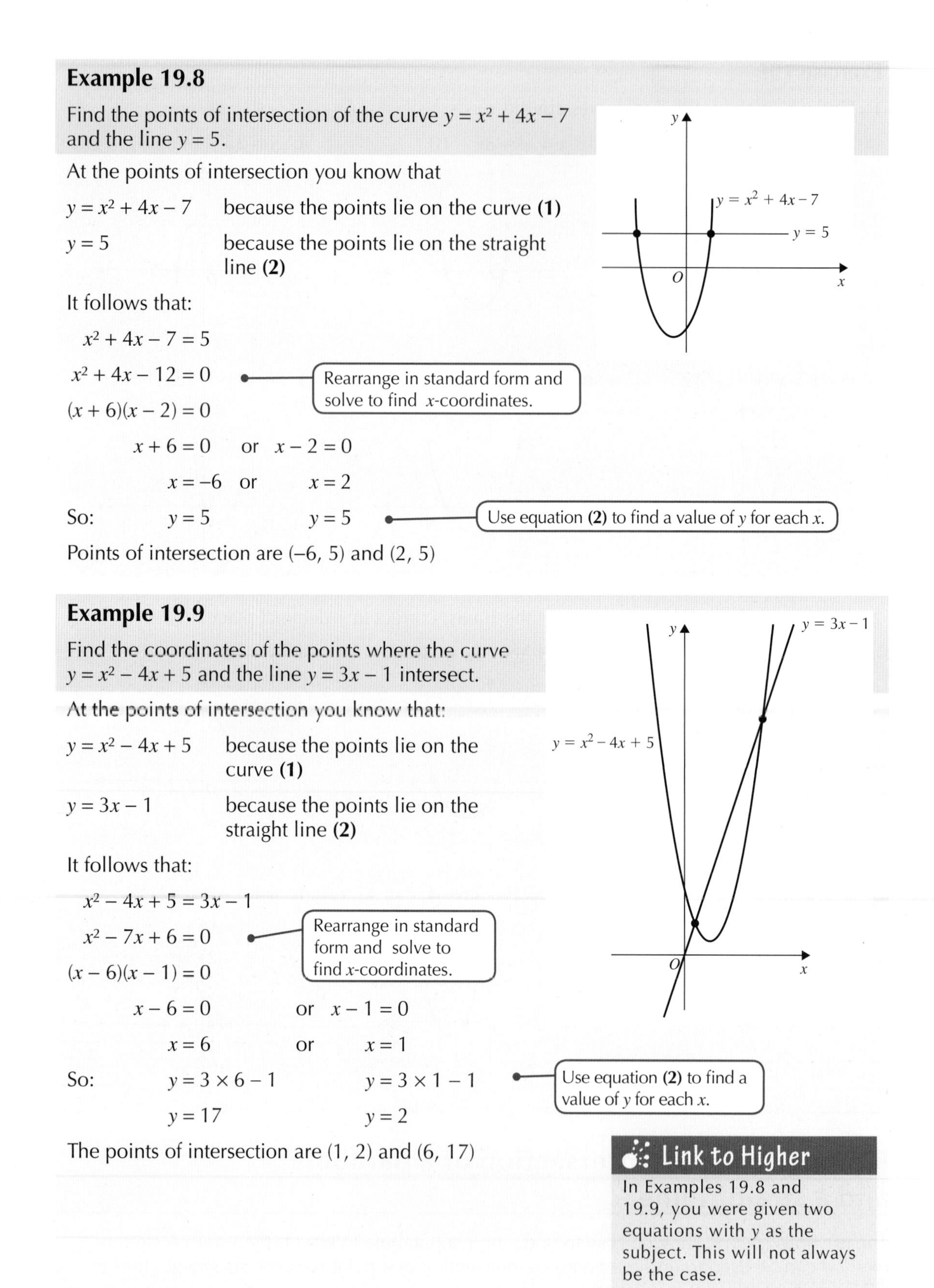

Example 19.8

Find the points of intersection of the curve $y = x^2 + 4x - 7$ and the line $y = 5$.

At the points of intersection you know that

$y = x^2 + 4x - 7$ because the points lie on the curve **(1)**

$y = 5$ because the points lie on the straight line **(2)**

It follows that:

$x^2 + 4x - 7 = 5$

$x^2 + 4x - 12 = 0$ — Rearrange in standard form and solve to find x-coordinates.

$(x + 6)(x - 2) = 0$

$x + 6 = 0$ or $x - 2 = 0$

$x = -6$ or $x = 2$

So: $y = 5$ $y = 5$ — Use equation **(2)** to find a value of y for each x.

Points of intersection are (–6, 5) and (2, 5)

Example 19.9

Find the coordinates of the points where the curve $y = x^2 - 4x + 5$ and the line $y = 3x - 1$ intersect.

At the points of intersection you know that:

$y = x^2 - 4x + 5$ because the points lie on the curve **(1)**

$y = 3x - 1$ because the points lie on the straight line **(2)**

It follows that:

$x^2 - 4x + 5 = 3x - 1$

$x^2 - 7x + 6 = 0$ — Rearrange in standard form and solve to find x-coordinates.

$(x - 6)(x - 1) = 0$

$x - 6 = 0$ or $x - 1 = 0$

$x = 6$ or $x = 1$

So: $y = 3 \times 6 - 1$ $y = 3 \times 1 - 1$ — Use equation **(2)** to find a value of y for each x.

$y = 17$ $y = 2$

The points of intersection are (1, 2) and (6, 17)

Link to Higher

In Examples 19.8 and 19.9, you were given two equations with y as the subject. This will not always be the case.

Example 19.10

Find the points of intersection of the quadratic curve with equation $y = 2x^2 - 4x + 5$ and the straight line with equation $2x + y = 9$.

The only difference is the form of the second equation so you must simply start by rearranging it:

$2x + y = 9$ can be rearranged to $y = 9 - 2x$

Once this has been done you proceed as before.

At the point of intersection you know that:

$y = 2x^2 - 4x + 5$ because the points lie on the curve **(1)**

$y = 9 - 2x$ because the points lie on the line **(2)**

It follows that:

$$2x^2 - 4x + 5 = 9 - 2x$$

$$2x^2 - 2x - 4 = 0$$

Rearrange in standard form and solve to find x-coordinates.

$$2(x^2 - x - 2) = 0$$

$$2(x - 2)(x + 1) = 0$$

$x - 2 = 0$ or $x + 1 = 0$

$x = 2$ or $x = -1$

So: $y = 9 - 2 \times 2$ $\quad$ $y = 9 - 2 \times (-1)$

$y = 5$ $\quad$ $y = 11$

Use equation **(2)** to find a value of y for each x.

The points of intersection are (2, 5) and (–1, 11).

Exercise 19F

1 Find the point(s) of intersection between each of the following.

a The curve $y = x^2 + 4x - 2$ and the line $y = 3$

b The line $y = 4$ and the curve $y = x^2 - 8x + 16$

c The curve $y = 3x^2 + 10x + 1$ and the line $y = 9$

d The curve $y = 7x - 2x^2$ and the line $y = 5$

e The line $y = 3$ and the curve $y = x^2 - 4x + 7$

★ **2** Find the point(s) of intersection between the following.

a The curve $y = x^2 - 3x + 3$ and the line $y = 2x - 3$

b The curve $y = 2x^2 - 5$ and the line $y = 2x + 3$

c The line $y = x + 3$ and the curve $y = 4 - 2x^2$

d The curve $y = 4x^2 - 5$ and the line $y = 4x - 6$

e The line $y = 8(1 - x)$ and the curve $y = 4 - 3x^2$

3 Find the point(s) of intersection between the following.

a The curve $y = x^2 + 10x + 12$ and the line $y - 2x = -3$

b The line $2x + y = 1$ and the curve $y = x^2 - 2x - 3$

c The curve $y = 2x^2 - 7x$ and the line $y - 2x = 5$

d The curve $y = 20 + 5x - x^2$ and $y - 3x - 5 = 0$

e The line $9x - y = 2$ and the curve $y = 4x^2 + x - 2$

Problem solving using quadratic equations

Many problems in mathematics, including real-life situations, can be modelled by forming equations. In some examples you are given an equation and must work with it. In others, you must recognise some form of equality in the question and use it to form an equation which you then go on to solve.

Example 19.11

The number of diagonals in an n-sided polygon is given by the formula:

$$d = \frac{n(n-3)}{2}$$

If a polygon has 27 diagonals, how many sides does it have?

It is given that:

$d = \frac{n(n-3)}{2}$ from the formula **(1)**

$d = 27$ the polygon has 27 diagonals **(2)**

It follows that:

$$\frac{n(n-3)}{2} = 27$$

You now have a quadratic equation which you can solve in the usual way.

$\frac{n(n-3)}{2} = 27$ — Multiply each side by 2 to remove the fraction.

$n(n-3) = 54$

$n^2 - 3n = 54$ — Multiply out the bracket.

$n^2 - 3n - 54 = 0$ — Rearrange in standard form.

$(n-9)(n+6) = 0$ — Factorise and solve.

$n - 9 = 0$ or $n + 6 = 0$

$n = 9$ or $n = -6$ — Check solutions against the problem.

While both 9 and –6 are roots of the equation, only 9 can be a solution to the problem as there cannot be a polygon with a negative number of sides.

The polygon has 9 sides.

Example 19.12

The area of the rectangle is 72 cm². Find the value of x.

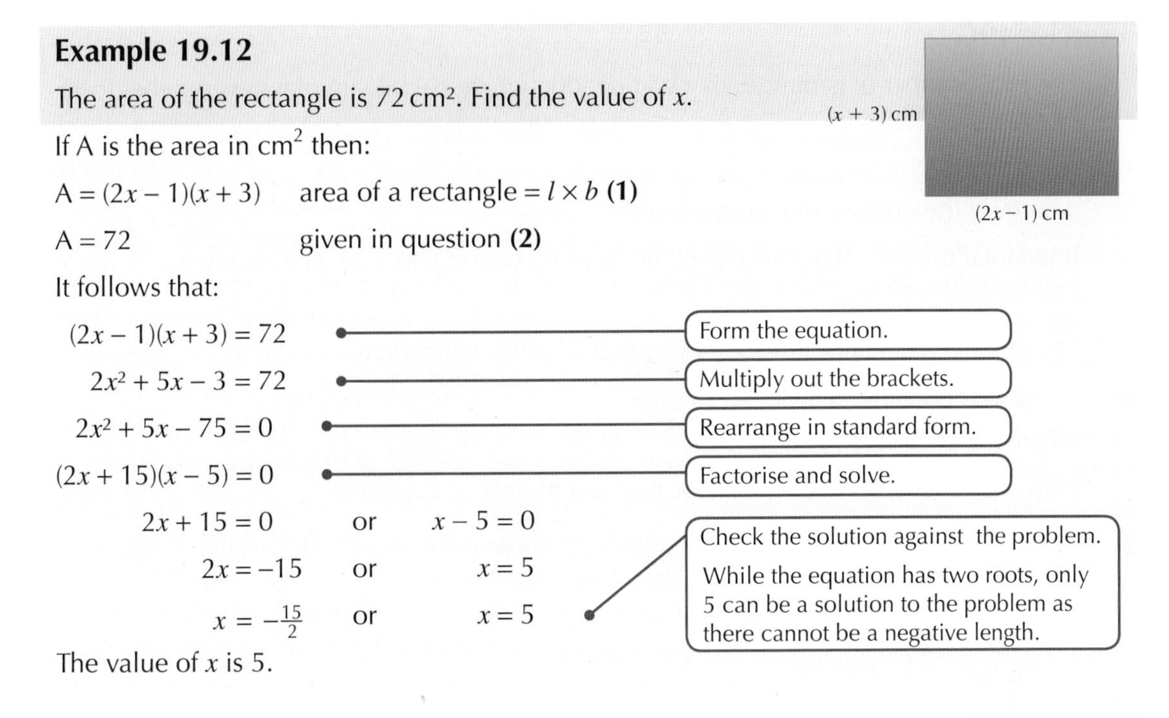

If A is the area in cm^2 then:

$A = (2x - 1)(x + 3)$ area of a rectangle $= l \times b$ **(1)**

$A = 72$ given in question **(2)**

It follows that:

$(2x - 1)(x + 3) = 72$ — Form the equation.

$2x^2 + 5x - 3 = 72$ — Multiply out the brackets.

$2x^2 + 5x - 75 = 0$ — Rearrange in standard form.

$(2x + 15)(x - 5) = 0$ — Factorise and solve.

$2x + 15 = 0$ or $x - 5 = 0$

$2x = -15$ or $x = 5$

$x = -\frac{15}{2}$ or $x = 5$

Check the solution against the problem. While the equation has two roots, only 5 can be a solution to the problem as there cannot be a negative length.

The value of x is 5.

Example 19.13

A rectangular lawn is surrounded by a path, 1 metre wide, on 3 sides as shown. The breadth of the lawn is x metres.

The length of the lawn is 4 metres more than its breadth

The area of the lawn is equal to the area of the path.

a Show that $x^2 + x + 6 = 0$.

b Calculate the dimensions of the lawn.

a Start with a labelled diagram to help calculate areas.

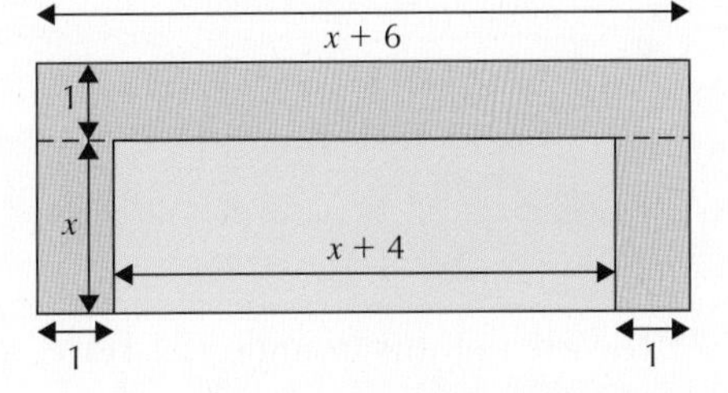

A(lawn) $= (x + 4) \times x = x(x + 4)$ equation **(1)**

A(path) $= 1 \times x + (x + 6) \times 1 + 1 \times x = 3x + 6$ equation **(2)**

area of the lawn = area of the path

$x(x + 4) = 3x + 6$

$x^2 + 4x = 3x + 6$

$x^2 + x - 6 = 0$

b $x^2 + x - 6 = 0$

$(x + 3)(x - 2) = 0$

$x + 3 = 0$ or $x - 2 = 0$

$x = -3$ or $x = 2$

While the equation has two roots, −3 and 2, we can only accept $x = 2$ as we cannot have a negative length.

The lawn is 6 m long and 2 m wide.

Exercise 19G

1 The total number of games, n, in a competition where each team plays every other team twice is given by:

$$n = t^2 - t$$

where t is the number of teams entered.

If the total number of games played in the competition was 110, how many teams were entered?

★ **2** The distance, x metres, travelled by a model train is given by the formula:

$$x = t^2 - 4t + 1$$

where t is the time in seconds.

How many seconds does it take for the train to travel 22 metres?

3 The number of handshakes, h, in a room of n people, if each pair shakes hands once, is given by the formula:

$$h = \tfrac{1}{2}n(n - 1), n > 1$$

If there are 36 handshakes, how many people are in the room?

4 John is 6 years older than Jim. The product of their ages is 135.

Let John's age = x years.

a Find an expression for Jim's age in terms of x.

b Find their respective ages algebraically.

5 Sarah is 4 years younger than Dave. The product of their ages is 320.

Let Sarah's age = x years.

Find their respective ages algebraically.

6 The dimensions and areas of the following rectangles are given. In each case find x.

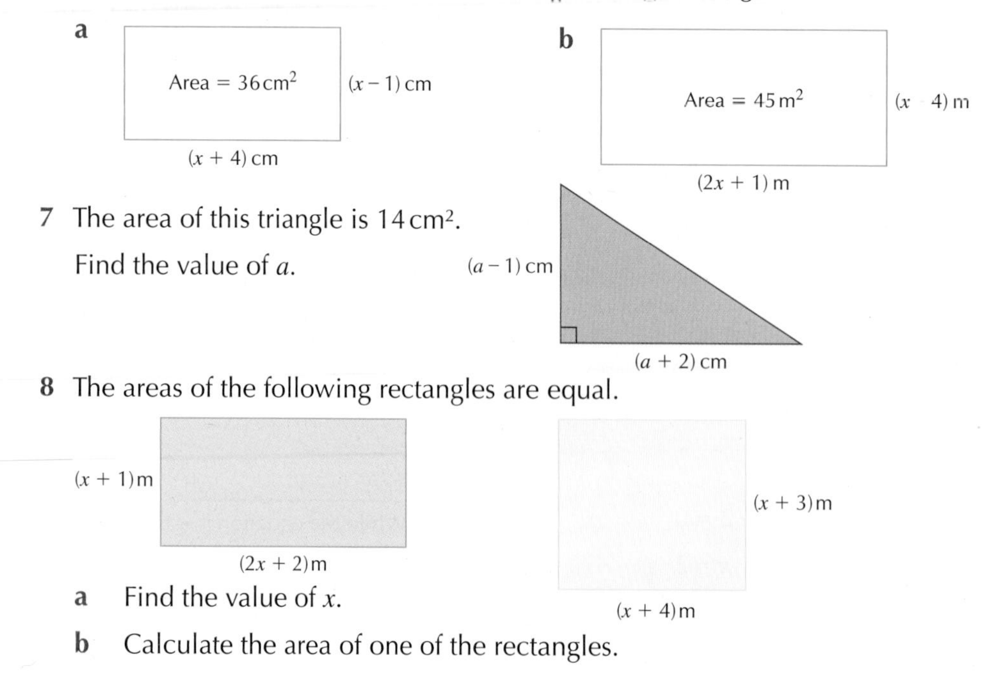

7 The area of this triangle is 14cm^2.

Find the value of a.

8 The areas of the following rectangles are equal.

a Find the value of x.

b Calculate the area of one of the rectangles.

9 The area of the square is $37\,\text{cm}^2$ bigger than the area of the triangle.

a Find the value of x.

b Find the area of each shape.

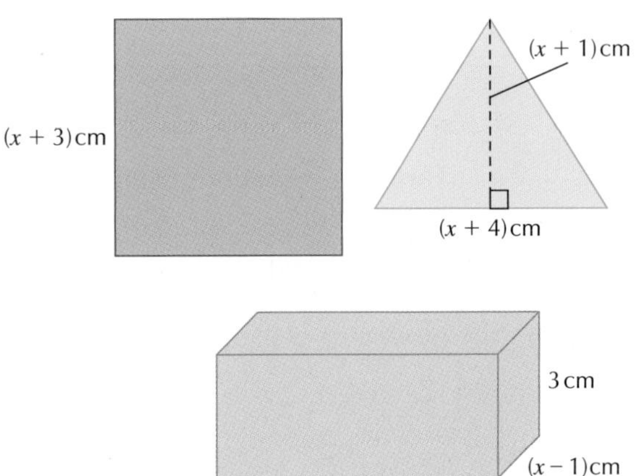

10 The volume of the following cuboid is $42\,\text{cm}^3$.

Find the value of x.

3 cm
(x − 1) cm
(x + 4) cm

11 By using Pythagoras' theorem, find the value of a in each of the following right-angled triangles.

a **b**

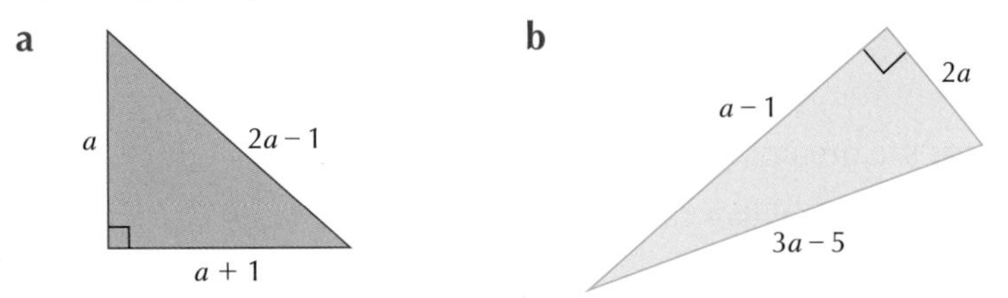

12 A rectangle measuring $12\,\text{cm} \times 9\,\text{cm}$ has four squares, each of length x cm, cut from its corners. The sides are then folded up to create a box of height x cm.

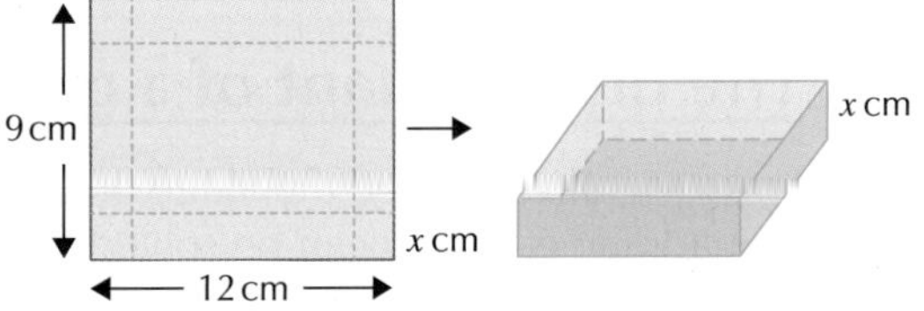

a The length l of the base of the box, in cm, is given by $l = 12 - 2x$. Find a similar expression for the breadth, b.

The area of the base of the box is $54\,\text{cm}^2$.

b Show that $2x^2 - 21x + 27 = 0$.

c Calculate the volume of the box.

13 A plan of Dave's garden is shown. The path is 2 ft wide.

a If the area of the lawn is twice the area of the path, show that $x^2 - 9 = 0$.

b Find the area of the lawn.

c Dave wants to re-pave the path at a cost of £3.50 per square foot. How much does the path cost?

(x + 5) ft
(3x + 1) ft

14 A school currently has a rectangular playing field measuring 150 metres by 100 metres. Owing to an extension to the main building, plans have been made to reduce both the length and the breadth by x metres.

The new playing field is 75% of the area of the original.

a Show that $x^2 - 250x + 3750 = 0$.

b Find the dimensions of the new playing field, correct to the nearest metre.

15 A pond is made from two squares with lengths x metres and y metres respectively.

It has an area of 52 m^2 and a perimeter of 32 m.

a Form two equations using the above information.

b Hence, find the values of x and y.

16 The perimeter of the following rectangle is equal to the perimeter of the square.

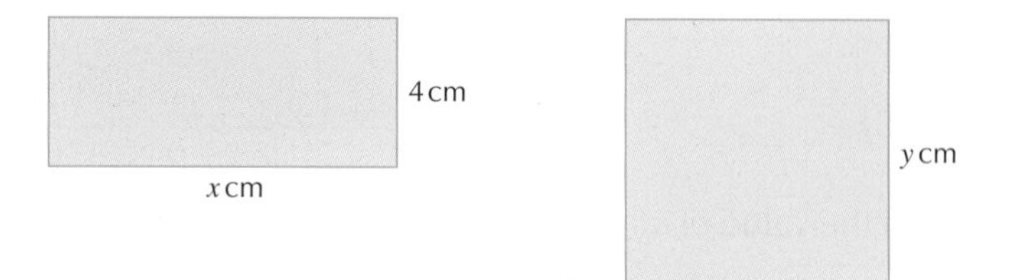

a Show that $x = 2y - 4$.

b Write an expression for the area of the rectangle in terms of y.

c The area of the square is 9 cm^2 bigger than the area of the rectangle.

Show that $y^2 - 8y + 7 = 0$.

d Find the dimensions of the square and the rectangle.

The discriminant of a quadratic equation

You have already used the value of the term $b^2 - 4ac$, when deciding whether or not a quadratic equation can be solved algebraically or when it is necessary to use the formula.

This quantity, $b^2 - 4ac$, is the **discriminant** of the equation. It can be used to tell us about the **nature of the roots** – the kind of roots that a quadratic equation has.

Value of discriminant	Roots	Nature of the roots
$b^2 - 4ac > 0$	$\frac{-b + \sqrt{b^2 - 4ac}}{2a}$ and $\frac{-b - \sqrt{b^2 - 4ac}}{2a}$	2 distinct real roots
$b^2 - 4ac = 0$	$\frac{-b + 0}{2a}$ and $\frac{-b - 0}{2a} = -\frac{b}{2a}$	2 equal real rational roots
$b^2 - 4ac < 0$	There is a problem – you cannot find the square root of a negative number.	no real roots

In the special case where $b^2 - 4ac > 0$ and is a perfect square, the two distinct real roots are also rational.

This is shown graphically using the fact that the roots correspond to where the graph of a quadratic function cuts the x-axis.

Consider the graph $y = ax^2 + bx + c$, $a > 0$:

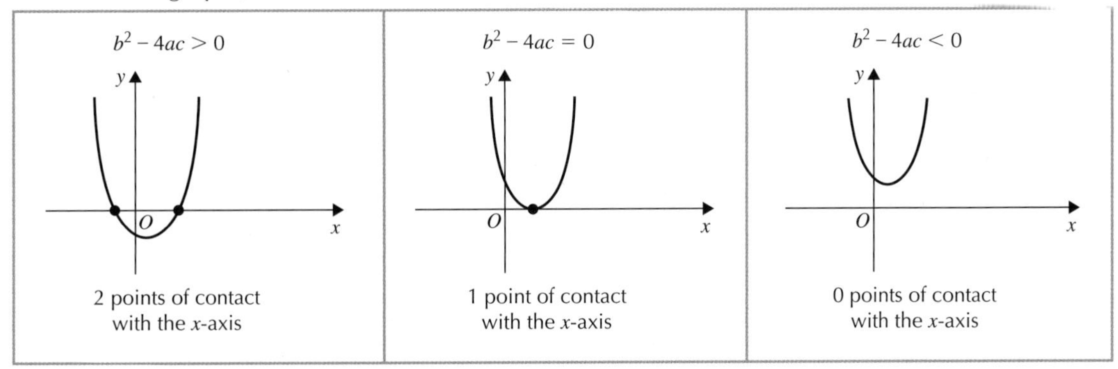

Notice that the same holds for the graph $y = ax^2 + bx + c$, $a < 0$:

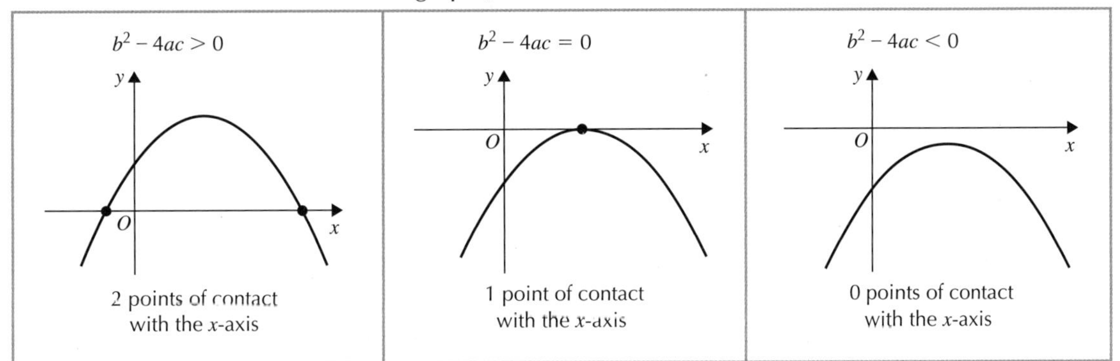

Example 19.14

By calculating the value of the discriminant, determine the nature of the roots of the following equations.

a $3x^2 + 2x + 1 = 0$ **b** $4x^2 - 12x + 9 = 0$ **c** $x^2 + 7x - 2 = 0$

a $a = 3, b = 2, c = 1$ — Write the values of a, b and c.

$b^2 - 4ac = 2^2 - 4 \times 3 \times 1$ — Calculate the value of the discriminant $b^2 - 4ac$.

$= 4 - 12$

$= -8$

$b^2 - 4ac < 0$ No real roots.

b $a = 4, b = -12, c = 9$

$b^2 - 4ac = (-12)^2 - 4 \times 4 \times 9$

$= 144 - 144$

$= 0$

$b^2 - 4ac = 0$ There are two roots which are real, rational and equal.

c $a = 1, b = 7, c = -2$

$b^2 - 4ac = 7^2 - 4 \times 1 \times (-2)$

$= 49 - (-8)$

$= 57$

$b^2 - 4ac > 0$ There are two roots which are real and distinct. They are not rational because 57 is not a perfect square.

Example 19.15

Show that the equation $5x^2 - 2x + 1 = 0$ has no real roots.

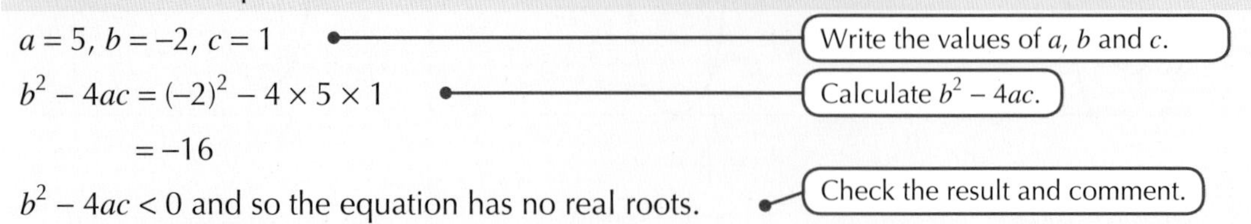
Write the values of a, b and c.
Calculate b² – 4ac.
Check the result and comment.

$a = 5, b = -2, c = 1$

$b^2 - 4ac = (-2)^2 - 4 \times 5 \times 1$

$= -16$

$b^2 - 4ac < 0$ and so the equation has no real roots.

Example 19.16

Find the range of values of p such that $2x^2 + 4x + p = 0$ has two real, distinct roots.

For real distinct roots, we need $b^2 - 4ac > 0$

$a = 2, b = 4, c = p$

$b^2 - 4ac > 0$

$4^2 - 4 \times 2 \times p > 0$

$16 - 8p > 0$

$16 > 8p$

$p < 2$

State the condition needed for real distinct roots.
Write the values of a, b and c.
Form the inequation and solve.

Example 19.17

Find the range of values of r such that $rx^2 - 3x + 2 = 0$ has no real roots.

For no real roots, we need $b^2 - 4ac < 0$

$a = r, b = -3, c = 2$

$b^2 - 4ac < 0$

$(-3)^2 - 4 \times r \times 2 < 0$

$9 - 8r < 0$

$9 < 8r$

$r > \frac{9}{8}$

Example 19.18

For what values of q does the equation $x^2 + qx = 3q$ have equal roots?

For equal roots, we need $b^2 - 4ac = 0$

$x^2 + qx = 3q$

$x^2 + qx - 3q = 0$

$a = 1, b = q, c = -3q$

$b^2 - 4ac = 0$

$q^2 - 4 \times 1 \times (-3q) = 0$

$q^2 + 12q = 0$

$q(q + 12) = 0$

$q = 0$ or $q = -12$

Exercise 19H

1 By calculating the value of the discriminant, determine the nature of the roots of the following quadratic equations.

a $x^2 + 6x + 9 = 0$ **b** $3x^2 - 4x + 2 = 0$ **c** $x^2 + 5x - 1 = 0$

d $5x^2 + 4x + 9 = 0$ **e** $4 - 2x - x^2 = 0$ **f** $9x^2 - 6x + 1 = 0$

★ **2** How many points of contact do graphs of the following quadratic functions have with the x-axis?

a $y = 3x^2 + 4x$ **b** $y = x^2 + \frac{1}{2}x + \frac{1}{4}$ **c** $y = 2x^2 - 9$

d $y = x^2 + 4x + 6$ **e** $y = 3x - x^2$ **f** $y = 5x^2 - 4x - 3$

★ **3** By first rearranging each quadratic equation to its standard form, determine the nature of the roots of the following.

a $x(x + 3) = 2x - 3$ **b** $(x - 2)^2 - 3 = 0$ **c** $4(2x - 4) = x^2$

d $(2x - 1)^2 - 3(x + 1) = 0$ **e** $x(4x + 3) + 6 = 5 - x$ **f** $(x + 3)^2 - 4x - 8 = 0$

4 Show that the equation $x^2 + 3x + 5 = 0$ has no real roots.

5 Show that the equation $x(x - 2) = 7 + 2x$ has real roots.

6 Find the range of values of k for which the equation $2x^2 + 4x + k = 0$ has real roots.

> Remember that real roots can be equal or distinct.

★ **7** Find the range of values of p for which the equation $px^2 - 2x + 1 = 0$ has no real roots.

8 Find the values of k for which $kx^2 - 10x + k = 0$ has equal roots.

9 Show that the graphs $y = x^2 - 3x + 2$ and $y = 2x - 9$ do not intersect.

10 For what values of k do the following equations have equal roots?

a $x^2 + (k - 5)x + 4k = 0$ **b** $x^2 = k(2x - 5)$

GO! Activity

Quadratic sequences

You will have met sequences like 3, 8, 13, 18, 23, … where the difference between consecutive terms is a constant. In this case, the constant is 5. The formula for the nth term (u_n) of this sequence is $u_n = 5n - 2$. A sequence like this is described as a **linear sequence**.

A linear sequence has a constant difference between consecutive terms and the formula for the nth term is of the form:

$$u_n = an + b$$

What if you are presented with a sequence which is not of this form and which you suspect is a quadratic sequence with the nth term of the form:

$$u_n = an^2 + bn + c$$

How would you verify this and how would you calculate the values of a, b and c?

(continued)

Consider the sequence $u_n = an^2 + bn + c$:

n	u_n	First difference	Second difference
1	$a+b+c$		
		$3a+b$	
2	$4a+2b+c$		$2a$
		$5a+b$	
3	$9a+3b+c$		$2a$
		$7a+b$	
4	$16a+4b+c$		$2a$
		$9a+b$	
5	$25a+5b+c$		

You will notice that it is not the first difference that is constant this time, it is the second.

So, to verify that the sequence is of the form $u_n = an^2 + bn + c$, and to calculate the values of a, b and c, we know that:

- a sequence is quadratic with $u_n = an^2 + bn + c$ if the second difference is constant
- a, b and c can be found by comparing with the template above.

This method for finding the formula for the nth term of a sequence can be extended to cubic sequences and beyond. It is called a **finite difference method**.

Consider the sequence 2, 7, 16, 29, 46, ...

Is it a quadratic sequence and, if it is, what is the formula for the nth term?

n	u_n	First difference	Second difference
1	2		
		5	
2	7		4
		9	
3	16		4
		13	
4	29		4
		17	
5	46		

The second difference is constant. This verifies that the formula for the nth term is of the form $u_n = an^2 + bn + c$.

Comparing with the template above:

$2a = 4$, so $a = 2$

$3a + b = 5 \rightarrow 6 + b = 5$, so $b = -1$

$a + b + c = 2 \rightarrow 2 + (-1) + c = 2$, so $c = 1$

The nth term $u_n = 2n^2 - n + 1$

(continued)

1 Find the formula for the nth term of the following quadratic sequences:

a 3, 7, 13, 21, 31, … b 4, 13, 28, 49, 76, … c 8, 31, 68, 119, 184, …

2 Now that you can deal with suspected quadratic sequences consider the next problem.

If three lines intersect then the maximum number of regions, as shown in the diagram, is 7. You will now investigate the relationship between the number of lines, n, and the maximum number of regions, R, which can be formed.

4 3 2 1 5 7 6

a By creating suitable diagrams complete a table showing the maximum number of regions for 1, 2, 3, 4 and 5 lines. Use that table to create a formula for R in terms of n.

b Use the formula to find the maximum number of regions created by 15 lines.

c Calculate the number of intersecting lines required to create a maximum of 46 regions.

d Using your knowledge of the discriminant of a quadratic equation explain why there is no value of n which will create a maximum of 35 regions.

- I can solve a quadratic equation algebraically when given in standard form, $ax^2 + bx + c = 0$. ★ Exercise 19A Q6
- I know that in order to solve a quadratic equation it must be given in standard form and I can deal with quadratic equations where this is not initially the case by first rearranging. ★ Exercise 19B Q2, Q3
- I can solve a quadratic equation where the quadratic expression does not factorise by using the quadratic formula. ★ Exercise 19C Q2
- I know when to use the quadratic formula. ★ Exercise 19D Q1
- I can recognise the relationship between the roots of a quadratic equation and the graph of its corresponding quadratic function. ★ Exercise 19E Q1 ★ Exercise 19H Q2
- I can find the points of intersection of a parabola and a straight line by solving a pair of simultaneous equations. ★ Exercise 19F Q2
- I can solve problems using quadratic equations. ★ Exercise 19G Q2, Q13
- I can work with the discriminant of a quadratic equation in questions involving the nature of its roots. ★ Exercise 19H Q3, Q7

For further assessment opportunities, see the Preparation for Assessment for Unit 2 on pages 288–296.

20 Applying Pythagoras' theorem

In this chapter you will learn how to:

- apply Pythagoras' theorem in complex 2D situations
- decide whether or not a triangle is right-angled using the converse of Pythagoras' theorem
- apply Pythagoras' theorem in 3D situations
- use Pythagoras' theorem to find the distance between two points on a 3D coordinate grid.

You should already know:

- how to find the length of a side of a right-angled triangle using Pythagoras' theorem
- how to use Pythagoras' theorem to find the distance between two points on a 2D coordinate grid.

Applying Pythagoras' theorem in complex 2D situations

Pythagoras' theorem is used to find the length of an unknown side in simple right-angled triangles using the equation $c^2 = a^2 + b^2$ where c is the hypotenuse (longest side). It can also be used in more complex geometrical problems in which right-angled triangles can be constructed. Sometimes you have to use Pythagoras' theorem more than once in a problem.

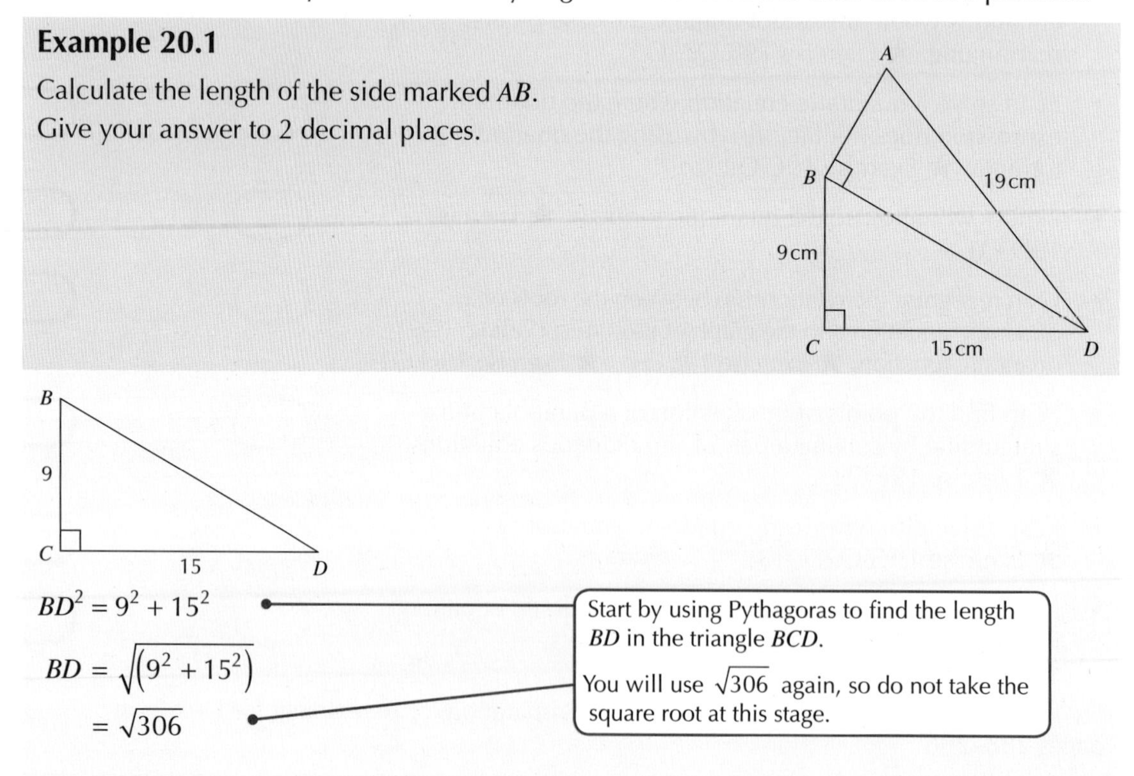

Example 20.1

Calculate the length of the side marked AB.
Give your answer to 2 decimal places.

$BD^2 = 9^2 + 15^2$

$BD = \sqrt{(9^2 + 15^2)}$

$= \sqrt{306}$

Start by using Pythagoras to find the length BD in the triangle BCD.

You will use $\sqrt{306}$ again, so do not take the square root at this stage.

(continued)

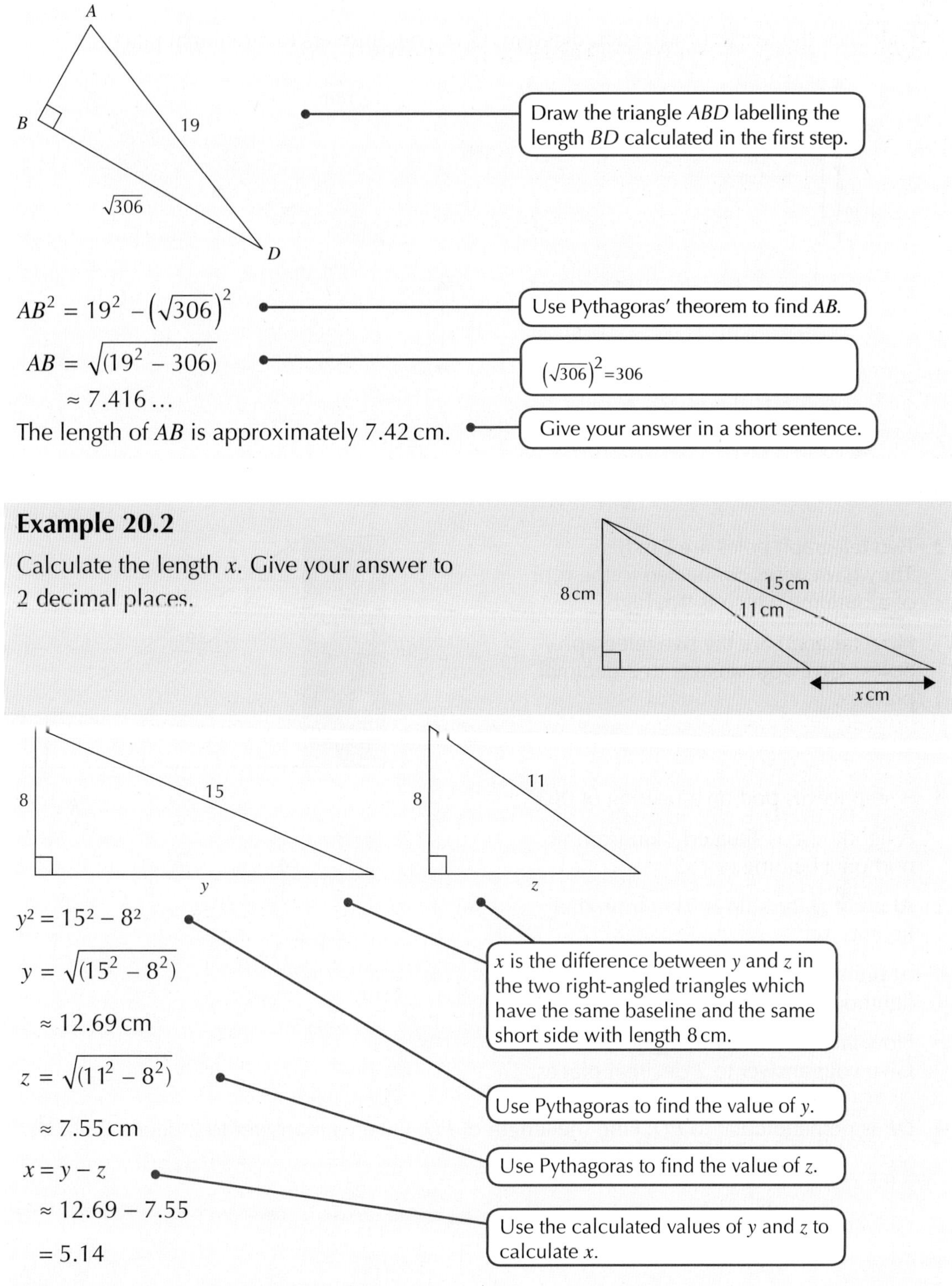

Draw the triangle ABD labelling the length BD calculated in the first step.

$AB^2 = 19^2 - \left(\sqrt{306}\right)^2$ — Use Pythagoras' theorem to find AB.

$AB = \sqrt{(19^2 - 306)}$ — $\left(\sqrt{306}\right)^2 = 306$

$\approx 7.416\ldots$

The length of AB is approximately 7.42 cm. — Give your answer in a short sentence.

Example 20.2

Calculate the length x. Give your answer to 2 decimal places.

$y^2 = 15^2 - 8^2$ — Use Pythagoras to find the value of y.

$y = \sqrt{(15^2 - 8^2)}$

≈ 12.69 cm

$z = \sqrt{(11^2 - 8^2)}$ — Use Pythagoras to find the value of z.

≈ 7.55 cm

$x = y - z$ — Use the calculated values of y and z to calculate x.

$\approx 12.69 - 7.55$

$= 5.14$

x is the difference between y and z in the two right-angled triangles which have the same baseline and the same short side with length 8 cm.

The length of x is approximately 5.14 cm.

Exercise 20A

1 Calculate the length of x in each diagram. Give your answers to 2 decimal places.

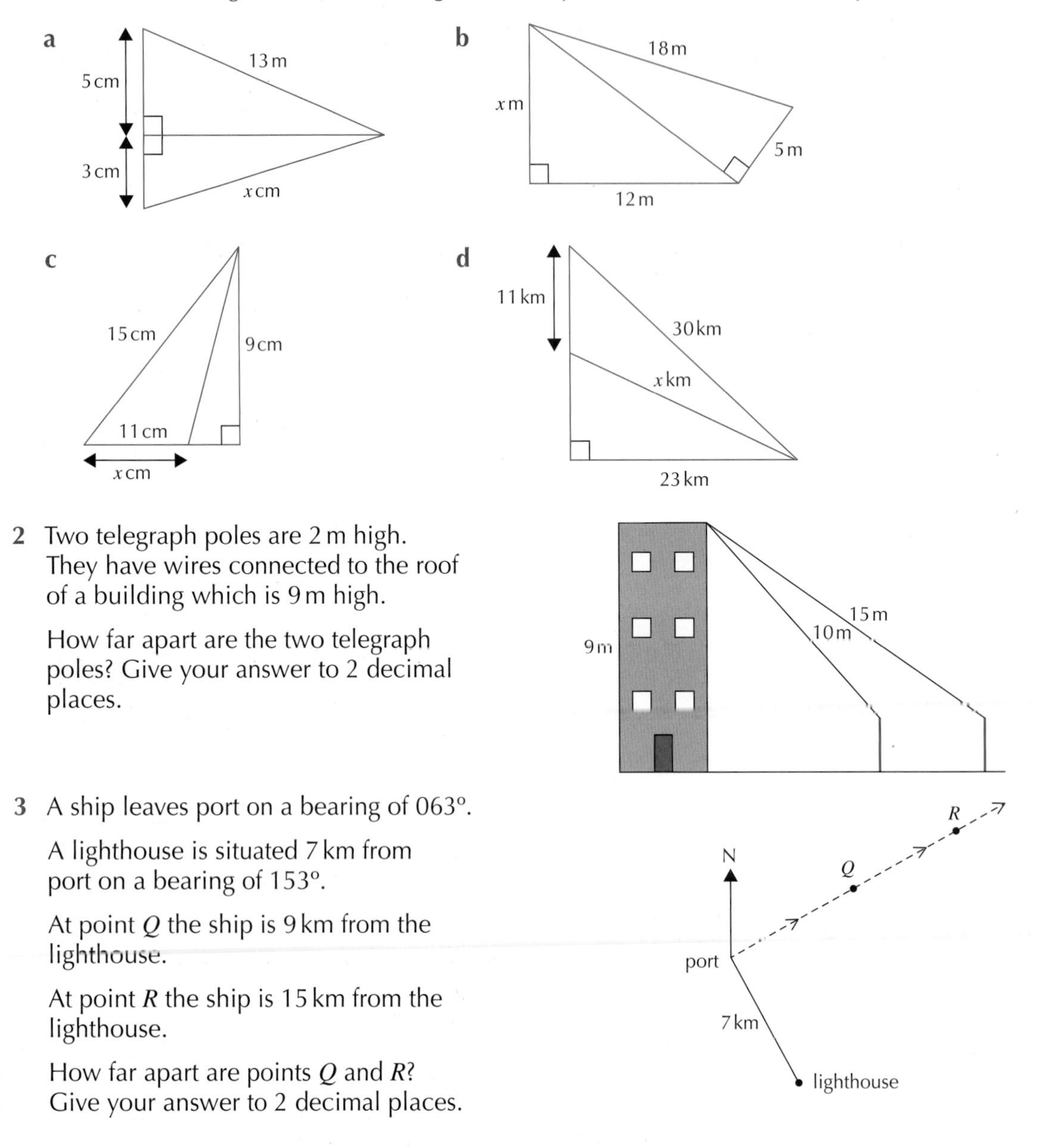

★ **2** Two telegraph poles are 2 m high. They have wires connected to the roof of a building which is 9 m high.

How far apart are the two telegraph poles? Give your answer to 2 decimal places.

3 A ship leaves port on a bearing of 063°.

A lighthouse is situated 7 km from port on a bearing of 153°.

At point Q the ship is 9 km from the lighthouse.

At point R the ship is 15 km from the lighthouse.

How far apart are points Q and R? Give your answer to 2 decimal places.

4 OP is perpendicular to PQ. Find the length of PQ. Give your answer to 2 decimal places.

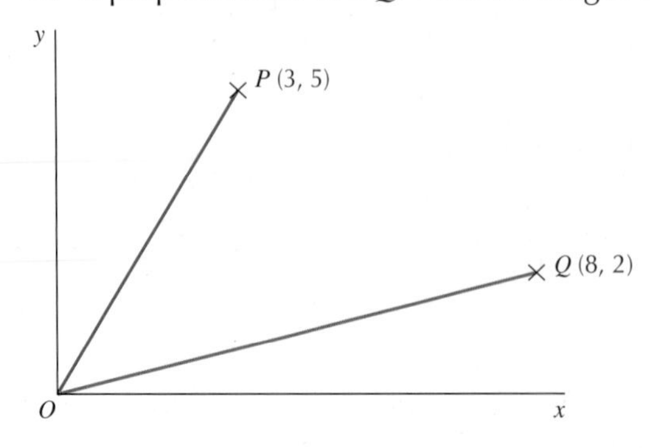

The converse of Pythagoras' theorem

Pythagoras' theorem is used to find the length of an unknown side of a right-angled triangle if you know the length of the other two sides. You can use the **converse** of Pythagoras' theorem as a tool for checking to see if a triangle is right-angled or not.

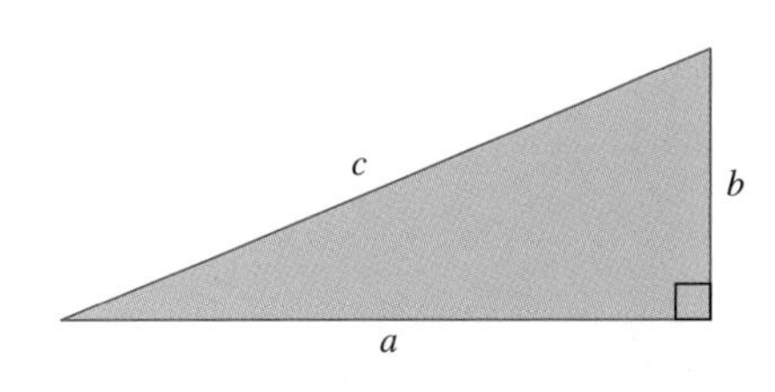

If c is the longest side of a triangle with a and b the other two sides, the converse of Pythagoras states:

if $c^2 = a^2 + b^2$ then the triangle is right-angled with c as its hypotenuse.

Example 20.3

Determine if this triangle is right-angled or not.

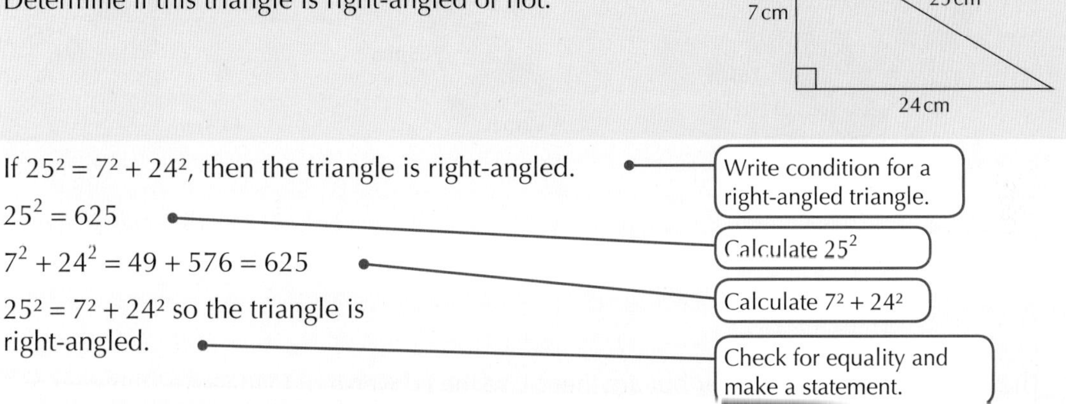

7 cm
25 cm
24 cm
Write condition for a right-angled triangle.
Calculate 25^2
Calculate $7^2 + 24^2$
Check for equality and make a statement.

If $25^2 = 7^2 + 24^2$, then the triangle is right-angled.

$25^2 = 625$

$7^2 + 24^2 = 49 + 576 = 625$

$25^2 = 7^2 + 24^2$ so the triangle is right-angled.

Example 20.4

A joiner is checking a window frame to ensure it has a 90° angle. The frame appears to be rectangular with a height of 1267 mm, a breadth of 349 mm and a diagonal of 1338 mm. Can the joiner allow the glazier to fit the glass?

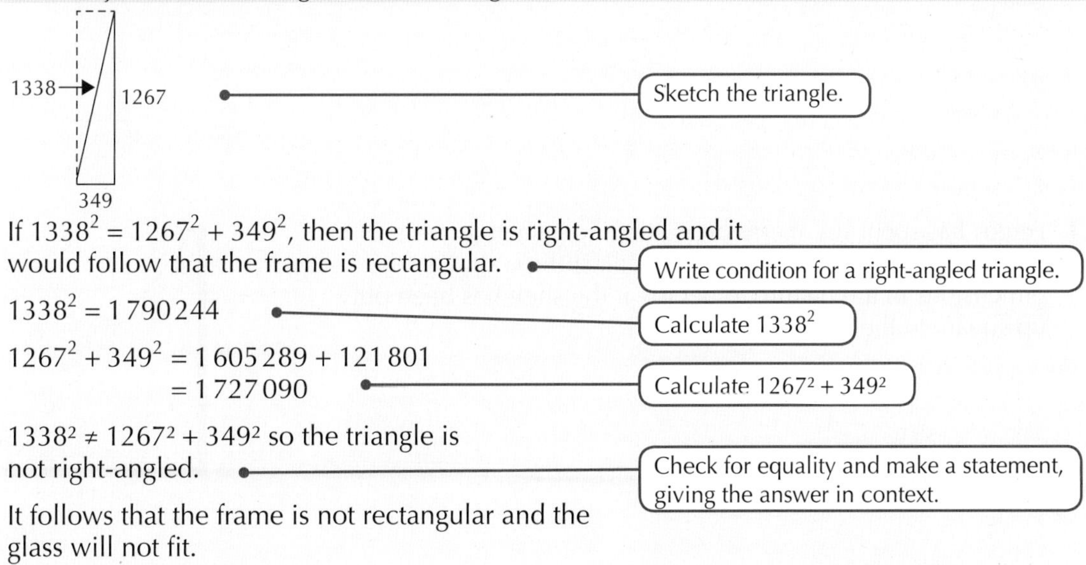

If $1338^2 = 1267^2 + 349^2$, then the triangle is right-angled and it would follow that the frame is rectangular.

$1338^2 = 1\,790\,244$

$1267^2 + 349^2 = 1\,605\,289 + 121\,801$
$= 1\,727\,090$

$1338^2 \neq 1267^2 + 349^2$ so the triangle is not right-angled.

It follows that the frame is not rectangular and the glass will not fit.

So the joiner will need to fix the frame and check it again before allowing the glazier to fit the glass.

Exercise 20B

1 Use the converse of Pythagoras' theorem to determine if these triangles are right-angled or not.

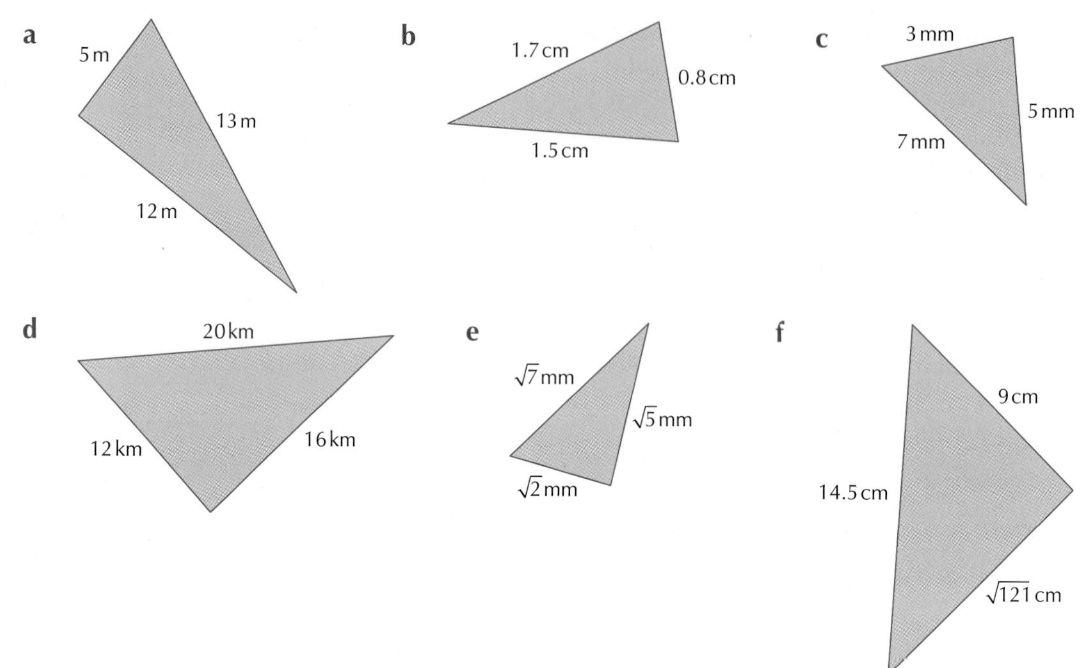

2 These shapes look rectangular but are they? Use the converse of Pythagoras' theorem to check.

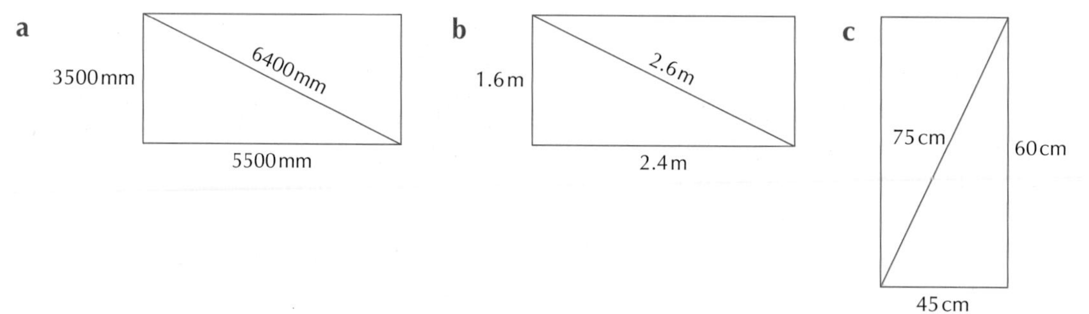

★ 3 Fergus has spent his afternoon putting up a shelf, but his brother James says it looks wrong against the wall. Use the dimensions in the picture to decide if the shelf has been put up square to the wall.

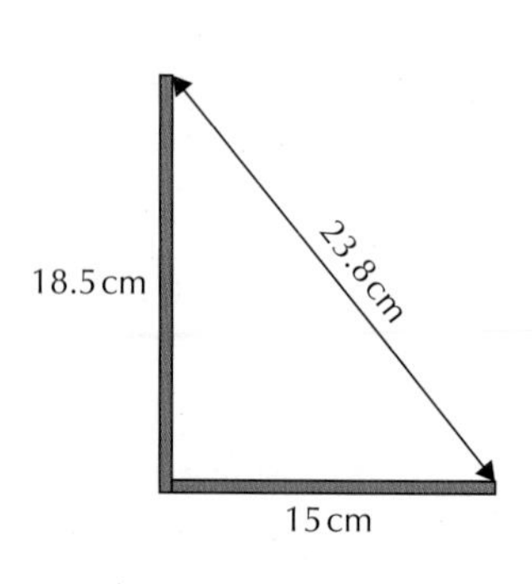

4 Two cafés share some pavement space in front for tables and chairs for customers. A line is painted on the ground to divide the space into two parts.

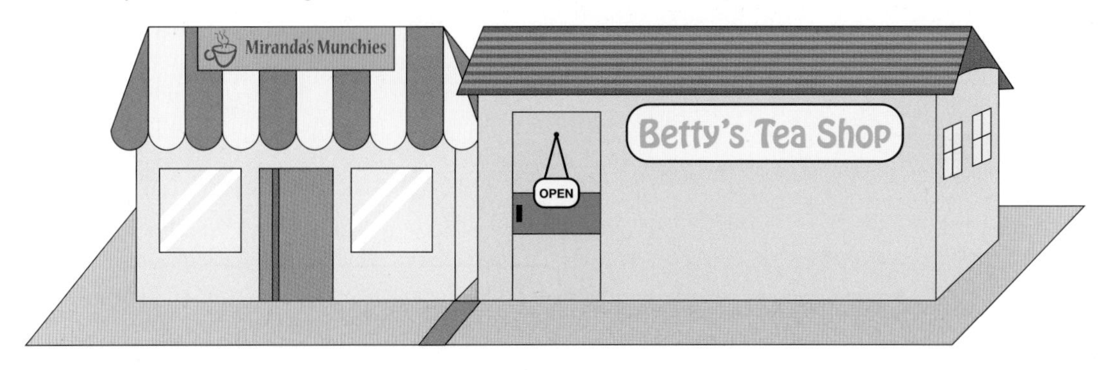

The diagram on the right shows the space in front of Miranda's Munchies.

'This is not fair!', says Miranda. 'The line is painted at an angle.' Betty does not agree. Using the dimensions shown in the diagram do you agree with Miranda or Betty?

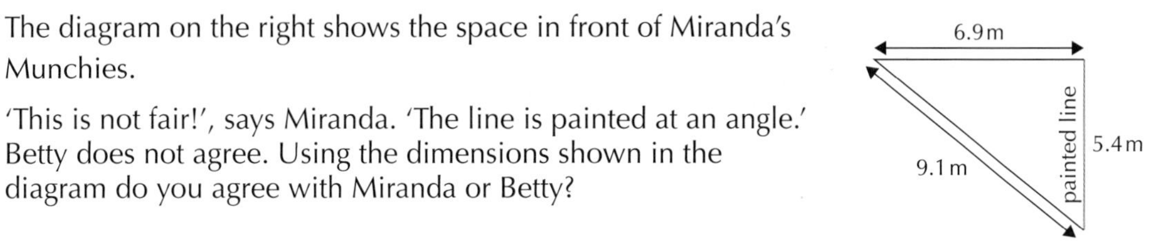

5 A company manufactures brackets for fitting in aeroplanes. It is important that the angles formed are right angles. The Quality Assurance department checks that the product meets the standard. Decide if the Quality Assurance department will send the bracket shown on the right to the aeroplane fitters.

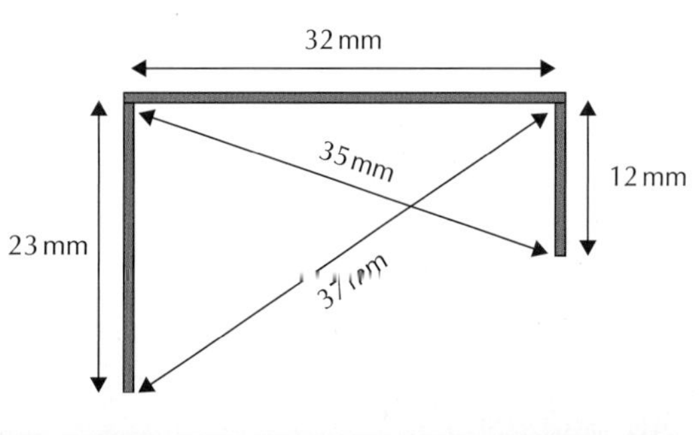

Activity

1. What is a Pythagorean triple?
   How many Pythagorean triples can you find where all three values are less than 40?
2. Rope stretchers in ancient Egypt were surveyors who used ropes knotted into Pythagorean triples. They used these tools to re-establish boundaries after floods. Use your list of Pythagorean triples to create your own rope set squares. Use your rope set squares to test the shapes of furniture for right angles. You may need some friends to help you keep the rope tight and accurate.

Applying Pythagoras' theorem in 3D problems

When working with 3D shapes there are two types of diagonal.

In this cuboid, there are:

- **face diagonals**, such as EG, which lie on a face of the shape
- **space diagonals**, such as AG, which join opposite vertices and are inside the shape.

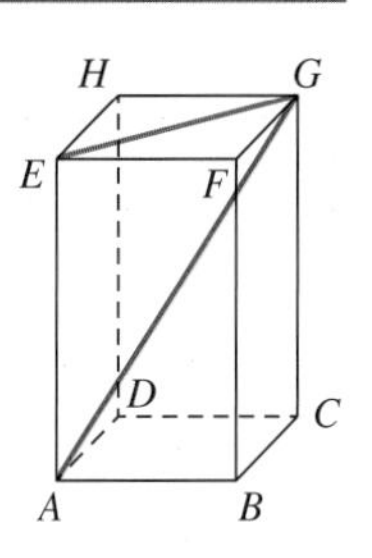

Example 20.5

Calculate the length of the space diagonal AG in the cuboid. Give your answer to 1 decimal place.

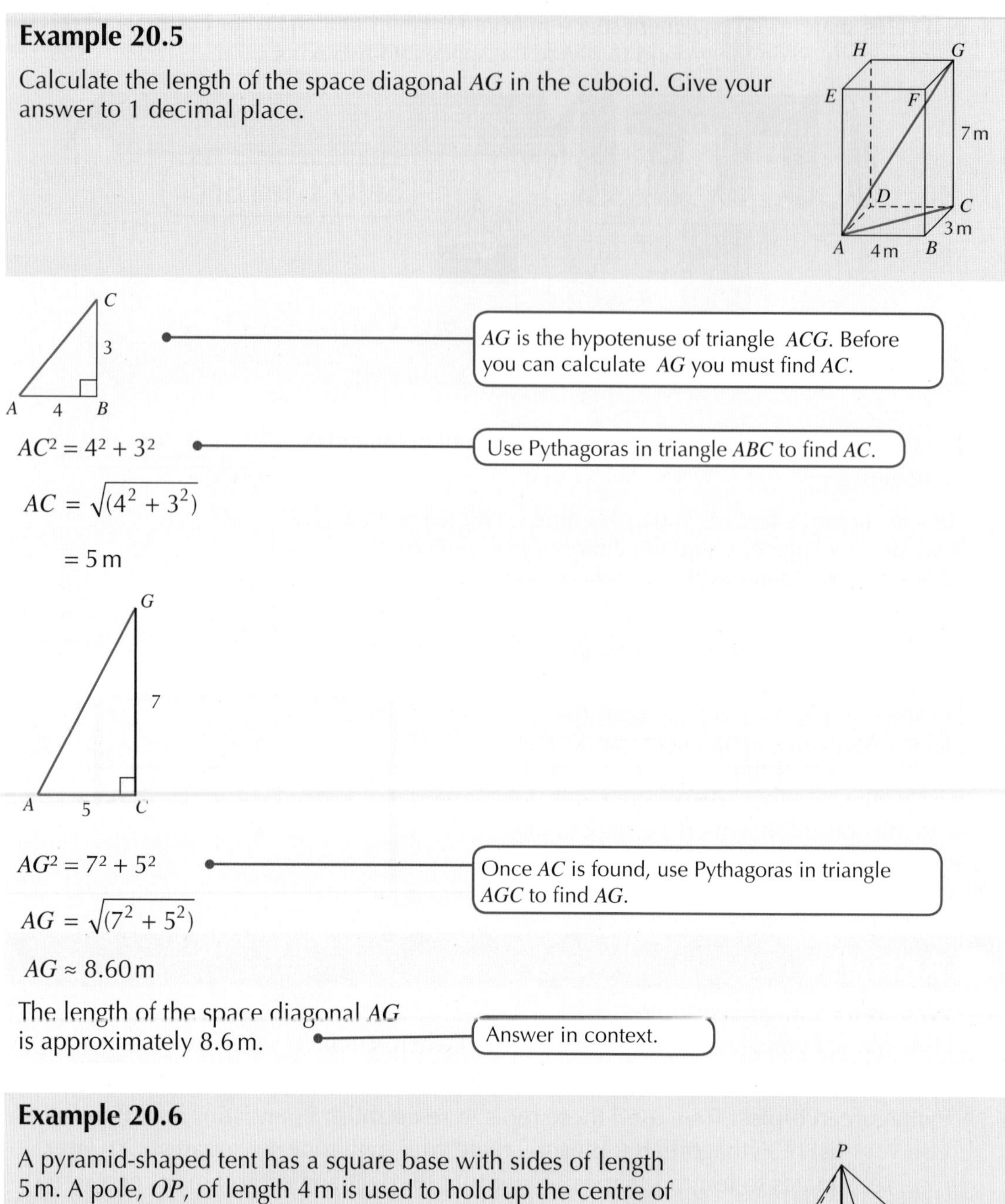

AG is the hypotenuse of triangle ACG. Before you can calculate AG you must find AC.

$AC^2 = 4^2 + 3^2$ — Use Pythagoras in triangle ABC to find AC.

$AC = \sqrt{(4^2 + 3^2)}$

$= 5\,\text{m}$

$AG^2 = 7^2 + 5^2$ — Once AC is found, use Pythagoras in triangle AGC to find AG.

$AG = \sqrt{(7^2 + 5^2)}$

$AG \approx 8.60\,\text{m}$

The length of the space diagonal AG is approximately 8.6 m. — Answer in context.

Example 20.6

A pyramid-shaped tent has a square base with sides of length 5 m. A pole, OP, of length 4 m is used to hold up the centre of the tent. Four poles are also used to connect the corners of the base to P. How long are these poles?

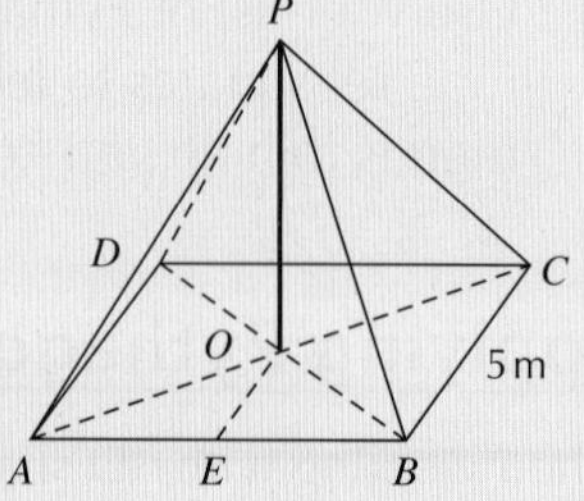

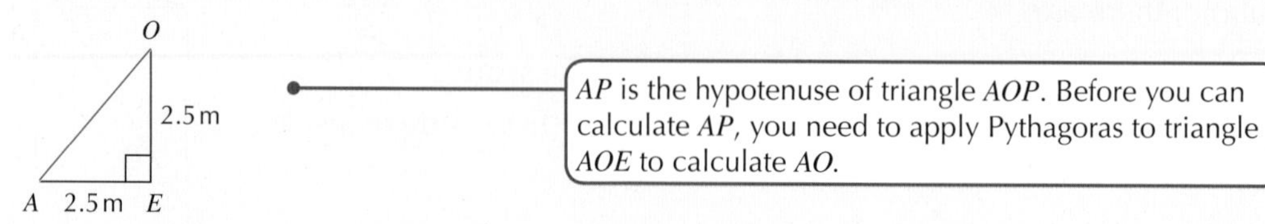

AP is the hypotenuse of triangle AOP. Before you can calculate AP, you need to apply Pythagoras to triangle AOE to calculate AO.

(continued)

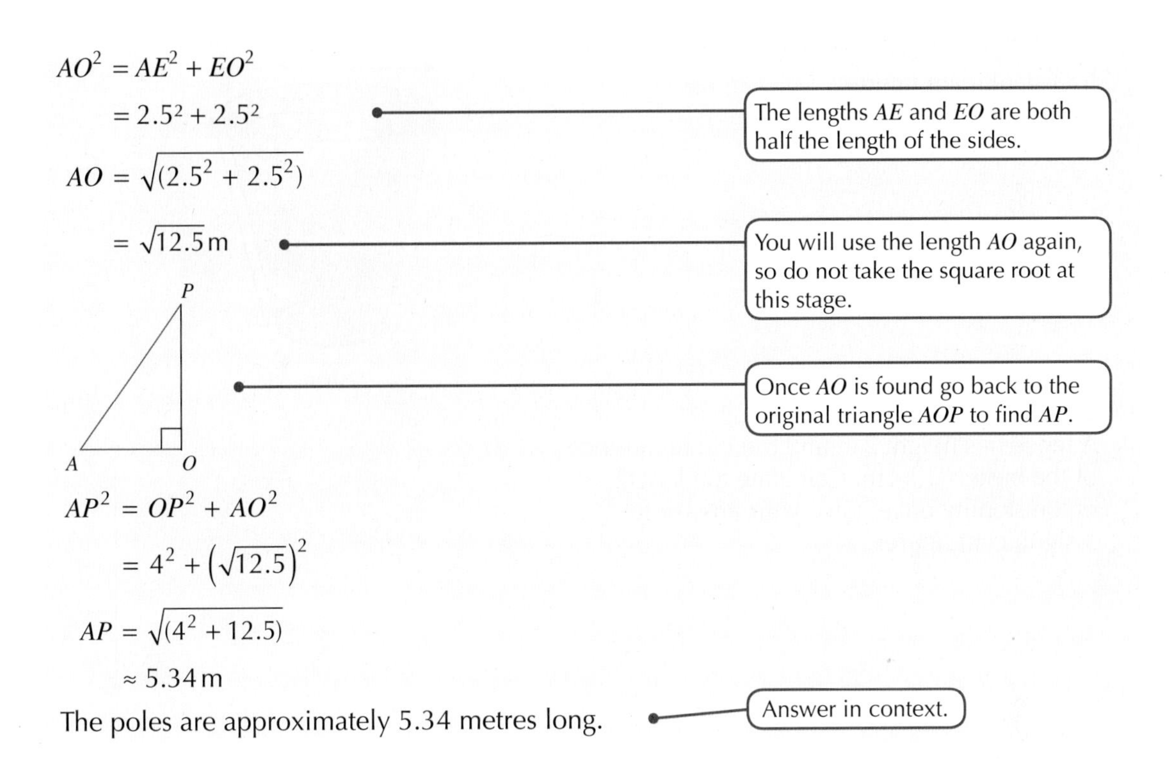

$AO^2 = AE^2 + EO^2$

$= 2.5^2 + 2.5^2$

$AO = \sqrt{(2.5^2 + 2.5^2)}$

$= \sqrt{12.5}\,\text{m}$

$AP^2 = OP^2 + AO^2$

$= 4^2 + \left(\sqrt{12.5}\right)^2$

$AP = \sqrt{(4^2 + 12.5)}$

$\approx 5.34\,\text{m}$

The poles are approximately 5.34 metres long.

The lengths AE and EO are both half the length of the sides.

You will use the length AO again, so do not take the square root at this stage.

Once AO is found go back to the original triangle AOP to find AP.

Answer in context.

Exercise 20C

1 For each of the following cuboids, calculate the space diagonal marked. Give your answers to 2 decimal places.

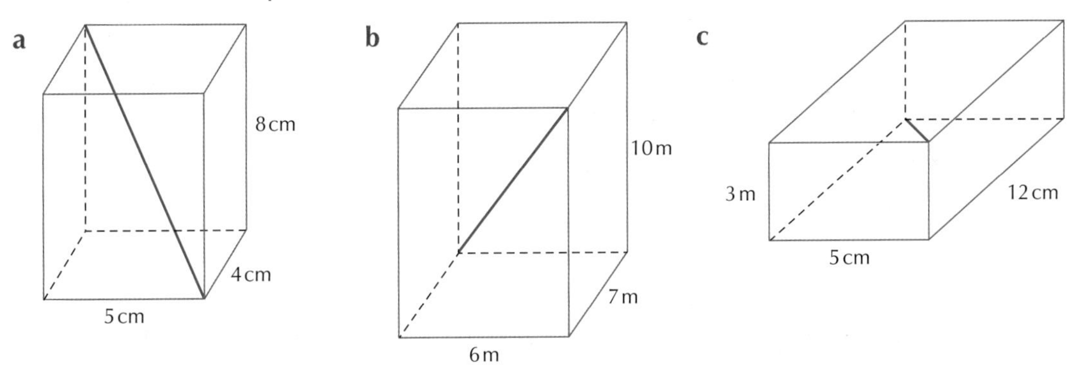

2 Calculate the length of the sloping edge on these pyramids. Give your answers to 2 decimal places.

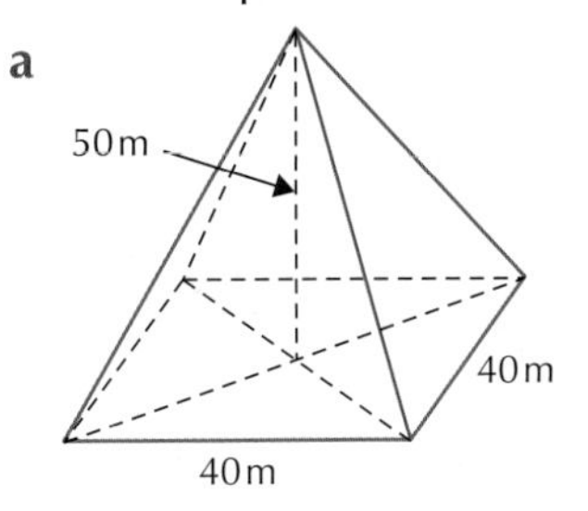

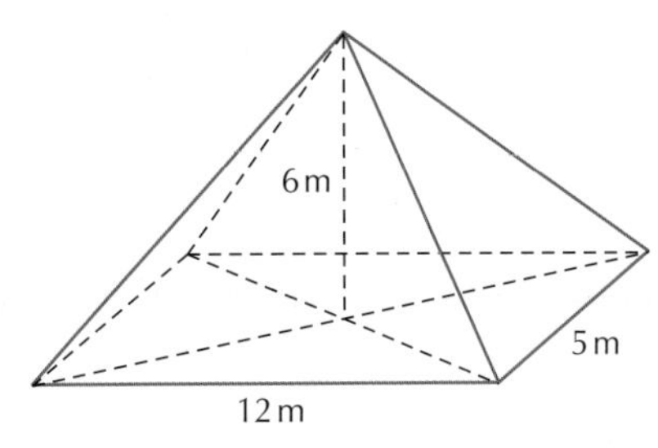

★ **3** How high is the bin? Give your answer to 3 significant figures.

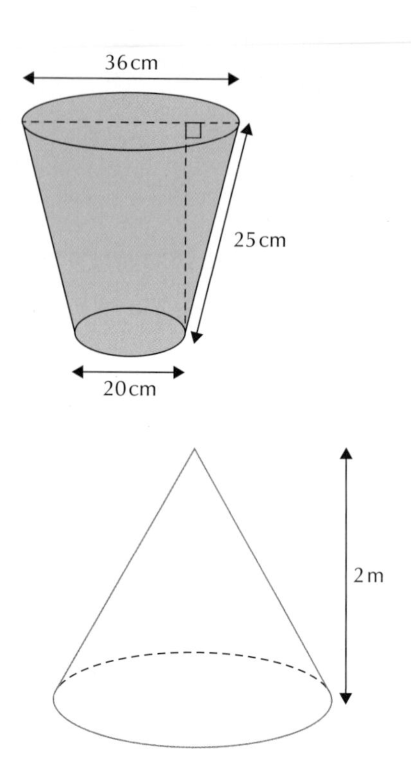

4 A tepee has height 2 m and the circumference of the base is 7.54 m. Calculate the length of the sloping edge. Give your answer to 3 significant figures.

5 Jason and Mark are arguing over whether they have enough room in their trailer to fit a metal pipe they need for a job. The pipe is 2.7 m long.

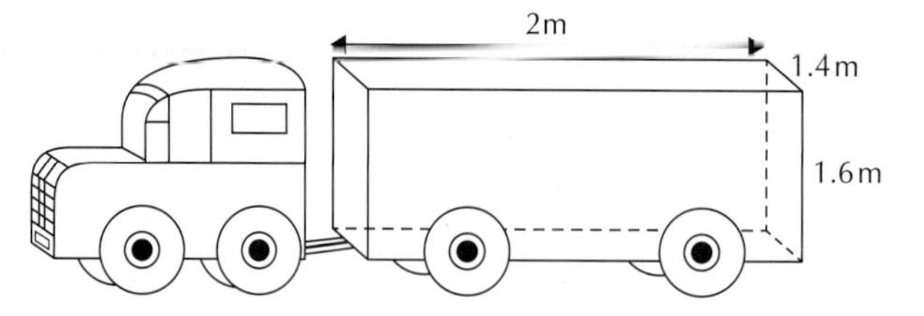

Jason claims that it will fit as the space diagonal is longer than the pipe.

Mark claims that it won't fit as the space diagonal is shorter than the pipe.

Who is correct? You must give a reason for your answer.

Distance between two points on a 3D coordinate grid

Pythagoras' theorem can be used to calculate the distance between two points on 3D coordinate diagrams, using the coordinate values of the two points.

Coordinates in 3D space

When working in two dimensions we use the x- and y-axes. When working in three dimensions we need to add a third axis, the z-axis, perpendicular to the xy plane as shown opposite.

The coordinates of a point P, correspond to the distances which must be travelled in the x, y and z directions, respectively, from the origin O to P.

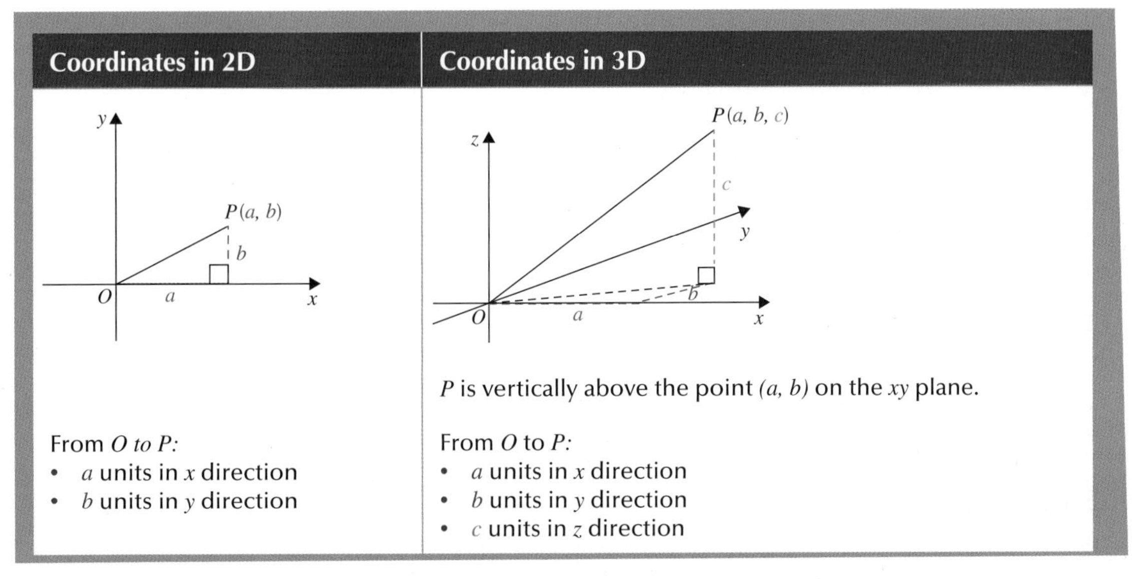

Coordinates in 2D
Coordinates in 3D
y
P(a, b)
b
O
a
x
P(a, b, c)
z
c
y
b
O
a
x
P is vertically above the point (a, b) on the xy plane.
From O to P:
a units in x direction
b units in y direction
From O to P:
a units in x direction
b units in y direction
c units in z direction

See Chapter 29 for more on 3D coordinates.

Distance between two points on the 2D coordinate plane

To develop a formula for the distance d between two points on the xy plane, consider two points, $A\ (x_1, y_1)$ and $B\ (x_2, y_2)$.

The distance d is found using Pythagoras' theorem:

$d^2 = (x_2 - x_1)^2 + (y_2 - y_1)^2$

So:

$$d = \sqrt{(x_2 - x_1)^2 + (y_2 - y_1)^2}$$

Example 20.7

Calculate the distance d between these pairs of points. Give your answer to 2 decimal places.

a A (1, 3) and B (8, 7) **b** C (−2, 5) and D (7, −3)

a $d = \sqrt{(x_2 - x_1)^2 + (y_2 - y_1)^2}$

$= \sqrt{(8 - 1)^2 + (7 - 3)^2}$ — Use the values of x and y for points A and B.

$= \sqrt{(7^2 + 4^2)}$

$\approx$ 8.06 units

b $d = \sqrt{(x_2 - x_1)^2 + (y_2 - y_1)^2}$

$= \sqrt{(7 - (-2))^2 + (-3 - 5)^2}$ — Use the values of x and y for points C and D.

$= \sqrt{(9^2 + (-8)^2)}$

$\approx$ 12.04 units

Distance between two points in 3D coordinate space

The formula for the distance between two points A (x_1, y_1, z_1) and B (x_2, y_2, z_2) in 3D coordinate space is a simple extension of the formula for the distance in 2D space.

Consider two points A (x_1, y_1, z_1) and B (x_2, y_2, z_2).

The formula for the distance between two points in 3D coordinate space is:

$$d = \sqrt{(x_2 - x_1)^2 + (y_2 - y_1)^2 + (z_2 - z_1)^2}$$

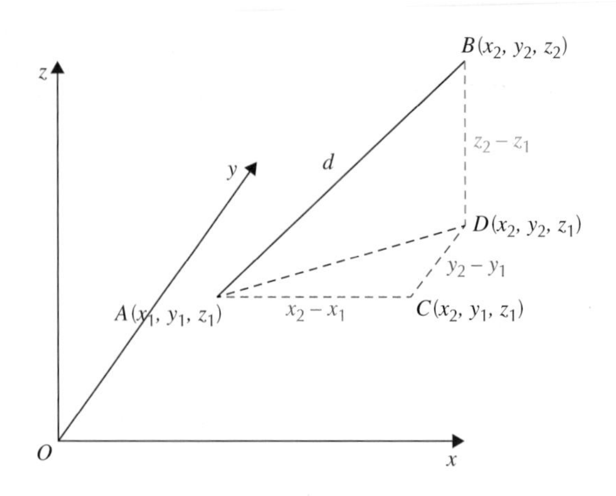

Example 20.8

Points A and B are (–2, 3, 4) and (3, 5, 7), respectively. Find the distance between A and B. Give your answer to 2 decimal places.

$$d = \sqrt{(x_2 - x_1)^2 + (y_2 - y_1)^2 + (z_2 - z_1)^2}$$

$$= \sqrt{(3 - (-2))^2 + (5 - 3)^2 + (7 - 4)^2}$$

$$= \sqrt{(5^2 + 2^2 + 3^2)}$$

$$= \sqrt{(25 + 4 + 9)}$$

$$= \sqrt{38}$$

≈ 6.16 units

Exercise 20D

1 Calculate the distance between the following pairs of points. Give your answers to 2 decimal places.

a A (3, 5, 2) and B (2, 4, 6)

b C (0, 2, 7) and D (3, 4, 3)

c E (–2, 3, 4) and F (2, 0, 5)

d G (5, –3, –4) and H (0, 6, –2)

2 Find the length of the line PQ when P is the point (0, 2, 0) and Q is the point (3, –3, 1). Give your answer to 2 decimal places.

3 A cuboid is drawn on a 3D coordinate grid.

The points A (0, 0, 0), B (4, 0, 5) and C (0, 3, 0) are shown on the diagram.

a Find the coordinates of D.

b Find the distance between A and D.

★ **4** A square-based pyramid is shown in the diagram.

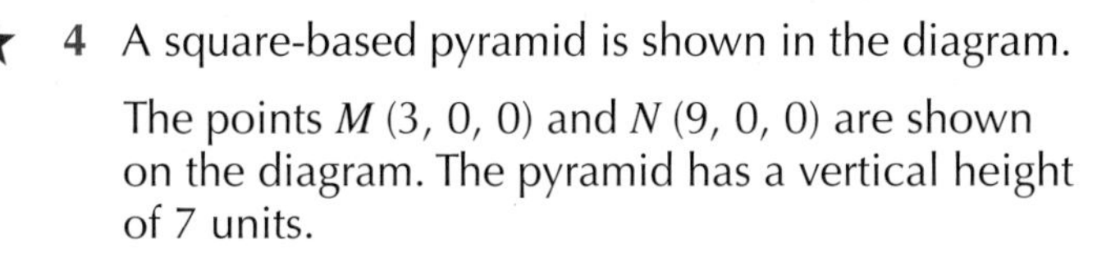

The points M (3, 0, 0) and N (9, 0, 0) are shown on the diagram. The pyramid has a vertical height of 7 units.

a Find the coordinates of P.

b Find the distance between M and P.

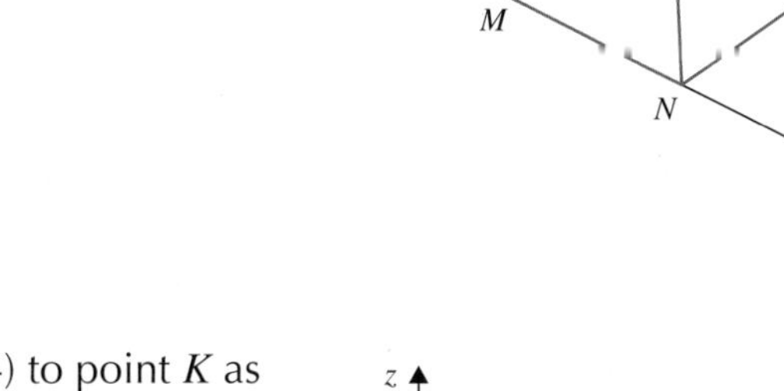

5 A line is drawn from point A (3, 2, 4) to point K as shown in the diagram. K is the midpoint of G (5, 7, 8) and H (7, 3, 4).

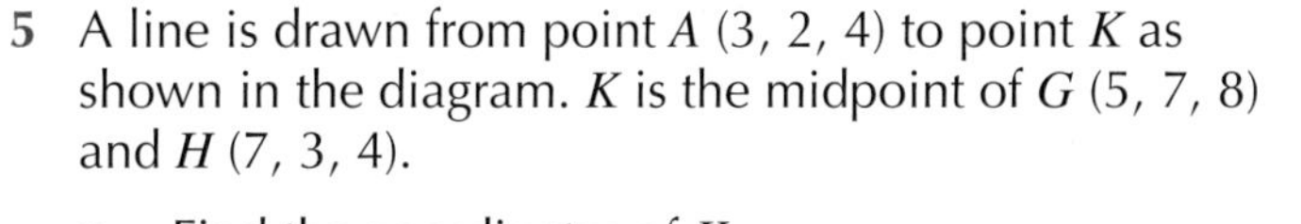

a Find the coordinates of K.

b Find the length of the line AK.

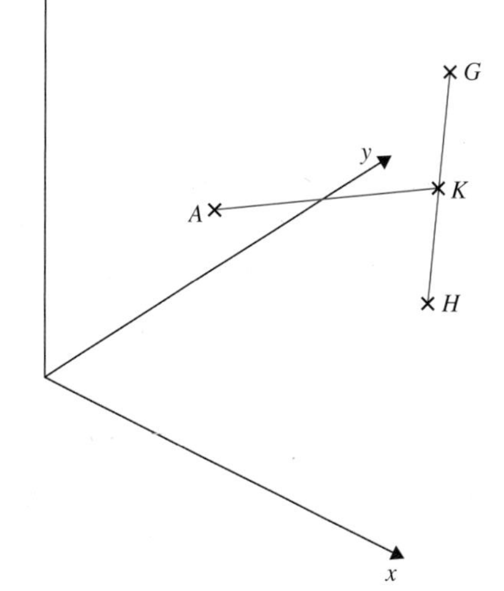

- I can solve problems which involve using Pythagoras' theorem more than once. ★ Exercise 20A Q2
- By applying my knowledge of the converse of Pythagoras' theorem, I can check whether or not a triangle is right-angled. ★ Exercise 20B Q3
- I can solve problems in 3D by creating right-angled triangles and using Pythagoras' theorem. ★ Exercise 20C Q3
- I have extended my knowledge of finding the distance between two points in 2D to finding the distance between two points in 3D. ★ Exercise 20D Q4

For further assessment opportunities, see the Preparation for Assessment for Unit 2 on pages 288–291.

21 Applying the properties of shapes to determine an angle involving at least two steps

In this chapter you will learn how to:

- find angles using combinations of the angle properties of triangles and common quadrilaterals
- find angles using combinations of angle properties of circles
- solve problems using symmetry properties of circles
- recognise polygons and find the sums of interior and exterior angles
- find the interior and exterior angles of regular polygons.

You should already know:

- basic angle properties (alternate, corresponding and vertically opposite angles)
- angle properties of triangles and common quadrilaterals
- how to find the side of a right-angled triangle using Pythagoras' theorem
- how to find a side or angle in a right-angled triangle using trigonometry
- angle properties of circles.

Finding angles using the angle properties of triangles and quadrilaterals

Angle properties of lines and 2D shapes are used to find unknown angles in geometric problems.

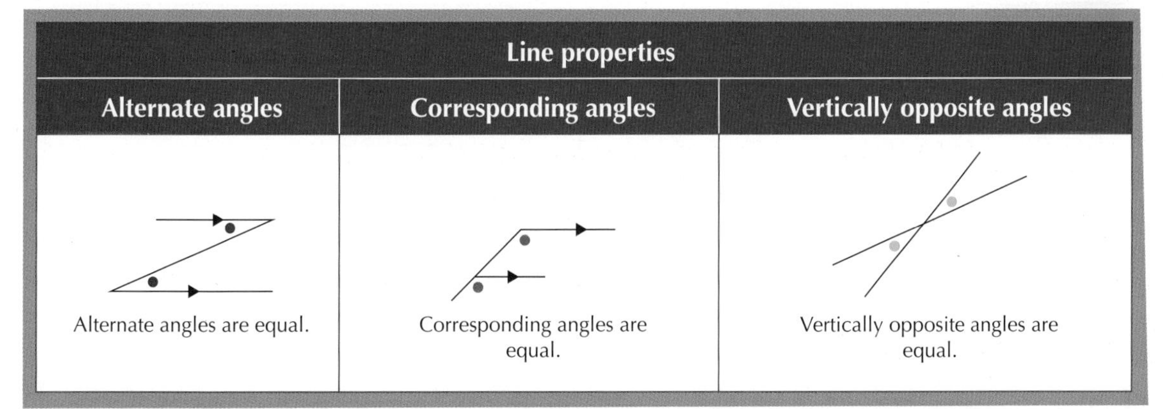

Line properties		
Alternate angles	Corresponding angles	Vertically opposite angles
Alternate angles are equal.	Corresponding angles are equal.	Vertically opposite angles are equal.

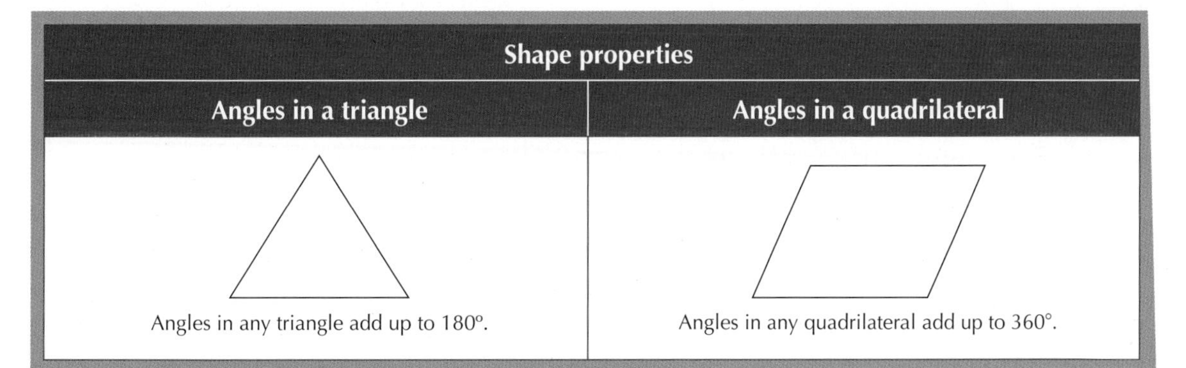

Shape properties	
Angles in a triangle	Angles in a quadrilateral
Angles in any triangle add up to 180°.	Angles in any quadrilateral add up to 360°.

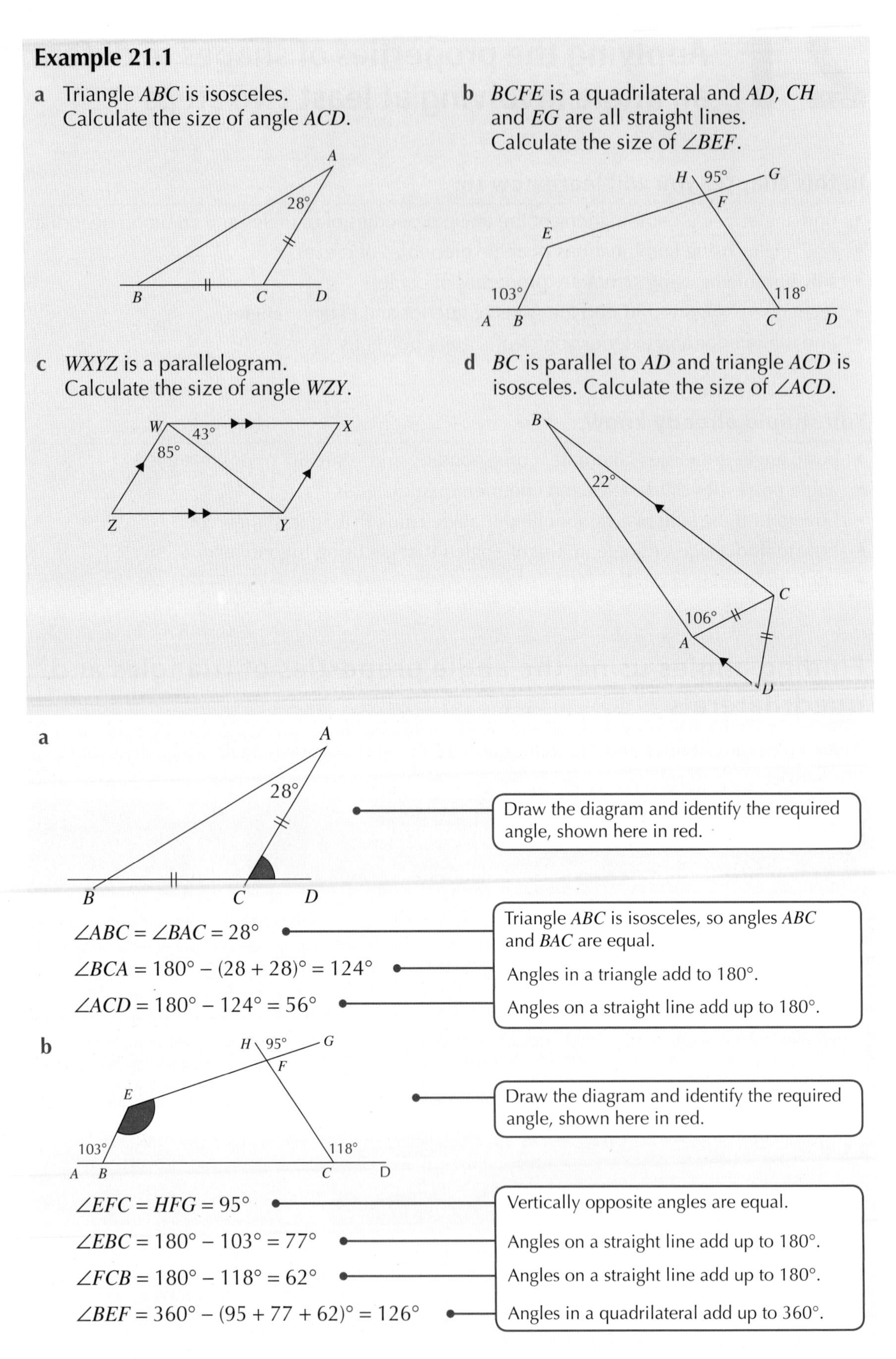

Example 21.1

a Triangle *ABC* is isosceles.
Calculate the size of angle *ACD*.

b *BCFE* is a quadrilateral and *AD*, *CH* and *EG* are all straight lines.
Calculate the size of $\angle BEF$.

c *WXYZ* is a parallelogram.
Calculate the size of angle *WZY*.

d *BC* is parallel to *AD* and triangle *ACD* is isosceles. Calculate the size of $\angle ACD$.

a

Draw the diagram and identify the required angle, shown here in red.

$\angle ABC = \angle BAC = 28°$ — Triangle *ABC* is isosceles, so angles *ABC* and *BAC* are equal.

$\angle BCA = 180° - (28 + 28)° = 124°$ — Angles in a triangle add to 180°.

$\angle ACD = 180° - 124° = 56°$ — Angles on a straight line add up to 180°.

b

Draw the diagram and identify the required angle, shown here in red.

$\angle EFC = HFG = 95°$ — Vertically opposite angles are equal.

$\angle EBC = 180° - 103° = 77°$ — Angles on a straight line add up to 180°.

$\angle FCB = 180° - 118° = 62°$ — Angles on a straight line add up to 180°.

$\angle BEF = 360° - (95 + 77 + 62)° = 126°$ — Angles in a quadrilateral add up to 360°.

(continued)

c

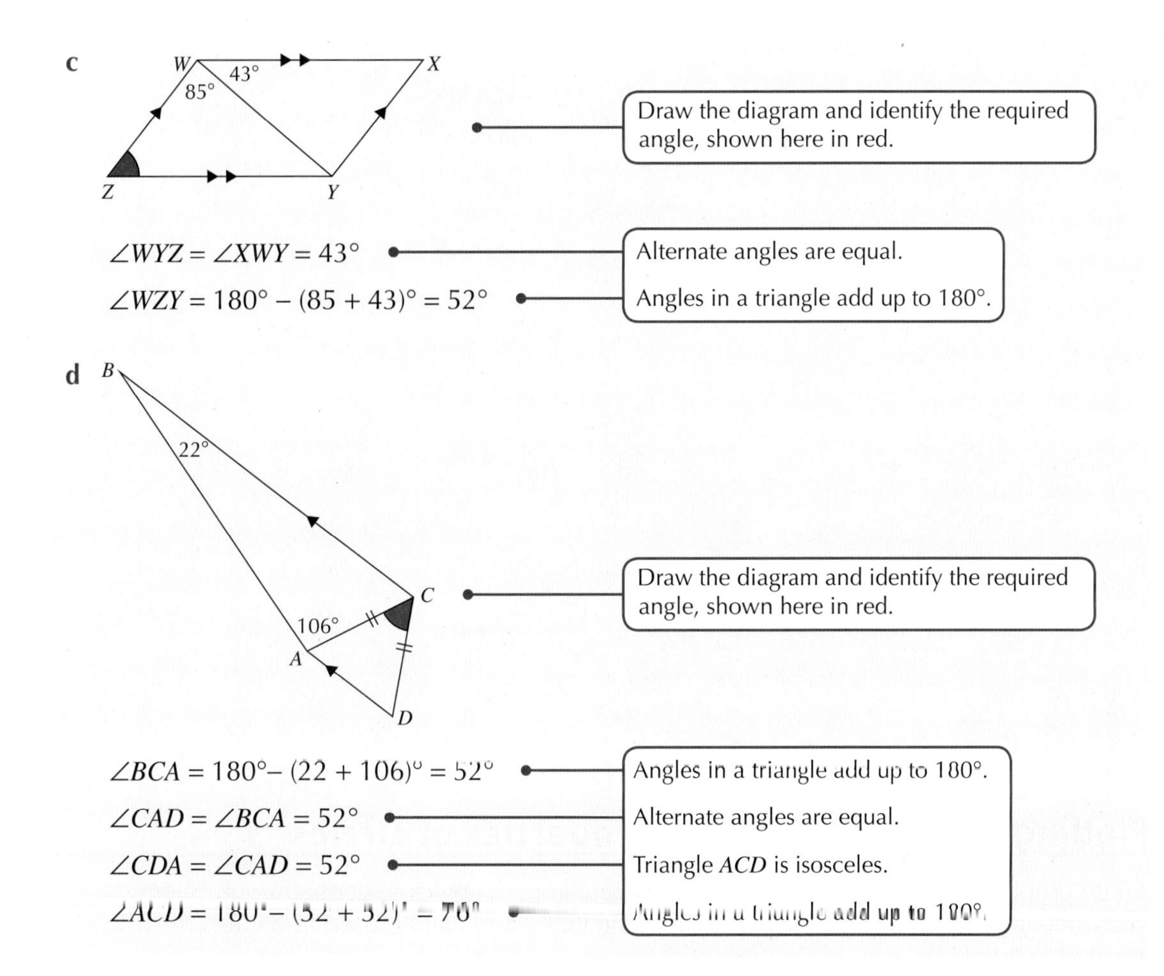

$\angle WYZ = \angle XWY = 43°$ — Alternate angles are equal.

$\angle WZY = 180° - (85 + 43)° = 52°$ — Angles in a triangle add up to 180°.

d

$\angle BCA = 180° - (22 + 106)° = 52°$ — Angles in a triangle add up to 180°.

$\angle CAD = \angle BCA = 52°$ — Alternate angles are equal.

$\angle CDA = \angle CAD = 52°$ — Triangle ACD is isosceles.

$\angle ACD = 180° - (52 + 52)° = 76°$ — Angles in a triangle add up to 180°.

Exercise 21A

★ **1** Find the size of the marked angles in the diagrams.

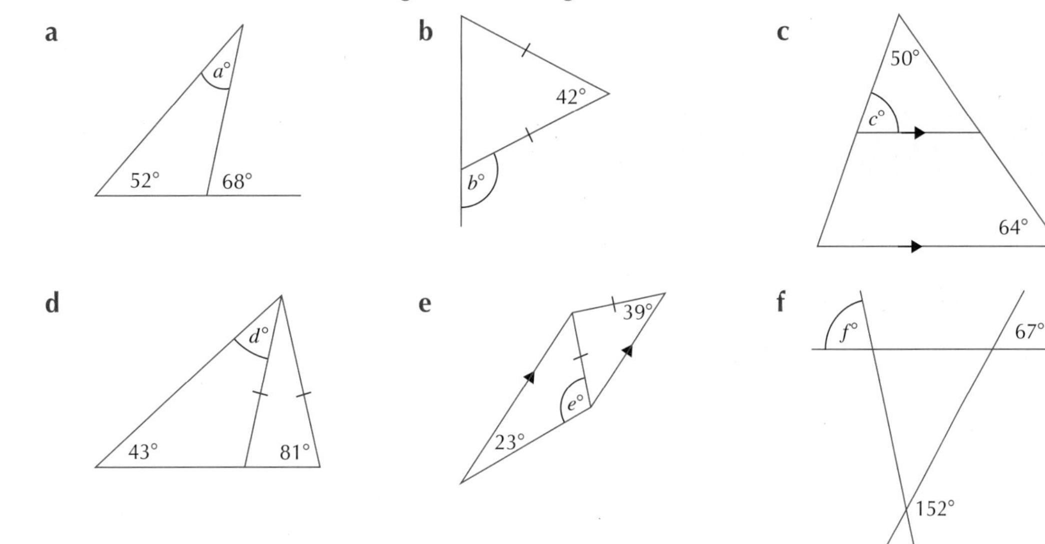

2 Find the size of the marked angle.

a

b $PQRS$ is a kite.

c

d

Finding angles using angle properties of circles

Angle properties within circles are used to find unknown angles in geometric problems. The three important properties are shown in the table, and can be used as a checklist when answering questions:

- is there an angle in a semi-circle?
- is there an angle between a tangent and a radius?
- is there an isosceles triangle formed by a chord and two radii?

When finding unknown angles in circles you can also use other sets of properties, such as those relating to parallel lines.

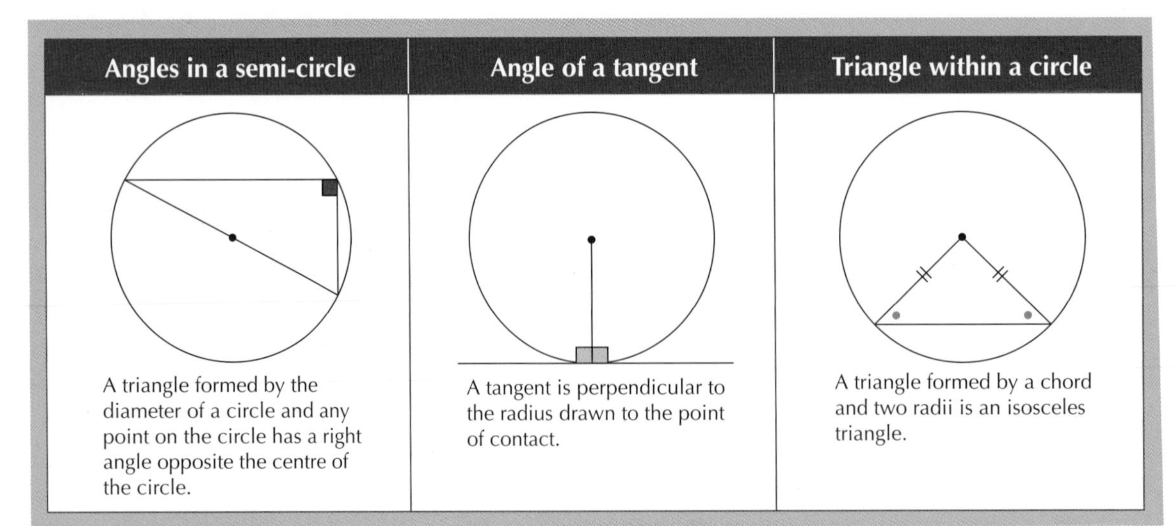

Angles in a semi-circle	Angle of a tangent	Triangle within a circle
A triangle formed by the diameter of a circle and any point on the circle has a right angle opposite the centre of the circle.	A tangent is perpendicular to the radius drawn to the point of contact.	A triangle formed by a chord and two radii is an isosceles triangle.

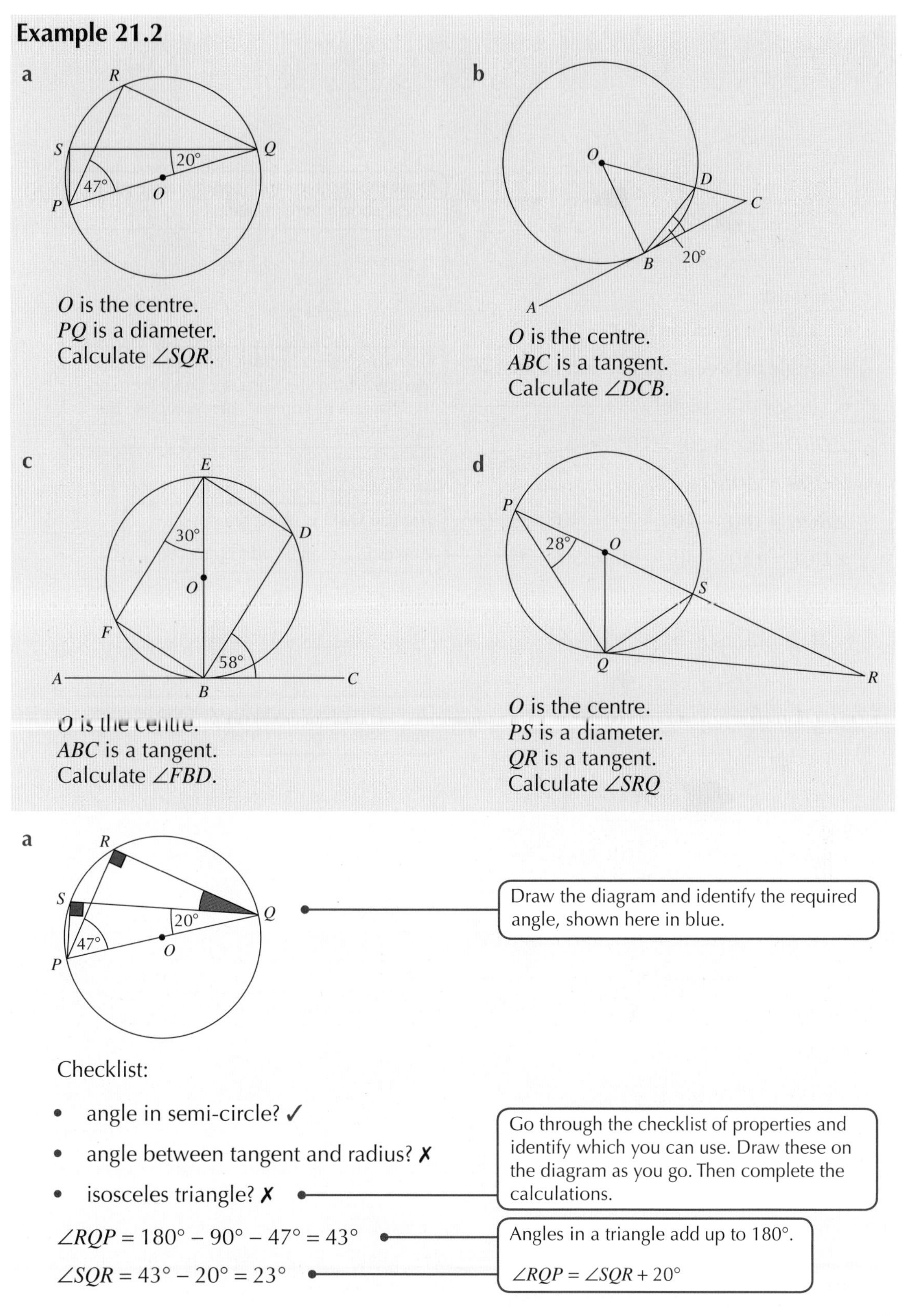

Example 21.2

a

O is the centre.
PQ is a diameter.
Calculate $\angle SQR$.

b

O is the centre.
ABC is a tangent.
Calculate $\angle DCB$.

c

O is the centre.
ABC is a tangent.
Calculate $\angle FBD$.

d

O is the centre.
PS is a diameter.
QR is a tangent.
Calculate $\angle SRQ$

a

Draw the diagram and identify the required angle, shown here in blue.

Checklist:

- angle in semi-circle? ✓
- angle between tangent and radius? ✗
- isosceles triangle? ✗

Go through the checklist of properties and identify which you can use. Draw these on the diagram as you go. Then complete the calculations.

$\angle RQP = 180° - 90° - 47° = 43°$

Angles in a triangle add up to 180°.

$\angle SQR = 43° - 20° = 23°$

$\angle RQP = \angle SQR + 20°$

(continued)

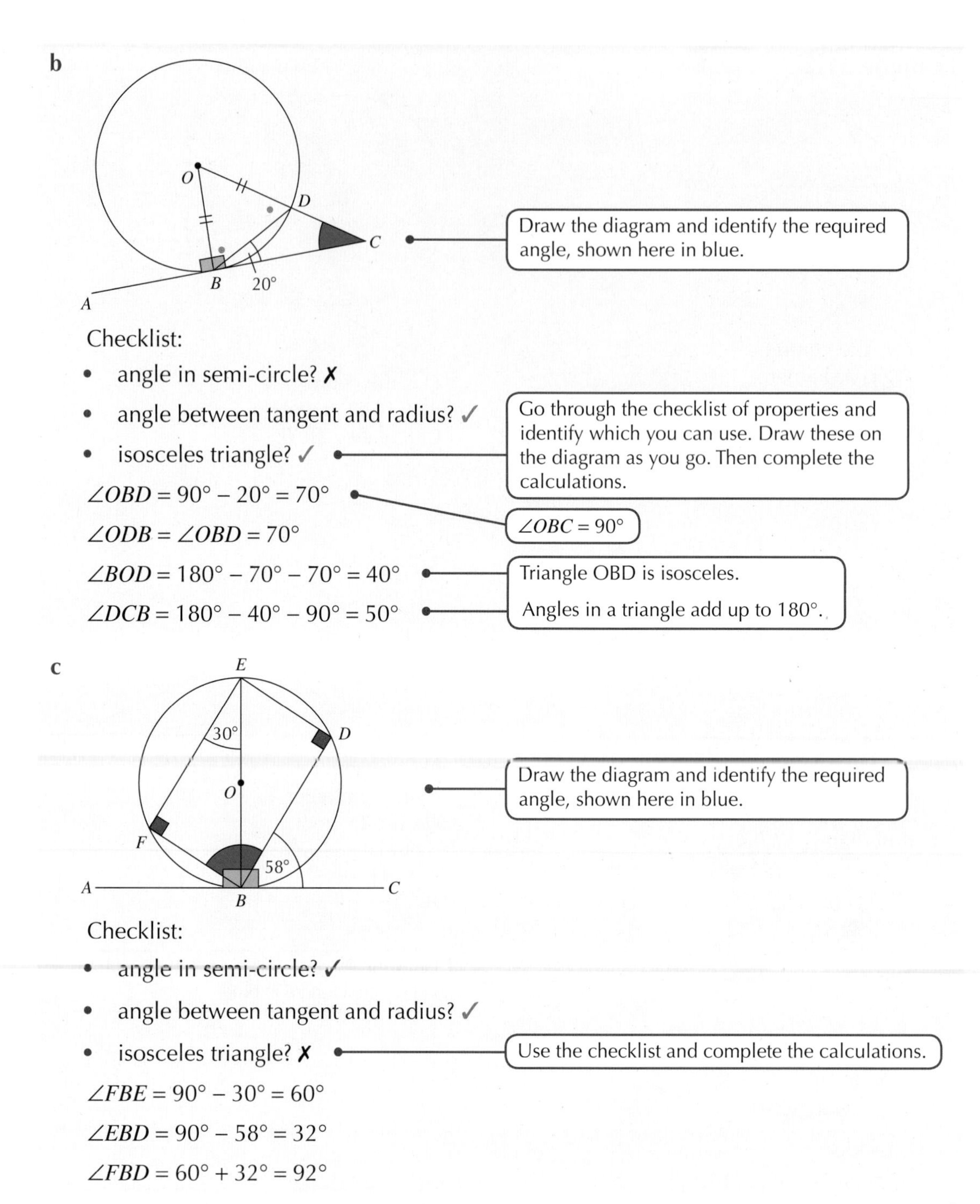

b

Draw the diagram and identify the required angle, shown here in blue.

Checklist:

- angle in semi-circle? ✗
- angle between tangent and radius? ✓
- isosceles triangle? ✓

Go through the checklist of properties and identify which you can use. Draw these on the diagram as you go. Then complete the calculations.

$\angle OBD = 90° - 20° = 70°$

$\angle OBC = 90°$

$\angle ODB = \angle OBD = 70°$

$\angle BOD = 180° - 70° - 70° = 40°$

Triangle OBD is isosceles.

$\angle DCB = 180° - 40° - 90° = 50°$

Angles in a triangle add up to 180°.

c

Draw the diagram and identify the required angle, shown here in blue.

Checklist:

- angle in semi-circle? ✓
- angle between tangent and radius? ✓
- isosceles triangle? ✗

Use the checklist and complete the calculations.

$\angle FBE = 90° - 30° = 60°$

$\angle EBD = 90° - 58° = 32°$

$\angle FBD = 60° + 32° = 92°$

(continued)

d

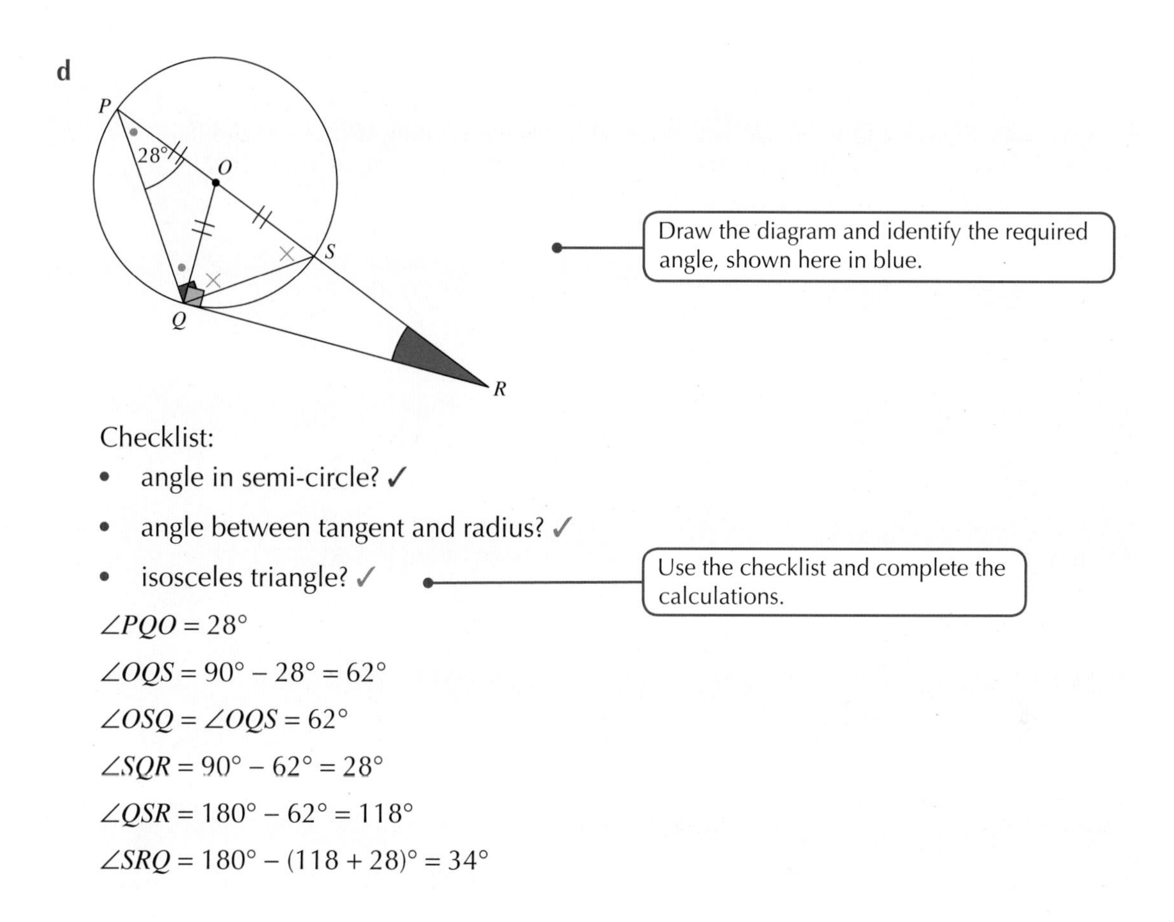

Draw the diagram and identify the required angle, shown here in blue.

Checklist:

- angle in semi-circle? ✓
- angle between tangent and radius? ✓
- isosceles triangle? ✓

Use the checklist and complete the calculations.

$\angle PQO = 28°$

$\angle OQS = 90° - 28° = 62°$

$\angle OSQ = \angle OQS = 62°$

$\angle SQR = 90° - 62° = 28°$

$\angle QSR = 180° - 62° = 118°$

$\angle SRQ = 180° - (118 + 28)° = 34°$

Example 21.3

O is the centre. *ABC* is a tangent. Calculate $\angle EOD$.

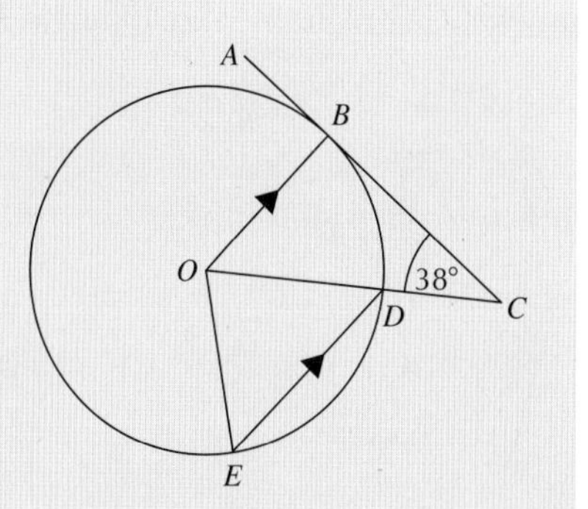

Checklist:

- angle in semi-circle? ✗
- angle between tangent and radius? ✓
- isosceles triangle? ✓

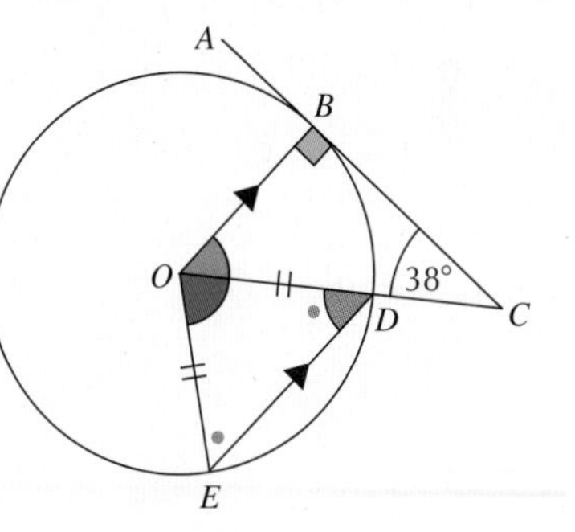

$\angle BOC = 90° - 38° = 52°$

$\angle EDO = \angle BOC = 52°$

Alternate angles are equal.

$\angle OED = \angle ODE = 52°$

$\angle EOD = 180° - (52 + 52)°$

$= 76°$

Exercise 21B

1 O is the centre. PQR is a tangent. QT is a diameter. Calculate the size of angle STQ.

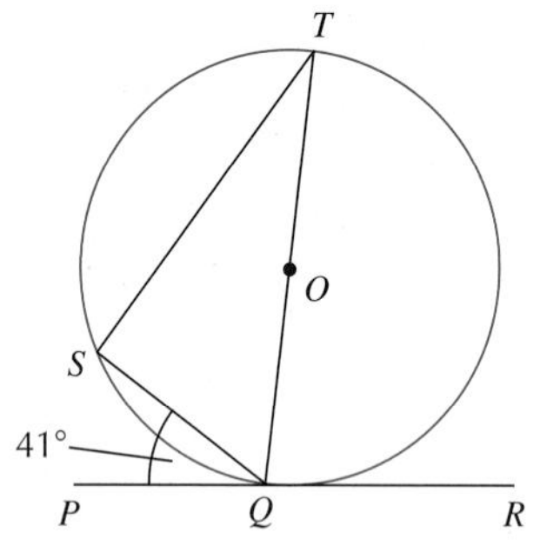

2 O is the centre. ABC is a tangent. Calculate the size of angle BOD.

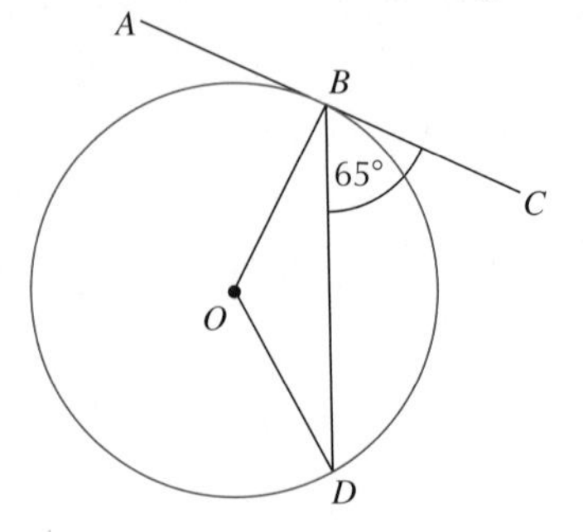

3 O is the centre. KLM is a tangent. Calculate the size of angle QNP.

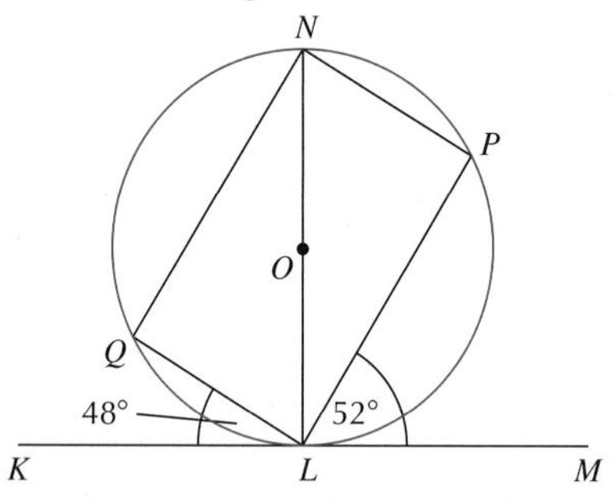

4 O is the centre. VWX is a tangent. Calculate the size of angle WXY.

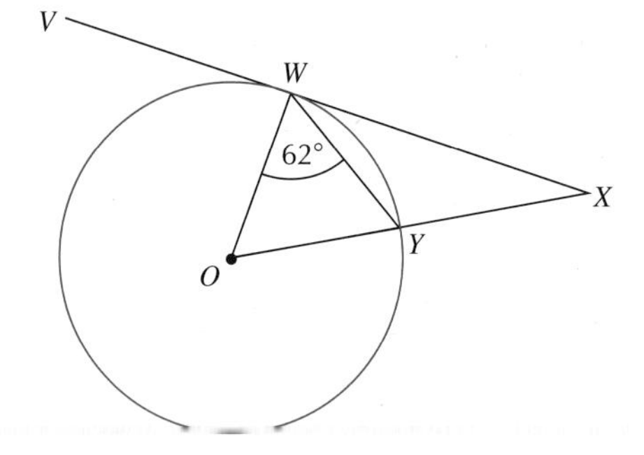

5 O is the centre. NPQ is a tangent. PR is a diameter. Calculate the size of angle PQO.

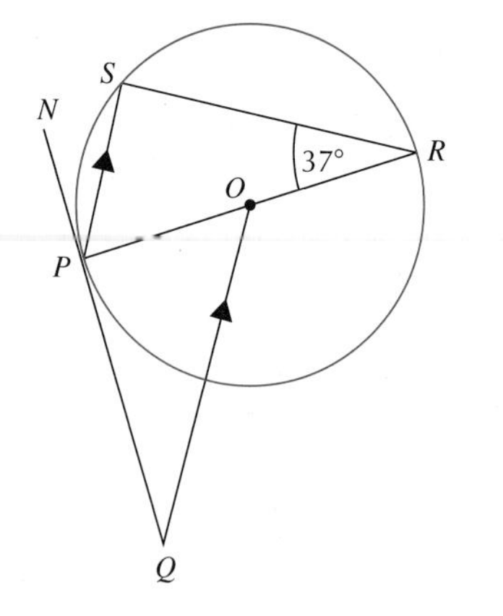

6 O is the centre. VW is a diameter. Calculate the size of angle YXW.

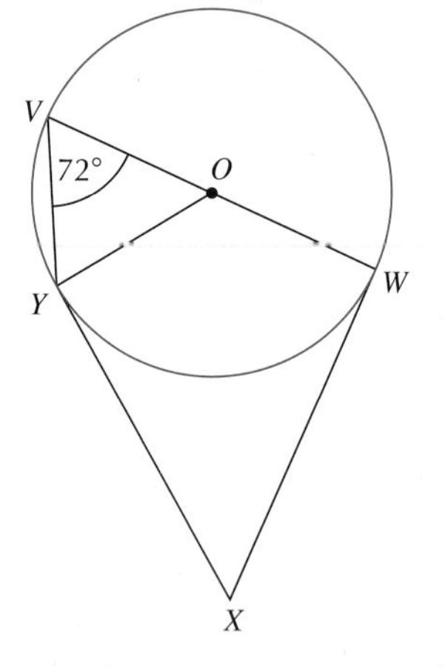

★ **7** O is the centre. ABC is a tangent. Calculate the size of angle OCB.

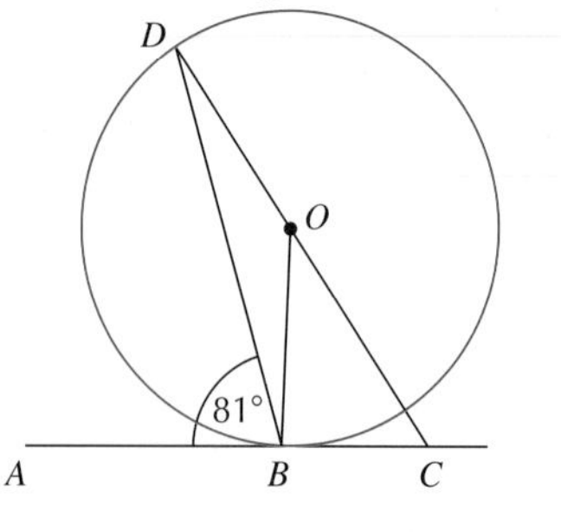

8 O is the centre. PQ is a diameter. Calculate the size of angle RQT.

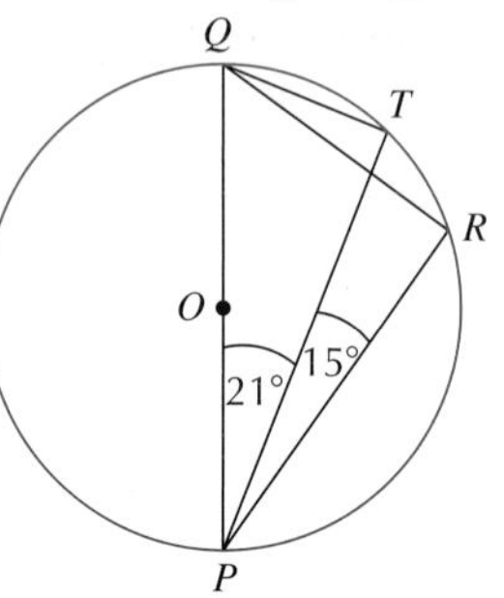

Symmetry in the circle

The triangle formed by two radii and a chord is isosceles, so it has all of the properties of an isosceles triangle. If an axis of symmetry is extended to form the diameter of the circle, there are more properties to note:

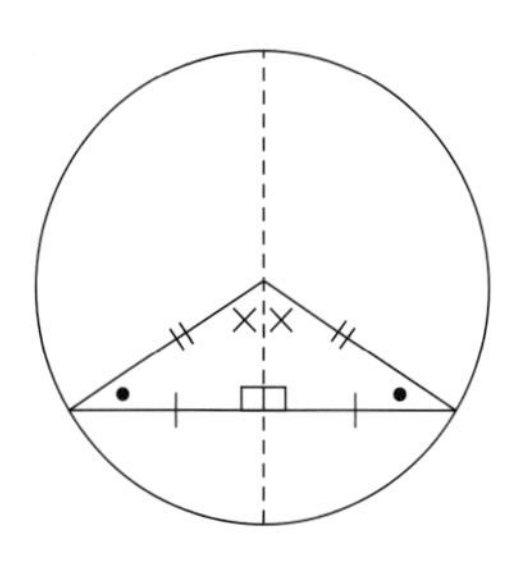

- the perpendicular bisector of a chord passes through the centre of the circle
- a diameter perpendicular to a chord bisects it
- a diameter which bisects a chord is perpendicular to it.

Pythagoras and trigonometry in circle calculations

When a triangle is formed by two radii and a chord of a circle, there are:

- two congruent right-angled triangles
- four equal radii.

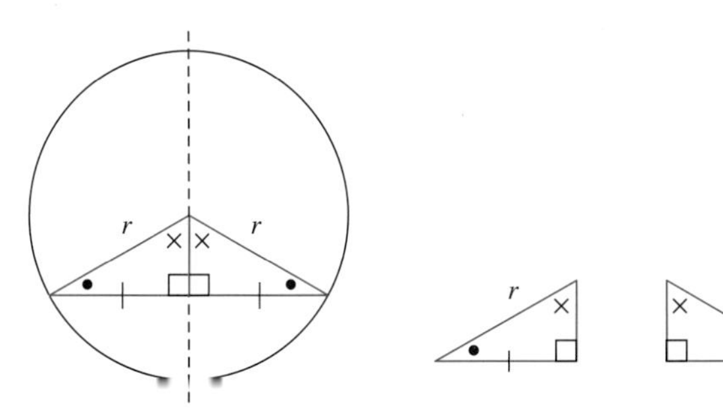

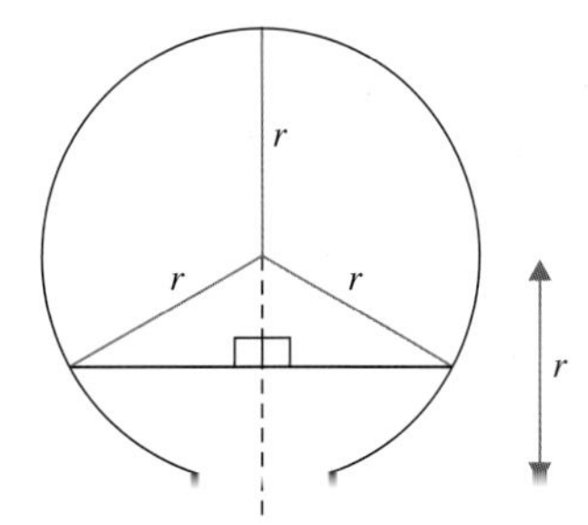

These two properties can be used to find unknown lengths based on symmetry in the circle.

Example 21.4

Calculate x to 1 decimal place.

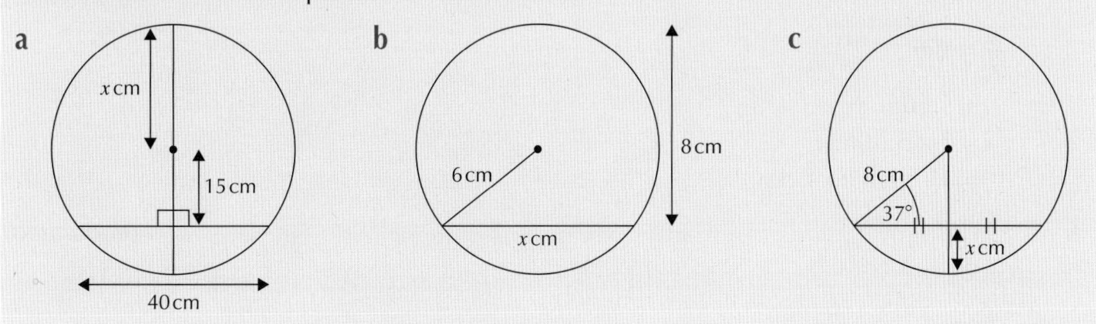

(continued)

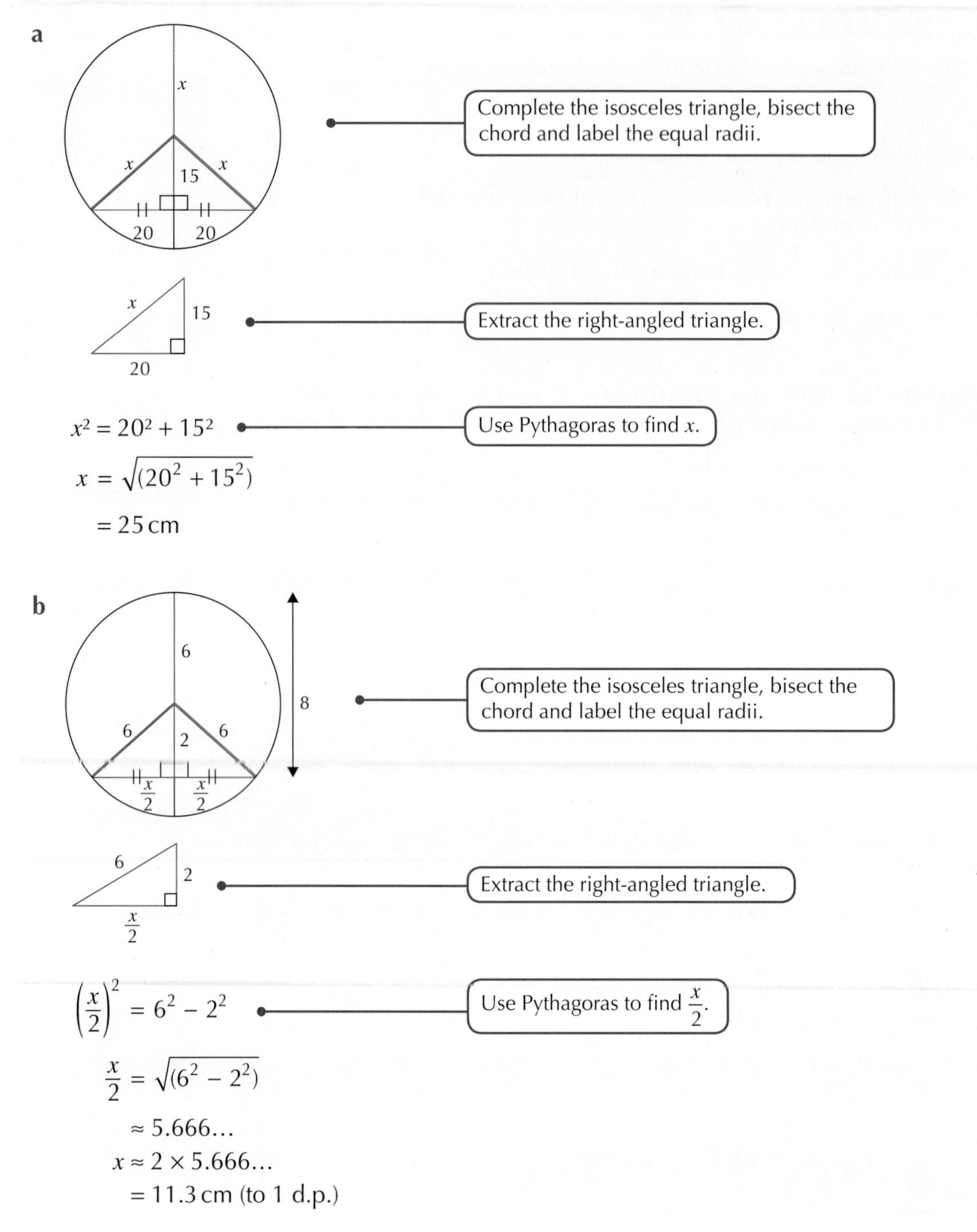
a
x
x
x
15
20
20
Complete the isosceles triangle, bisect the chord and label the equal radii.
x
15
20
Extract the right-angled triangle.
$x^2 = 20^2 + 15^2$
Use Pythagoras to find x.
$x = \sqrt{(20^2 + 15^2)}$
$= 25\,\text{cm}$
b
6
6
6
2
8
$\frac{x}{2}$
$\frac{x}{2}$
Complete the isosceles triangle, bisect the chord and label the equal radii.
6
2
$\frac{x}{2}$
Extract the right-angled triangle.
$\left(\frac{x}{2}\right)^2 = 6^2 - 2^2$
Use Pythagoras to find $\frac{x}{2}$.
$\frac{x}{2} = \sqrt{(6^2 - 2^2)}$
$\approx 5.666\ldots$
$x \approx 2 \times 5.666\ldots$
$= 11.3\,\text{cm}$ (to 1 d.p.)

(continued)

c

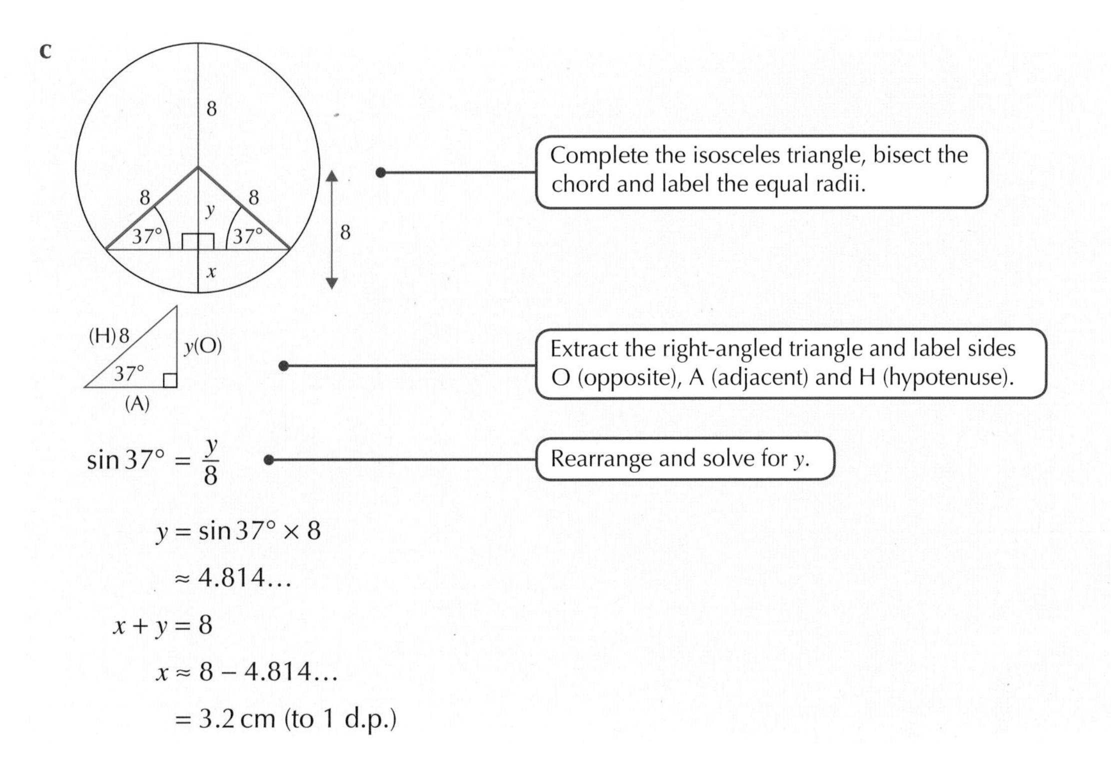

$\sin 37° = \dfrac{y}{8}$

$y = \sin 37° \times 8$

$\approx 4.814...$

$x + y = 8$

$x \approx 8 - 4.814...$

$= 3.2$ cm (to 1 d.p.)

Exercise 21C

★ **1** Calculate the value of x. Give your answers to 2 decimal places where necessary.

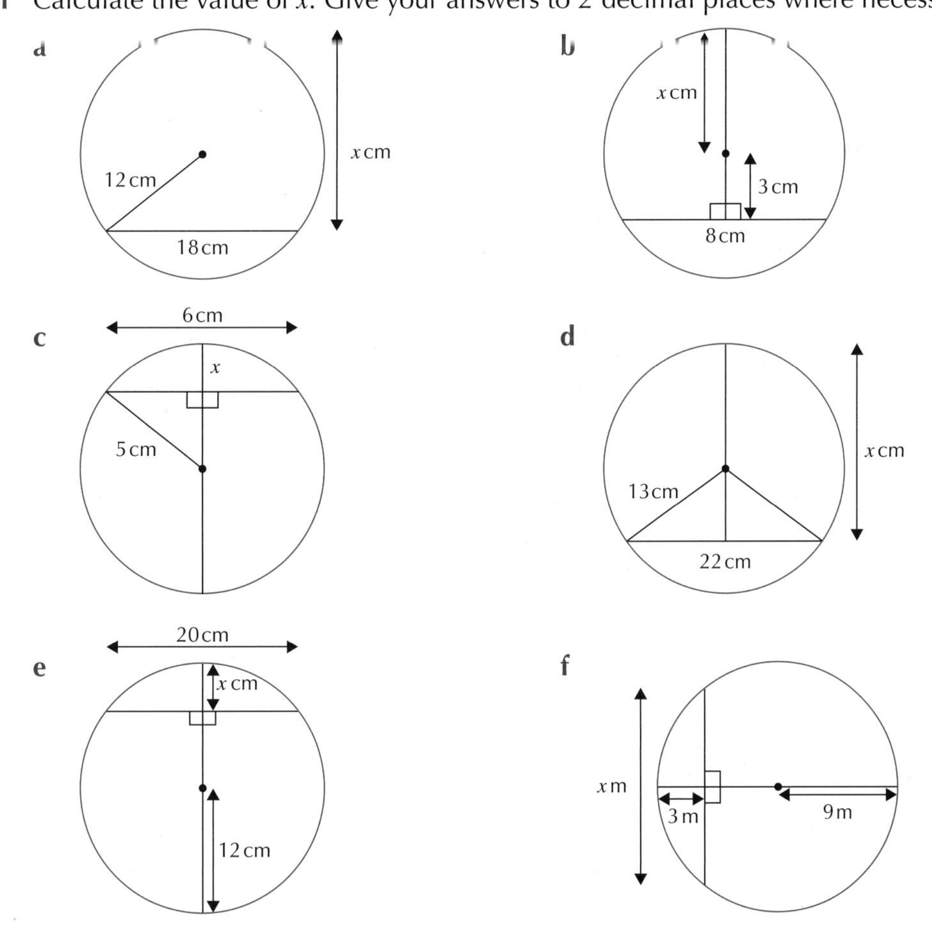

★ **2** Find the value of x.

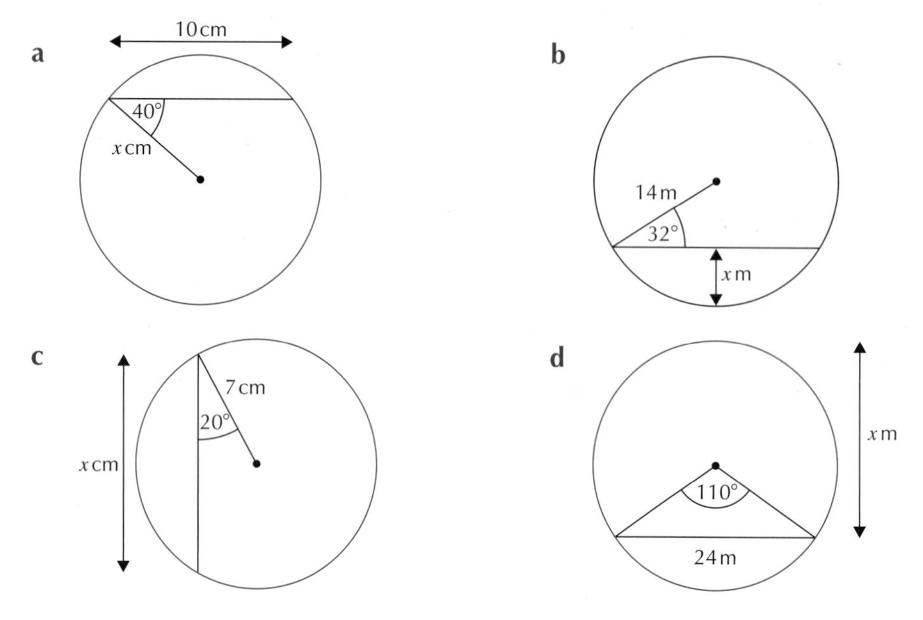

Problem solving based on symmetry in the circle

Properties based on symmetry in a circle can be used to solve problems. Most problems will also involve the use of Pythagoras and trigonometry. It is useful to be able to visualise parts of circles in ways that let you use the properties you have been working with.

Typical scenarios are:

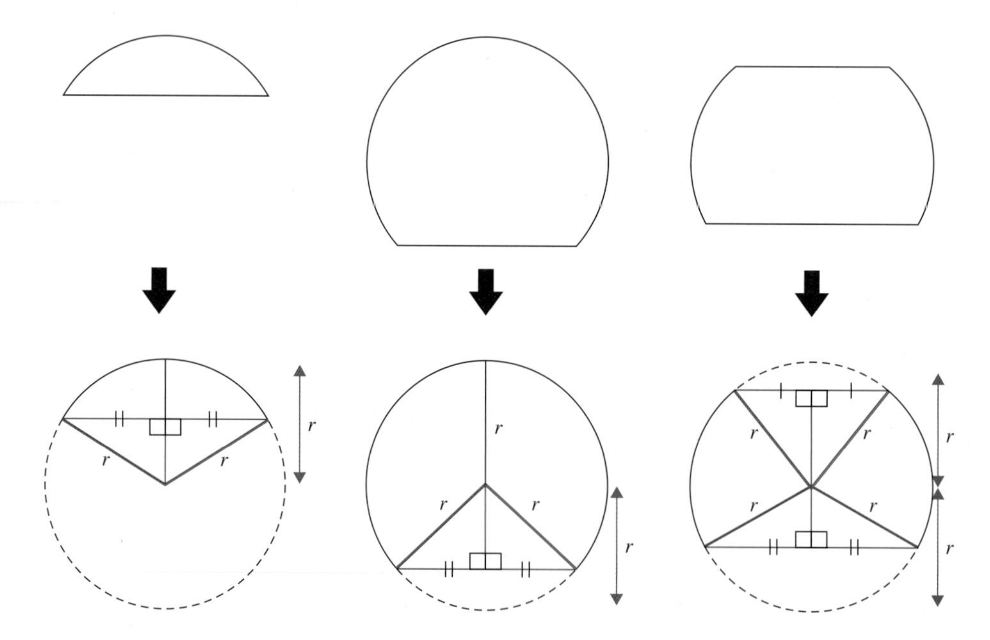

Example 21.5

The diagram shows the cross-section of a tunnel.

It has a horizontal floor AB, and the curved part is the arc of a circle, centre O and radius OA. $AB = 2.8$ m and $OA = 3.2$ m. Calculate the height of the tunnel.

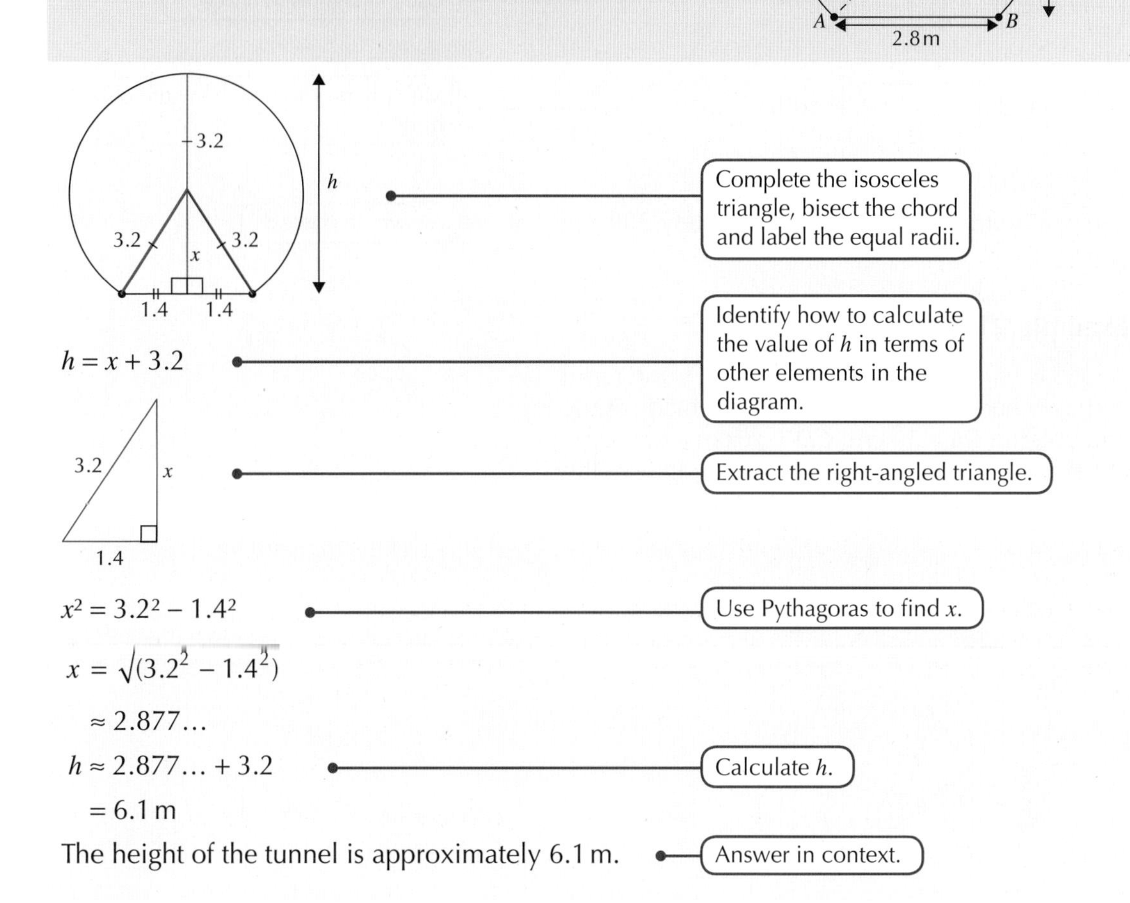

$h = x + 3.2$ — Identify how to calculate the value of h in terms of other elements in the diagram.

Extract the right-angled triangle.

$x^2 = 3.2^2 - 1.4^2$ — Use Pythagoras to find x.

$x = \sqrt{(3.2^2 - 1.4^2)}$

$\approx 2.877\ldots$

$h \approx 2.877\ldots + 3.2$ — Calculate h.

$= 6.1$ m

The height of the tunnel is approximately 6.1 m. — Answer in context.

Example 21.6

The curved part of a bridge is formed by an arc of a circle with radius 200 m. The height of the bridge is 27 m. Calculate its width w metres.

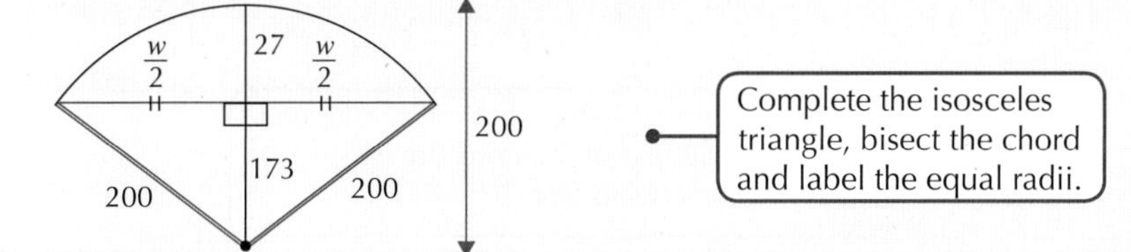

Complete the isosceles triangle, bisect the chord and label the equal radii.

(continued)

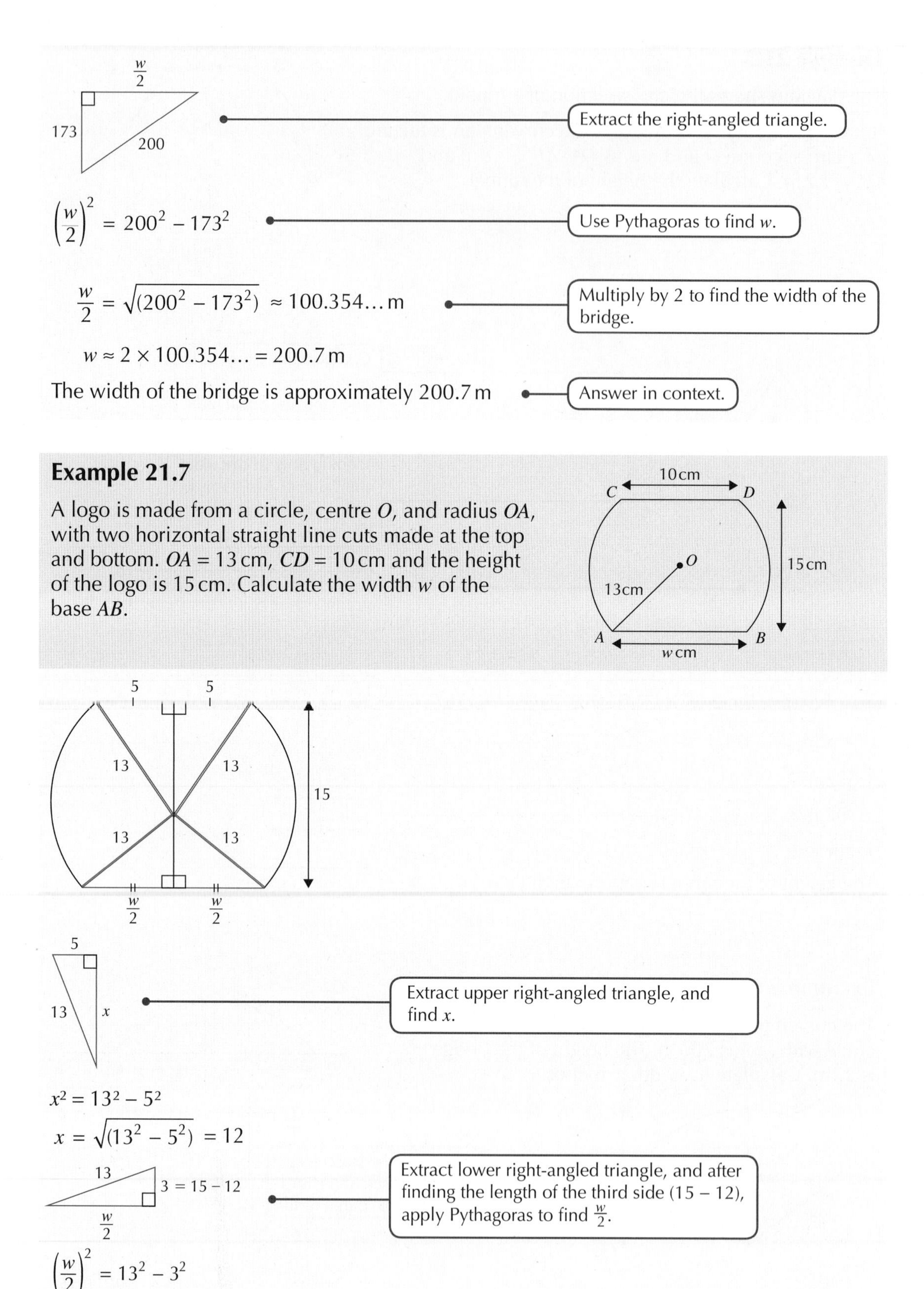

Extract the right-angled triangle.

$$\left(\frac{w}{2}\right)^2 = 200^2 - 173^2$$

Use Pythagoras to find w.

$$\frac{w}{2} = \sqrt{(200^2 - 173^2)} \approx 100.354\ldots \text{ m}$$

Multiply by 2 to find the width of the bridge.

$$w \approx 2 \times 100.354\ldots = 200.7 \text{ m}$$

The width of the bridge is approximately 200.7 m

Answer in context.

Example 21.7

A logo is made from a circle, centre O, and radius OA, with two horizontal straight line cuts made at the top and bottom. $OA = 13$ cm, $CD = 10$ cm and the height of the logo is 15 cm. Calculate the width w of the base AB.

Extract upper right-angled triangle, and find x.

$$x^2 = 13^2 - 5^2$$

$$x = \sqrt{(13^2 - 5^2)} = 12$$

Extract lower right-angled triangle, and after finding the length of the third side (15 − 12), apply Pythagoras to find $\frac{w}{2}$.

$$\left(\frac{w}{2}\right)^2 = 13^2 - 3^2$$

$$\frac{w}{2} = \sqrt{(13^2 - 3^2)} \approx 12.649\ldots$$

$$w \approx 2 \times 12.649\ldots = 25.3 \text{ cm}$$

Exercise 21D

1 A cylindrical water tank with diameter 140 cm is partly filled with water. The depth of the water is 30 cm. Calculate the width of AB.

2 A set of park gates form part of a circle with radius 4 m. The width of the gates is 6 m. Calculate the height h of the gates.

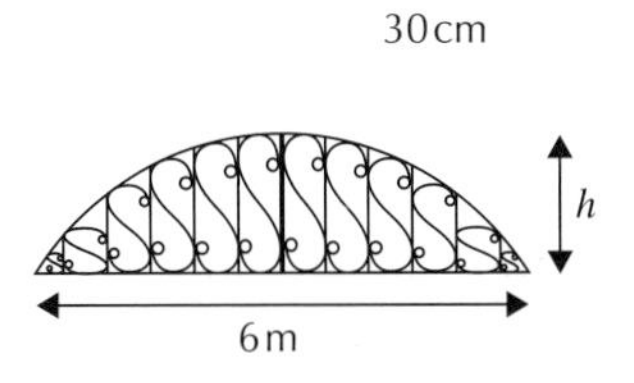

3 A radar on ship A covers a radius of 60 km. Ship B travels on a straight course through the points P and Q. The length of PQ is 100 km. Ship A has a warning alarm which is activated if another ship passes within 30 km of it. Will the alarm be activated as ship B passes?

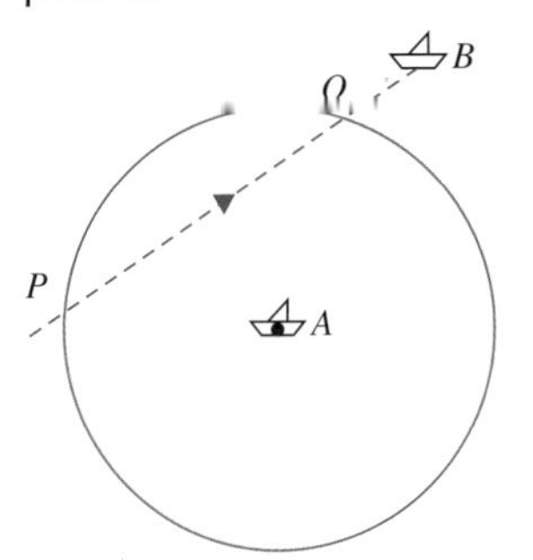

4 A circular table with diameter 2 m has foldable sides as shown in the diagram. When the sides are folded, the table sides are 1.8 m in length. What is its width w?

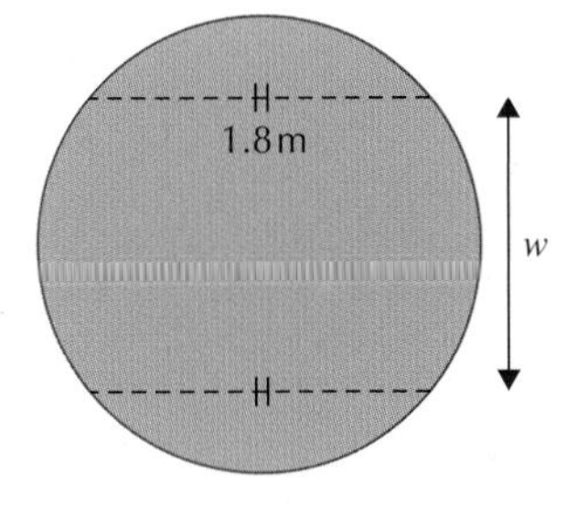

5 The cross-section of a lampshade is part of a circle with radius 15 cm. The width of the top is 9 cm and the height is 22 cm. What is the width w of the bottom?

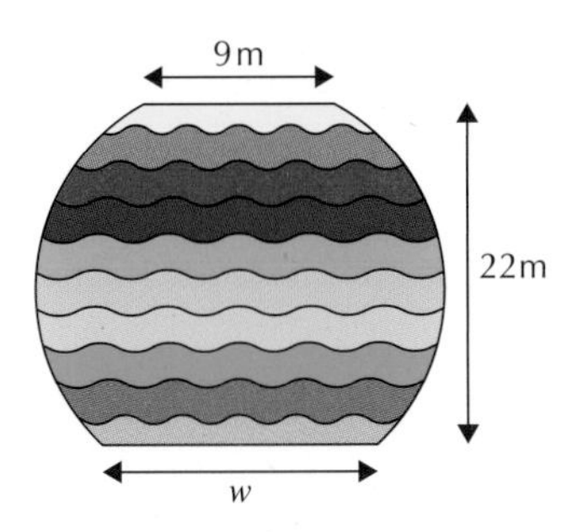

6 A fruit bowl is part of the cross-section of a circle with radius 20 cm. The width of the top is 36 cm and the width of the base is 12 cm. Calculate the height h of the bowl.

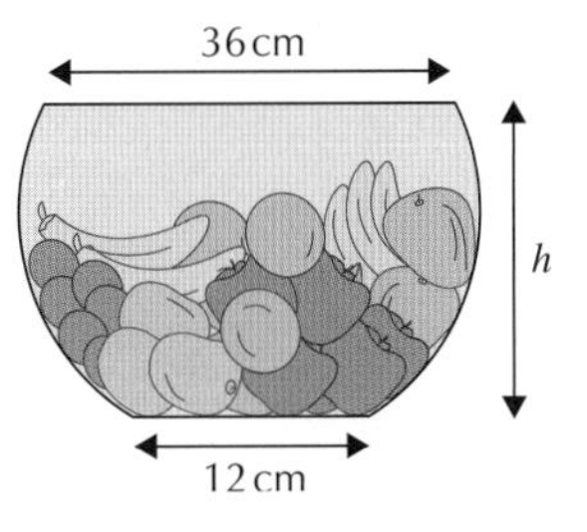

7 A doorway is made from a rectangle and part of a circle with radius 70 cm. Calculate its width w.

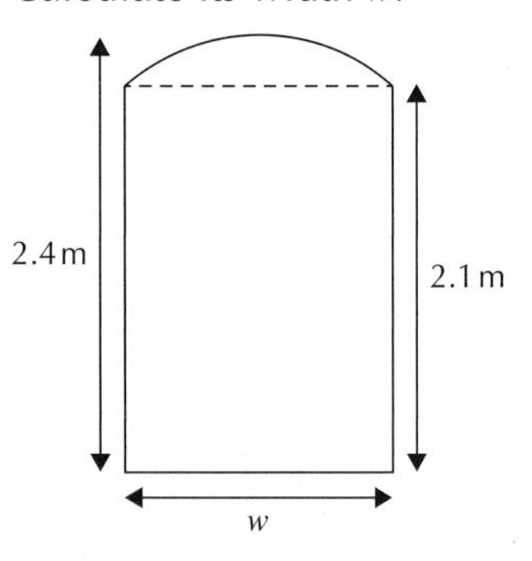

8 A drinking trough is part of the cross-section of a circle with diameter 70 cm. The width of the top is 46 cm. What is the depth d of the water when the trough is full?

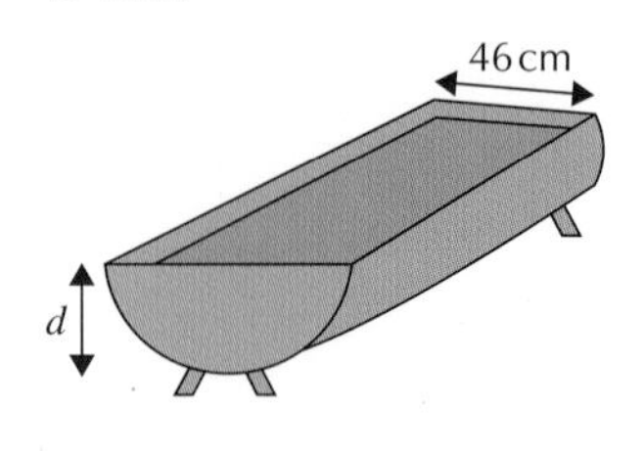

9 A badge for employees of Beta Corp is shown.

The badge is made from an equilateral triangle, with sides of length 3 cm, which touches the circumference of the circle at P, Q and R.

a Calculate the size of angle OPQ.

b Calculate the length of the radius of the circle OP.

★ **10** A factory security light positioned at point O covers a distance of 40 m in all directions. The length of the factory wall AB is 60 m. Calculate the size of angle AOB.

Polygons and their interior and exterior angles

A **polygon** is a 2D shape with straight sides. The simplest polygons are the triangle and the quadrilateral.

Name	Number of sides
Pentagon	5
Hexagon	6
Heptagon	7
Octagon	8
Nonagon	9
Decagon	10
Dodecagon	12

A polygon is said to be **regular** if all of its angles are equal and all of its sides are equal. If it is not regular it said to be **irregular.**

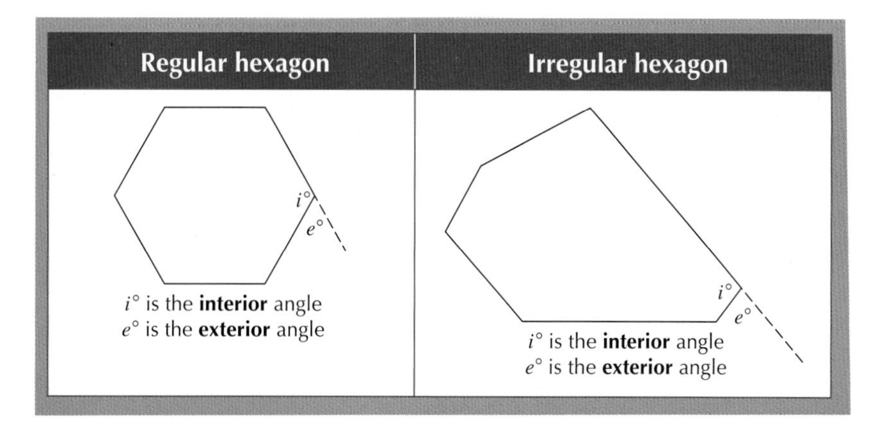
Regular hexagon
Irregular hexagon
i°
e°
i° is the interior angle
e° is the exterior angle
i°
e°
i° is the interior angle
e° is the exterior angle

In each hexagon above, the angle labelled $i°$, inside the shape, is called an **interior** angle. The **exterior** angle $e°$ is formed by one side of the polygon and a line extended from an adjacent side as illustrated. Each pair of interior and exterior angles adds up to 180°.

In a regular polygon all of the interior angles are equal and all of the exterior angles are equal. This is not the case with irregular polygons.

Finding formulae for the sum of the interior angles and the sum of the exterior angles of a polygon

Activity

The interior angles of any polygon, whether regular or irregular, can be found by splitting it into triangles as shown here for a pentagon.

A pentagon splits into 3 triangles.

The angles in any triangle add up to 180°, so it follows that:
the sum of the angles of a pentagon = $3 \times 180° = 540°$

1 a Copy and complete the table.

b Use your results to find a formula for the sum of the interior angles of an n-sided polygon.

Name of polygon	Number of sides	Sum of the interior angles
Quadrilateral	4	
Pentagon	5	$3 \times 180° = 540°$
Hexagon	6	
Heptagon	7	
Octagon	8	
Nonagon	9	
Decagon	10	
Dodecagon	12	

Each interior and exterior angle pair adds up to 180°. This means that in an n-sided figure, the sum of all of the interior and exterior angles added together is $n \times 180° = 180n°$.

2 Using the formula which you have just developed for the sum of the interior angles of an n-sided polygon, find a formula for the sum of the exterior angles.

Finding formulae for the interior and exterior angles of a regular polygon

In a regular polygon all of the interior angles are equal and all of the exterior angles are equal which means that it is possible to develop a formula for finding each.

The interior and exterior angles of a regular polygon can be found by splitting it into isosceles triangles as illustrated here for a pentagon.

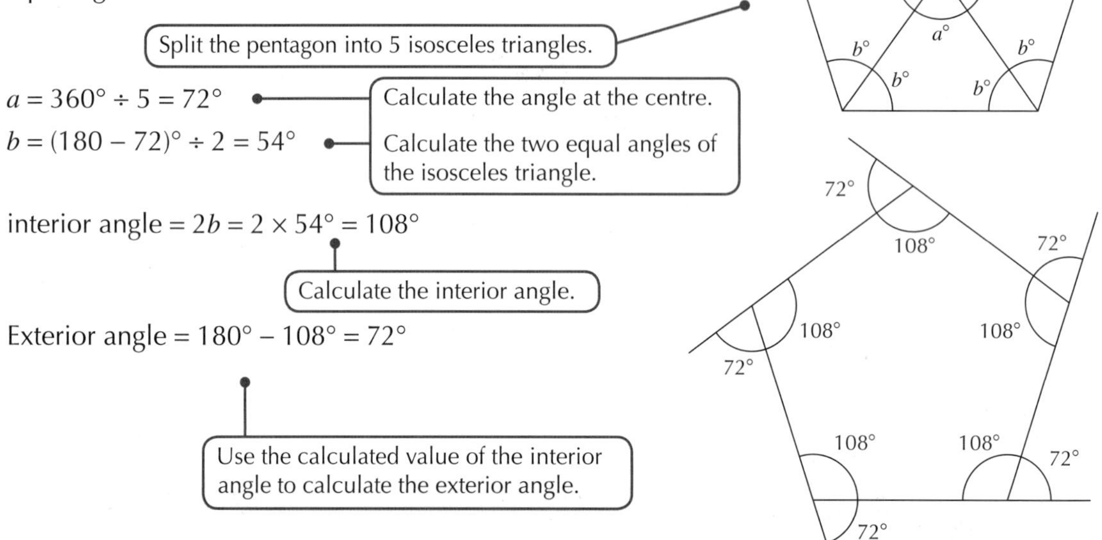

Split the pentagon into 5 isosceles triangles.
$a = 360° \div 5 = 72°$
Calculate the angle at the centre.
$b = (180 - 72)° \div 2 = 54°$
Calculate the two equal angles of the isosceles triangle.
interior angle = $2b = 2 \times 54° = 108°$
Calculate the interior angle.
Exterior angle = $180° - 108° = 72°$
Use the calculated value of the interior angle to calculate the exterior angle.
a°
b°
72°
108°

Activity

1 Copy and complete the table for different regular polygons.

Regular polygon	Number of sides	Angle at the centre	Interior angle	Exterior angle
Square	4			
Pentagon	5	$\frac{360°}{5} = 72°$	108°	72°
Hexagon	6			
Octagon	8			

2 Compare the exterior angle to the angle at the centre of the circle and use this to find a formula for the exterior angle of an n-sided regular polygon.

3 If each interior and exterior angle pair adds to 180°, what is the formula for the interior angle of an n-sided polygon?

4 If the interior angle of a regular polygon is 165°, how many sides does the polygon have?

Look for the link between the exterior angle and the angle at the centre.

Finding angles using the angle properties of polygons

Angle properties for polygons include:

- for any n-sided polygon, the sum S of the interior angles is given by $S = 180(n - 2)°$
- for any regular n-sided polygon,

  the exterior angle = angle at the centre = $\frac{360°}{n}$
- for any regular n-sided polygon,

  the interior angle = $180° - \frac{360°}{n}$

  $= 180° -$ exterior angle

> Learn how these formulae were derived so you can recall them more readily when required.

Example 21.8

ABCDE is a pentagon. Calculate the size of angle *EDC*.

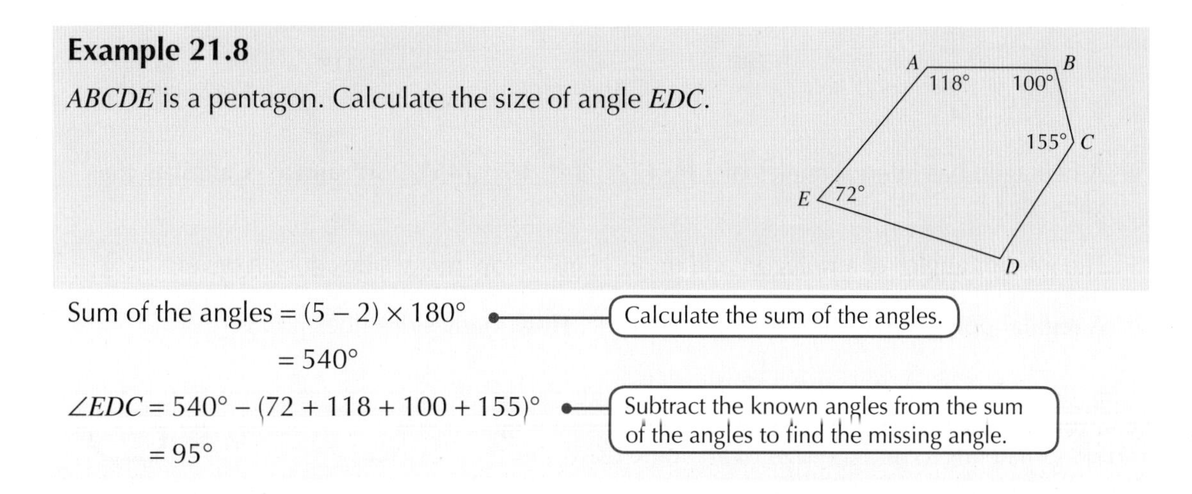

Sum of the angles = $(5 - 2) \times 180°$ — Calculate the sum of the angles.

$= 540°$

$\angle EDC = 540° - (72 + 118 + 100 + 155)°$ — Subtract the known angles from the sum of the angles to find the missing angle.

$= 95°$

Example 21.9

John added up the angles of a regular polygon. The answer came to 2340. What was the size of each interior angle?

Sum = 2340

$(n - 2) \times 180° = 2340$ — Use the sum of the angles to find the number of sides.

$n - 2 = \frac{2340}{180}$

$= 13$

$n = 15$

exterior angle = $\frac{360°}{15} = 24°$ — Use the number of sides to find the exterior angle.

interior angle = $180° - 24° = 156°$ — Use the exterior angle to find the interior angle.

Exercise 21E

★ **1** Calculate the size of the interior and exterior angles of regular polygons with these numbers of sides.

a 15 **b** 18 **c** 72

2 What is the sum of the interior angles of regular polygons with these numbers of sides?

a 12 **b** 14 **c** 22

★ **3** Calculate the size of angle $x°$ in each of the irregular polygons.

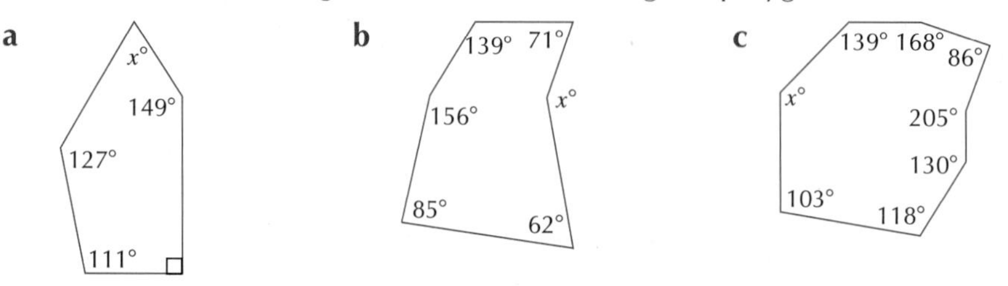

4 A heptagon has interior angles of 73°, 122°, 34°, 15°, 145°, 230° and $x°$. Calculate the value of x.

5 The interior angles of a polygon add up to 1260°. How many sides does it have?

6 A regular polygon has interior angles of 144°. How many sides does it have?

- I can use my knowledge of the angle properties of triangles and quadrilaterals to find the size of an angle using more than one step. ★ Exercise 21A Q1
- I can use my knowledge of angles in circles to find the size of an angle involving more than one step. ★ Exercise 21B Q7
- I know the symmetry properties of a circle and can use this knowledge to find lengths using Pythagoras or trigonometry. ★ Exercise 21C Q1, Q2
- I can solve real-life problems using my knowledge of the symmetry of a circle. ★ Exercise 21D Q2, Q6, Q10
- I can find formulae for the sums of the interior angles and the exterior angles of a polygon. *Activity* p. 233
- I can find formulae for the interior angle and the exterior angle of a regular polygon. *Activity* p. 234
- I can use my knowledge of angles in polygons to find the size of an angle using more than one step. ★ Exercise 21E Q1, Q3

For further assessment opportunities, see the Preparation for Assessment for Unit 2 on pages 288–291.

22 Using similarity

In this chapter you will learn how to:

- recognise and work with **similar** triangles formed by parallel lines
- use **scale factors** for length, area and volume of similar shapes in problem solving.

You should already know:

- how to recognise and use the properties of alternate, corresponding and vertically opposite angles
- how to recognise similar shapes
- how to calculate a scale factor
- how to use a scale factor to enlarge or reduce a shape.

Similar triangles formed by parallel lines

You will already know that geometrical shapes are **similar** if:

- corresponding angles are equal (**equiangular**)
- corresponding sides are in the same ratio (have the same **scale factor**).

In triangles, these two properties are mutually dependent, so if one is true then the other must also be true. If triangles have equal corresponding angles, making them similar, then the corresponding sides of the similar triangles must be in proportion. Instead of saying that corresponding angles in similar triangles are equal you can simply state that the triangles are **equiangular**.

When using scale factors, use ESF as an abbreviation for 'enlargement scale factor' and RSF for a 'reduction scale factor'.

Example 22.1

Calculate x in each of the following. Give your answers to 2 decimal places.

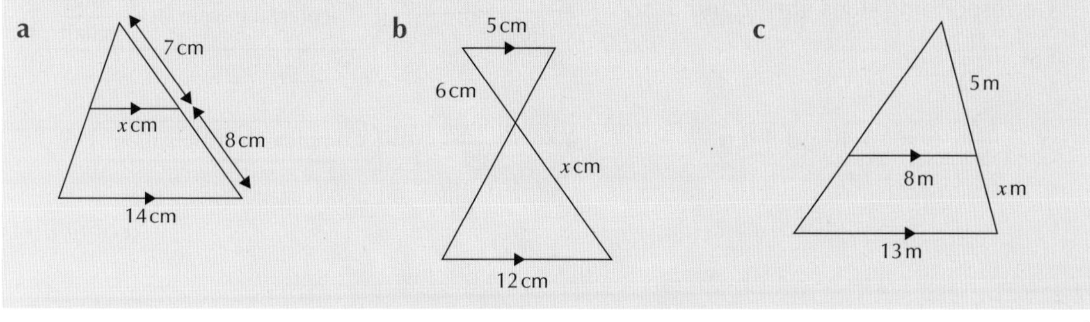

(*continued*)

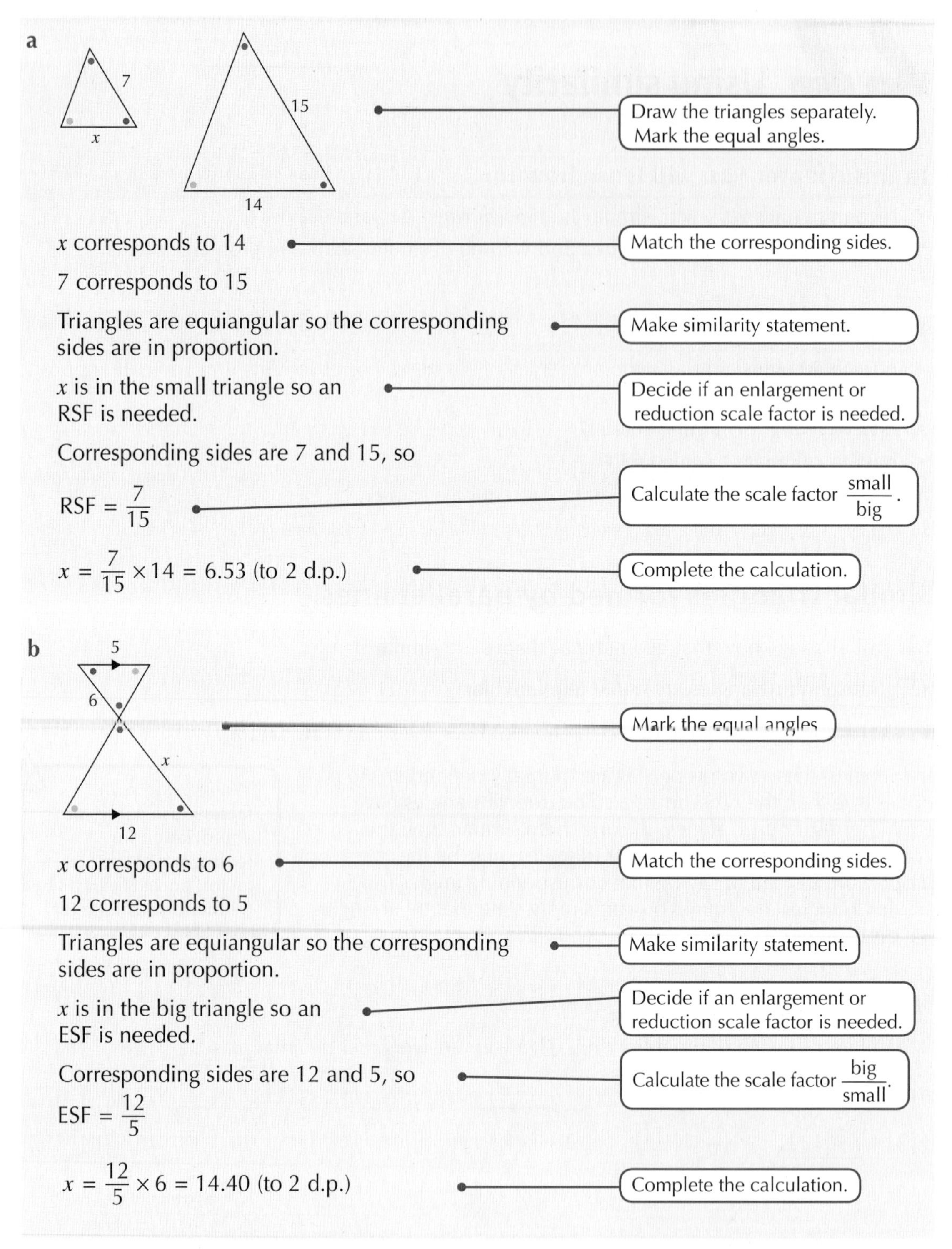

a

x corresponds to 14 — Match the corresponding sides.

7 corresponds to 15

Triangles are equiangular so the corresponding sides are in proportion. — Make similarity statement.

x is in the small triangle so an RSF is needed. — Decide if an enlargement or reduction scale factor is needed.

Corresponding sides are 7 and 15, so

$\text{RSF} = \dfrac{7}{15}$ — Calculate the scale factor $\dfrac{\text{small}}{\text{big}}$.

$x = \dfrac{7}{15} \times 14 = 6.53$ (to 2 d.p.) — Complete the calculation.

Draw the triangles separately. Mark the equal angles.

b

Mark the equal angles

x corresponds to 6 — Match the corresponding sides.

12 corresponds to 5

Triangles are equiangular so the corresponding sides are in proportion. — Make similarity statement.

x is in the big triangle so an ESF is needed. — Decide if an enlargement or reduction scale factor is needed.

Corresponding sides are 12 and 5, so

$\text{ESF} = \dfrac{12}{5}$ — Calculate the scale factor $\dfrac{\text{big}}{\text{small}}$.

$x = \dfrac{12}{5} \times 6 = 14.40$ (to 2 d.p.) — Complete the calculation.

(*continued*)

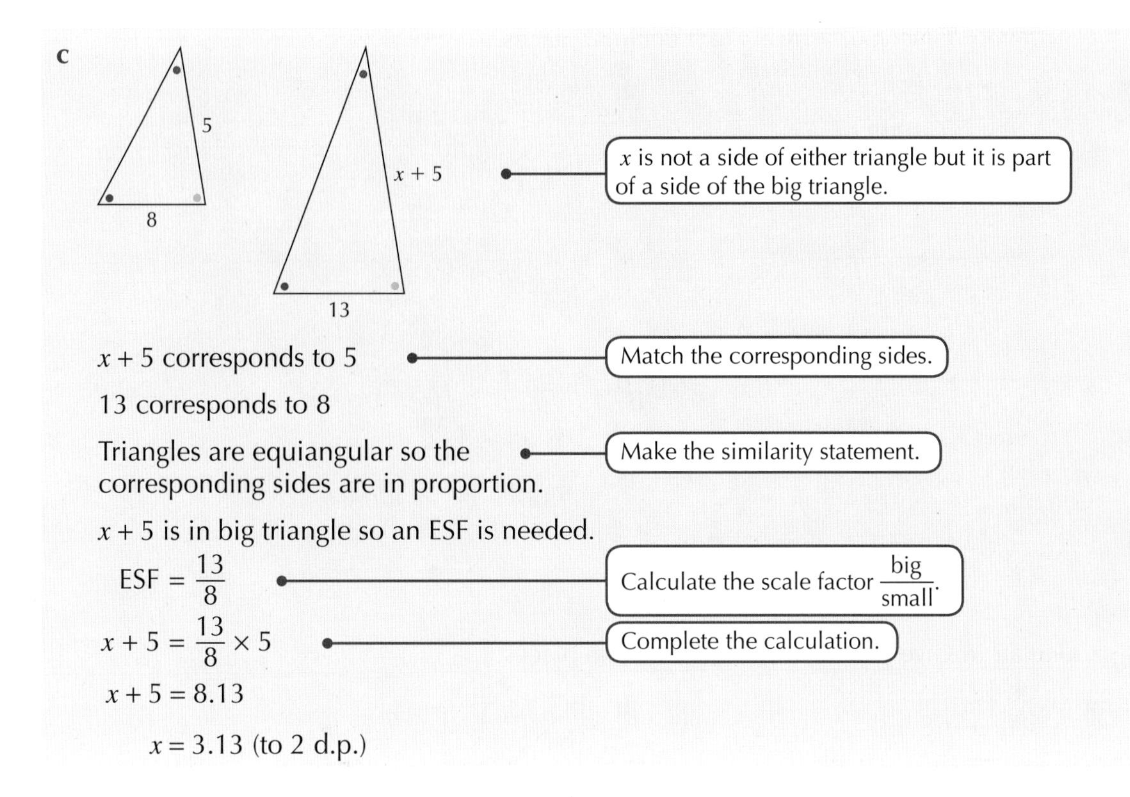

c
5
8
x + 5
13
x is not a side of either triangle but it is part of a side of the big triangle.
Match the corresponding sides.
Make the similarity statement.
Calculate the scale factor big/small.
Complete the calculation.

$x + 5$ corresponds to 5

13 corresponds to 8

Triangles are equiangular so the corresponding sides are in proportion.

$x + 5$ is in big triangle so an ESF is needed.

$\text{ESF} = \frac{13}{8}$

$x + 5 = \frac{13}{8} \times 5$

$x + 5 = 8.13$

$x = 3.13$ (to 2 d.p.)

Exercise 22A

1 Calculate x. Give your answers to 2 decimal places.

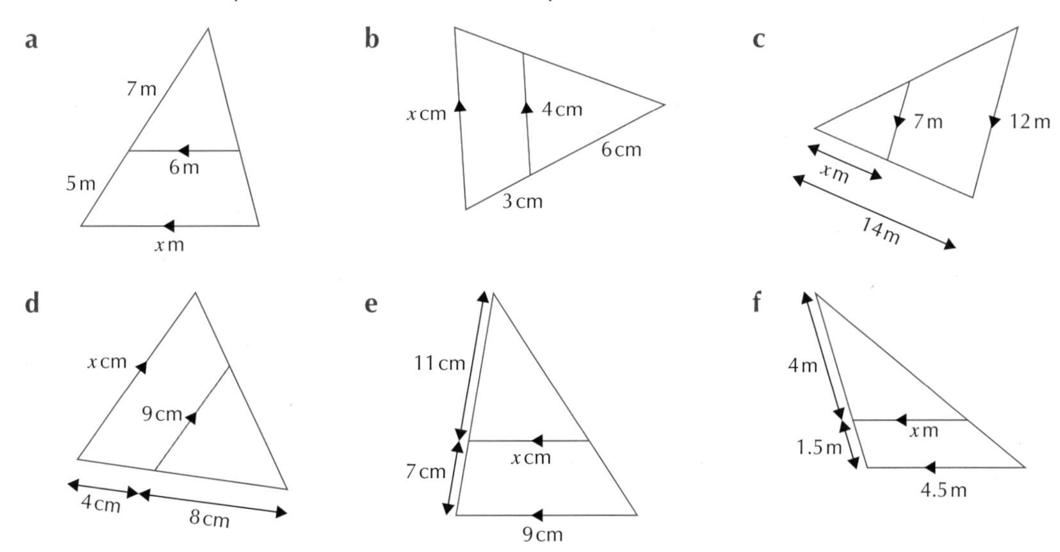

2 Calculate x. Give your answers to 2 decimal places.

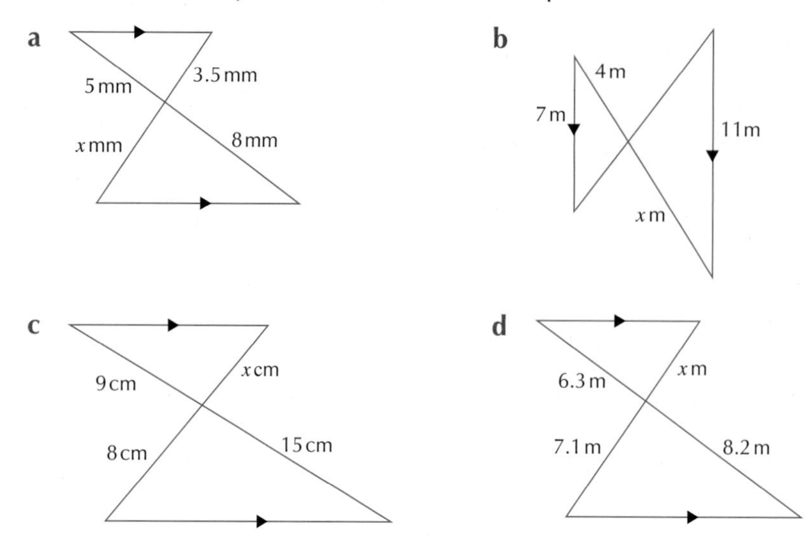

3 Calculate x. Give your answers to 2 decimal places.

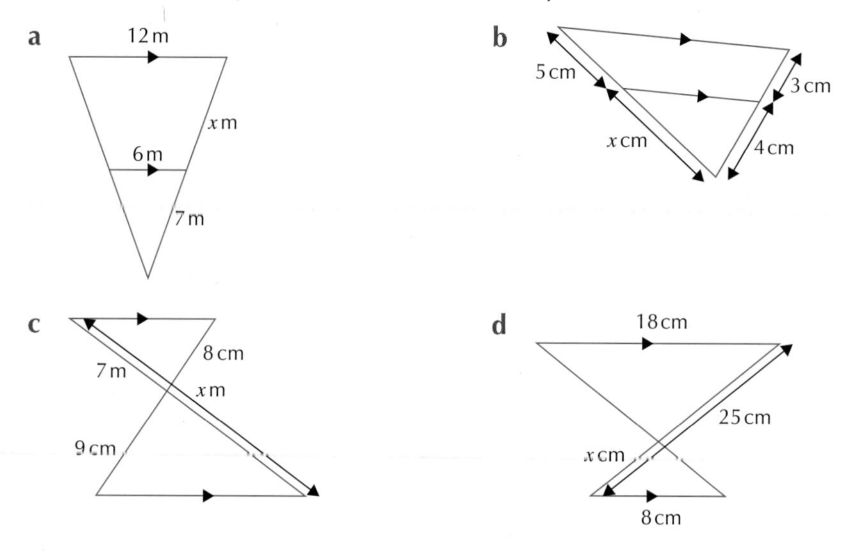

4 A side-on view of a table is shown. The table-top overlaps the legs by 5 cm on each side. The feet of the legs are 70 cm apart. Calculate the length of the table top. Give your answer to 3 significant figures.

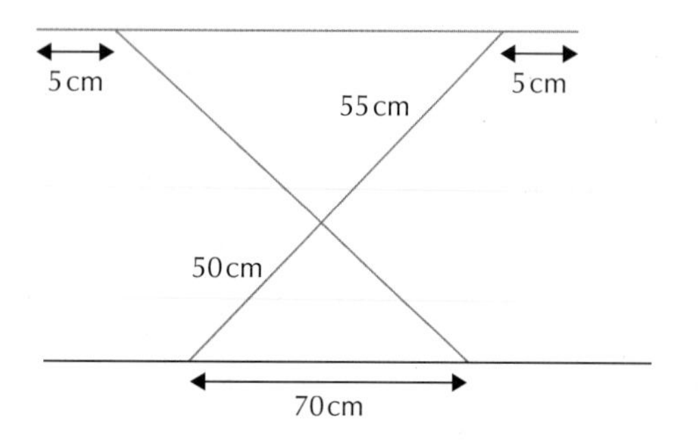

★ 5 Jake stands 3 m from the base of a street lamp which is 5 m tall. The length of his shadow is 1 m. How tall is Jake? Give your answer to 3 significant figures.

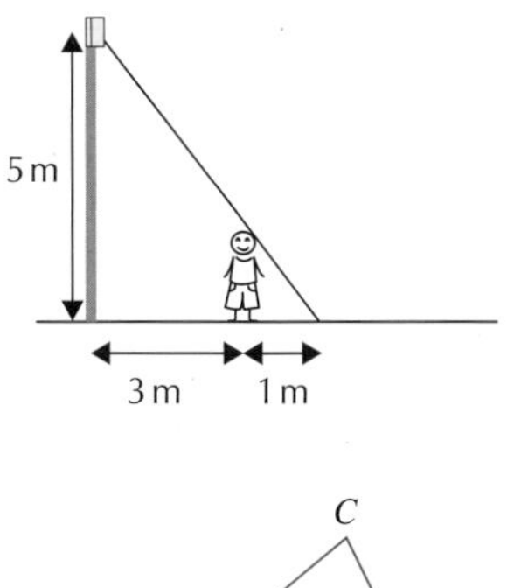

★ 6 A roof truss ABC is shown. A support beam, parallel to AB, is added between points D and E.

The length of DC is 1.2 m. The lengths of AB and DE are 6 m and 2 m, respectively. Calculate the length of AD. Give your answer to 3 significant figures.

7 When Sarah stands 3 metres from her garden wall, she can just see the top of her house over the wall. The distance from the wall to the house is 9 metres.

If the wall is 2.3 metres tall, and Sarah's eye level is at 1.4 metres, calculate the height of the house. Give your answer to 3 significant figures.

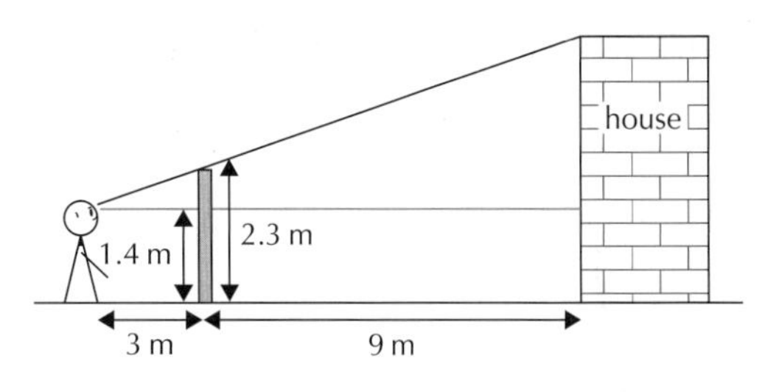

8 John wants to measure the distance across a river from A to B but has no practical means of doing so directly.

Instead he walks 15 metres along the riverbank from A to C and marks the spot. He then walks a further a further 20 metres to D which he also marks. From D he walks away from the river bank at right angles to it until he reaches a point E from where ECB is a straight line. He then measures the distance ED to be 23 metres.

Use this information to find the distance from A to B. Give your answer to 3 significant figures.

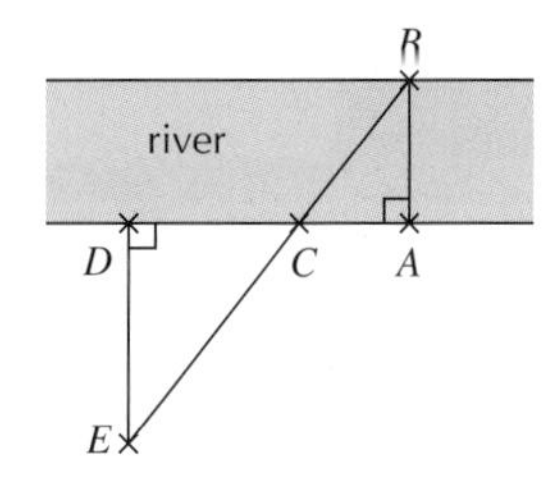

Activity

There are many situations where direct measurement of the distance between two points is impossible. There may be a natural obstruction, or the distance may simply be too great to measure physically. The method used in Question 8 in Exercise 22A overcomes this difficulty.

Working in pairs or as a group, select a distance to be calculated. Using a suitable measuring instrument and cones or sticks to mark appropriate points, use the method demonstrated in Question 8 to measure the distance.

You will not be using accurate measuring tools for either distances or angles, so your answer will only be an approximation. Repeat the task several times using different distances for the equivalent of A to C and C to D before deciding on your final result.

Areas and volumes of similar shapes

If you double the length of a cube, what happens to the area of a face? What happens to the volume? Enlargement and reduction scale factors for length are related to the scale factors for area and volume of similar shapes, and can be used to solve problems.

Consider the following lines, 2D shapes and 3D objects, with all lengths in centimetres.

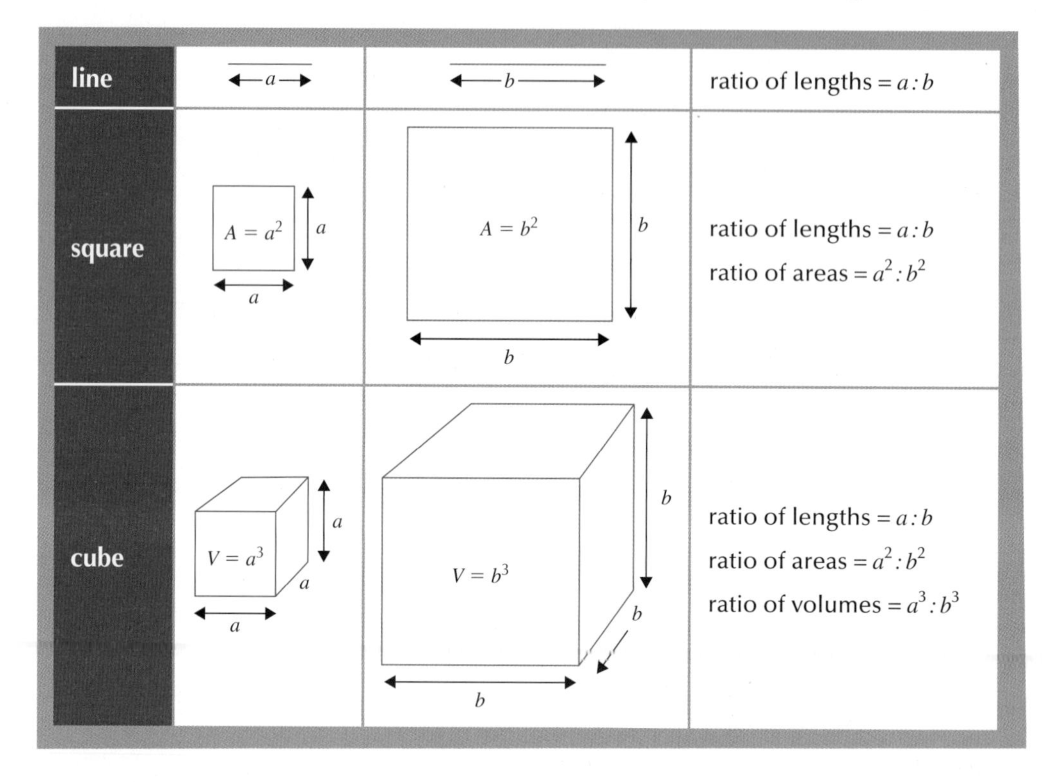

line
a
b
ratio of lengths = $a:b$
square
$A = a^2$
a
a
$A = b^2$
b
b
ratio of lengths = $a:b$
ratio of areas = $a^2:b^2$
cube
$V = a^3$
a
a
a
$V = b^3$
b
b
b
ratio of lengths = $a:b$
ratio of areas = $a^2:b^2$
ratio of volumes = $a^3:b^3$

There is an obvious relationship between linear, area and volume ratios. Although it has only been examined using a cube here, the result is the same for any similar shapes.

Whether dealing with enlargement or reduction:

- area scale factor = (linear scale factor)^2
- volume scale factor = (linear scale factor)^3.

Example 22.2

The rugs shown are mathematically similar.

The large rug has an area of 3.6 m².

Calculate the area of the small rug correct to 1 decimal place.

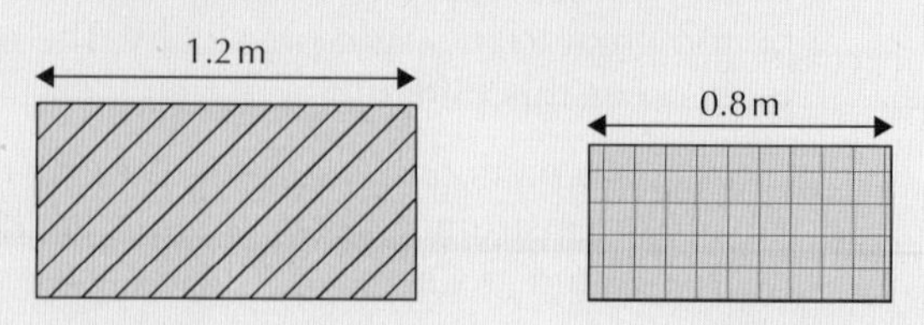

(continued)

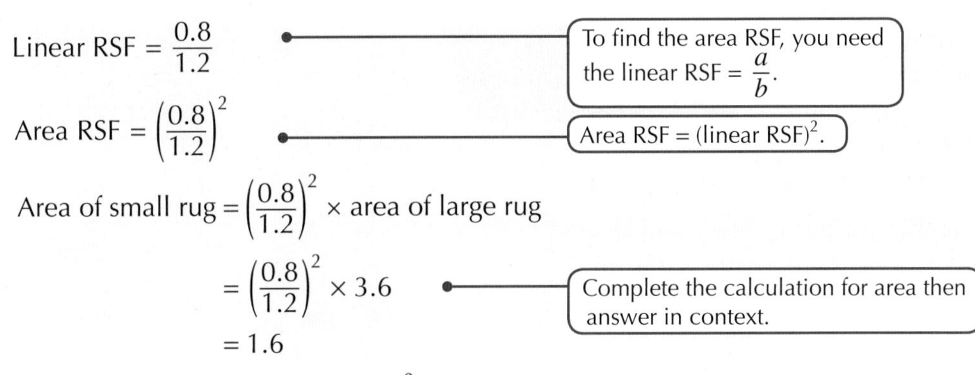

Linear RSF $= \frac{0.8}{1.2}$ — To find the area RSF, you need the linear RSF $= \frac{a}{b}$.

Area RSF $= \left(\frac{0.8}{1.2}\right)^2$ — Area RSF = (linear RSF)^2.

Area of small rug $= \left(\frac{0.8}{1.2}\right)^2 \times$ area of large rug

$= \left(\frac{0.8}{1.2}\right)^2 \times 3.6$ — Complete the calculation for area then answer in context.

$= 1.6$

The area of the small rug is 1.6 m^2.

Example 22.3

The following vases are mathematically similar. The small vase has a volume of 250 ml.

Calculate the volume of the large vase. Give your answer to the nearest millilitre.

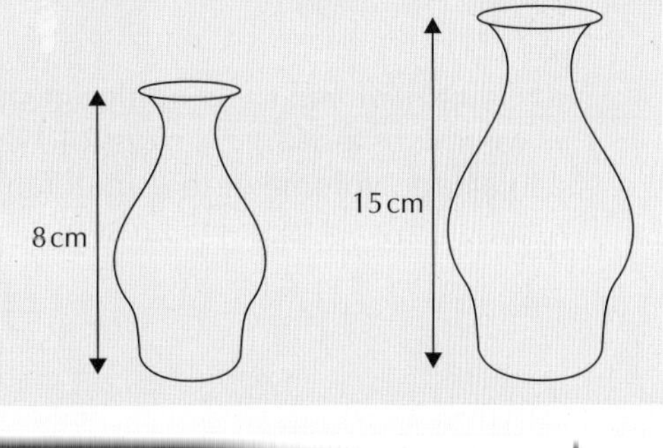

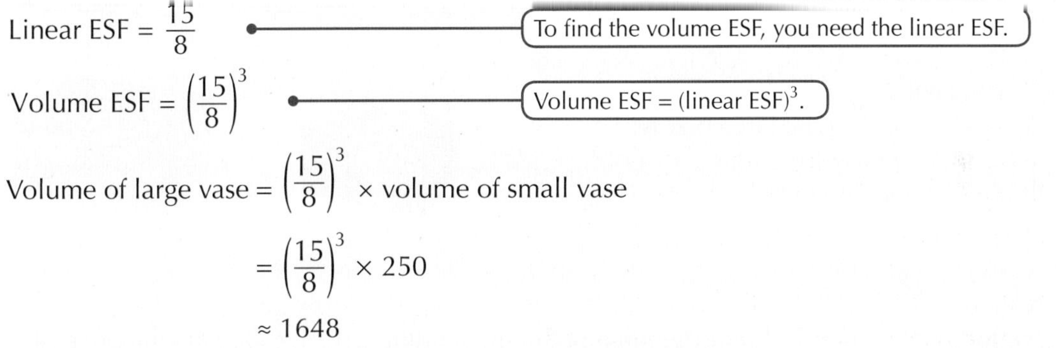

Linear ESF $= \frac{15}{8}$ — To find the volume ESF, you need the linear ESF.

Volume ESF $= \left(\frac{15}{8}\right)^3$ — Volume ESF = (linear ESF)^3.

Volume of large vase $= \left(\frac{15}{8}\right)^3 \times$ volume of small vase

$= \left(\frac{15}{8}\right)^3 \times 250$

≈ 1648

The volume of the large vase is 1648 ml to the nearest millilitre.

Exercise 22B

1 The following photographs are mathematically similar. The area of the smaller photograph is 450 cm^2. Calculate the area of the larger photograph. Give your answer to 3 significant figures.

30 cm

2 Two tabletops are similar in shape. The area of the large tabletop is 8000 cm². Calculate the area of the small tabletop, correct to the nearest cm².

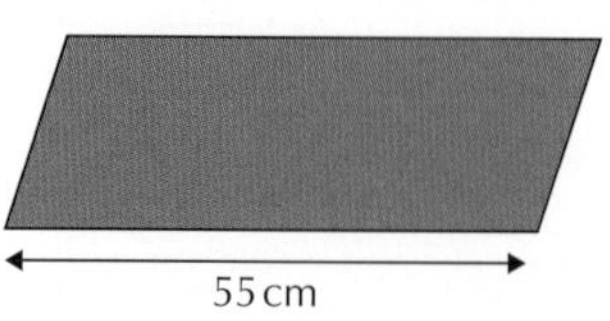

★ **3** The two bottles shown are similar in shape. The volume of the small bottle is 250 ml. Calculate the volume of the large bottle. Give your answer to the nearest 10 ml.

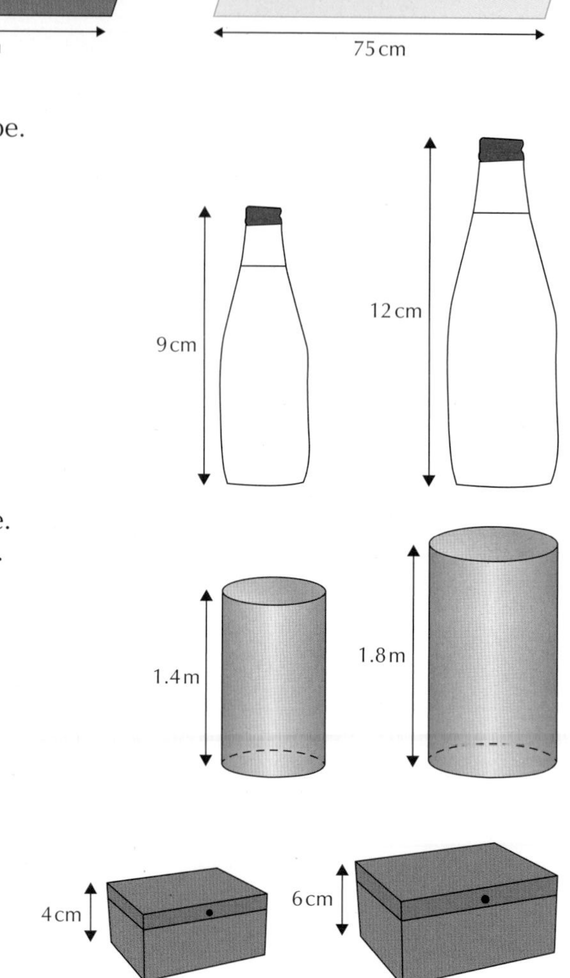

4 Two cylindrical tanks are similar in shape. The volume of the large tank is 240 litres. Calculate the volume of the small tank. Give your answer to the nearest litre

★ **5** Two metal boxes are similar in shape. The surface area of the large box is 140 cm². Calculate the surface area of the small box, correct to the nearest cm².

6 A shop sells bottles of perfume in similar-sized bottles. The prices of the bottles are proportional to their volume. The price of the small bottle is £25. Calculate the price of the large bottle.

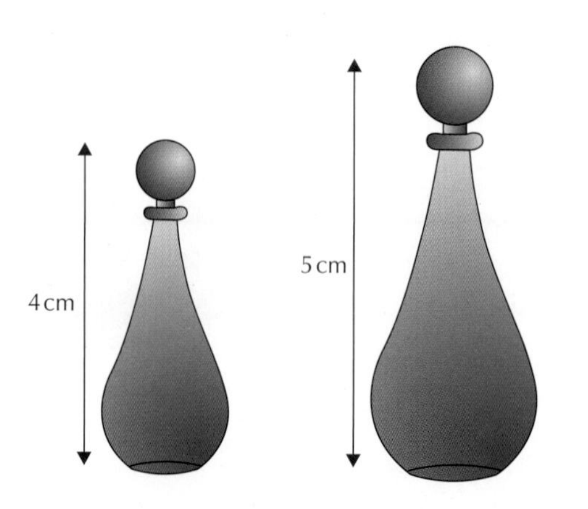

7 The price of a rug is proportional to its area. A circular rug with diameter 1.2 m costs £80. How much would a similar rug with diameter 1.5 m cost?

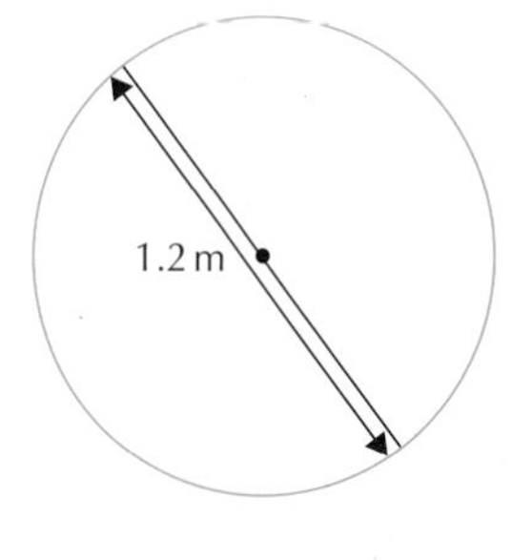

8 A company rents out skips. The cost is based on the volume of the skip. Two skips which are similar in shape are shown.

1.2 m 1.4 m

The rental price of the small skip is £80. How much would it cost to rent the large skip?

9 A box of cereal costs £1.62.

15 cm CEREAL

a How much would a similar-shaped box cost if it is 20 cm high?

b A similar-shaped box of height 25 cm costs £7.50. Is this cost proportional to the volume? Give a reason for your answer.

10 An ironmonger sells firelighters in two different sizes. The boxes are similar in shape. The small box is 6 cm high and costs £2.40. The large box is 8 cm high and costs £5.80.

The shop claims that the prices of the boxes are proportional to their volume. Is this correct? You must justify your answer.

More complex problems involving area and volume

In straightforward questions, you can always find the linear scale factor which is then used to find areas and volumes. This will not always be the case, and you need to be able to use indices and powers to work out scale factors.

In particular you need to know these rules:

- $\sqrt[n]{a} = a^{\frac{1}{n}}$
- $(\sqrt[n]{a})^m = a^{\frac{m}{n}}$

See Chapter 2 for more on indices and powers.

These rules give the following relationships between linear scale factors and scale factors for area and volume:

- linear scale factor = $\sqrt{\text{area scale factor}}$ = (area scale factor)$^{\frac{1}{2}}$
- linear scale factor = $\sqrt[3]{\text{volume scale factor}}$ = (volume scale factor)$^{\frac{1}{3}}$

Example 22.4

The paper cones shown are similar in shape. The small cone is 8 cm high and uses 180 cm² of paper. What will be the height of the larger cone if it is made from 290 cm² of card? Give your answer to 2 decimal places.

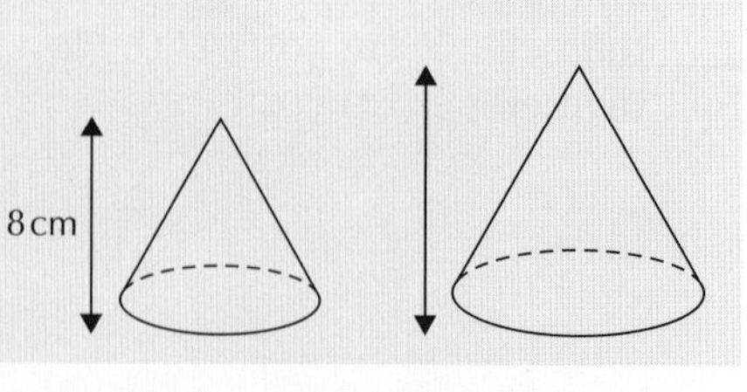

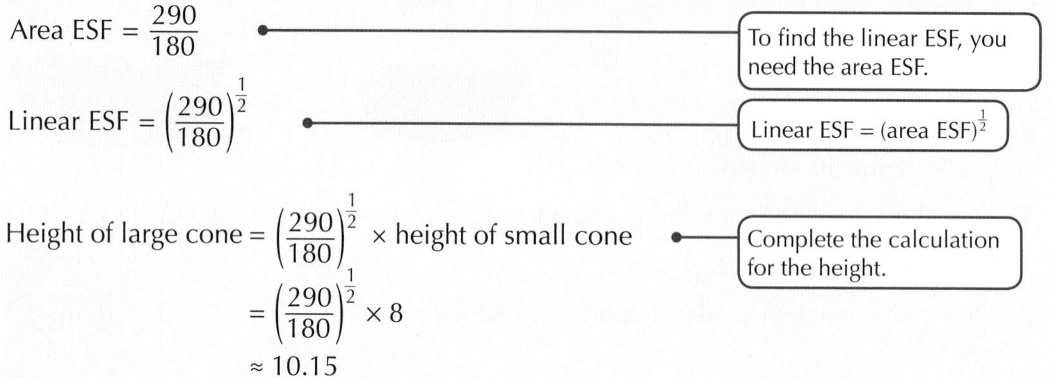

To find the linear ESF, you need the area ESF.
Linear ESF = (area ESF)$^{\frac{1}{2}}$
Complete the calculation for the height.

Area ESF $= \dfrac{290}{180}$

Linear ESF $= \left(\dfrac{290}{180}\right)^{\frac{1}{2}}$

Height of large cone $= \left(\dfrac{290}{180}\right)^{\frac{1}{2}} \times$ height of small cone

$= \left(\dfrac{290}{180}\right)^{\frac{1}{2}} \times 8$

≈ 10.15

The height of the larger cone will be approximately 10.15 cm.

Example 22.5

Two tins of paint are similar in shape as are their labels. They hold 1.4 litres and 2 litres, respectively.

The area of the label on the small tin is 80 cm². Calculate the area of the label on the large tin. Give your answer to the nearest cm².

Volume ESF $= \dfrac{2}{1.4}$

> To find the area ESF, you need the volume ESF.

Area ESF $= \left(\dfrac{2}{1.4}\right)^{\frac{2}{3}}$

> Linear ESF = (volume ESF)$^{\frac{1}{3}}$
>
> So, area ESF = (linear ESF)^2 = (volume ESF)$^{\frac{2}{3}}$

Area of large label $= \left(\dfrac{2}{1.4}\right)^{\frac{2}{3}} \times$ area of small label

$= \left(\dfrac{2}{1.4}\right)^{\frac{2}{3}} \times 80$

≈ 101

The area of the label on the larger tin is approximately 101 cm².

Exercise 22C

★ **1** A piece of card, 800 cm^2 in area, will make a tube 9 cm long. What is the length of a similar tube made from a similar piece of card with an area of 1000 cm^2? Give your answer to 1 decimal place.

2 The containers are similar-shaped.

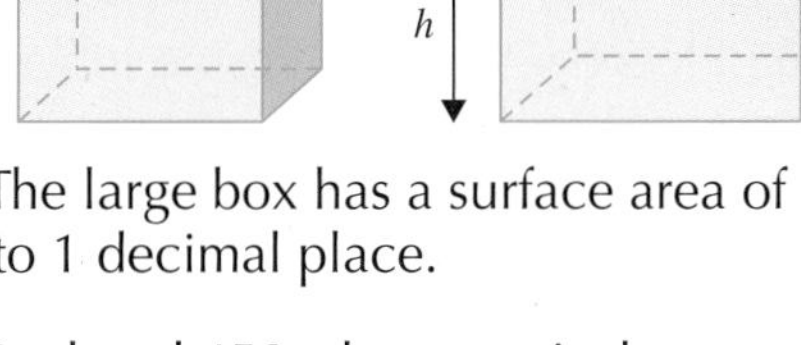

The total surface area of the small box is 400 cm^2. The large box has a surface area of 900 cm^2. Calculate its height, h. Give your answer to 1 decimal place.

3 The volume of two similar-shaped canisters are 300 ml and 450 ml, respectively. The height of the small canister is 12 cm. Calculate the height of the large canister. Give your answer to 1 decimal place.

4 Two water jugs are similar-shaped. They hold 500 ml and 700 ml, respectively.

The area of the base of the large jug is 35 cm^2. Calculate the area of the base of the small jug. Give your answer to 1 decimal place.

500 ml 700 ml

★ **5** Two cereal boxes are similar in shape.

The areas of the logos on the front of the box are 25 cm^2 and 30 cm^2, respectively. The volume of the large box is 1200 cm^3. Calculate the volume of the small box. Give your answer to 1 decimal place.

6 Two candles are similar-shaped. They have volumes of 60 cm^3 and 75 cm^2, respectively. The area of the base of the small candle is 4 cm^2. Calculate the area of the base of the large candle. Give your answer to 1 decimal place.

- I can use angle properties to establish the similarity of triangles before finding a scale factor and using it in calculations. ★ Exercise 22A Q5, Q6
- I can derive the enlargement or reduction scale factors for the area and volume of similar shapes from the linear scale factor and can use the result in problem solving. ★ Exercise 22B Q3, Q5
- I can work between the scale factors for length, area and volume, being able to derive any one from another and use the result in problem solving. ★ Exercise 22C Q1, Q5

For further assessment opportunities, see the Preparation for Assessment for Unit 2 on pages 288–291.

23 Working with the graphs of trigonometric functions

In this chapter you will learn how to:

- sketch and identify the **graphs** of the sine, cosine and tangent functions
- sketch and identify sine and cosine functions with **different amplitudes**
- sketch and identify sine, cosine and tangent functions with **different periods** and note the link with **multiple angles**
- sketch and identify sine, cosine and tangent functions which have undergone a **vertical translation**
- sketch and identify sine, cosine, and tangent functions which have undergone a **horizontal translation** and know what is meant by a **phase angle**
- solve problems involving sine and cosine graphs.

You should already know:

- how to find a side or angle in a right-angled triangle using the sine, cosine and tangent ratios.

Sine, cosine and tangent ratios

The three main trigonometric ratios are the sine, cosine and tangent. In right-angled triangles, they are defined as:

- $\sin\theta° = \dfrac{\text{Opposite}}{\text{Hypotenuse}} = \dfrac{\text{O}}{\text{H}}$
- $\cos\theta° = \dfrac{\text{Adjacent}}{\text{Hypotenuse}} = \dfrac{\text{A}}{\text{H}}$
- $\tan\theta° = \dfrac{\text{Opposite}}{\text{Adjacent}} = \dfrac{\text{O}}{\text{A}}$

H
O
$\theta°$
A

These definitions are only valid for angles between 0° and 90°. By transferring the triangle onto the coordinate plane we can get a more general definition valid for all angles.

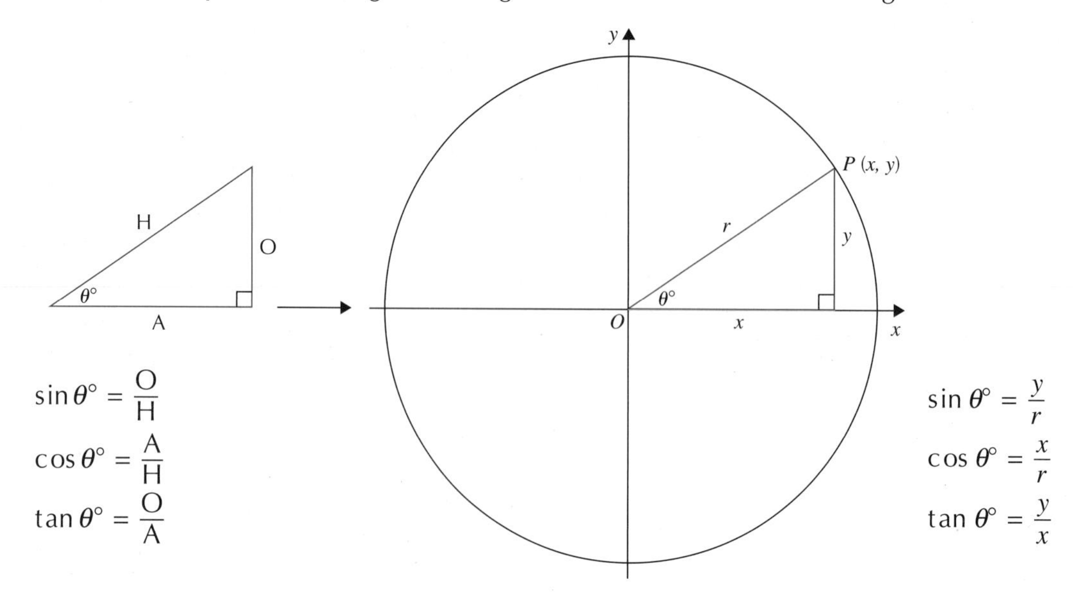

As the angle $\theta°$ changes, the point $P\ (x, y)$ moves round the circle, with a unique point P for each angle $\theta°$.

The angle $\theta°$ is measured from the positive x-axis.

- For $\theta > 0$, OP rotates anti-clockwise.
- For $\theta < 0$, OP rotates clockwise.

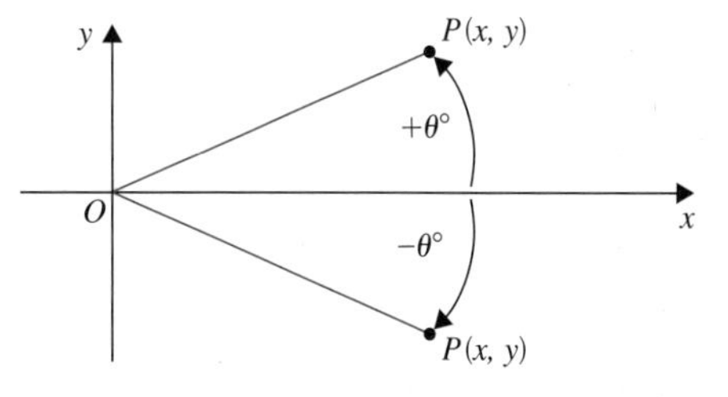

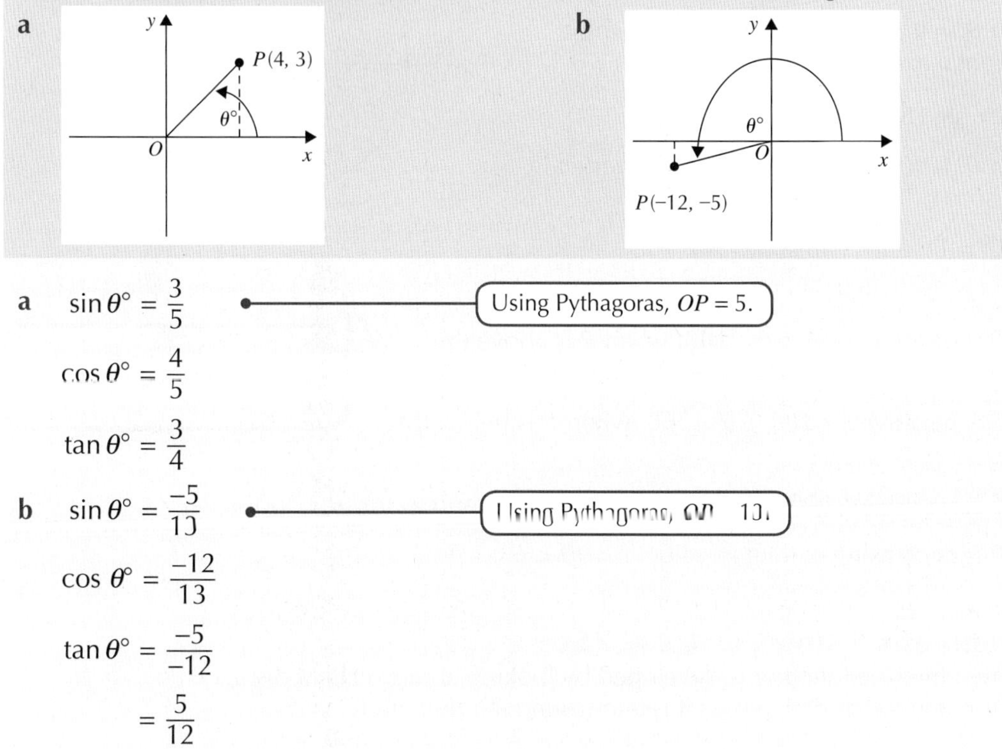

Example 23.1

Write down the value of sin, cos and tan of $\theta°$ in each of the following.

a $\sin\theta° = \frac{3}{5}$ — Using Pythagoras, $OP = 5$.

$\cos\theta° = \frac{4}{5}$

$\tan\theta° = \frac{3}{4}$

b $\sin\theta° = \frac{-5}{13}$ — Using Pythagoras, $OP = 13$.

$\cos\theta° = \frac{-12}{13}$

$\tan\theta° = \frac{-5}{-12}$

$= \frac{5}{12}$

The graphs of the sine and cosine functions

The definitions of the three trigonometric functions can be used to sketch the graphs of the sine and cosine functions.

A simple way of working out the graphs is by considering the point P travelling around the circle centred on the origin with radius 1 unit. This is referred to as the **unit circle**. The ratios are fixed for each angle, regardless of the size of the circle because of the properties of similarity.

As $r = 1$ in this unit circle, it follows that:

- $\sin\theta° = \frac{y}{1} = y$
- $\cos\theta° = \frac{x}{1} = x$

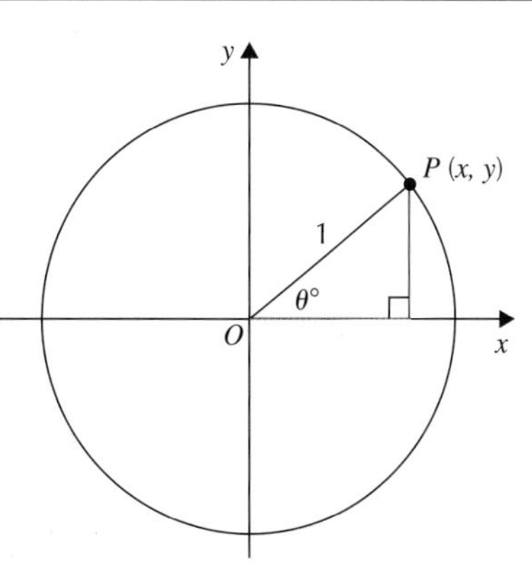

It follows that, for any angle $\theta°$:

- $\sin\theta°$ = the **vertical displacement** of the corresponding point P from the **horizontal axis**
- $\cos\theta°$ = the **horizontal displacement** of the corresponding point P from the **vertical axis.**

This gives us valuable information about the graphs of these functions.

The key points on the unit circle

By noting the vertical displacement of the point P_θ from the horizontal axis as θ increases from 0 to 360, we can see that:

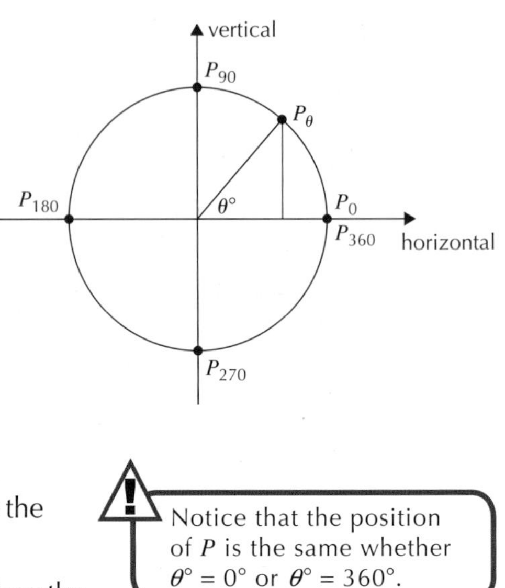

- when $\theta° = 0°$, $180°$ or $360°$, P lies on the horizontal axis, so:

  $\sin 0° = \sin 180° = \sin 360° = 0$

- when $\theta° = 90°$ or $270°$, P lies on the vertical axis, so:

  $\sin 90° = 1$ and $\sin 270° = -1$

It can also be seen that:

- $\sin\theta°$ is **positive** for $0 < \theta < 180$, when P is above the x-axis
- $\sin\theta°$ is **negative** for $180 < \theta < 360$, when P is below the x-axis
- $\sin\theta°$ is **increasing** (rising in value) for $0 < \theta < 90$ and $270 < \theta < 360$
- $\sin\theta°$ is **decreasing** (falling in value) for $90 < \theta < 270$.

Notice that the position of P is the same whether $\theta° = 0°$ or $\theta° = 360°$.

The graph of $y = \sin x°$, $0 \leqslant x \leqslant 360$

The graph of the sine function is developed by looking at the **vertical** displacement of the point P from the horizontal axis as it rotates round the unit circle.

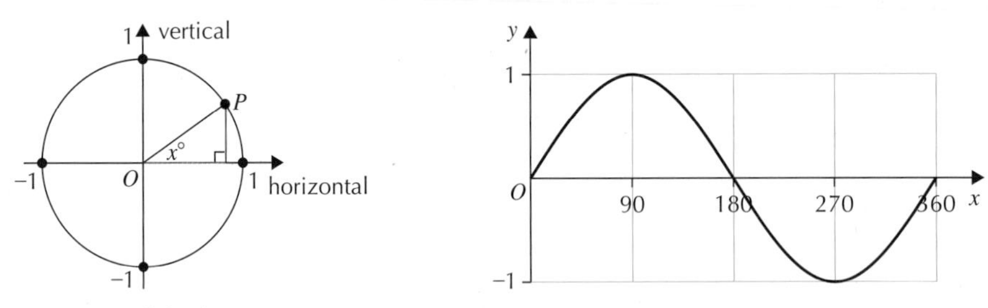

$\sin x°$ = vertical displacement

The key points to remember on the graph $y = \sin x°$, $0 \leqslant x \leqslant 360$ are:

(0°, 0), (90°, 1), (180°, 0), (270°, −1) and (360°, 0).

The graph of $y = \sin x°$, $x \in \mathbb{R}$

The rotating line OP is not restricted to moving through the angles between 0° and 360°. It can rotate through any anti-clockwise (positive) angle or any clockwise (negative) angle but P will return to the same spot every 360°, so it can be seen that the graph repeats every 360°.

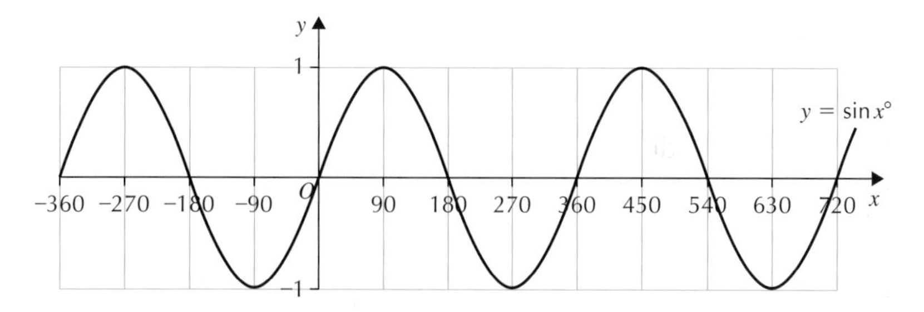

The graph of $y = \sin x°$ is described as having:

- **amplitude** = 1 (maximum distance above and below the horizontal axis)
- **period** = 360° (the graph repeats every 360°).

The graph of $y = \cos x°$, $0 \leqslant x \leqslant 360$

The graph of the cosine function is developed in a similar way to that of the sine function. For the cosine function, it is the **horizontal** displacement of the point P from the horizontal axis as it travels anti-clockwise round the unit circle which defines the graph.

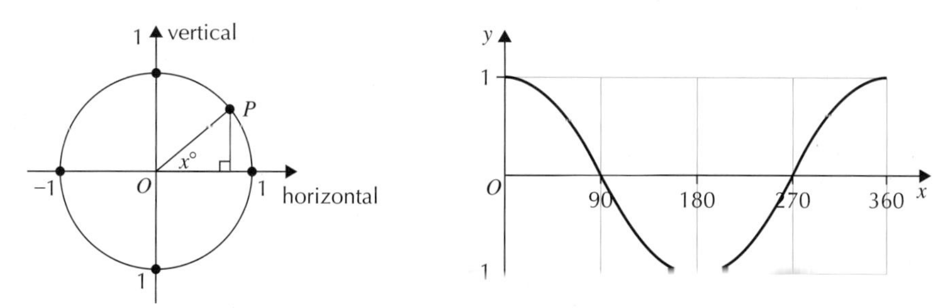

$\cos x°$ = horizontal displacement

The key points to remember on the graph $y = \cos x°$, $0 \leqslant x \leqslant 360$ are:
(0°, 1), (90°, 0), (180°, –1), (270°, 0) and (360°, 1).

The graph of $y = \cos x°$, $x \in \mathbb{R}$

As for sine, the graph of the cosine function is not restricted to angles between 0°and 360°.

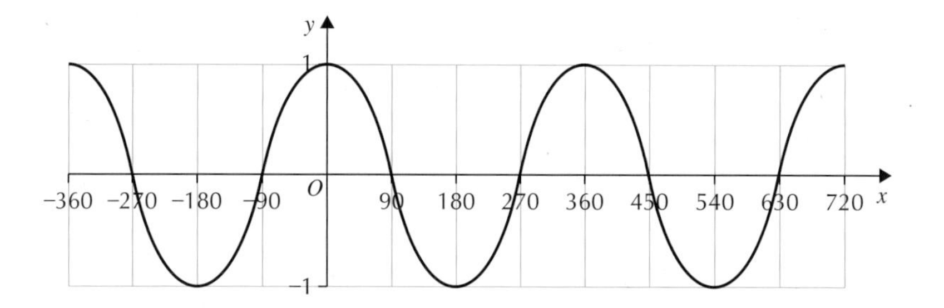

The graph of $y = \cos x°$ is also described as having:

- **amplitude** = 1 (maximum distance above and below the horizontal axis)
- **period** = 360° (the graph repeats every 360°).

The graphs of sine and cosine functions with different amplitudes: $y = a\sin x°$ and $y = a\cos x°$

In the graphs of $y = a\sin x°$ and $y = a\cos x°$ for different values of a, multiplying by a only affects the amplitude of the graph, and does not change the period.

The graphs of $y = a\sin x°$ and $y = a\cos x°$ have:

- amplitude = $|a|$
- period = 360°.

> $|a|$ is the **absolute value** of a. This is the positive difference between a and zero.

Example 23.2

Sketch the graphs of:

a $y = 3\sin x°$, $0 \leqslant x \leqslant 360$ **b** $y = \frac{1}{2}\cos x°$, $0 \leqslant x \leqslant 360$ **c** $y = -2\sin x°$, $0 \leqslant x \leqslant 360$

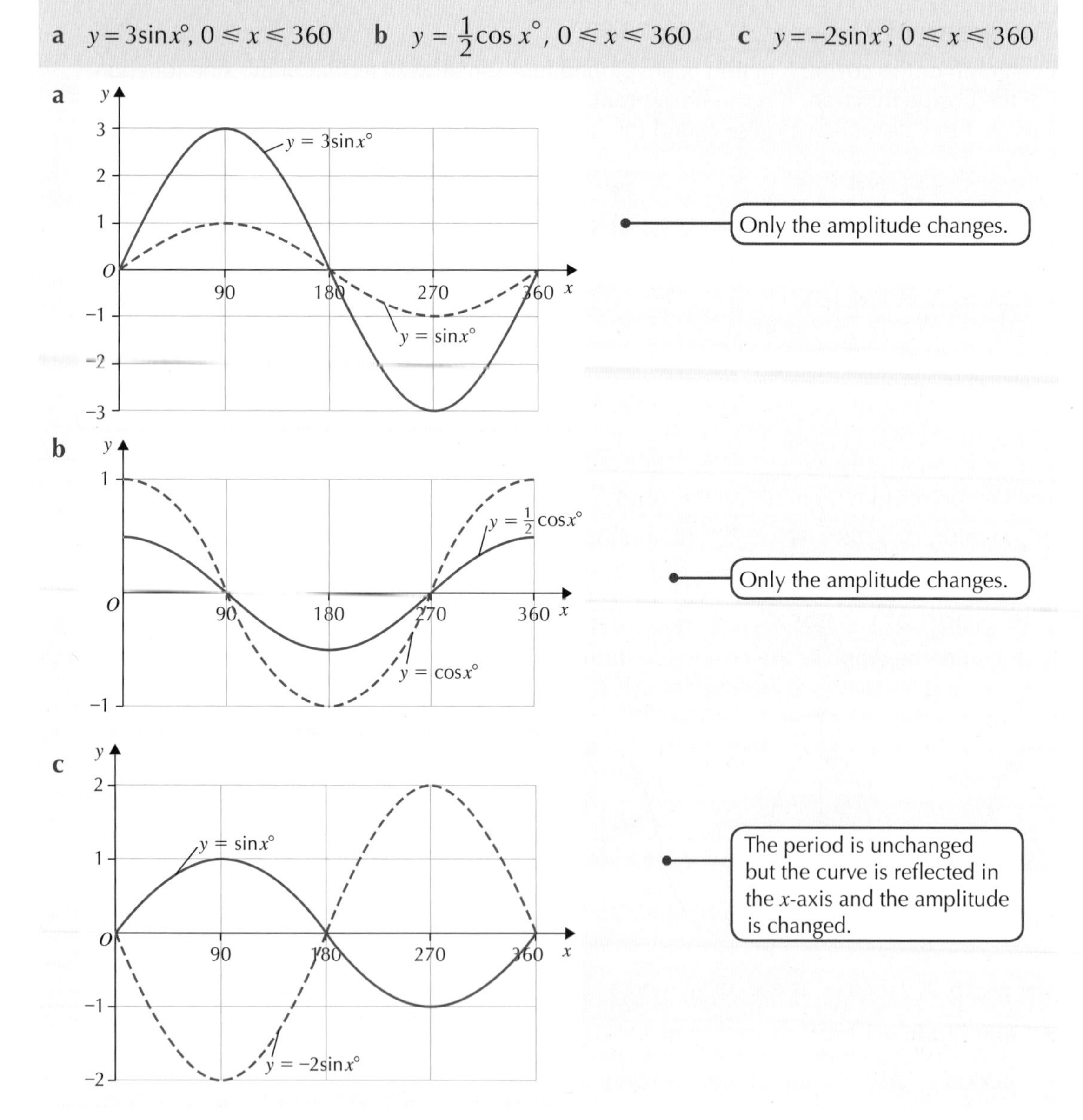

Exercise 23A

★ **1** Sketch the graphs of the following for $0 \leqslant x \leqslant 360$.

a $y = 5\sin x°$ **b** $y = \frac{3}{4}\cos x°$ **c** $y = -4\sin x°$

d $y = 3\cos x°$ **e** $y = \frac{1}{2}\sin x°$ **f** $y = -2\cos x°$

★ **2** State the equations of each of the following graphs.

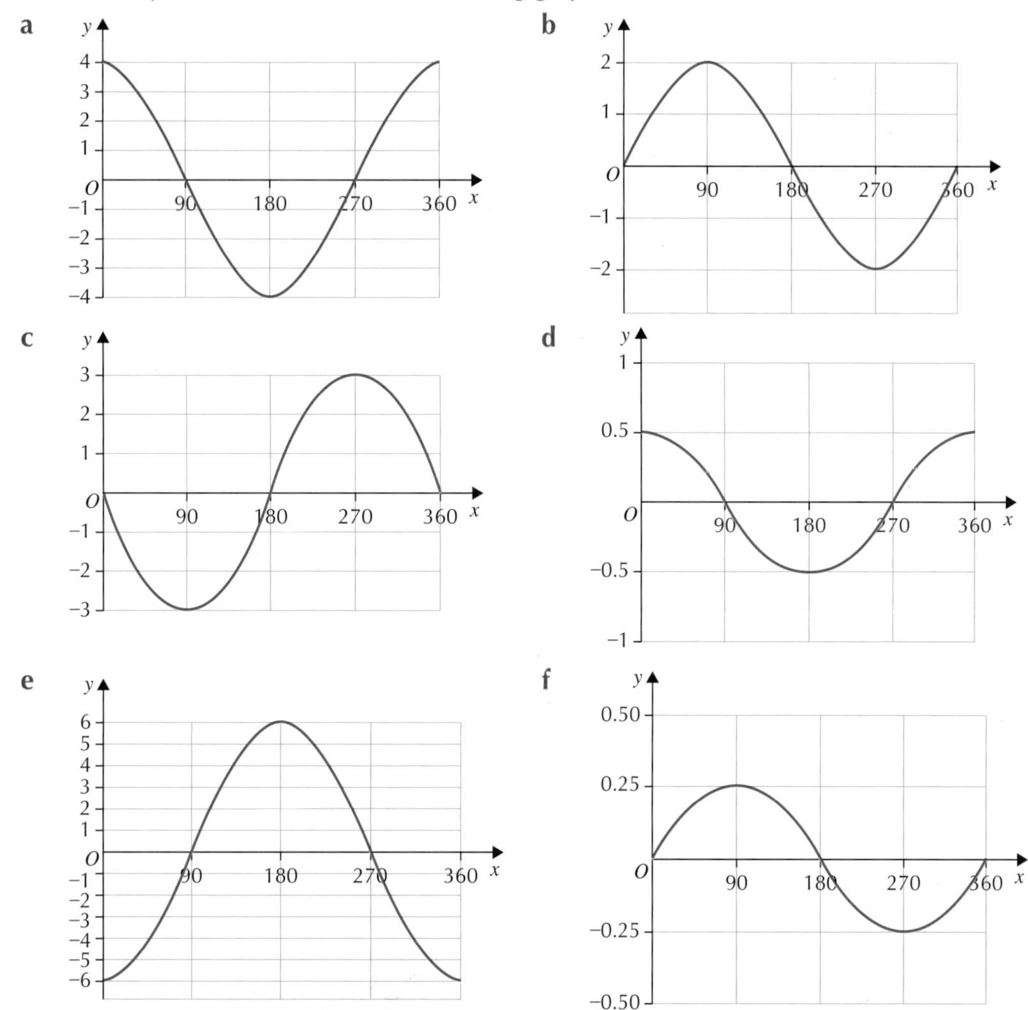

The graphs of sine and cosine functions with different periods: $y = \sin bx°$ and $y = \cos bx°$, $b > 0$

In the graphs of $y = \sin bx°$ and $y = \cos bx°$ for different values of b, multiplying by b only affects the period of the graph, and does not change the amplitude. If $y = \sin bx°$ or $y = \cos bx°$ there are b cycles in 360°, giving one cycle every $\left(\frac{360}{b}\right)°$.

The graphs of $y = \sin bx°$ and $y = \cos bx°$ where $b > 0$ have:

- amplitude = 1
- period = $\left(\frac{360}{b}\right)°$

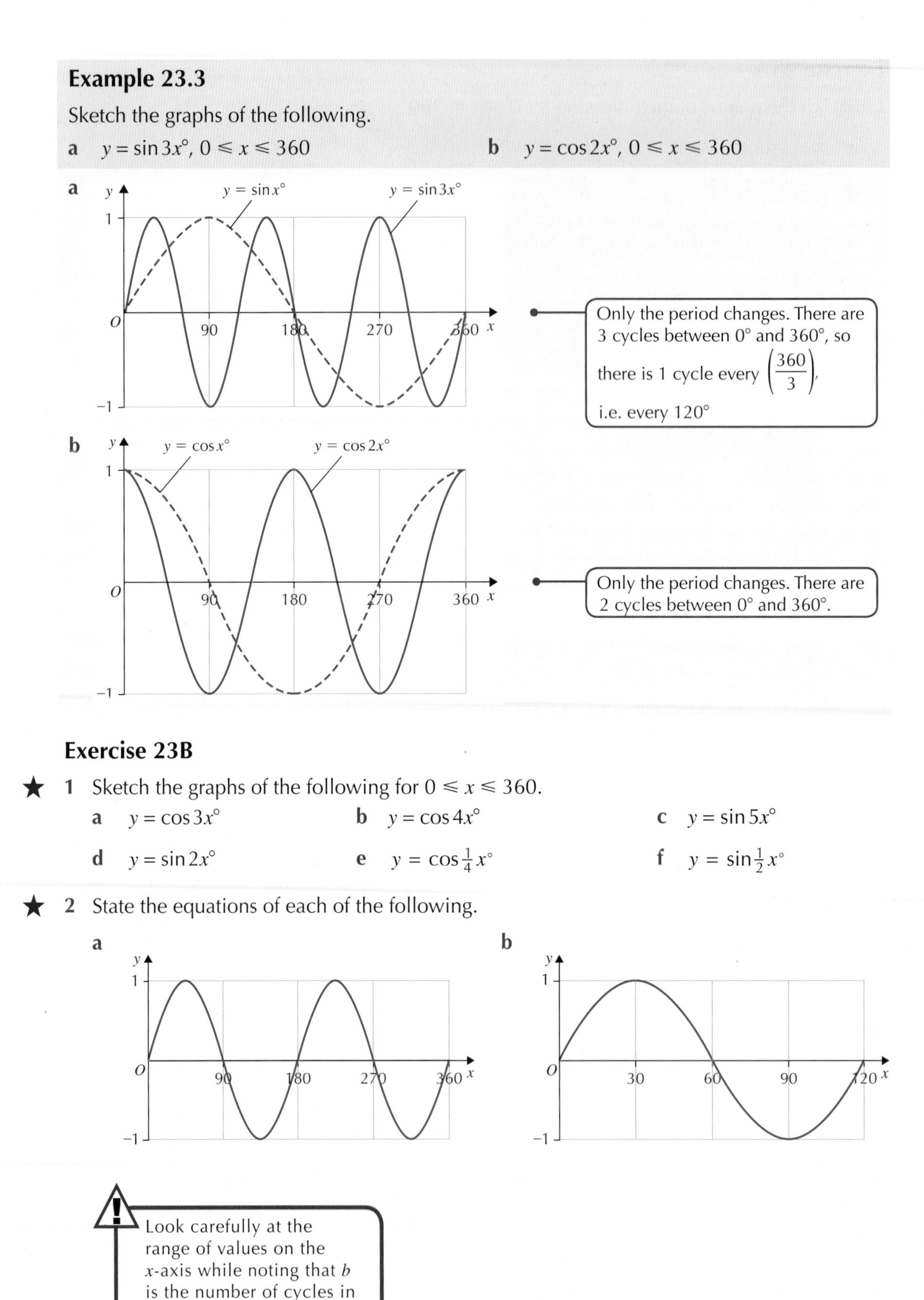

Example 23.3

Sketch the graphs of the following.

a $y = \sin 3x°$, $0 \leqslant x \leqslant 360$ **b** $y = \cos 2x°$, $0 \leqslant x \leqslant 360$

Exercise 23B

★ **1** Sketch the graphs of the following for $0 \leqslant x \leqslant 360$.

a $y = \cos 3x°$ **b** $y = \cos 4x°$ **c** $y = \sin 5x°$

d $y = \sin 2x°$ **e** $y = \cos \frac{1}{4}x°$ **f** $y = \sin \frac{1}{2}x°$

★ **2** State the equations of each of the following.

Look carefully at the range of values on the x-axis while noting that b is the number of cycles in 360°.

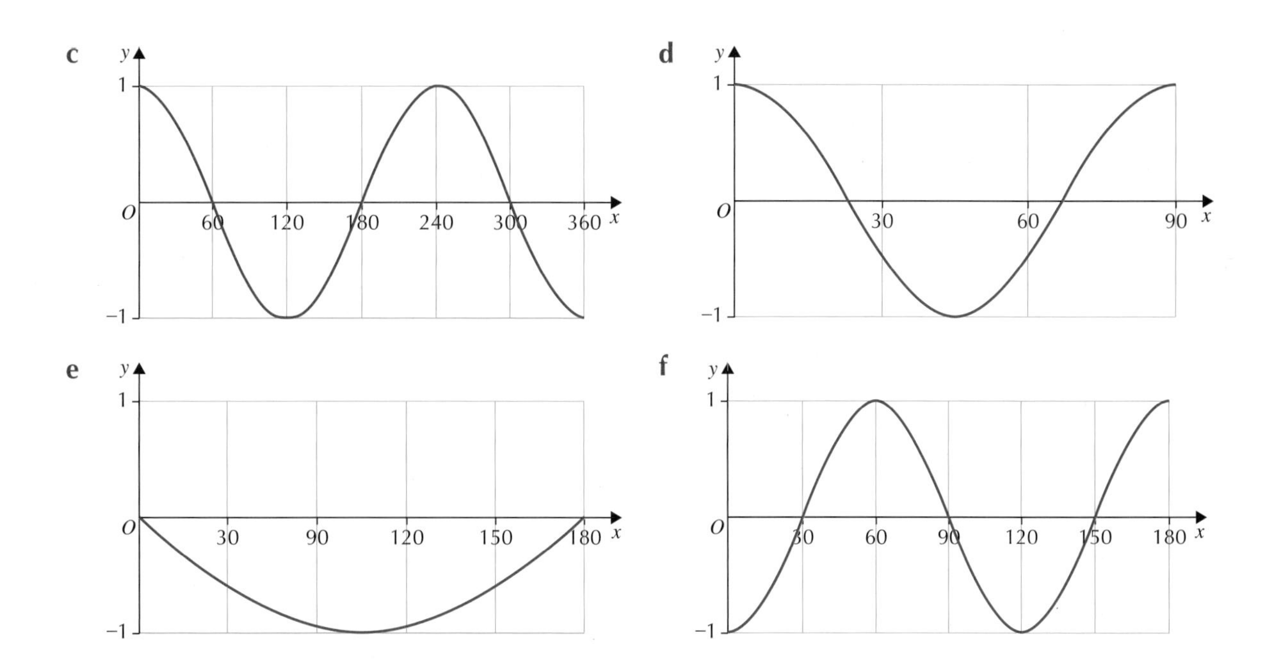
c
d
e
f

The graphs of sine and cosine functions with different amplitudes and periods: $y = a\sin bx°$ and $y = a\cos bx°$

The graphs of $y = a\sin bx°$ and $y = a\cos bx°$ have:

- amplitude = $|a|$
- period = $\left(\frac{360}{b}\right)^\circ$ (b cycles in 360°).

Example 23.4

Sketch the graphs of the following.

a $y = 3\sin 2x°$, $0 \leqslant x \leqslant 360$ **b** $y = 5\cos 3x°$, $0 \leqslant x \leqslant 360$

c $y = -2\cos 2x°$, $0 \leqslant x \leqslant 360$ **d** $y = 3\sin 4x°$, $0 \leqslant x \leqslant 180$

a $y = 3\sin 2x°$, $0 \leqslant x \leqslant 360$

$a = 3$, so amplitude = 3

$b = 2$, so 2 cycles in 360°, giving 1 cycle in 180°

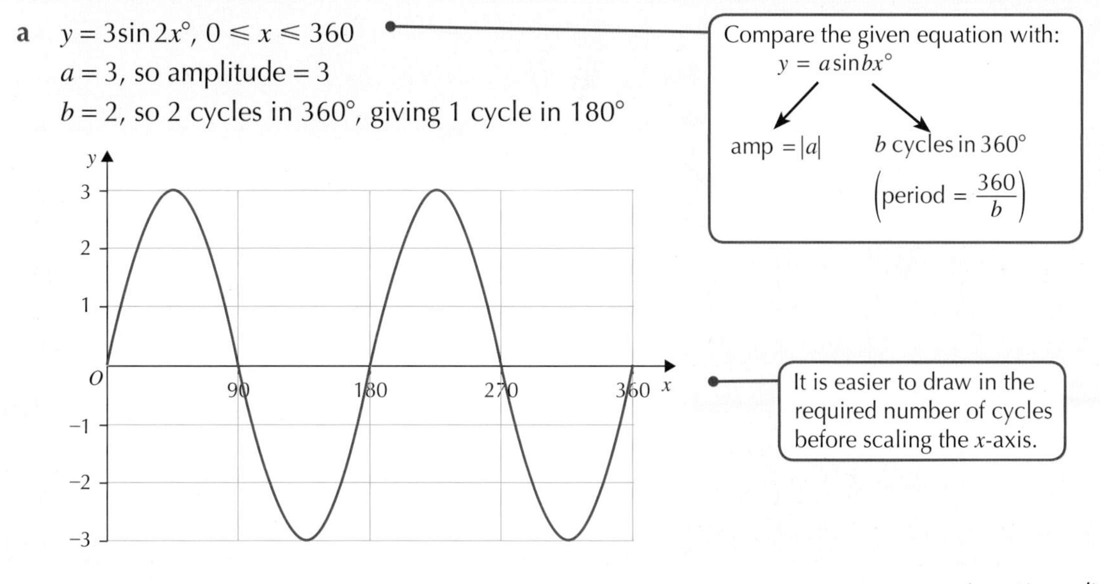
Compare the given equation with:
y = a sin bx°
amp = |a|
b cycles in 360°
(period = 360/b)
It is easier to draw in the required number of cycles before scaling the x-axis.

(continued)

b $y = 5\cos 3x°$, $0 \leqslant x \leqslant 360$

$a = 5$, so amplitude $= 5$

$b = 3$, so 3 cycles in 360°, giving 1 cycle in 120°

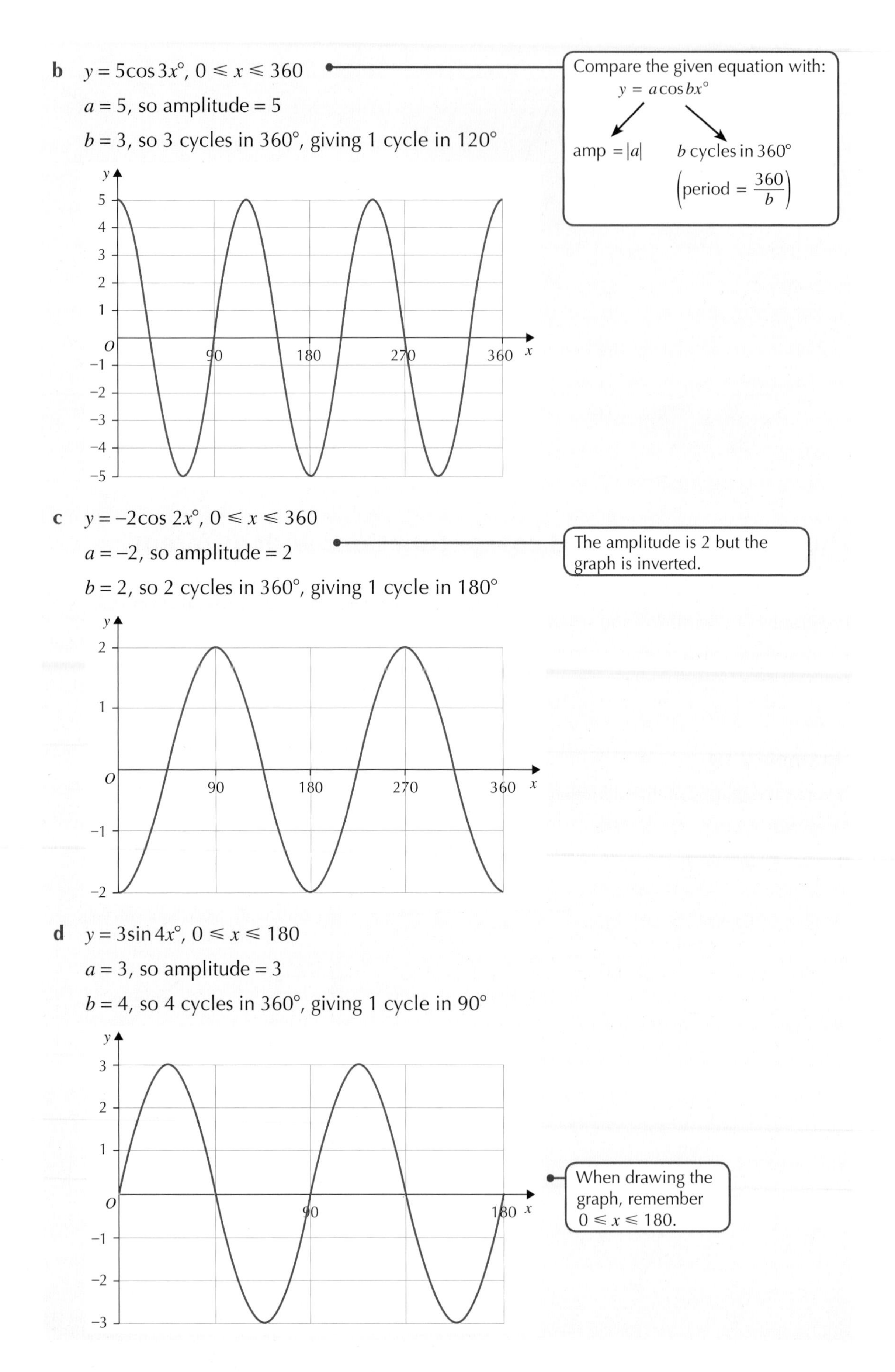

Compare the given equation with:
$y = a\cos bx°$
amp $= |a|$
b cycles in 360°
(period $= \frac{360}{b}$)
The amplitude is 2 but the graph is inverted.
When drawing the graph, remember $0 \leqslant x \leqslant 180$.

c $y = -2\cos 2x°$, $0 \leqslant x \leqslant 360$

$a = -2$, so amplitude $= 2$

$b = 2$, so 2 cycles in 360°, giving 1 cycle in 180°

d $y = 3\sin 4x°$, $0 \leqslant x \leqslant 180$

$a = 3$, so amplitude $= 3$

$b = 4$, so 4 cycles in 360°, giving 1 cycle in 90°

Example 23.5

State the equation of each of the following graphs.

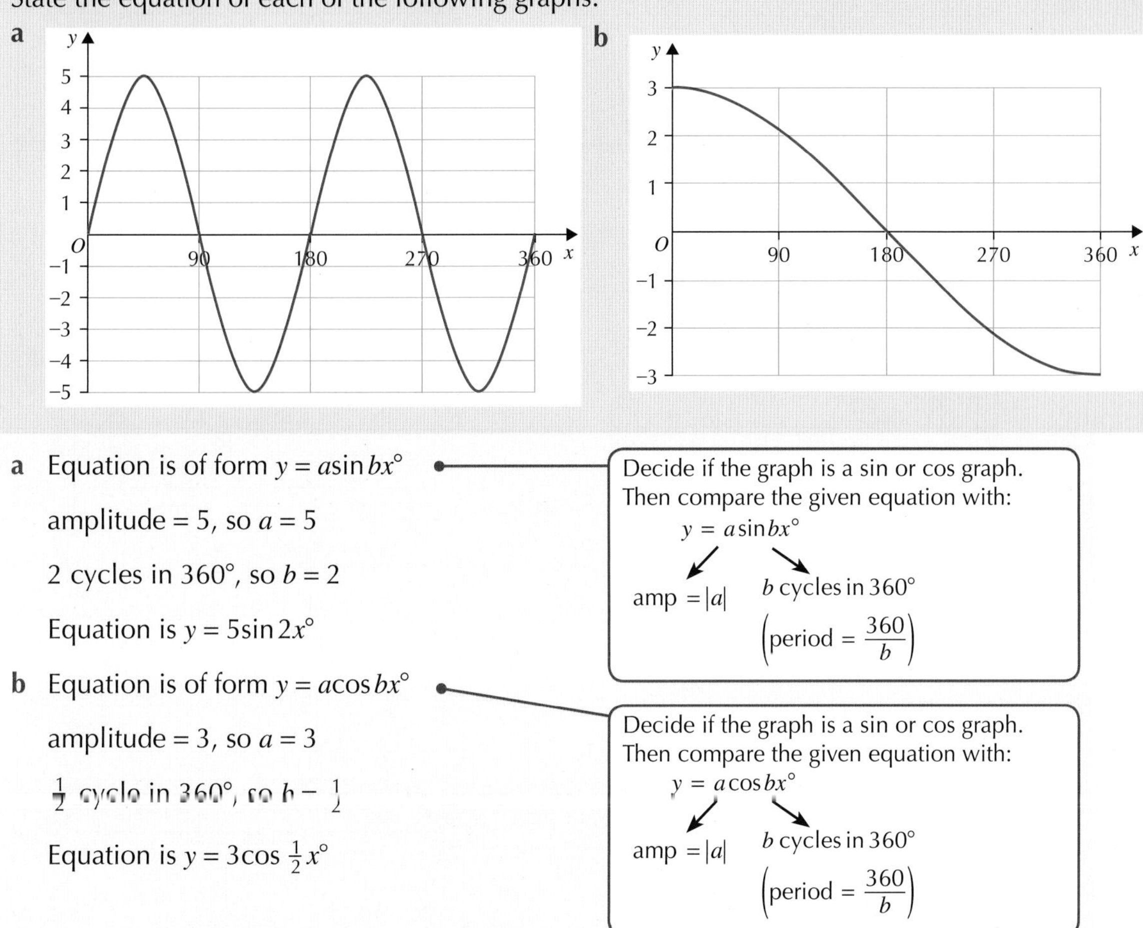

a Equation is of form $y = a\sin bx°$

amplitude = 5, so $a = 5$

2 cycles in 360°, so $b = 2$

Equation is $y = 5\sin 2x°$

b Equation is of form $y = a\cos bx°$

amplitude = 3, so $a = 3$

$\frac{1}{2}$ cycle in 360°, so $b = \frac{1}{2}$

Equation is $y = 3\cos \frac{1}{2}x°$

Decide if the graph is a sin or cos graph. Then compare the given equation with:
$y = a\sin bx°$
amp $= |a|$; b cycles in 360° $\left(\text{period} = \frac{360}{b}\right)$

Decide if the graph is a sin or cos graph. Then compare the given equation with:
$y = a\cos bx°$
amp $= |a|$; b cycles in 360° $\left(\text{period} = \frac{360}{b}\right)$

Exercise 23C

★ **1** Sketch the graphs of the following for $0 \leqslant x \leqslant 360$.

a $y = 3\sin 2x°$ **b** $y = 5\cos 3x°$ **c** $y = 2\cos \frac{1}{2}x°$ **d** $y = 2\sin 4x°$

e $y = -2\cos 3x°$ **f** $y = -3\sin 2x°$ **g** $y = 2\cos 5x°$ **h** $y = 3\sin \frac{1}{2}x°$

2 Sketch the following.

a $y = 3\cos 2x°$, $0 \leqslant x \leqslant 180$ **b** $y = 4\sin 3x°$, $0 \leqslant x \leqslant 180$

c $y = 2\sin 3x°$, $0 \leqslant x \leqslant 720$ **d** $y = 2\cos 2x°$, $0 \leqslant x \leqslant 540$

★ **3** State the equations of each of the following graphs.

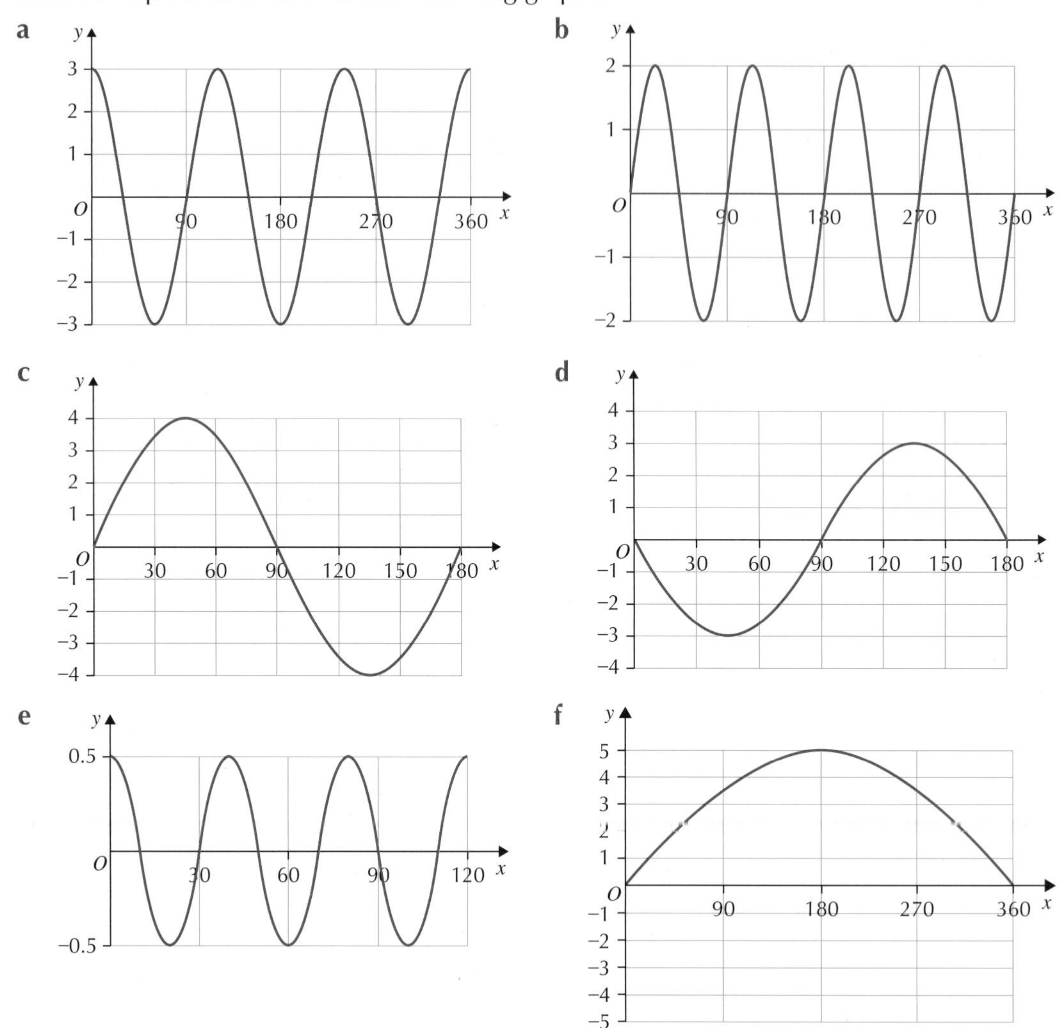

The graphs of sine and cosine functions which have undergone a vertical translation: $y = a\sin bx° + c$ and $y = a\cos bx° + c$

In the graphs of $y = a\sin bx° + c$ and $y = a\cos bx° + c$ for different values of c, adding c translates the graphs vertically up or down by $|c|$ according to whether c is positive or negative.

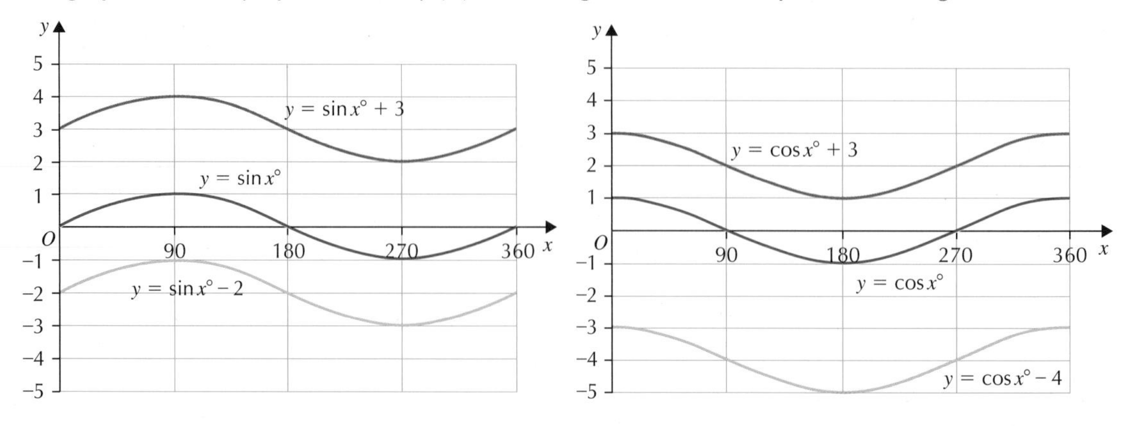

Example 23.6

Sketch each of the following where $0 \leqslant x \leqslant 360$.

a $y = \sin x° + 5$ **b** $y = \cos 3x° - 2$ **c** $y = 2\cos x° + 3$ **d** $y = 1 + 4\sin 3x°$

a $y = \sin x° + 5$

Compare the given equation with: $y = a\sin bx \pm c$
amp = $|a|$ up/down c
b cycles in 360°

$a = 1$, so amplitude = 1

$b = 1$, so 1 cycle in 360°

$c = 5$, which moves the graph up 5, so maximum value = 5 + 1 = 6 and minimum value = 5 + −1 = 4

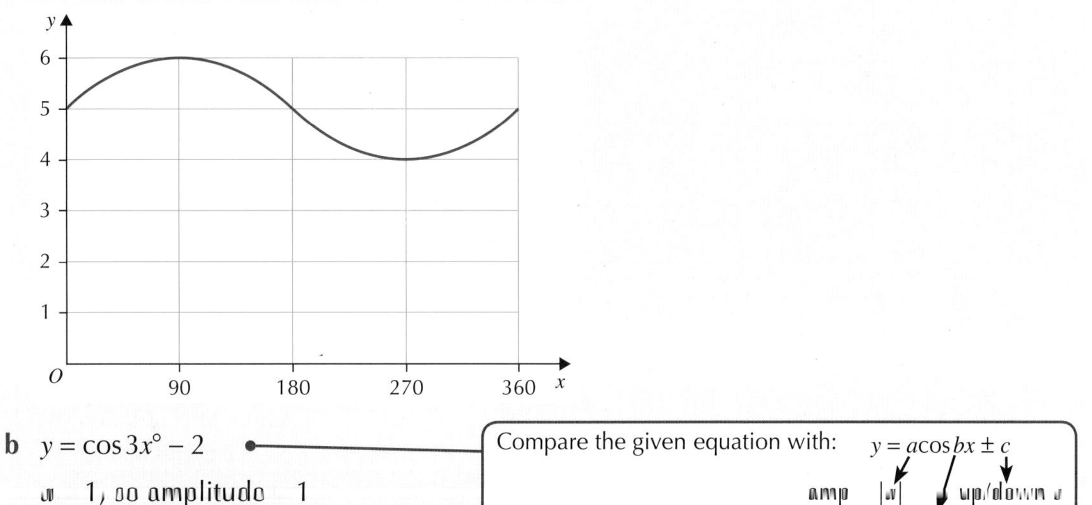

b $y = \cos 3x° - 2$

Compare the given equation with: $y = a\cos bx \pm c$
amp = $|a|$ up/down c
b cycles in 360°

$a = 1$, so amplitude = 1

$b = 3$, so 3 cycles in 360°, giving 1 cycle in 120°

$c = -2$, which moves the graph down 2, so maximum value = −2 + 1 = −1 and minimum value = −2 − 1 = −3

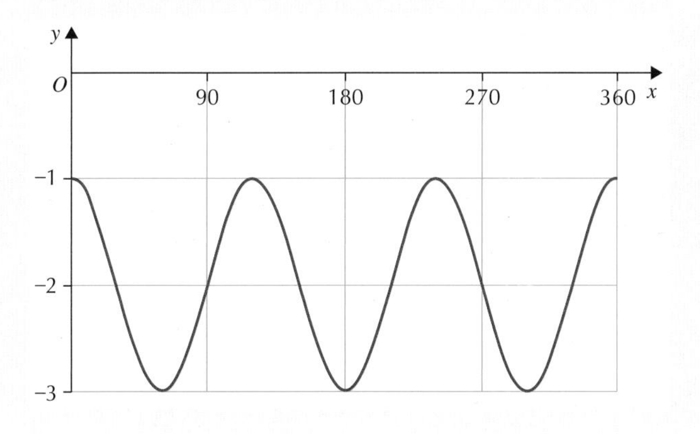

(continued)

c $y = 2\cos x° + 3$

Compare the given equation with: $y = a\cos bx \pm c$

amp = $|a|$ up/down c

b cycles in 360°

$a = 2$, so amplitude = 2

$b = 1$, so 1 cycle in 360°

$c = 3$, which moves the graph up 3, so maximum value = 3 + 2 = 5 and minimum value = 3 + −2 = 1

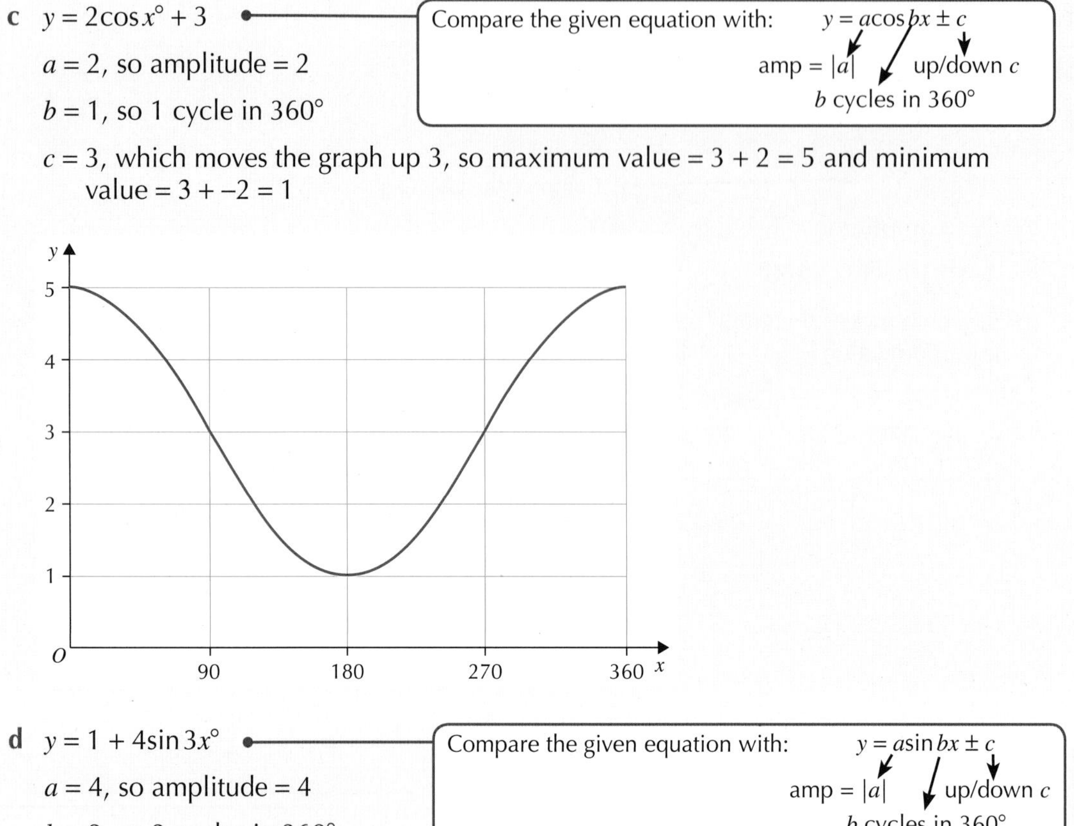

d $y = 1 + 4\sin 3x°$

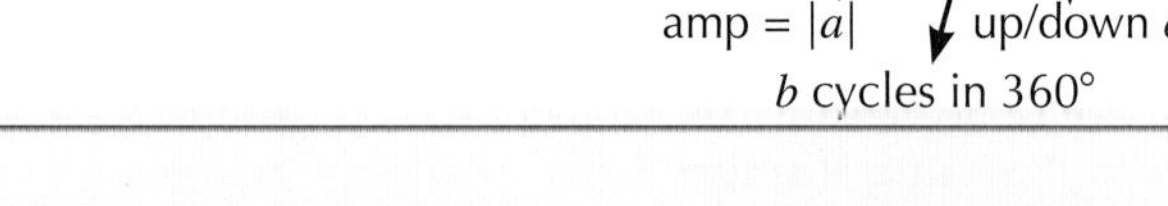

$a = 4$, so amplitude = 4

$b = 3$, so 3 cycles in 360°, giving 1 cycle in 120°

$c = 1$, which moves the graph up 1, so maximum value = 1 + 4 = 5 and minimum value = 1 + −4 = −3

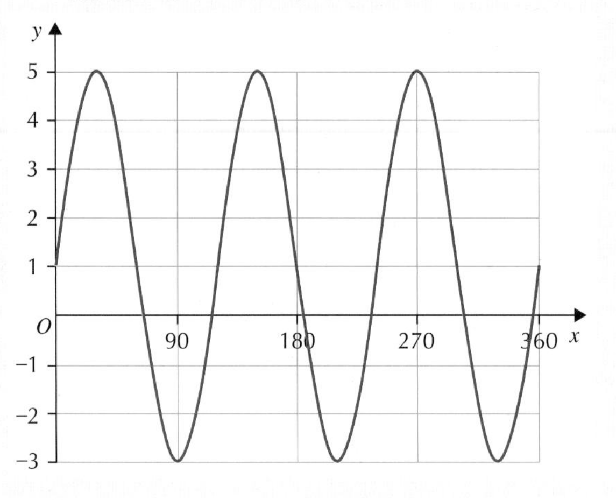

Example 23.7

Find the equation of each of the following.

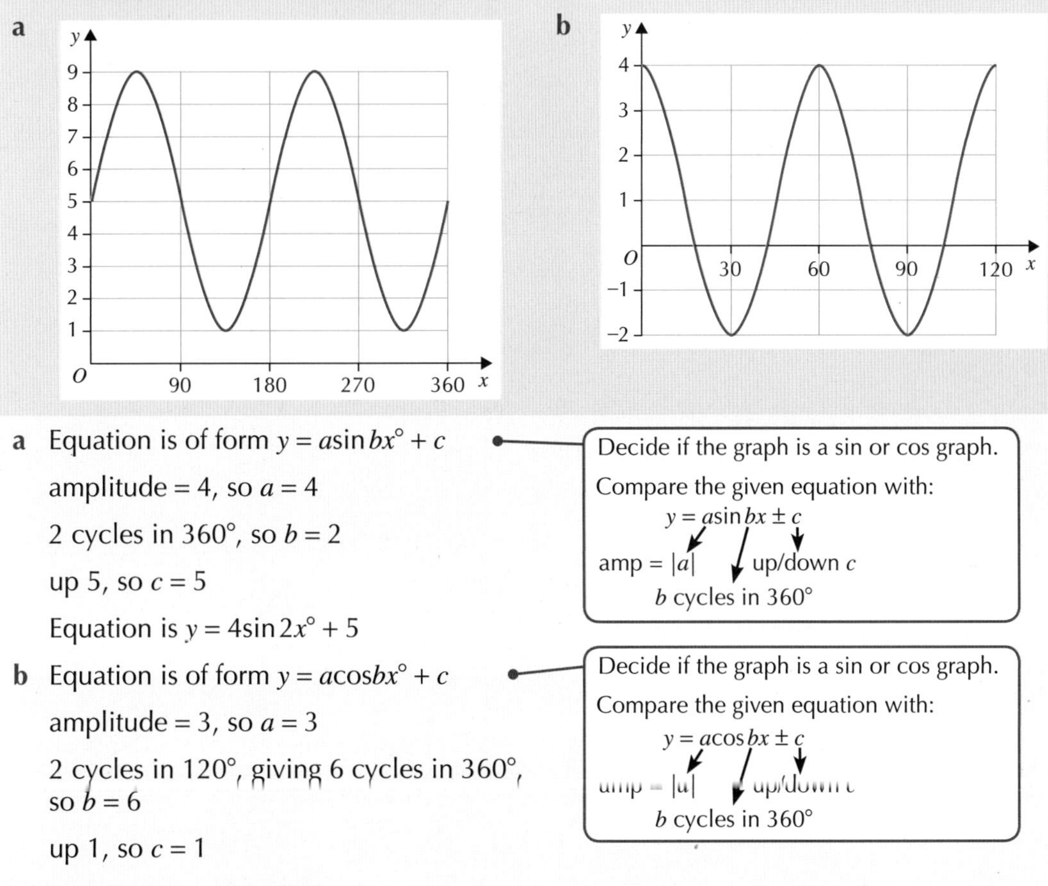

a Equation is of form $y = a\sin bx° + c$

amplitude = 4, so $a = 4$

2 cycles in 360°, so $b = 2$

up 5, so $c = 5$

Equation is $y = 4\sin 2x° + 5$

> Decide if the graph is a sin or cos graph.
> Compare the given equation with:
> $y = a\sin bx \pm c$
> amp = $|a|$ up/down c
> b cycles in 360°

b Equation is of form $y = a\cos bx° + c$

amplitude = 3, so $a = 3$

2 cycles in 120°, giving 6 cycles in 360°, so $b = 6$

up 1, so $c = 1$

Equation is $y = 3\cos 6x° + 1$

> Decide if the graph is a sin or cos graph.
> Compare the given equation with:
> $y = a\cos bx \pm c$
> amp = $|a|$ up/down c
> b cycles in 360°

Exercise 23D

★ **1** Sketch the graphs of the following for $0 \leqslant x \leqslant 360$.

a $y = \sin 2x° + 3$ **b** $y = 4\cos 3x° - 1$ **c** $y = 3\cos 2x° + 5$

d $y = 3\sin x° - 5$ **e** $y = 2\sin 3x° - 1$ **f** $y = 5\cos\frac{1}{2}x° + 2$

g $y = 4\sin 2x° + 3$ **h** $y = 5 + 3\cos x°$ **i** $y = 3 + \cos 2x°$

2 Sketch the following.

a $y = \cos 2x° + 3,\ 0 \leqslant x \leqslant 180$

b $y = 2\sin x° - 3,\ 0 \leqslant x \leqslant 180$

c $y = 5 + 3\sin 2x°,\ 0 \leqslant x \leqslant 720$

★ **3** Write down the equations of the following graphs.

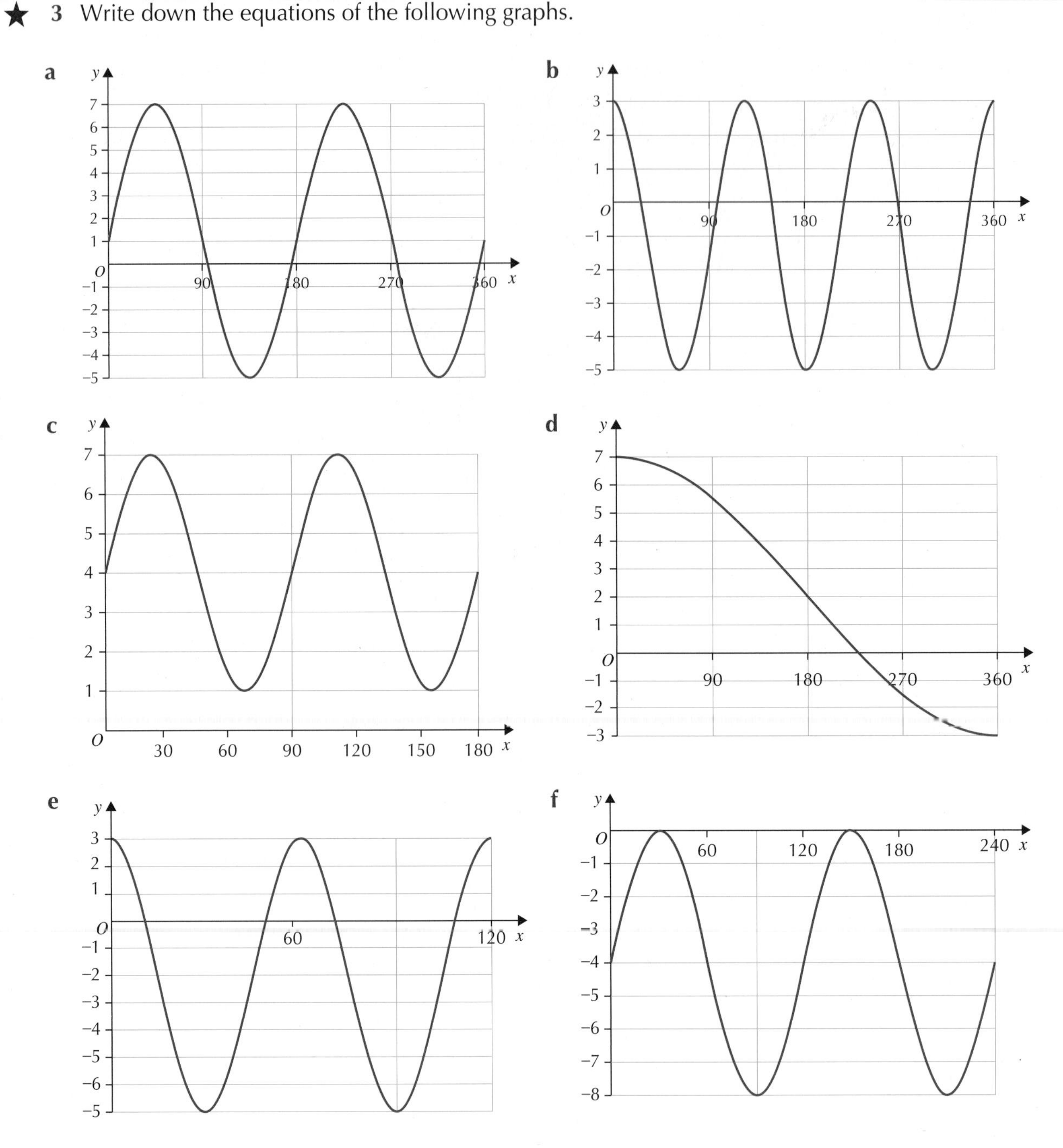

Problems involving the sine and cosine graphs

Many real-life problems involving circular motion can be solved by using sine and cosine graphs.

Example 23.8

The level of water in a harbour rises and falls. The depth, d metres, is given by the formula

$$d = 15 + 9\sin(30t)°$$

where t is the number of hours after midnight.

a What is the depth of the water at 1 pm?

b Draw the graph of $d = 15 + 9\sin(30t)°$.

c Use your graph to find the depth of water at high tide.

d Use your graph to find the time of the first low tide.

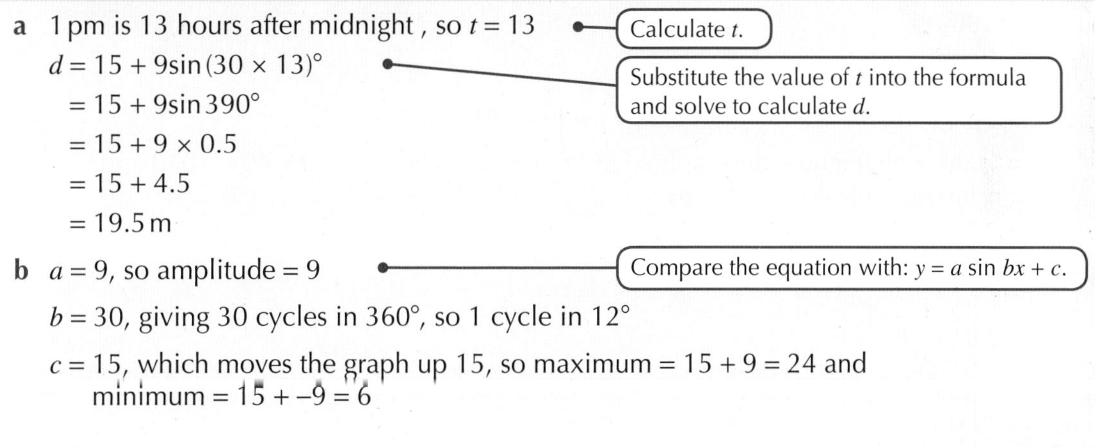

Calculate t.
Substitute the value of t into the formula and solve to calculate d.
Compare the equation with: $y = a \sin bx + c$.

a 1 pm is 13 hours after midnight , so $t = 13$

$d = 15 + 9\sin(30 \times 13)°$

$= 15 + 9\sin 390°$

$= 15 + 9 \times 0.5$

$= 15 + 4.5$

$= 19.5$ m

b $a = 9$, so amplitude $= 9$

$b = 30$, giving 30 cycles in 360°, so 1 cycle in 12°

$c = 15$, which moves the graph up 15, so maximum $= 15 + 9 = 24$ and minimum $= 15 + -9 = 6$

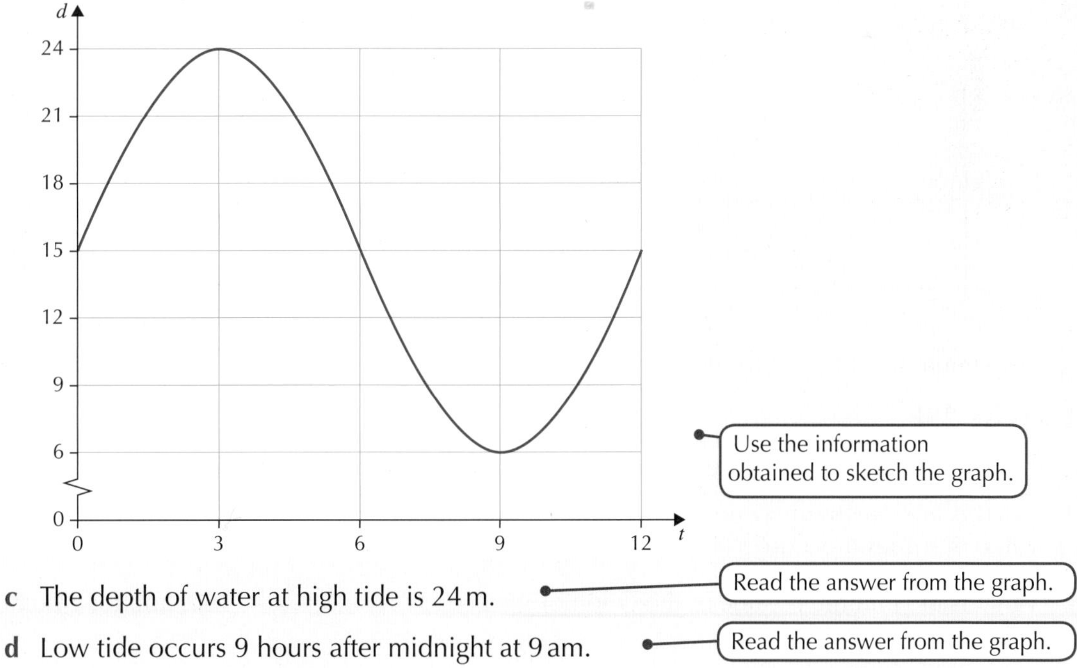

d
24
21
18
15
12
9
6
0
0
3
6
9
12
t
Use the information obtained to sketch the graph.

c The depth of water at high tide is 24 m. — Read the answer from the graph.

d Low tide occurs 9 hours after midnight at 9 am. — Read the answer from the graph.

Example 23.9

The distance, d km, of a space capsule above the Earth, t hours after timing began, is given by the formula

$$d = 40\sin(12t)° + 1000$$

a What was the height when timing began?

b Draw the graph of $d = 40\sin(12t)° + 1000$.

c Use your graph to calculate the maximum height and how long the capsule would take to reach it for the first time.

a When timing began, $t = 0$ — Calculate t.

$d = 40\sin(12t)° + 1000$ — Substitute the value of t into the formula and solve to calculate d.

$= 40\sin(12 \times 0)° + 1000$

$= 1000$ m

b $a = 40$, so amplitude $= 40$

$b = 12$, giving 12 cycles in 360°, so 1 cycle in 30°

$c = 1000$, which moves the graph up 1000, so maximum $= 1000 + 40 = 1040$ and minimum $= 1000 + -40 = 960$

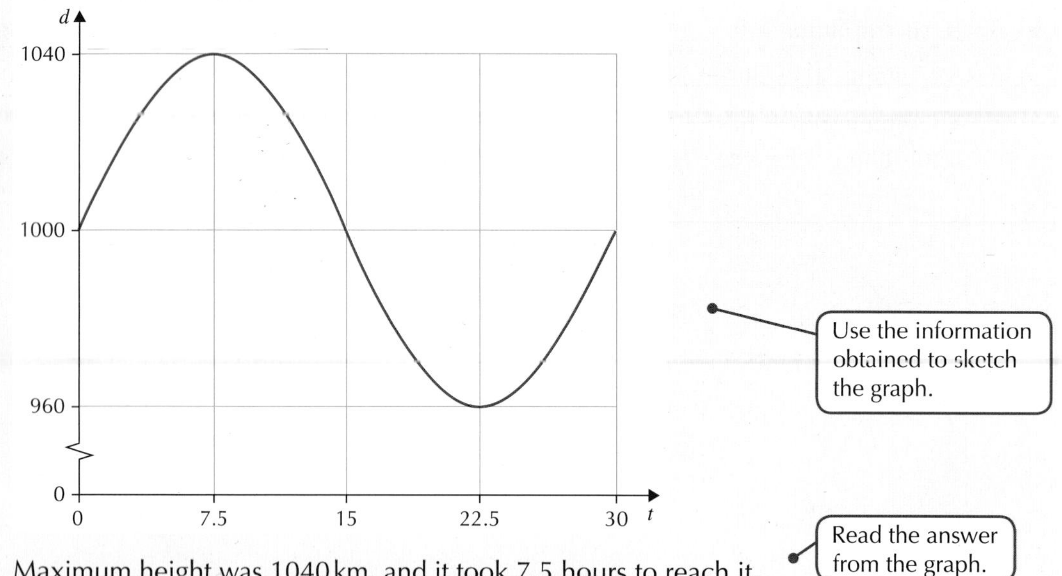

c Maximum height was 1040 km, and it took 7.5 hours to reach it. — Read the answer from the graph.

Exercise 23E

Use the graphs of the appropriate functions to answer the following questions.

1 What is the maximum value of the graph $y = 5 + 3\sin 4x°$, $0 \leqslant x \leqslant 360$, and for which value of x does it occur?

2 What is the minimum value of the graph of $y = 2\cos 3x° + 4$, $0 \leqslant x \leqslant 360$, and for which value of x does it occur?

★ **3** A sensor is attached to the tip of a turbine blade. Its height, h metres, is given by the equation

$$h = 3\cos(24t)° + 11$$

where t is the time in seconds after starting.

a How high is the sensor after 10 seconds?

b What is the maximum height of the sensor and after how many seconds does it first occur?

c When is the sensor at its lowest height?

4 A weight is attached to a spring as shown in the diagram.

It is released from the point P and allowed to oscillate.

The height, h cm, of the weight above the ground after t seconds is given by the formula

$$h = 70 - 50\cos(60t)°$$

a What is the height of the weight from the ground after 20 seconds?

b After how many seconds does the weight first return to the point P?

c i What is the furthest the weight gets from the ground?

ii When does this first occur?

5 The London Eye is a large Ferris wheel situated by the River Thames. The height, h metres, of a capsule t minutes after getting on is given by the formula

$$h = 75 - 60\sin(24t)°$$

a How high would a capsule be after 8 minutes?

b How many minutes does it take to reach the top?

6 The height, h metres, of water in a harbour is given by the formula

$$h = 10 + 4\sin(15t)°$$

where t is the time in hours after 6 am.

a What is the height of the water at 3 pm?

b When is high tide?

c It is not deemed to be safe for a ship to enter the harbour if the water level falls below a height of 7 metres. Will this ever happen? You must explain your answer.

The graph of the tangent function

You used the unit circle to investigate the sine and cosine functions and establish their graphs. You will now use the unit circle to develop the graph of the tangent function.

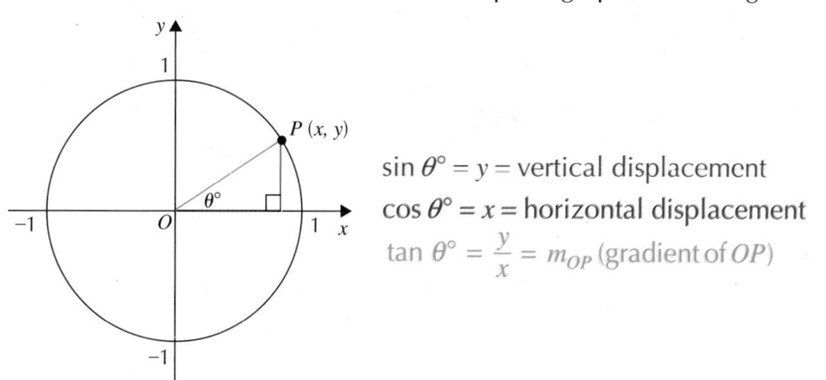

Several points can be noted as $\theta°$ increases from 0° to 360° and OP rotates anti-clockwise from the positive x-axis.

- When $\theta° = 0°$, P is on the x-axis, meaning $m_{OP} = 0$, so $\tan 0° = 0$.
- When $\theta° = 45°$, $m_{OP} = 1$, so $\tan 45° = 1$.
- When $\theta° = 90°$, P is on the y-axis, meaning m_{OP} is undefined, so $\tan 90°$ is undefined.

For $\theta° = 90°$, $\tan\theta°$ is undefined. This is shown on the diagram by the dotted vertical line. This line is called an **asymptote**, which is defined as a line such that the distance between the curve and the asymptote approaches zero as the curve approaches infinity.

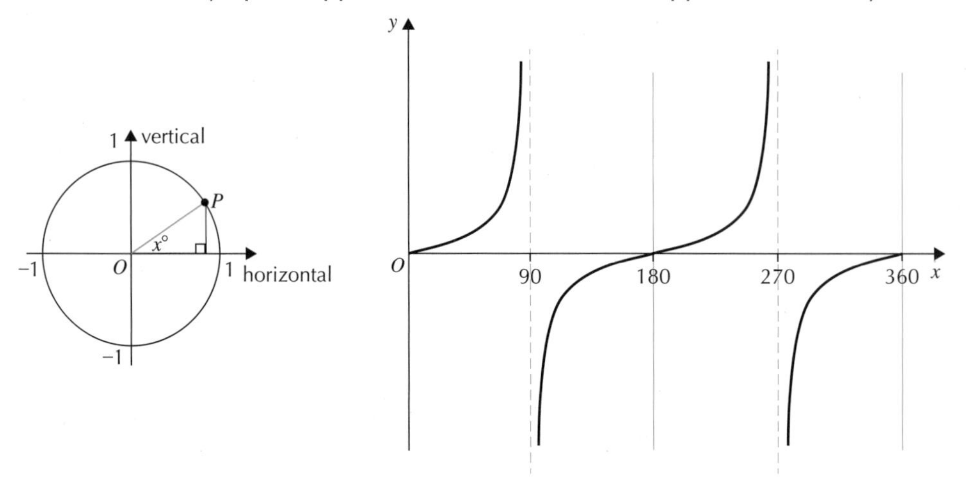

The graph of $y = \tan x°$, $x \in \mathbb{R}$

The graph of $y = \tan x°$ is not like the graphs of the sine and cosine functions. It is not continuous for all values of x, and it has a period of 180°.

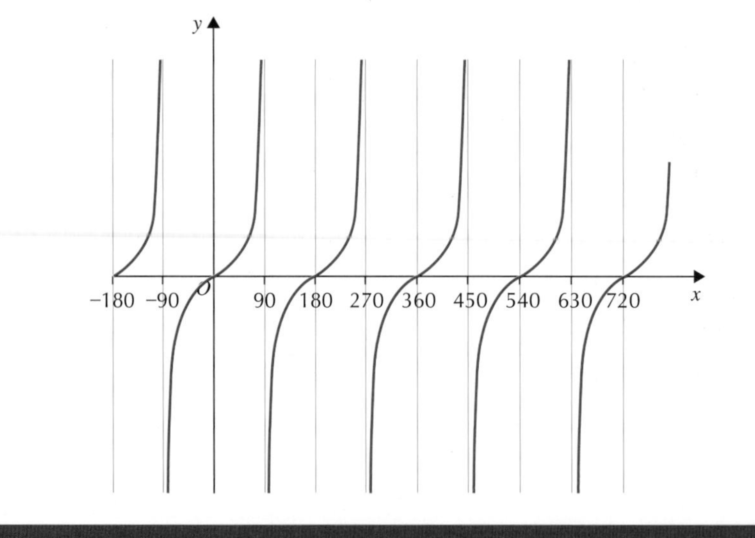

Activity

1 Investigate the graphs of
- $y = \tan(bx)°$
- $y = \tan(bx)° + c$

for different values of b and c and comment on your findings.

2 Use your results to sketch the graphs of the following for $0 \leqslant x \leqslant 360$.

a $y = \tan(2x)°$ **b** $y = \tan x° + 3$ **c** $y = \tan(2x)° - 1$

(continued)

3 Now identify the equations of the following graphs.

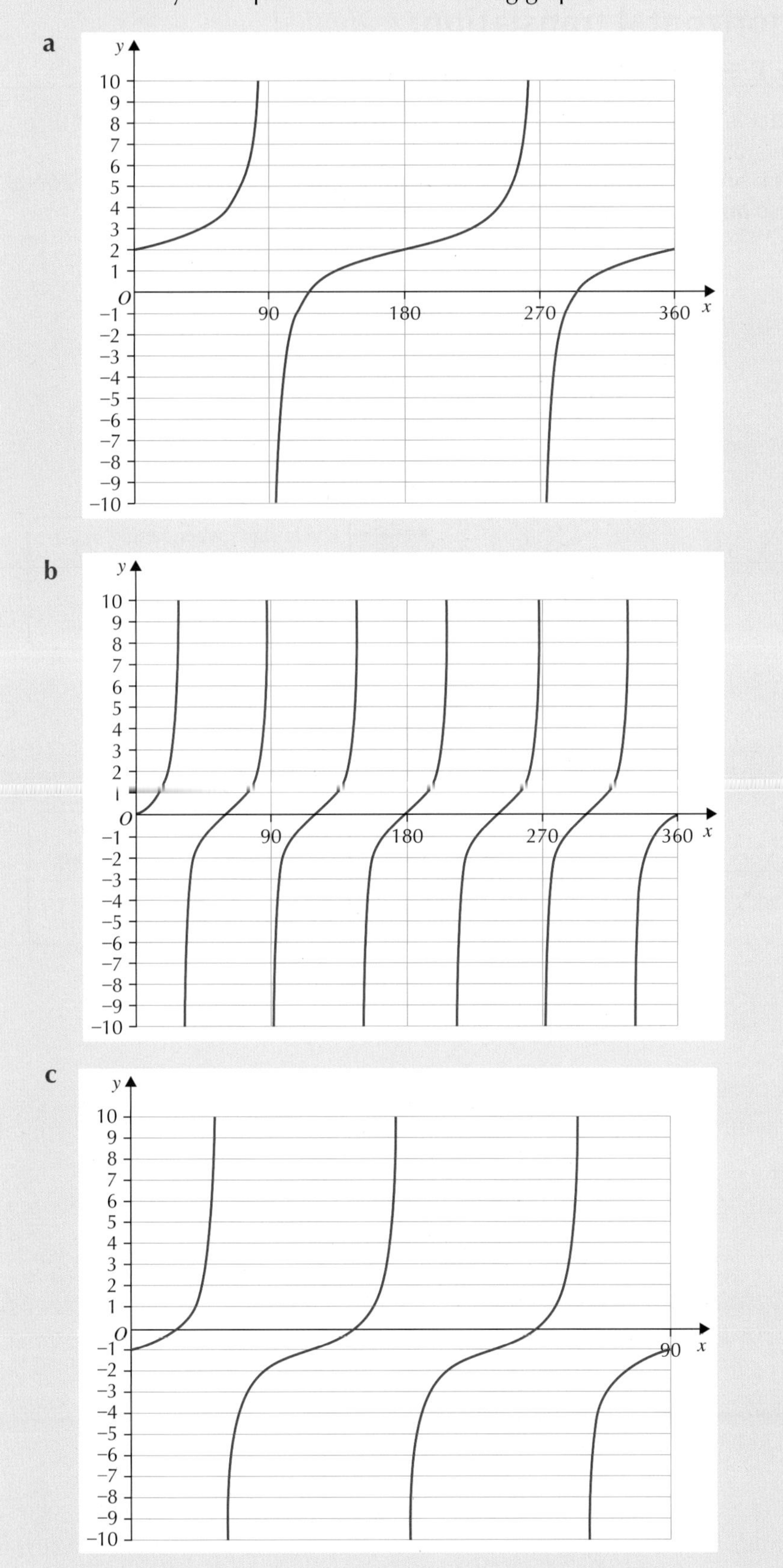

Graphs of sine, cosine and tangent functions which have undergone a horizontal translation: $y = \sin(x + d)°$, $y = \cos(x + d)°$ and $y = \tan(x + d)°$

In the graphs of $y = \sin(x + d)°$, $y = \cos(x + d)°$, and $y = \tan(x + d)°$ for different values of d, adding d translates the graphs horizontally left or right by $|d|$, according to whether d is positive or negative. The angle $d°$, which gives the magnitude and direction of this horizontal shift, is called the **phase angle** of the function.

Example 23.10

Sketch the graphs of the following functions.

a $y = \sin(x + 30)°$, $\leqslant x \leqslant 360$ **b** $y = \cos(x - 50)°$, $0 \leqslant x \leqslant 360$

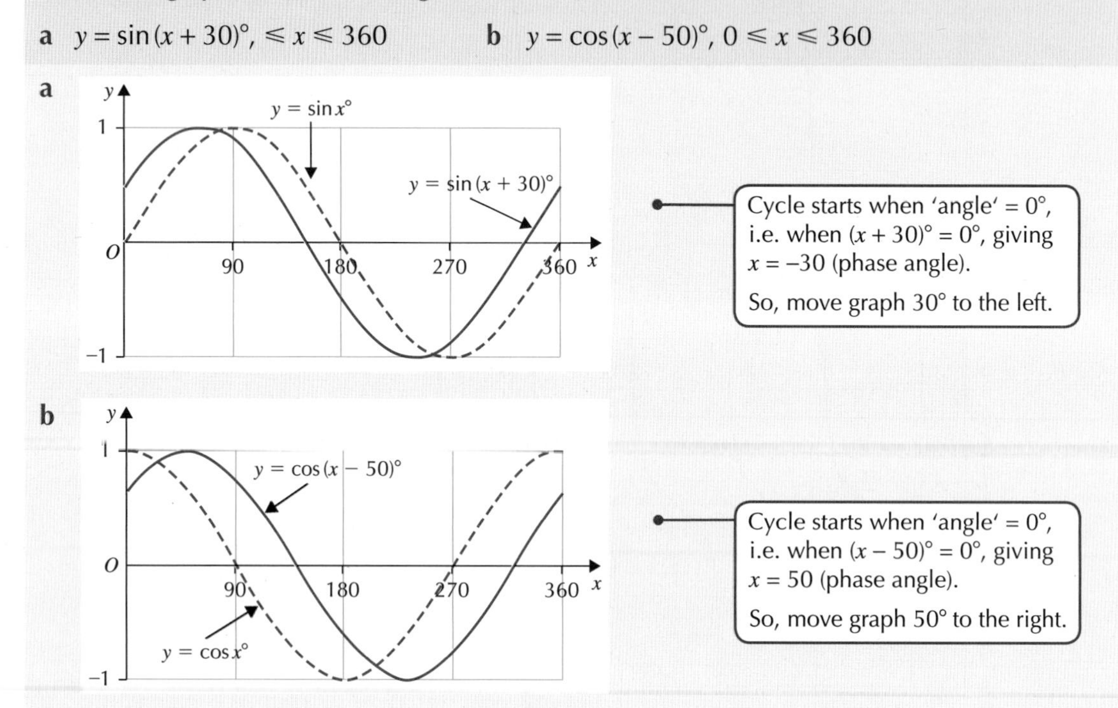

a
y
1
y = sin x°
y = sin (x + 30)°
O
90
180
270
360
x
−1
Cycle starts when 'angle' = 0°, i.e. when (x + 30)° = 0°, giving x = −30 (phase angle).
So, move graph 30° to the left.
b
y
1
y = cos (x − 50)°
O
90
180
270
360
x
y = cos x°
−1
Cycle starts when 'angle' = 0°, i.e. when (x − 50)° = 0°, giving x = 50 (phase angle).
So, move graph 50° to the right.

Exercise 23F

★ **1** Sketch the graphs of the following for $0 \leqslant x \leqslant 360$.

a $y = \sin(x + 45)°$ **b** $y = \cos(x + 30)°$ **c** $y = \cos(x - 15)°$

d $y = \sin(x - 60)°$ **e** $y = \tan(x - 90)°$ **f** $y = \tan(x + 30)°$

★ **2** The following graphs are of the form $y = \sin(x + d)°$, $y = \cos(x + d)°$ or $y = \tan(x + d)°$, $-90 < d < 90$. Write down their equations.

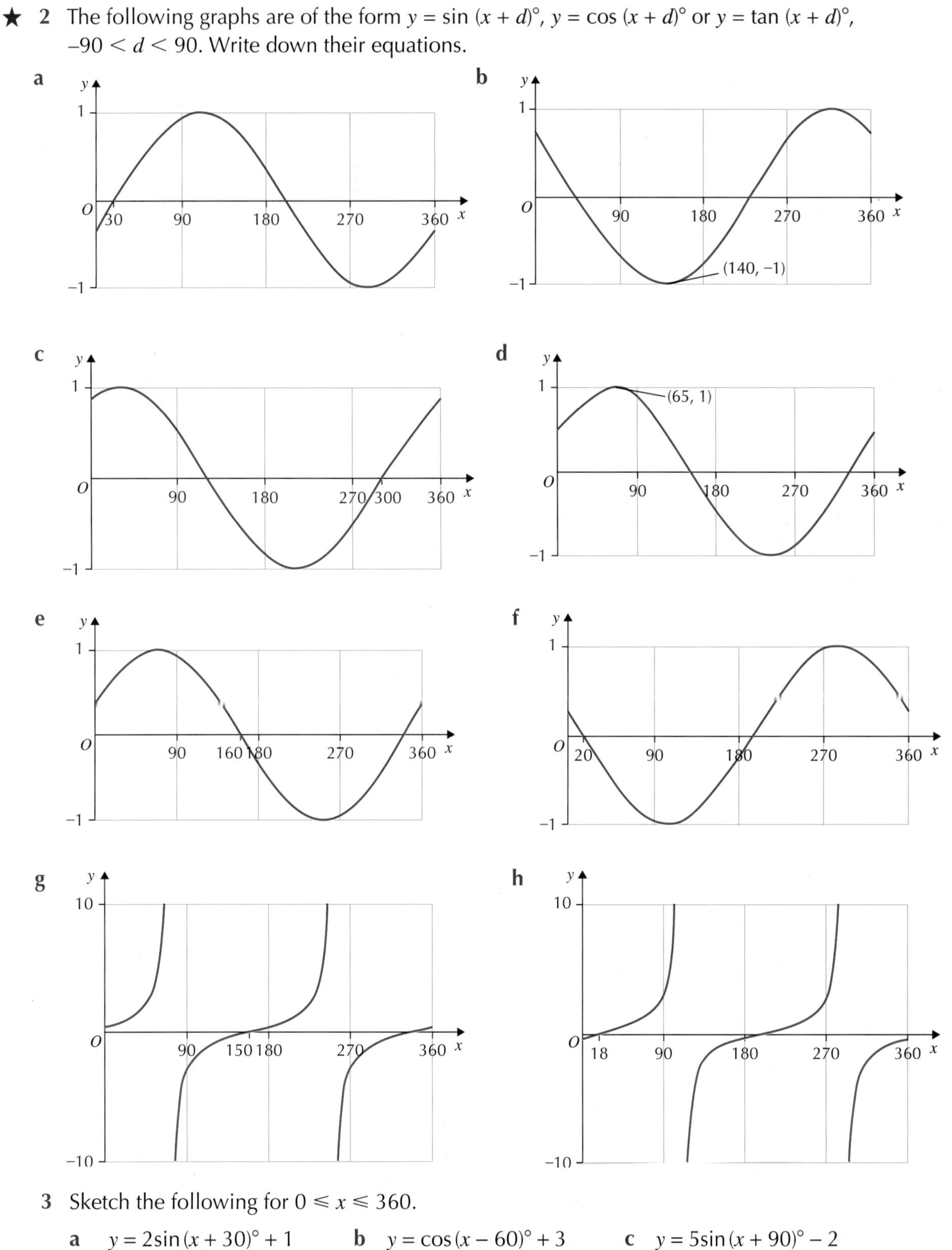

3 Sketch the following for $0 \leqslant x \leqslant 360$.

a $y = 2\sin(x + 30)° + 1$ **b** $y = \cos(x - 60)° + 3$ **c** $y = 5\sin(x + 90)° - 2$

d $y = 3\sin(x - 40)°$ **e** $y = 2\cos(x - 45)° - 1$ **f** $y = 4\cos(x + 60)° + 3$

Activity

Copy and complete the table to summarise the following transformations.

Function	Transformation
$y = f(x) + c$	If c is +ve, the graph moves up c units If c is –ve, the graph moves down c units
$y = af(x)$	
$y = f(bx)$	
$y = f(x + d)$	

- I can sketch and identify sine and cosine functions with different amplitudes. ★ Exercise 23A Q1, Q2
- I can sketch and identify sine and cosine functions with different periods noting the link with multiple angles. ★ Exercise 23B Q1, Q2
- I can sketch and identify sine and cosine functions with different amplitudes and periods. ★ Exercise 23C Q1, Q3
- I can sketch and identify sine and cosine functions which have undergone a vertical translation. ★ Exercise 23D Q1, Q3
- I can apply my knowledge of sine and cosine graphs to solve real-life problems. ★ Exercise 23E Q3
- I can sketch and identify sine, cosine and tangent functions which have undergone a horizontal translation and know what is meant by a phase angle. ★ Exercise 23F Q1, Q2

For further assessment opportunities, see the Preparation for Assessment for Unit 2 on pages 288–291.

24 Working with trigonometric relationships in degrees

In this chapter you will learn how to:

- identify related angles
- solve basic trigonometric equations
- calculate where trigonometric graphs and straight lines intersect
- work with **exact value** ratios
- work with **trigonometric identities**.

You should already know:

- how to find a side or angle in a right-angled triangle using the sine, cosine and tangent ratios
- the key features of the graphs of $y = \sin x°$, $y = \cos x°$ and $y = \tan x°$.

Key points on the trigonometric graphs

Using trigonometric ratios to find the values of angles when side lengths are known is a straightforward process. For example, a straightforward calculation will give x in the triangle when the opposite and hypotenuse are known, as shown below.

$$\sin x° = \tfrac{1}{2}$$
$$x° = \sin^{-1}\left(\tfrac{1}{2}\right)$$
$$= 30°$$

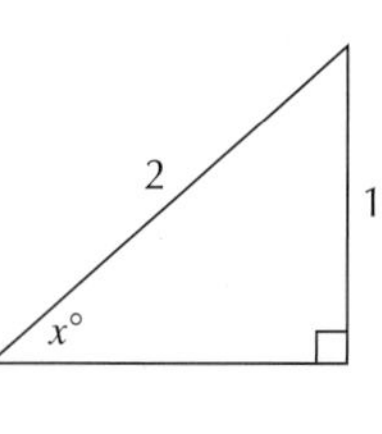

In the restricted context of the right-angled triangle, the equation $\sin x° = \frac{1}{2}$ has only one solution.

However, the periodicity of the sine curve means there are many values of x for which $\sin x° = \frac{1}{2}$.

See Chapter 23 for more about the periodicity of sine and cosine graphs.

You can see from the graph that for values of $x°$ between 0° and 360° (in the range $0 \leqslant x \leqslant 360$), there are two solutions for $x°$ for which $\sin x° = \frac{1}{2}$: $x° = 30°$ or $x° = 150°$.

When we extend the graph for values of $x°$ between 0° and 720° (in the range $0 \leqslant x \leqslant 720$), there are four solutions for $x°$ for which $\sin x° = \frac{1}{2}$: $x° = 30°, 150°, 390°, 510°$.

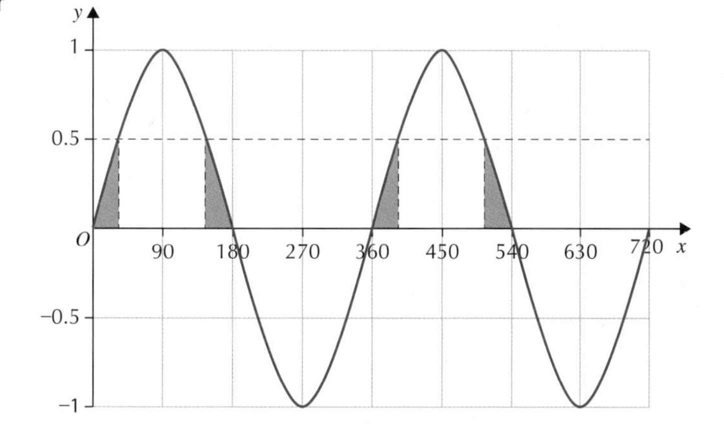

Trig graphs can be used to solve equations, but it is not the most effective method unless you are dealing with the key points linked to angles 0°, 90°, 180°, 270° and 360°.

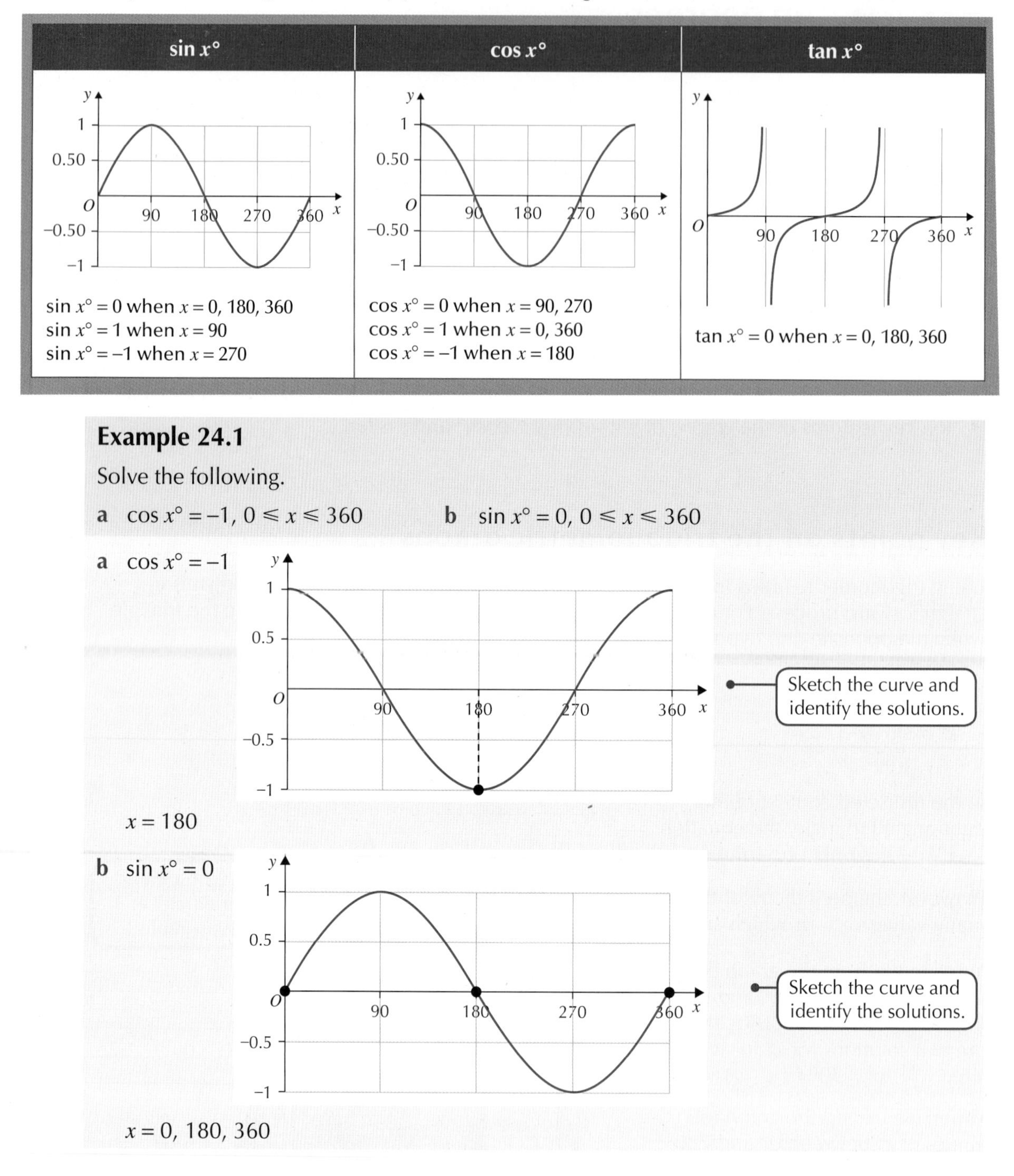
sin x°
cos x°
tan x°
Sketch the curve and identify the solutions.
Sketch the curve and identify the solutions.

sin x°	cos x°	tan x°
sin x° = 0 when x = 0, 180, 360 sin x° = 1 when x = 90 sin x° = −1 when x = 270	cos x° = 0 when x = 90, 270 cos x° = 1 when x = 0, 360 cos x° = −1 when x = 180	tan x° = 0 when x = 0, 180, 360

Example 24.1

Solve the following.

a $\cos x° = -1,\ 0 \leqslant x \leqslant 360$ **b** $\sin x° = 0,\ 0 \leqslant x \leqslant 360$

a $\cos x° = -1$

$x = 180$

b $\sin x° = 0$

$x = 0, 180, 360$

Exercise 24A

★ **1** Solve for $0 \leqslant x \leqslant 360$.

a $\sin x° = 1$ **b** $\cos x° = 0$ **c** $\tan x° = 0$ **d** $\sin x° = -1$ **e** $\cos x° = 1$

2 Solve for $0 \leqslant x \leqslant 720$.

a $\sin x° = 0$ **b** $\cos x° = 1$

Working with related angles

The angles 0°, 90°, 180°, 270° and 360° are special cases. To solve equations for other angles, it is necessary to look again at the definitions of sine, cosine and tangent as defined using the unit circle (see Chapter 23).

Consider the position of a general point P as it starts on the positive x-axis, with angle 0°, and travels with increasing angle (anti-clockwise) round the unit circle.

1st quadrant	2nd quadrant	3rd quadrant	4th quadrant
$P_1\ (x, y)$	$P_2\ (-x, y)$	$P_3\ (-x, -y)$	$P_4\ (x, -y)$
$\sin\theta° = y$	$\sin(180 - \theta)° = y$	$\sin(180 + \theta)° = -y$	$\sin(360 - \theta)° = -y$
$\cos\theta° = x$	$\cos(180 - \theta)° = -x$	$\cos(180 + \theta)° = -x$	$\cos(360 - \theta)° = x$
$\tan\theta° = m_{op}$	$\tan(180 - \theta)° = m_{op}$	$\tan(180 + \theta)° = m_{op}$	$\tan(360 - \theta)° = m_{op}$
sin +ve	sin −ve	sin −ve	sin −ve
cos +ve	cos −ve	cos −ve	cos +ve
tan +ve	tan −ve	tan +ve	tan −ve

From the diagrams above you can see there is a group of related angles:

$\theta°$, $(180 - \theta)°$, $(180 + \theta)°$, $(360 - \theta)°$

which share the same absolute value for each of the sine, cosine and tangent ratios.
The following **quadrant diagram** shows where each ratio is **positive**.

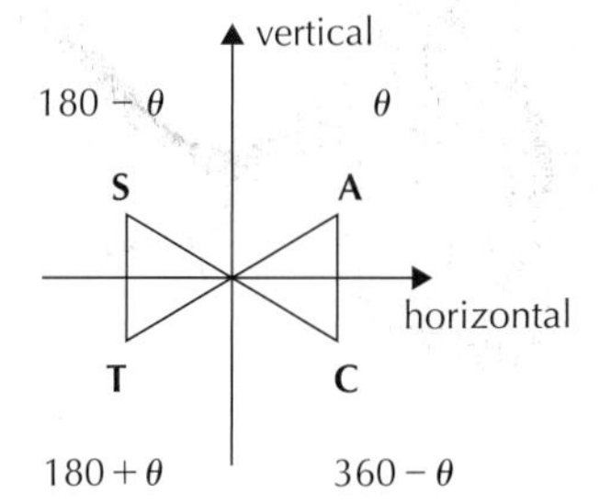

All positive in 1st quadrant
Sin positive in 2nd quadrant
Tan positive in 3rd quadrant
Cos positive in 4th quadrant

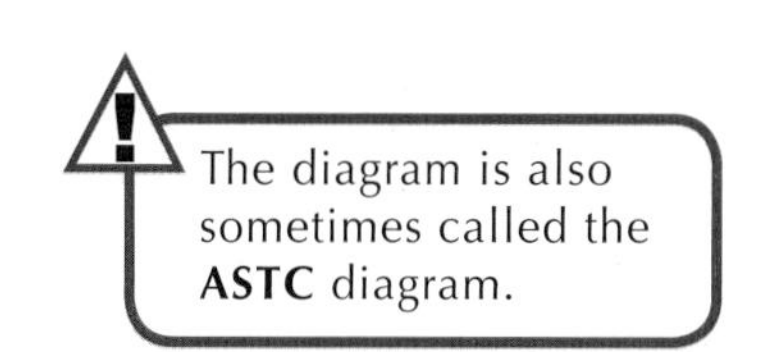

The diagram is also sometimes called the ASTC diagram.

By extension, we can also find the related angles for those greater than 360° or less than 0°.

Example 24.2

Without using your calculator, state whether the following ratios are positive or negative.

a sin 340° **b** cos 200° **c** tan 240°

a 340° is in 4th quadrant; sin 340° is –ve

b 200° is in 3rd quadrant; cos 200° is –ve

c 240° is in 3rd quadrant; tan 240° is +ve

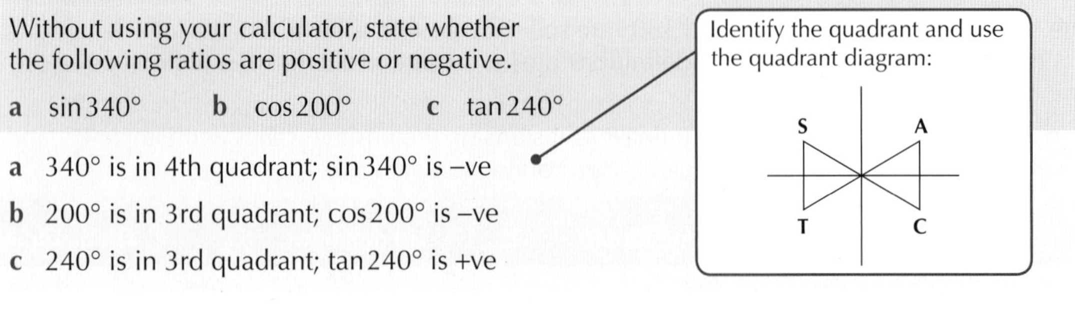

Identify the quadrant and use the quadrant diagram:
S
A
T
C

Example 24.3

Give three angles between 0° and 360° which are related to:

a 30° **b** 120° **c** 320°

a 30° is related to:

2nd quadrant: (180 – 30)° = 150°

3rd quadrant: (180 + 30)° = 210°

4th quadrant: (360 – 30)° = 330°

30° is in the 1st quadrant.

b 120° is related to:

1st quadrant: 60°

3rd quadrant: (180 + 60)° = 240°

4th quadrant: (360 – 60)° = 300°

120° is in the 2nd quadrant.

120° = (180 – 60)°

c 320° is related to:

1st quadrant: 40°

2nd quadrant: (180 – 40)° = 140°

3rd quadrant: (180 + 40)° = 220°

320° is in the 4th quadrant.

320° = (360 – 40)°

Exercise 24B

1 Without using a calculator, state whether the following ratios are positive or negative.

a cos 120°	**b** sin 60°	**c** tan 300°	**d** cos 315°
e sin 156°	**f** cos 244°	**g** tan 129°	**h** sin 143°
i tan 210°	**j** sin 317°	**k** cos 295°	**l** tan 282°

★ **2** Give three angles between 0° and 360° which are related to these angles.

a 45°	**b** 60°	**c** 150°	**d** 140°
e 320°	**f** 215°	**g** 165°	**h** 350°
i 200°	**j** 115°	**k** 205°	**l** 345°

Solving trigonometric equations

You can use your knowledge of related angles and the quadrant diagram to solve trigonometric equations. The solutions to trigonometric equations are always found by using the acute angle in the 1st quadrant and then applying the conditions for the 2nd, 3rd and 4th quadrants, so the essential first stage in solving trigonometric equations is to find the 1st quadrant angle.

Example 24.4

Solve for $0 \leqslant x \leqslant 360$, giving your answer to the nearest degree.

a $\sin x° = 0.835$ **b** $\cos x° = 0.231$ **c** $\tan x° = 3.42$

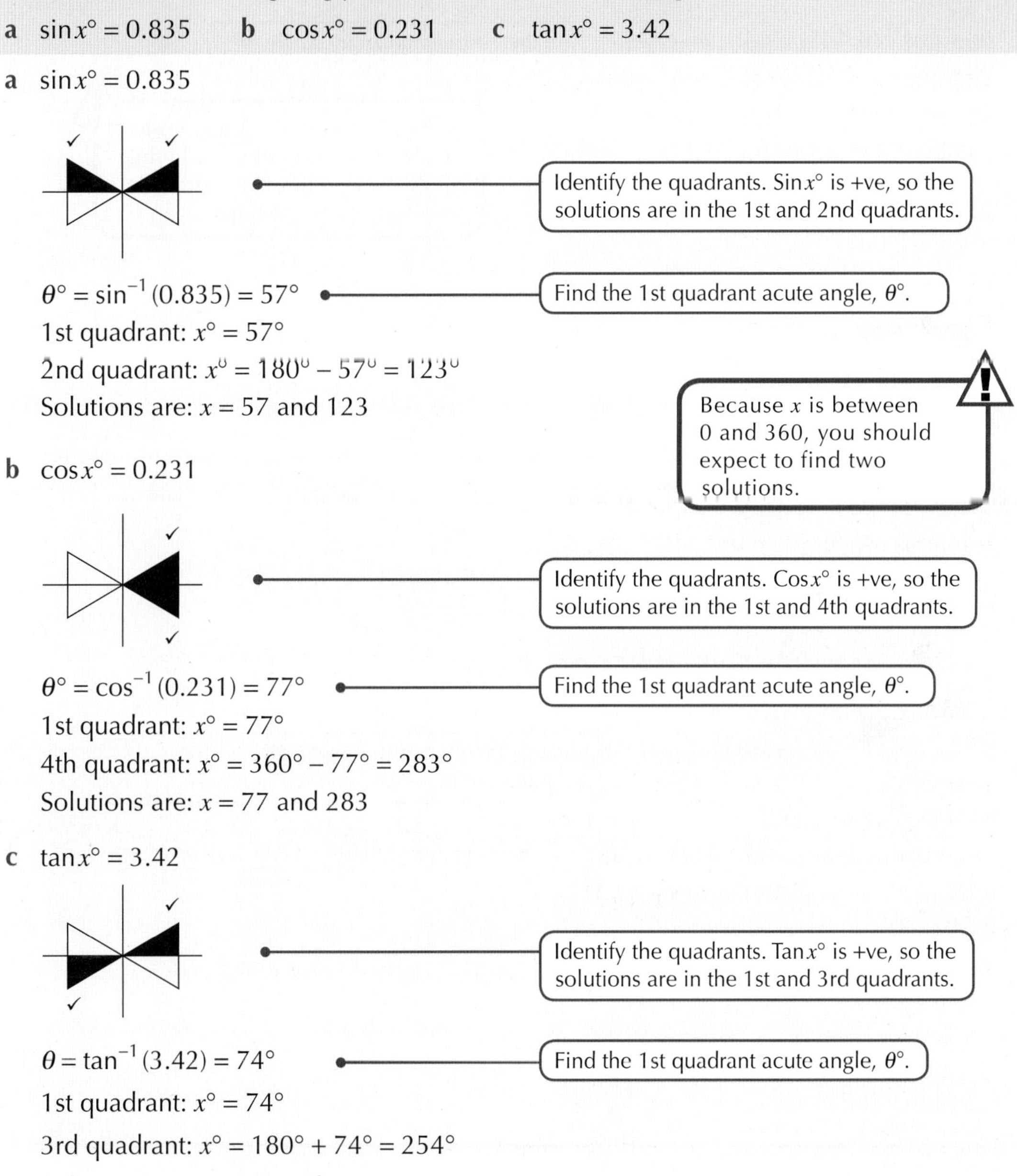

a $\sin x° = 0.835$

Identify the quadrants. Sin$x°$ is +ve, so the solutions are in the 1st and 2nd quadrants.

$\theta° = \sin^{-1}(0.835) = 57°$ — Find the 1st quadrant acute angle, $\theta°$.

1st quadrant: $x° = 57°$

2nd quadrant: $x° = 180° - 57° = 123°$

Solutions are: $x = 57$ and 123

Because x is between 0 and 360, you should expect to find two solutions.

b $\cos x° = 0.231$

Identify the quadrants. Cos$x°$ is +ve, so the solutions are in the 1st and 4th quadrants.

$\theta° = \cos^{-1}(0.231) = 77°$ — Find the 1st quadrant acute angle, $\theta°$.

1st quadrant: $x° = 77°$

4th quadrant: $x° = 360° - 77° = 283°$

Solutions are: $x = 77$ and 283

c $\tan x° = 3.42$

Identify the quadrants. Tan$x°$ is +ve, so the solutions are in the 1st and 3rd quadrants.

$\theta = \tan^{-1}(3.42) = 74°$ — Find the 1st quadrant acute angle, $\theta°$.

1st quadrant: $x° = 74°$

3rd quadrant: $x° = 180° + 74° = 254°$

Solutions are: $x = 74$ and 254

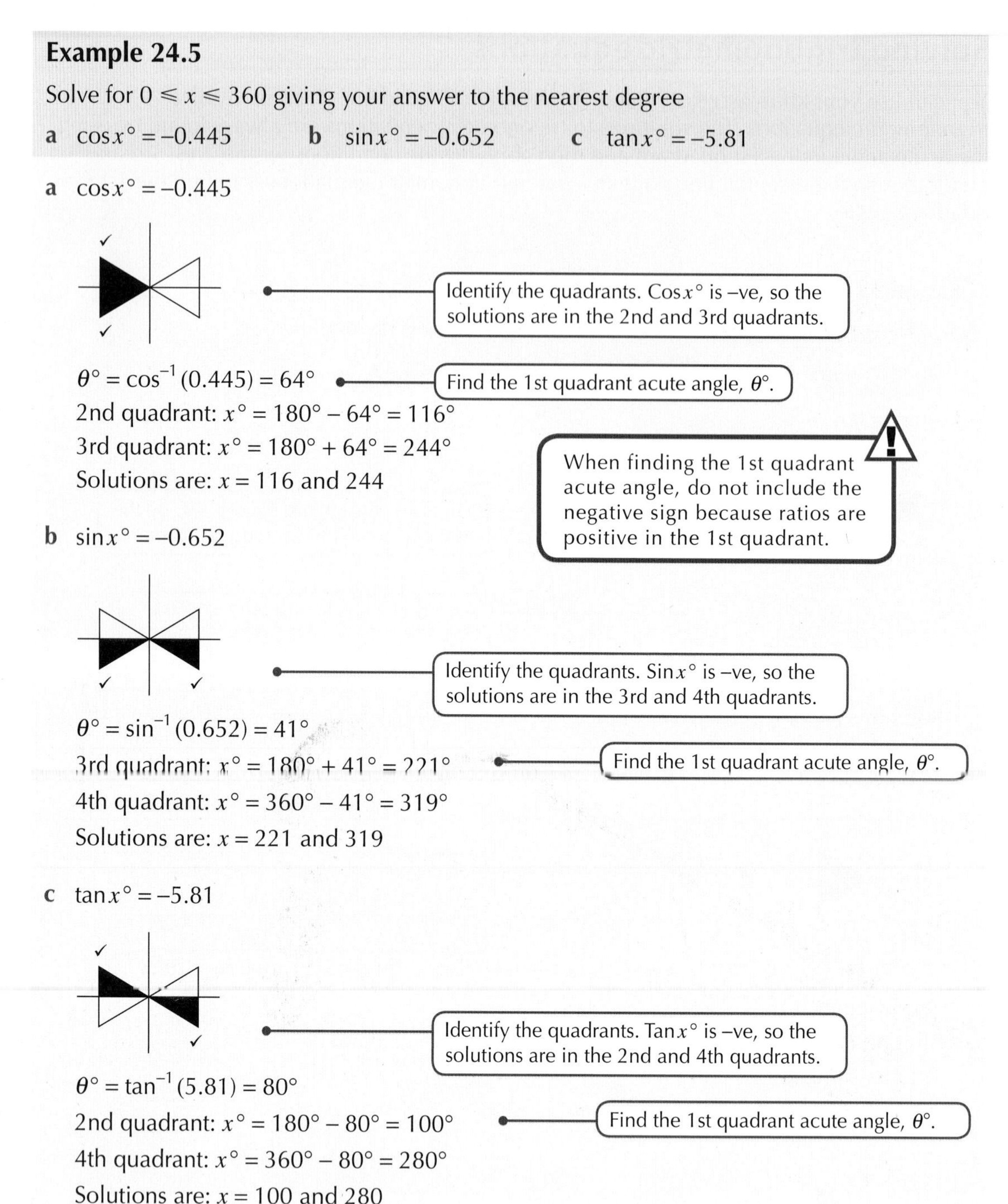

Example 24.5

Solve for $0 \leqslant x \leqslant 360$ giving your answer to the nearest degree

a $\cos x° = -0.445$ **b** $\sin x° = -0.652$ **c** $\tan x° = -5.81$

a $\cos x° = -0.445$

Identify the quadrants. $\text{Cos}\,x°$ is –ve, so the solutions are in the 2nd and 3rd quadrants.

$\theta° = \cos^{-1}(0.445) = 64°$

Find the 1st quadrant acute angle, $\theta°$.

2nd quadrant: $x° = 180° - 64° = 116°$

3rd quadrant: $x° = 180° + 64° = 244°$

Solutions are: $x = 116$ and 244

When finding the 1st quadrant acute angle, do not include the negative sign because ratios are positive in the 1st quadrant.

b $\sin x° = -0.652$

Identify the quadrants. $\text{Sin}\,x°$ is –ve, so the solutions are in the 3rd and 4th quadrants.

$\theta° = \sin^{-1}(0.652) = 41°$

Find the 1st quadrant acute angle, $\theta°$.

3rd quadrant: $x° = 180° + 41° = 221°$

4th quadrant: $x° = 360° - 41° = 319°$

Solutions are: $x = 221$ and 319

c $\tan x° = -5.81$

Identify the quadrants. $\text{Tan}\,x°$ is –ve, so the solutions are in the 2nd and 4th quadrants.

$\theta° = \tan^{-1}(5.81) = 80°$

Find the 1st quadrant acute angle, $\theta°$.

2nd quadrant: $x° = 180° - 80° = 100°$

4th quadrant: $x° = 360° - 80° = 280°$

Solutions are: $x = 100$ and 280

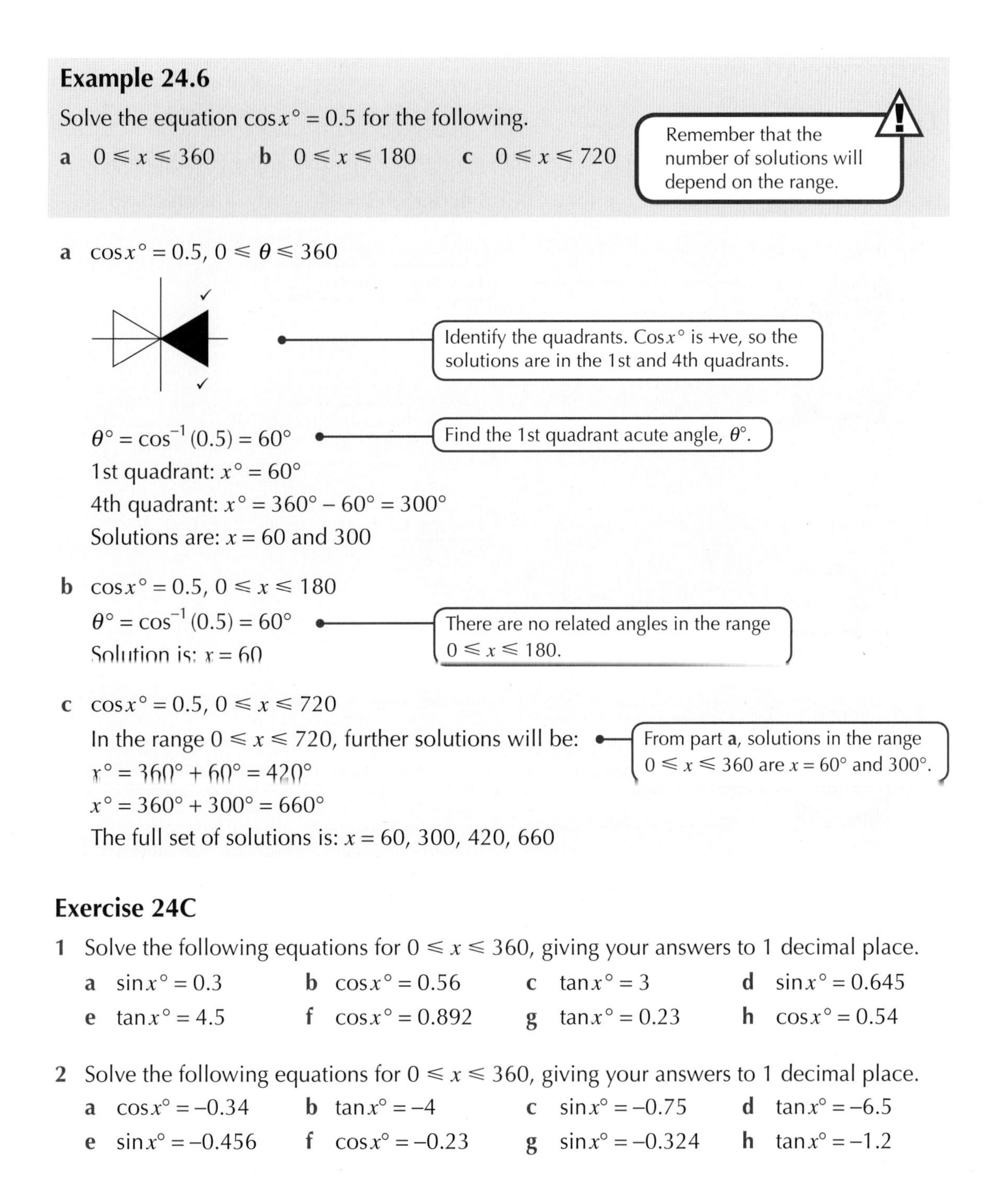

Identify the quadrants. Cos$x°$ is +ve, so the solutions are in the 1st and 4th quadrants.
Find the 1st quadrant acute angle, $\theta°$.
There are no related angles in the range $0 \leqslant x \leqslant 180$.
From part **a**, solutions in the range $0 \leqslant x \leqslant 360$ are $x = 60°$ and $300°$.
Remember that the number of solutions will depend on the range.

Example 24.6

Solve the equation $\cos x° = 0.5$ for the following.

a $0 \leqslant x \leqslant 360$ **b** $0 \leqslant x \leqslant 180$ **c** $0 \leqslant x \leqslant 720$

a $\cos x° = 0.5, 0 \leqslant \theta \leqslant 360$

$\theta° = \cos^{-1}(0.5) = 60°$

1st quadrant: $x° = 60°$

4th quadrant: $x° = 360° - 60° = 300°$

Solutions are: $x = 60$ and 300

b $\cos x° = 0.5, 0 \leqslant x \leqslant 180$

$\theta° = \cos^{-1}(0.5) = 60°$

Solution is: $x = 60$

c $\cos x° = 0.5, 0 \leqslant x \leqslant 720$

In the range $0 \leqslant x \leqslant 720$, further solutions will be:

$x° = 360° + 60° = 420°$

$x° = 360° + 300° = 660°$

The full set of solutions is: $x = 60, 300, 420, 660$

Exercise 24C

1 Solve the following equations for $0 \leqslant x \leqslant 360$, giving your answers to 1 decimal place.

a $\sin x° = 0.3$ **b** $\cos x° = 0.56$ **c** $\tan x° = 3$ **d** $\sin x° = 0.645$

e $\tan x° = 4.5$ **f** $\cos x° = 0.892$ **g** $\tan x° = 0.23$ **h** $\cos x° = 0.54$

2 Solve the following equations for $0 \leqslant x \leqslant 360$, giving your answers to 1 decimal place.

a $\cos x° = -0.34$ **b** $\tan x° = -4$ **c** $\sin x° = -0.75$ **d** $\tan x° = -6.5$

e $\sin x° = -0.456$ **f** $\cos x° = -0.23$ **g** $\sin x° = -0.324$ **h** $\tan x° = -1.2$

3 Solve the following equations for $0 \leqslant p \leqslant 360$. Round your answers to 2 decimal places.

a $\sin p° = 0.5$ **b** $\cos p° = -0.2$ **c** $\tan p° = -3.4$ **d** $\cos p° = 0.443$

e $\sin p° = -0.17$ **f** $\tan p° = 12$ **g** $\sin p° = 0.9$ **h** $\cos p° = -0.205$

★ **4** Solve the following, giving your answers to the nearest degree.

a $\cos x° = 0.8, 0 \leqslant x \leqslant 180$ **b** $\sin x° = 0.62, 0 \leqslant x \leqslant 720$

c $\tan x° = 4, 180 \leqslant x \leqslant 360$ **d** $\sin x° = -0.72, 0 \leqslant x \leqslant 720$

e $\cos x° = -0.55, 0 \leqslant x \leqslant 180$ **f** $\tan x° = -6.5, 180 \leqslant x \leqslant 540$

Further equations

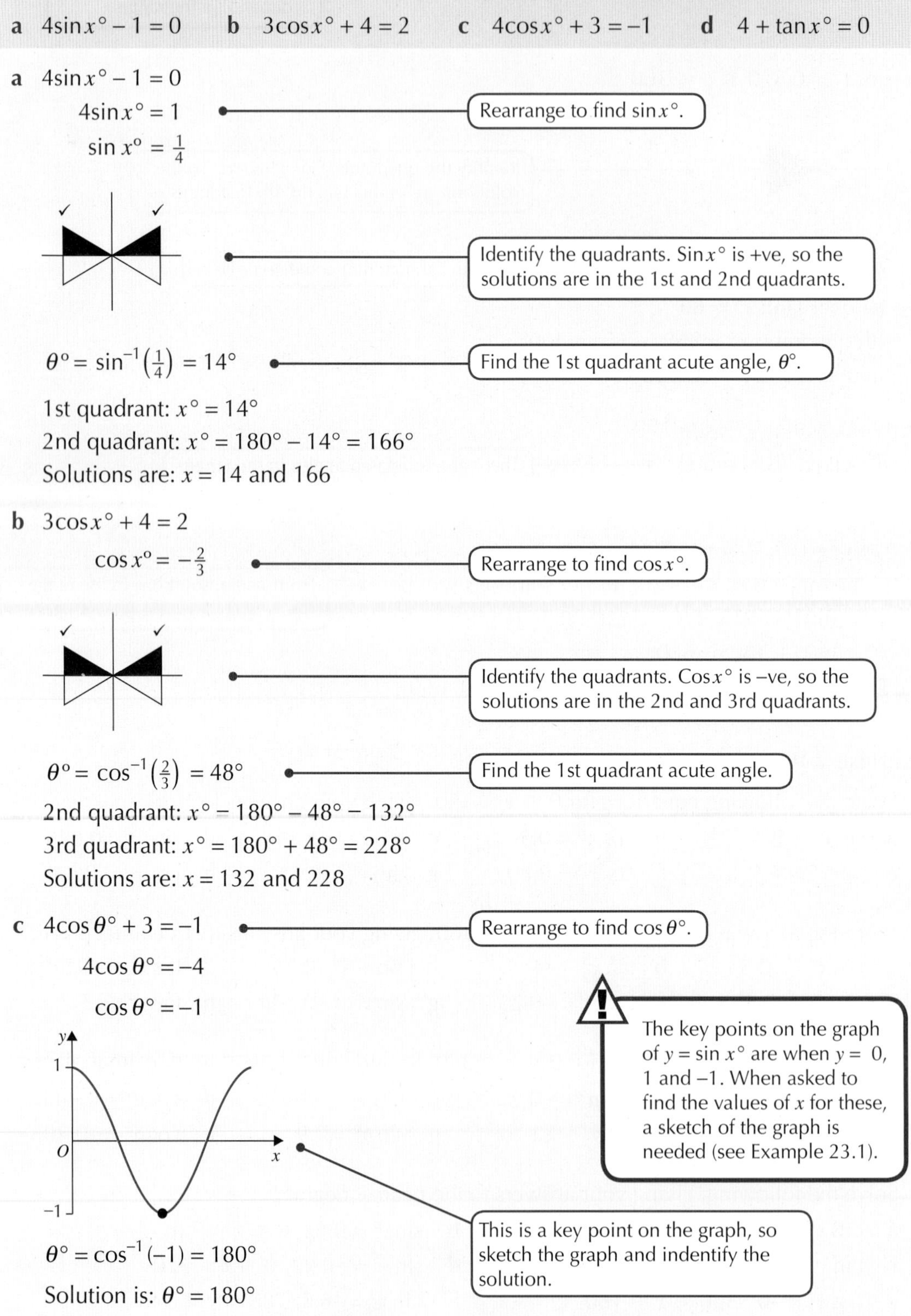

Example 24.7

Solve the following equations for $0 \leqslant x \leqslant 360$, giving your answers to the nearest degree.

a $4\sin x° - 1 = 0$ **b** $3\cos x° + 4 = 2$ **c** $4\cos x° + 3 = -1$ **d** $4 + \tan x° = 0$

a $4\sin x° - 1 = 0$

$4\sin x° = 1$ — Rearrange to find $\sin x°$.

$\sin x° = \frac{1}{4}$

Identify the quadrants. $\sin x°$ is +ve, so the solutions are in the 1st and 2nd quadrants.

$\theta° = \sin^{-1}\left(\frac{1}{4}\right) = 14°$ — Find the 1st quadrant acute angle, $\theta°$.

1st quadrant: $x° = 14°$

2nd quadrant: $x° = 180° - 14° = 166°$

Solutions are: $x = 14$ and 166

b $3\cos x° + 4 = 2$

$\cos x° = -\frac{2}{3}$ — Rearrange to find $\cos x°$.

Identify the quadrants. $\cos x°$ is −ve, so the solutions are in the 2nd and 3rd quadrants.

$\theta° = \cos^{-1}\left(\frac{2}{3}\right) = 48°$ — Find the 1st quadrant acute angle.

2nd quadrant: $x° = 180° - 48° = 132°$

3rd quadrant: $x° = 180° + 48° = 228°$

Solutions are: $x = 132$ and 228

c $4\cos\theta° + 3 = -1$ — Rearrange to find $\cos\theta°$.

$4\cos\theta° = -4$

$\cos\theta° = -1$

The key points on the graph of $y = \sin x°$ are when $y = 0$, 1 and −1. When asked to find the values of x for these, a sketch of the graph is needed (see Example 23.1).

This is a key point on the graph, so sketch the graph and indentify the solution.

$\theta° = \cos^{-1}(-1) = 180°$

Solution is: $\theta° = 180°$

(continued)

d $4 + \tan x° = 0$

$\tan x° = -4$

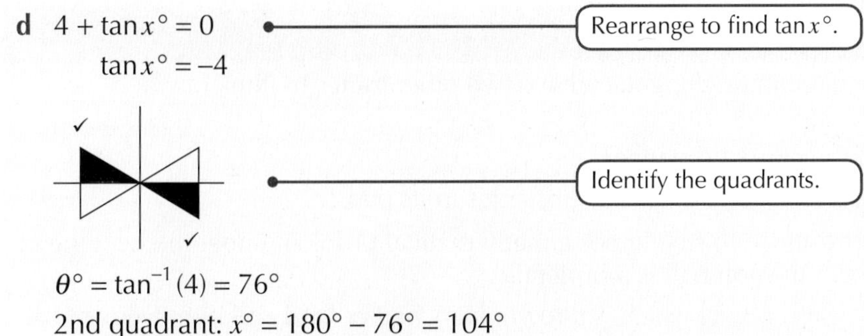

$\theta° = \tan^{-1}(4) = 76°$

2nd quadrant: $x° = 180° - 76° = 104°$

4th quadrant: $x° = 360° - 76° = 284°$

Solutions are: $x = 104$ and 284

Exercise 24D

★ **1** Solve for $0 \leqslant x \leqslant 360$, giving your answers to 1 decimal place.

a $3\sin x° - 2 = 0$	**b** $4\cos x° + 1 = 2$	**c** $5\tan x° - 4 = 3$
d $4\cos x° + 3 = 0$	**e** $8 + 3\tan x° = 4$	**f** $7 - 3\cos x° = 6$
g $4\sin x° - 5 = -1$	**h** $6\cos x° + 2 = 7$	**i** $3 + 5\sin x° = 7$
j $2\tan x° - 3 = 5$	**k** $2\cos x° + 3 = 1$	**l** $5 \quad 4\sin x° - 8$

2 The height, h metres, of a carriage on a Ferris wheel is given by the equation

$h = 35 + 30\sin t°$

where t is the time in seconds after starting.

a Calculate the height after 45 seconds.

b **i** After how many seconds is the carriage first at a height of 50 m?

ii When does it next reach this height?

3 In the diagram, the point L represents a lift.

The height, h metres, of the lift above the ground is given by the equation

$h = 10\tan x° + 1.7$

where x is the angle of elevation from P.

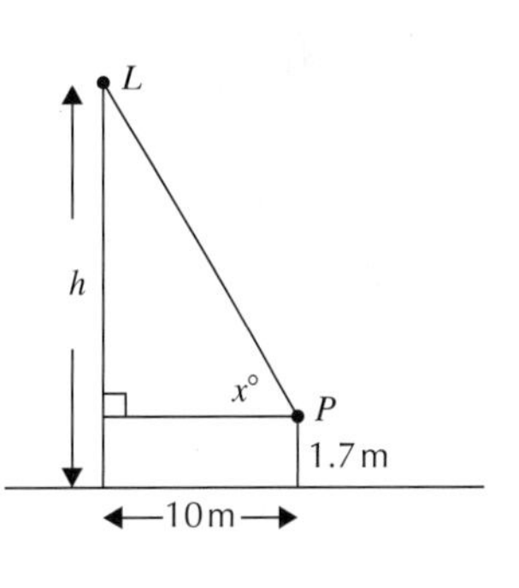

a Calculate the height of the lift when the angle of elevation is 25°.

b Calculate the angle of elevation at point P when the lift is at a height of 6 metres.

4 The depth, d metres, of water in a reservoir is measured over the course of a 6-hour period. The depth is given by the formula

$d = 12 + 7\cos t°, 0 \leqslant t \leqslant 360$

where t is the number of minutes from the start.

a What is the depth of water at the start?

b Calculate the depth of water after 2 hours.

c At which two times will the water have a depth of 8 m?

Activity

Apart from getting the direction right, golfers have two other things to think of when preparing for a shot:

- will the ball travel the correct distance?
- will the ball go high enough to clear any obstacles in its path?

Ignoring spin and air resistance this becomes a mathematical problem under the heading of 'projectile motion' where the golf ball is a projectile.

A projectile is defined as any object which is thrown, launched or fired and its movement can be modelled using mathematical equations. Ignoring any forces other than gravity, the path of a projectile depends on two variables. In the case of the golfer they are:

- the initial velocity of the golf ball
- the initial angle at which it is struck.

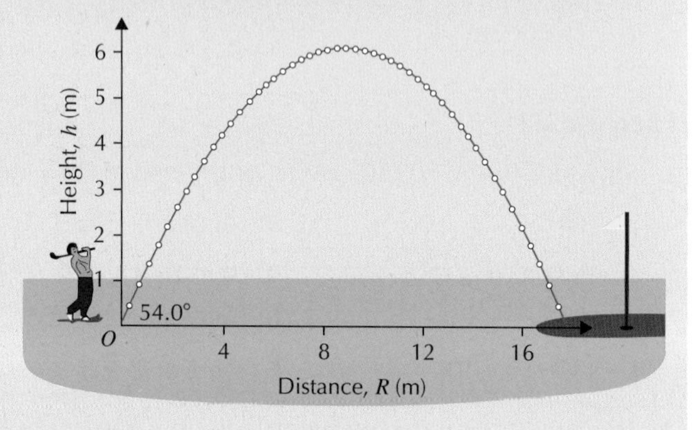

A golfer's swing controls both of these.

If the initial velocity of the ball is u m/s, the initial angle is $\theta°$ and we take the acceleration due to gravity as g m/s^2, then, assuming that the path of the ball starts and ends at the same height, the study of ball motion would show the following results:

- the time, t seconds, taken for the ball to reach its maximum height is:

  $t = \dfrac{u\sin\theta°}{g}$

- the maximum height, h metres, reached by the ball is: $h = \dfrac{u^2\sin^2\theta°}{2g}$

  which can also be expressed as $h = \dfrac{u^2(\sin\theta°)^2}{2g}$

- the time, T seconds, taken for the ball to land is: $T = \dfrac{2u\sin\theta°}{g}$

- the total distance, R metres travelled horizontally is: $R = \dfrac{u^2\sin(2\theta)°}{g}$

Use $g = 9.8$ m/s^2 as an approximation for acceleration due to gravity to answer the following questions.

1 A golf ball is launched at an angle of 23° with an initial velocity of 60 m/s.
  - **a** How long is the golf ball in the air?
  - **b** Find its maximum height.
  - **c** How far does the golf ball travel horizontally?

2 A golf ball driven at a speed of 80 m/s hits the ground 270 metres from the starting point. At what angle was it struck?

3 John is close to the green and wants his shot to travel 40 metres. He hits the ball at a speed of 28 m/s. Assuming the ball travels the desired distance, calculate the angle at which the ball starts off.

You must remember that these calculations do not take into account factors such as air resistance, spin or how far the ball bounces or rolls after landing.

Calculating where trigonometric graphs and straight lines intersect

Points of intersection between trigonometric graphs and straight lines can be calculated in similar ways to the methods used to find the points of intersection of a straight line and a parabola. (See Chapter 19.)

Example 24.8

The diagram shows part of the graph of $y = \sin x°$ and the straight line $y = 0.4$. The straight line intersects the sine curve at points A and B. Find the coordinates of A and B.

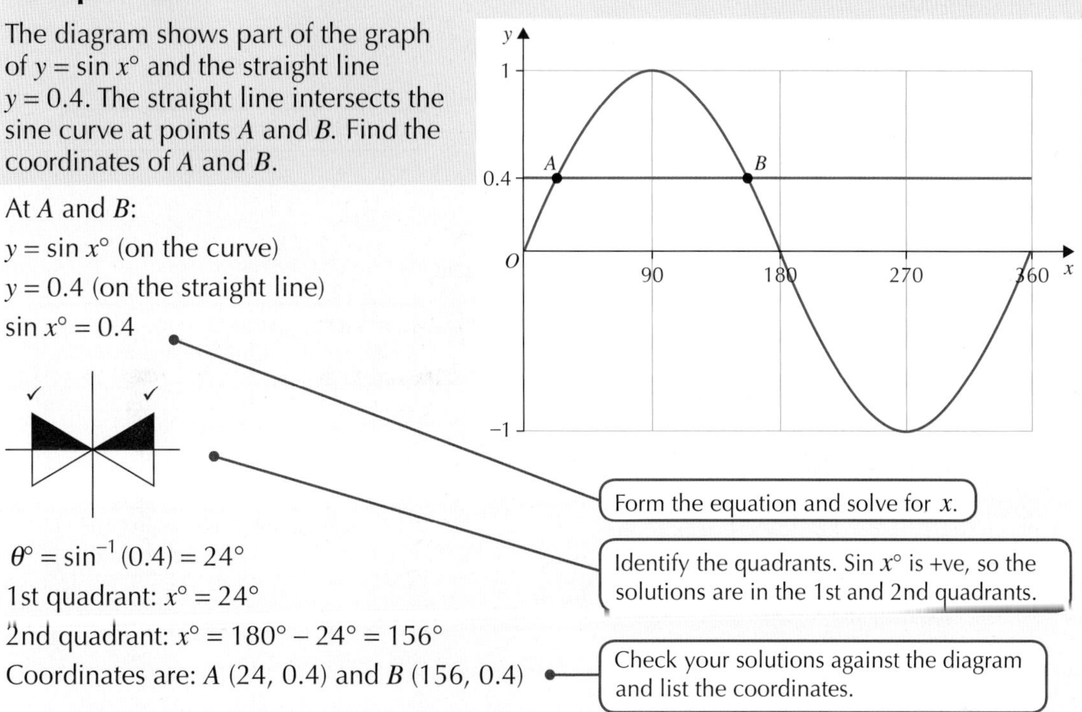

At A and B:

$y = \sin x°$ (on the curve)

$y = 0.4$ (on the straight line)

$\sin x° = 0.4$

$\theta° = \sin^{-1}(0.4) = 24°$

1st quadrant: $x° = 24°$

2nd quadrant: $x° = 180° - 24° = 156°$

Coordinates are: A (24, 0.4) and B (156, 0.4)

Check your solutions against the diagram and list the coordinates.

Example 24.9

The graph shows part of a graph of the form $y = a\sin bx° + c$.

a Find the values of a, b and c and hence write down the equation of the graph.

b The line with equation $y = 1.5$ intersects the graph at P and Q. Find the coordinates of P and Q.

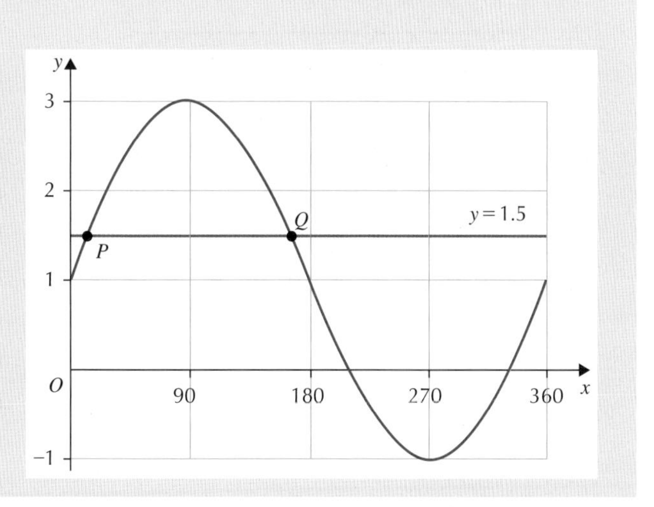

(continued)

a $y = a\sin bx° + c$

amplitude = 2, so $a = 2$

1 cycle in 360°, so $b = 1$

graph moved up 1, so $c = 1$

Equation is $y = 2\sin x° + 1$

Compare with: $y = a\sin bx° \pm c$ — amp = $|a|$; b cycles in 360°; up/down c

b At P and Q:

$y = 2\sin x° + 1$ (on the curve)

$y = 1.5$ (on the straight line)

$2\sin x° + 1 = 1.5$

$2\sin x° = 0.5$

$\sin x° = 0.25$

Form the equation and solve for x.

✓ ✓

Identify the quadrants. $\sin x°$ is +ve, so the solutions are in the 1st and 2nd quadrants.

$\theta° = \sin^{-1}(0.25) = 14°$

1st quadrant: $x° = 14°$

2nd quadrant: $x° = 180° - 14° = 166°$

Coordinates are: A (14, 1.5) and B (166, 1.5)

Check your solutions against the diagram and list the coordinates.

Exercise 24E

1 Find the coordinates of the marked points.

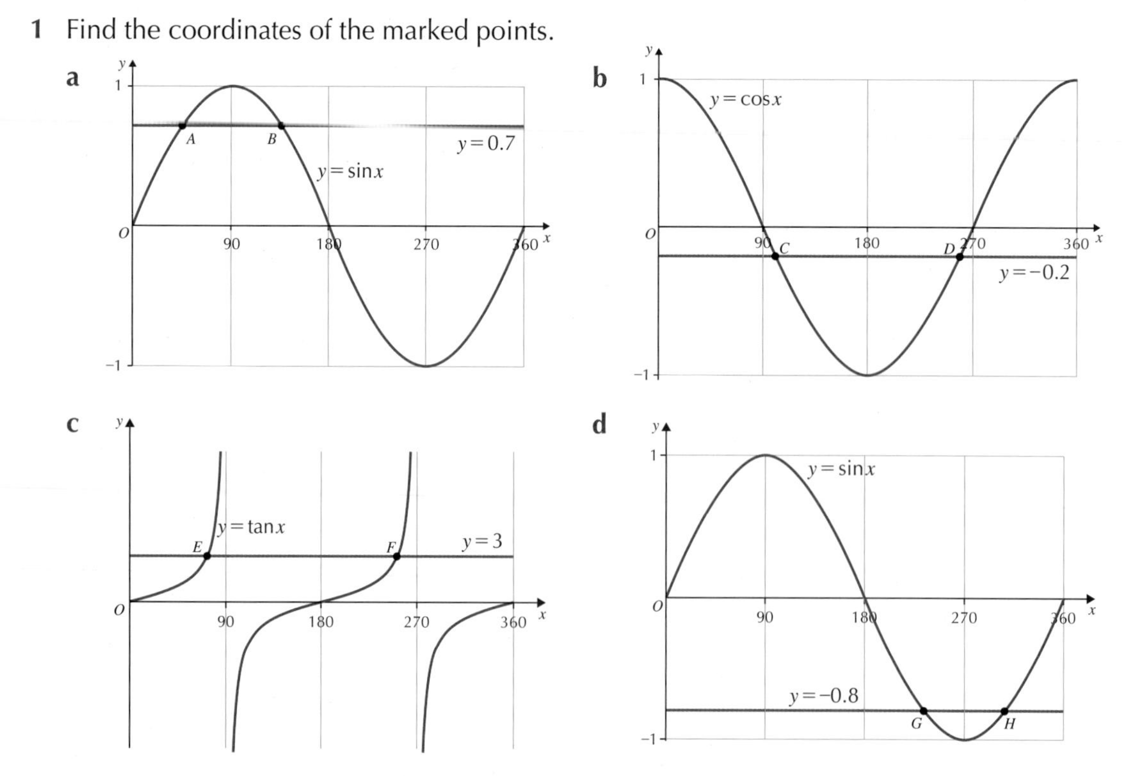

e

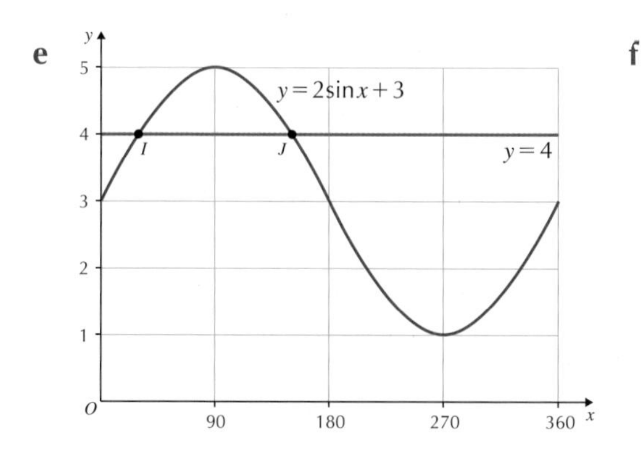

f

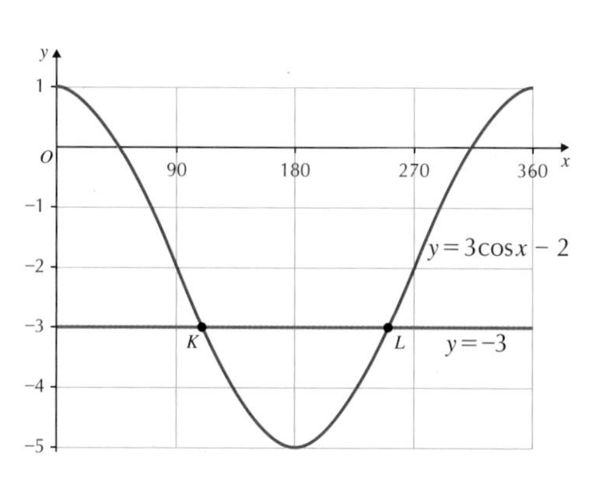

2 The graph shown is of the form $y = p\sin x° + q$.

a Find the values of p and q.

b The line with equation $y = 5$ intersects the graph at the points A and B. Find the coordinates of A and B.

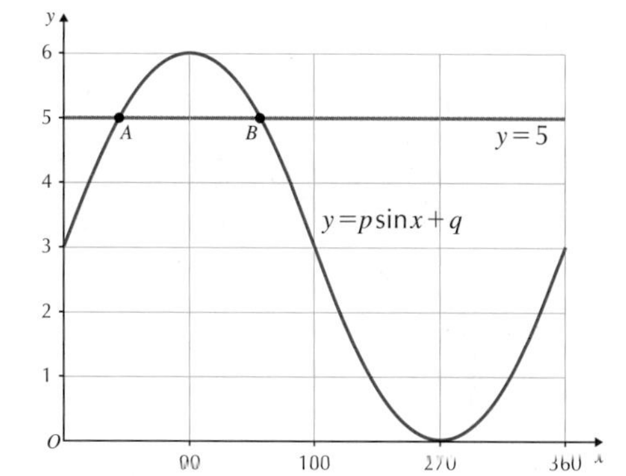

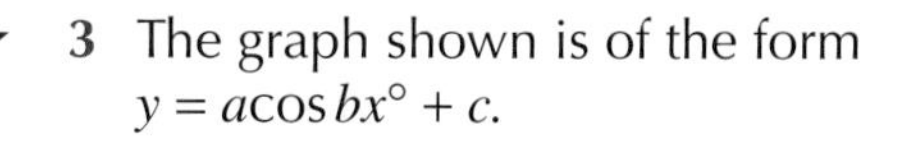

★ **3** The graph shown is of the form $y = a\cos bx° + c$.

a Find the values of a, b and c.

b The line with equation $y = -1$ intersects the graph at the points P and Q. Find the coordinates of P and Q.

c Write down the coordinates of the next two points of intersection.

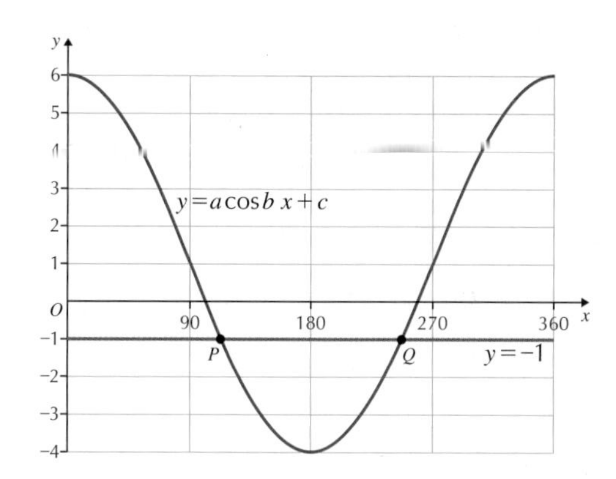

Learning and working with exact value ratios

There is a limited number of angles for which we can give an **exact** value for sin, cos and tan. For other angles, we can only get an approximate value.

The most obvious angles for which we can obtain exact values are those for which the exact values are 0, 1, or −1.

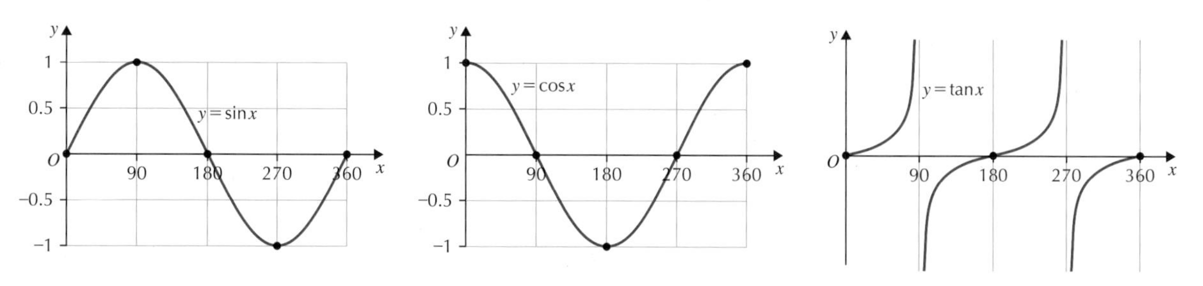

For example, we know that the **exact** value of $\cos 270° = 0$.

The acute angles 30°, 45° and 60° (and their related angles) also have exact value ratios.

Exact values for 30° and 60°

Consider an equilateral triangle with sides of length 2 units. This can be shown as two right-angled triangles with hypotenuse 2 and sides 1 and $\sqrt{3}$ (found by Pythagoras).

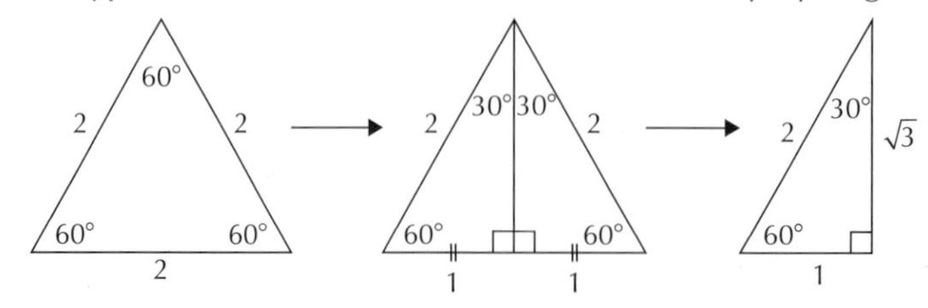

From this, the exact values for sin, cos and tan can be found, as shown in the table.

30°	60°
$\sin 30° = \frac{1}{2}$	$\sin 60° = \frac{\sqrt{3}}{2}$
$\cos 30° = \frac{\sqrt{3}}{2}$	$\cos 60° = \frac{1}{2}$
$\tan 30° = \frac{1}{\sqrt{3}}$	$\tan 60° = \sqrt{3}$

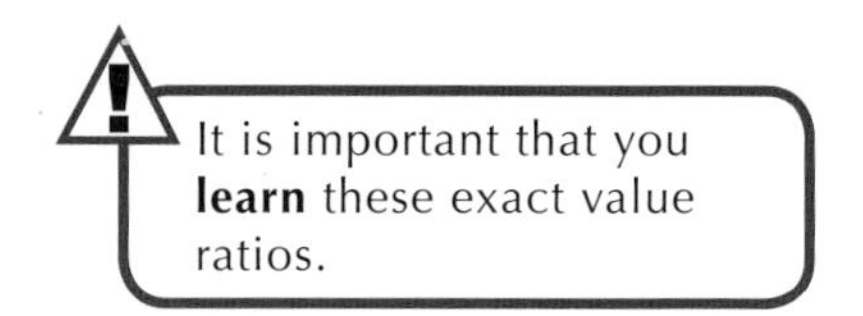

It is important that you **learn** these exact value ratios.

Exact values for 45°

Consider a right-angled isosceles triangle with shorter sides of length 1 unit. This has a hypotenuse of $\sqrt{2}$ (found by Pythagoras).

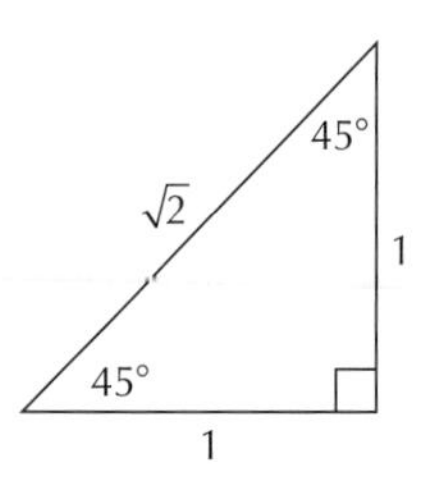

Again, exact values for sin, cos and tan can be found.

45°
$\sin 45° = \frac{1}{\sqrt{2}}$
$\cos 45° = \frac{1}{\sqrt{2}}$
$\tan 45° = 1$

Example 24.10

Without using a calculator solve the following equations for $0 \leqslant x \leqslant 360$.

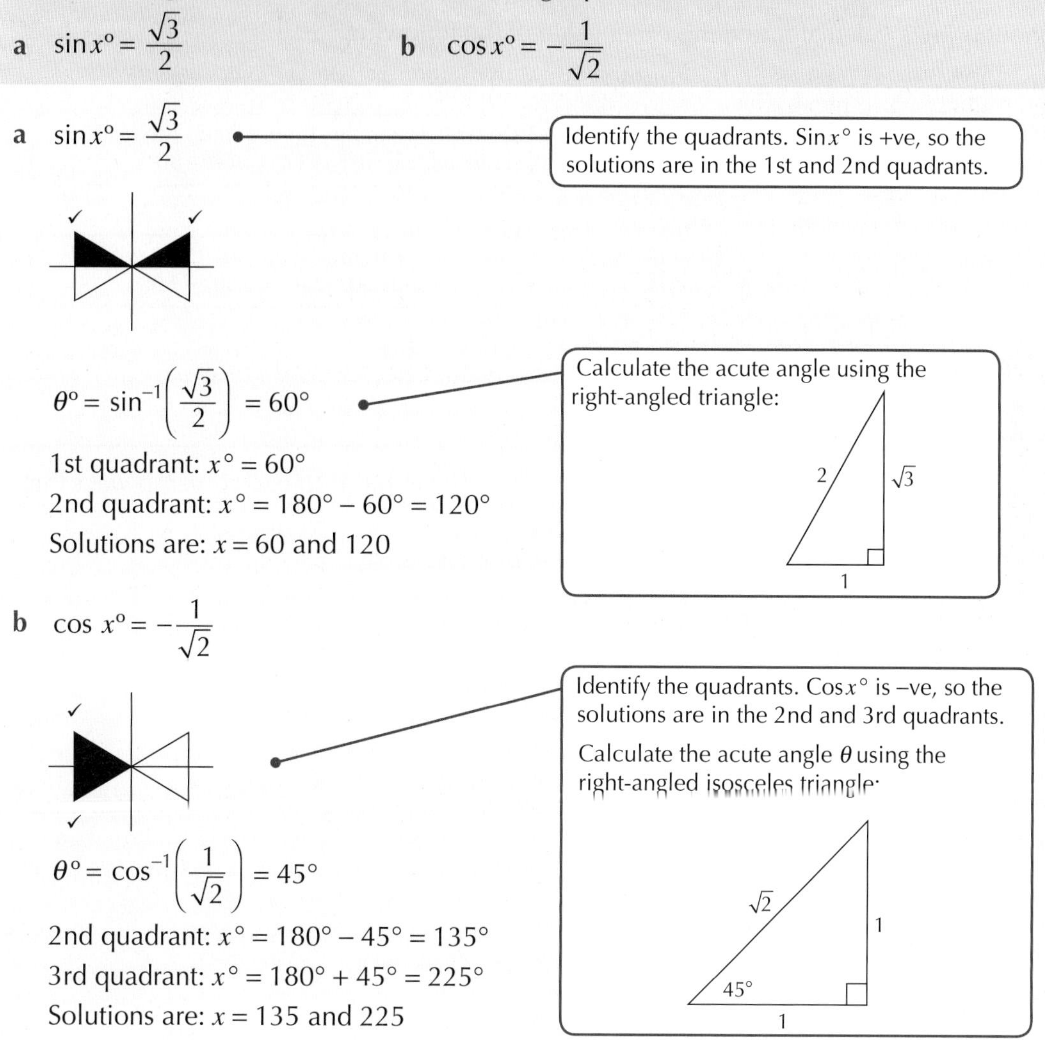

a $\sin x^\circ = \frac{\sqrt{3}}{2}$ **b** $\cos x^\circ = -\frac{1}{\sqrt{2}}$

a $\sin x^\circ = \frac{\sqrt{3}}{2}$

Identify the quadrants. $\sin x^\circ$ is +ve, so the solutions are in the 1st and 2nd quadrants.

$\theta^\circ = \sin^{-1}\left(\frac{\sqrt{3}}{2}\right) = 60^\circ$

Calculate the acute angle using the right-angled triangle:

1st quadrant: $x^\circ = 60^\circ$

2nd quadrant: $x^\circ = 180^\circ - 60^\circ = 120^\circ$

Solutions are: $x = 60$ and 120

b $\cos x^\circ = -\frac{1}{\sqrt{2}}$

Identify the quadrants. $\cos x^\circ$ is −ve, so the solutions are in the 2nd and 3rd quadrants.

Calculate the acute angle θ using the right-angled isosceles triangle:

$\theta^\circ = \cos^{-1}\left(\frac{1}{\sqrt{2}}\right) = 45^\circ$

2nd quadrant: $x^\circ = 180^\circ - 45^\circ = 135^\circ$

3rd quadrant: $x^\circ = 180^\circ + 45^\circ = 225^\circ$

Solutions are: $x = 135$ and 225

Exercise 24F

1 Without using a calculator, solve the following equations for angles between 0° and 360°.

a $\sin x^\circ = \frac{1}{\sqrt{2}}$ **b** $\tan p^\circ = \frac{1}{\sqrt{3}}$ **c** $\cos q^\circ = \frac{\sqrt{3}}{2}$

d $\tan a^\circ = -\sqrt{3}$ **e** $\sin x^\circ = -\frac{\sqrt{3}}{2}$ **f** $\cos b^\circ = -\frac{1}{2}$

★ 2 Without using a calculator, solve the following equations for $0 \leqslant x \leqslant 360$.

a $2\sin x^\circ - 1 = 0$ **b** $3\tan x^\circ = 3$ **c** $\sqrt{2}\cos x^\circ - 1 = 0$

d $\tan x^\circ + \sqrt{3} = 0$ **e** $2\cos x^\circ + \sqrt{3} = 0$ **f** $\sqrt{2}\sin x^\circ + 2 = 1$

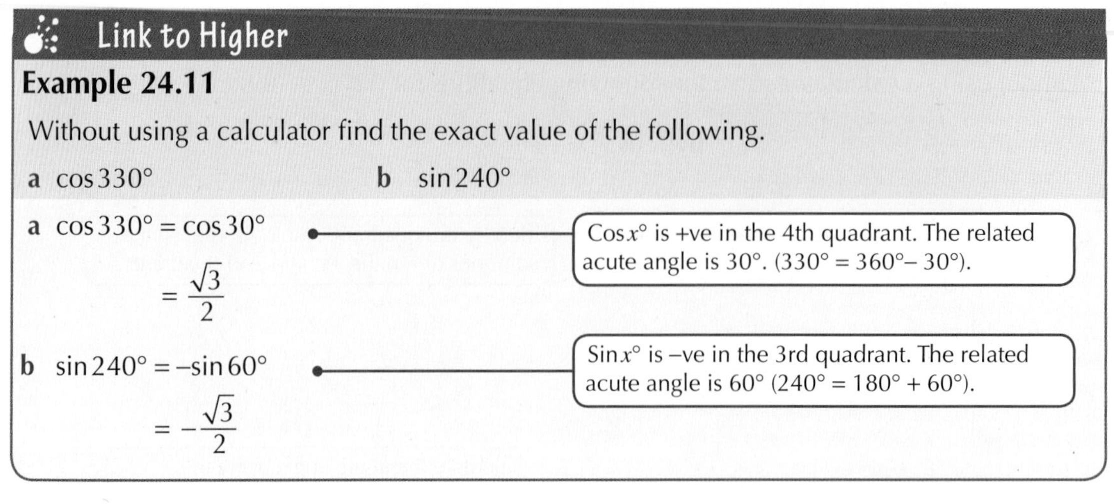

Link to Higher

Example 24.11

Without using a calculator find the exact value of the following.

a $\cos 330°$ **b** $\sin 240°$

a $\cos 330° = \cos 30°$

$= \frac{\sqrt{3}}{2}$

Cos$x$° is +ve in the 4th quadrant. The related acute angle is 30°. (330° = 360°– 30°).

b $\sin 240° = -\sin 60°$

$= -\frac{\sqrt{3}}{2}$

Sin$x$° is –ve in the 3rd quadrant. The related acute angle is 60° (240° = 180° + 60°).

The trigonometric identities

Trigonometric identities relate the three basic trigonometric ratios to each other in ways that can help with the solution of complex equations.

The trigonometric identities are derived using the unit circle.

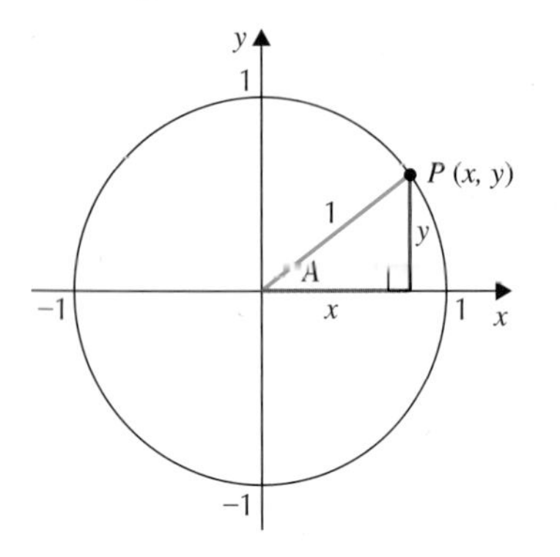

- $\sin A = \frac{y}{1} = y$
- $\cos A = \frac{x}{1} = x$
- $\tan A = \frac{y}{x}$

Using Pythagoras:

$$y^2 + x^2 = 1$$

$$(\sin A)^2 + (\cos A)^2 = 1$$

$$\sin^2 A + \cos^2 A = 1 \text{ for any angle } A$$

In trigonometry the convention is to write, for example, $(\sin A)^2$ as $\sin^2 A$.

Using the definition of $\tan A$

$$\tan A = \frac{y}{x}$$

$$\tan A = \frac{\sin A}{\cos A}$$

So, for any angle A the trigonometric identities are:

$$\mathbf{\sin^2 A + \cos^2 A = 1}$$

$$\mathbf{\tan A = \frac{\sin A}{\cos A}}$$

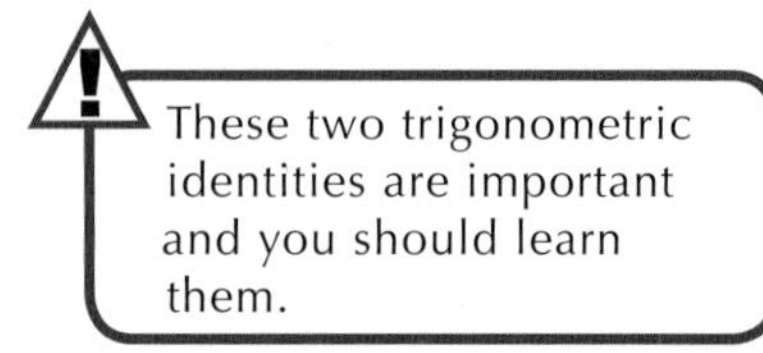

These two trigonometric identities are important and you should learn them.

Example 24.12

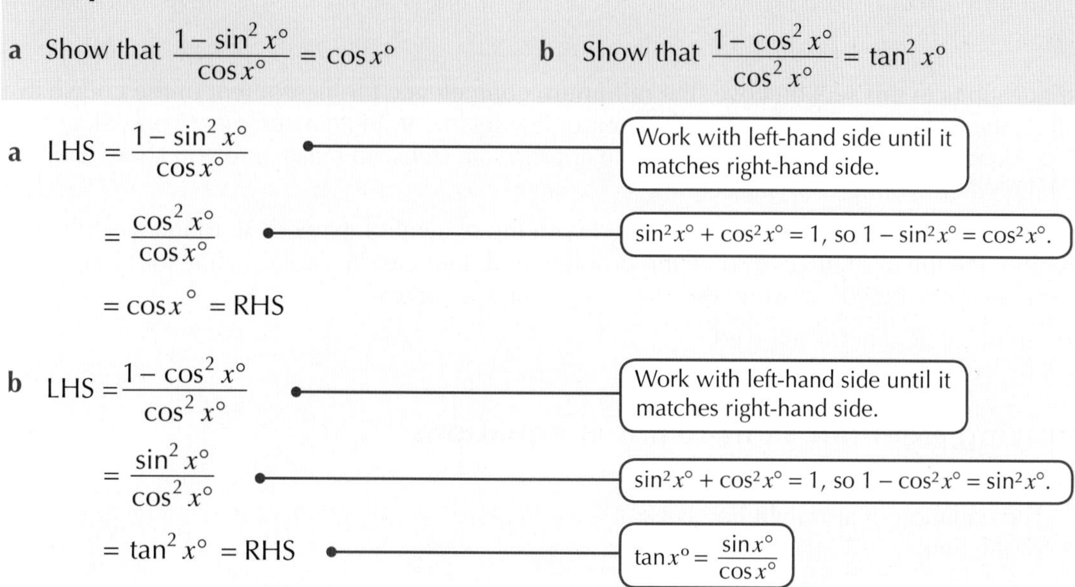

a Show that $\dfrac{1-\sin^2 x°}{\cos x°} = \cos x°$ **b** Show that $\dfrac{1-\cos^2 x°}{\cos^2 x°} = \tan^2 x°$

a $\text{LHS} = \dfrac{1-\sin^2 x°}{\cos x°}$ — Work with left-hand side until it matches right-hand side.

$= \dfrac{\cos^2 x°}{\cos x°}$ — $\sin^2 x° + \cos^2 x° = 1$, so $1 - \sin^2 x° = \cos^2 x°$.

$= \cos x° = \text{RHS}$

b $\text{LHS} = \dfrac{1-\cos^2 x°}{\cos^2 x°}$ — Work with left-hand side until it matches right-hand side.

$= \dfrac{\sin^2 x°}{\cos^2 x°}$ — $\sin^2 x° + \cos^2 x° = 1$, so $1 - \cos^2 x° = \sin^2 x°$.

$= \tan^2 x° = \text{RHS}$ — $\tan x° = \dfrac{\sin x°}{\cos x°}$

Exercise 24G

1 Show that $\sin x° \tan x° = \dfrac{\sin^2 x°}{\cos x°}$

2 Show that $\sin^3 x° + \sin x° \cos^2 x° = \sin x°$

3 Show that $3\sin^2 x° + 3\cos^2 x° = 3$

4 Show that $\dfrac{\sin x°}{\tan x°} = \cos x°$

5 Show that $5 - 5\cos^2 x° = 5\sin^2 x°$

★ 6 Show that $\dfrac{\sin^2 x°}{1-\cos^2 x°} = 1$

★ 7 Show that $\dfrac{\sin x° \cos x°}{\cos^2 x°} = \tan x°$

- I can solve simple trigonometric equations. ★ Exercise 24A Q1
- I know how to identify related angles when working with trigonometric functions. ★ Exercise 24B Q2
- I know how to solve basic trigonometric equations. ★ Exercise 24C Q4 ★ Exercise 24D Q1
- I know how to calculate where trigonometric graphs and straight lines intersect. ★ Exercise 24E Q3
- I can use exact value ratios to solve trigonometric equations without a calculator. ★ Exercise 24F Q2
- I know the trigonometric identities, $\tan A = \dfrac{\sin A}{\cos A}$ and $\sin^2 A + \cos^2 A = 1$, and can use them to simplify algebraic expressions. ★ Exercise 24G Q6, Q7

For further assessment opportunities, see the Preparation for Assessment for Unit 2 on pages 288–291.

Preparation for Assessment: Unit 2

The questions in this section cover the minimum competence for the content of the course in Unit 2. They are a good preparation for your unit assessment. In an assessment you will get full credit only if your solution includes the appropriate steps, so make sure you show your thinking when writing your answers.

Remember that reasoning questions marked with the ⚙ symbol expect you to interpret the situation, identify a strategy to solve the problem and then clearly explain your solution. If a context is given you must relate your answer to this context.

The use of a calculator is allowed.

Applying algebraic skills to linear equations (Chapters 12, 13, 14 and 15)

1 The equation of a straight line is $y = 4 - 3x$. Write down the gradient and y-intercept of the line.

2 A straight line with gradient 4 passes through the point (5, −7). Determine the equation of the straight line.

3 Solve the equation $2(x - 6) + 3 = 4 - (x + 2)$.

4 Solve the inequation $4t + 6 \geqslant 2t - 9$.

5 $f(x) = 2x - 4$ and $g(x) = 8 - x$.

a Draw the graphs of $y = f(x)$ and $y = g(x)$.

b At what point do they cross?

6 Solve the following system of equations algebraically:

$$x - 5y = 15$$

$$3x - 7y = 17$$

7 The attendance at a local football match was 1200 people. The entrance fee for adults was £6.50 and children were charged half the adult price. The total amount taken for entry tickets for the game was £6630.

a Write two equations to represent this information.

b Solve your equations and find the number of adults and the number of children who attended the game.

8 This formula is used to calculate the area of a trapezium.

$$A = \tfrac{1}{2}(a + b)h$$

where h is the perpendicular height and a and b are the lengths of the parallel sides.

Change the subject of the formula to a.

9 This formula is used in physics to calculate the displacement s of a body given its initial speed u, its acceleration a and the time taken t.

$$s = ut + \tfrac{1}{2}at^2$$

Change the subject of the formula to a.

Applying algebraic skills to graphs of quadratic relationships (Chapters 16, 17 and 18)

10 The diagram shows the parabola with equation $y = kx^2$. What is the value of k?

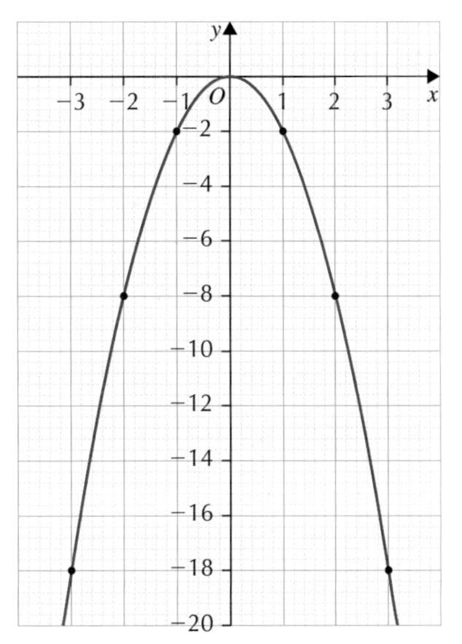

11 The diagram shows the graph of the equation $y = (x + p)^2 + q$. What are the values of p and q?

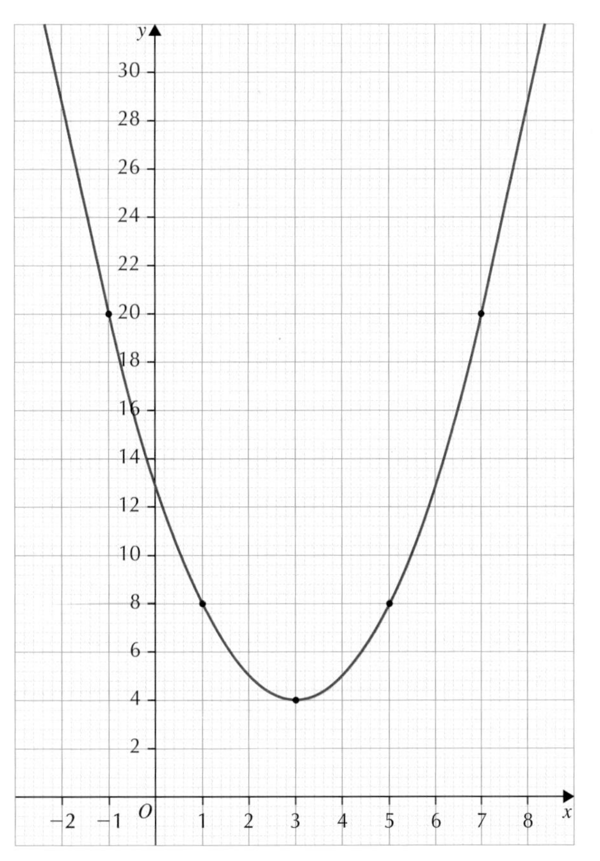

12 Sketch the graph $y = (x - 3)(x + 1)$ on plain paper.
Mark clearly where the graph crosses the axes and state the coordinates of the turning point.

13 **a** Sketch the graph $y = (x - 5)^2 + 2$. Mark clearly where the graph crosses the axes and state the coordinates of the turning point.

b State the equation of the axis of symmetry.

c State the nature of the turning point.

Applying algebraic skills to graphs of quadratic relationships (Chapter 19)

14 Solve the equation $x^2 + 2x - 24 = 0$.

15 The graph of the equation $y = 10 + 3x - x^2$ is shown in the diagram. Use the graph to write down the solutions to this equation.

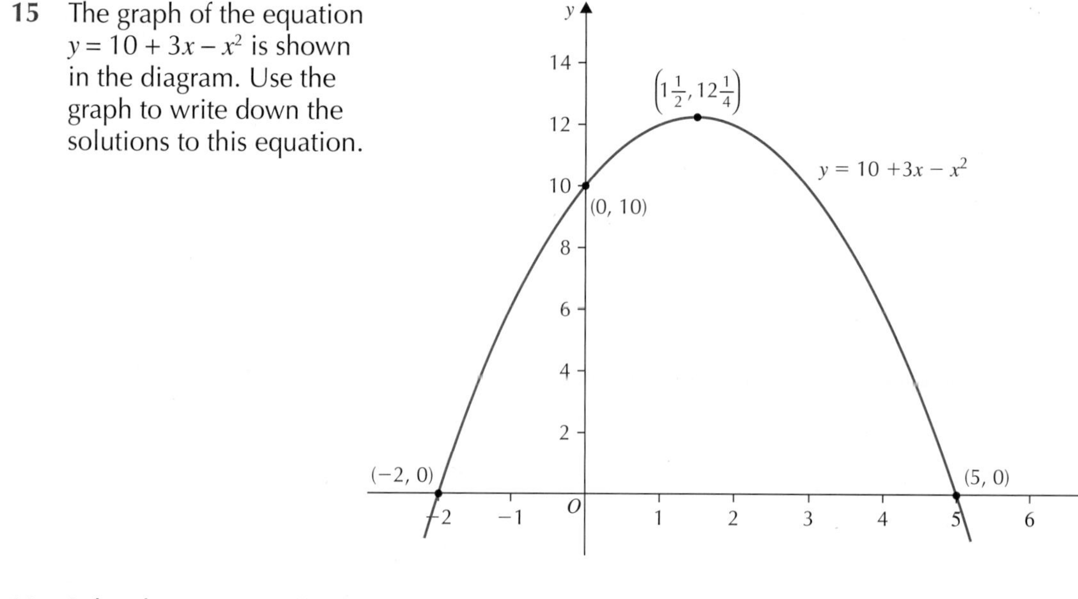

16 Solve the equation $x^2 - 5x - 3 = 0$ using the quadratic formula. Give your answer correct to 1 decimal place.

17 Solve the equation $x^2 - 9x + 20 = 0$ and state the roots.

18 Determine the nature of the roots of the equation $3x^2 + 2x - 3 = 0$.

Applying geometric skills to lengths, angles and similarity (Chapters 20, 21 and 22)

19 The council instructs a building firm to construct a new sports hall. The council insists that the footprint of the building must be a rectangle. The diagram shows the footprint of the new building. Determine if the council will be satisfied or not.

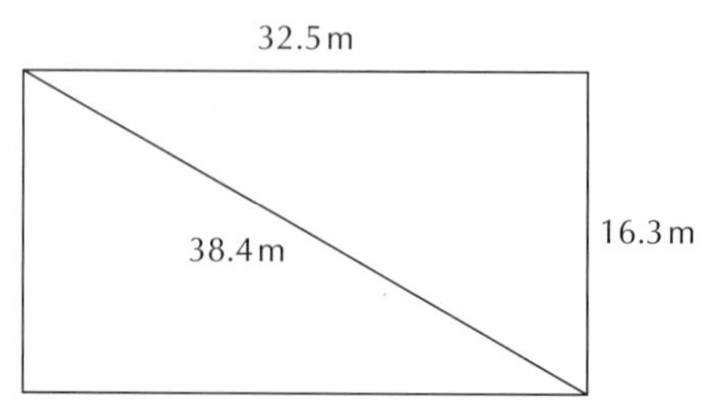

20 The diagram shows a regular heptagon divided into seven congruent triangles. Find the size of angle a and angle b in the triangles.

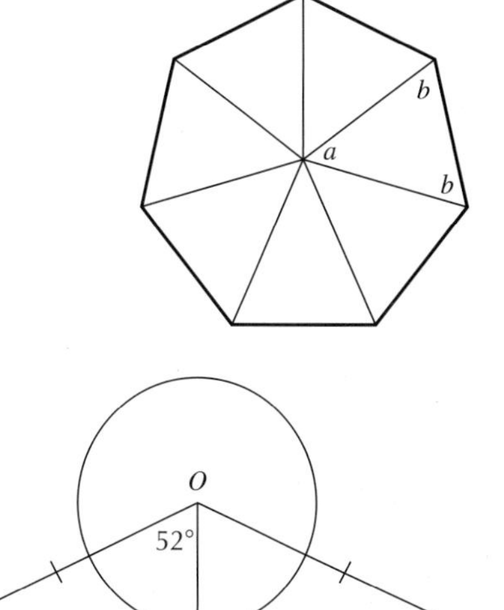

21 The picture shows a bicycle stand. AOB is an isosceles triangle and the wheel is a circle with centre O. AB is a tangent to the circle and M is the midpoint of AB.

Given that AOM is 52°, calculate angle MBO.

22 A supermarket sells baked beans in two sizes of cylindrical tins, standard and family size. The family size tin holds 670 ml and is 12 cm high. The standard size tin is mathematically similar to the larger family size. If the standard baked bean tin is 8 cm high, what volume does it hold?

Applying trigonometric skills to graphs and identities (Chapters 23 and 24)

23 Sketch the graphs of:

a $y = \sin x°$ for $0 \leqslant x \leqslant 360$

b $y = \cos x°$ for $0 \leqslant x \leqslant 360$

c $y = \tan x°$ for $0 \leqslant x \leqslant 360$.

24 Sketch the graph of $y = 2\cos x°$ for $0 \leqslant x \leqslant 360$.

25 Sketch the graph of $y = \sin x° - 2$ for $0 \leqslant x \leqslant 360$.

26 Write down the amplitude and period of the graph of the equation $y = 3\sin 4x°$.

27 Solve the equation $2\tan x° - 3 = 0$ for $0 \leqslant x \leqslant 180$.

28 Solve the equation $3\sin x° + 2 = 0$ for $0 \leqslant x \leqslant 360$.

25 Calculating the area of a triangle using trigonometry

In this chapter you will learn how to:

- use trigonometry to work out the area of non-right-angled triangles using the formula **area = $\frac{1}{2}ab \sin C$**
- carry out calculations in context and calculate missing sides or angles given the area.

You should already know:

- how to use your calculator to carry out trigonometric calculations
- how to substitute values into algebraic expressions
- how to calculate the area of a triangle given breadth and height.

Calculating the area of a triangle using trigonometry

You have already calculated the area of a triangle given perpendicular dimensions using the formula area = $\frac{1}{2}bh$. However, if you are not given the base and height of the triangle you can use trigonometry to calculate the area of any triangle using the formula:

$$\textbf{area} = \tfrac{1}{2}ab\sin C$$

where a, b and C are labelled in the diagram.

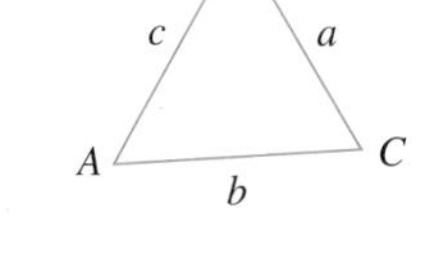

You can work out that this is the same as:

$$\textbf{area} = \tfrac{1}{2}bc\sin A = \tfrac{1}{2}ac\sin B$$

Example 25.1

Calculate the area of the triangle shown giving your answer to 1 decimal place.

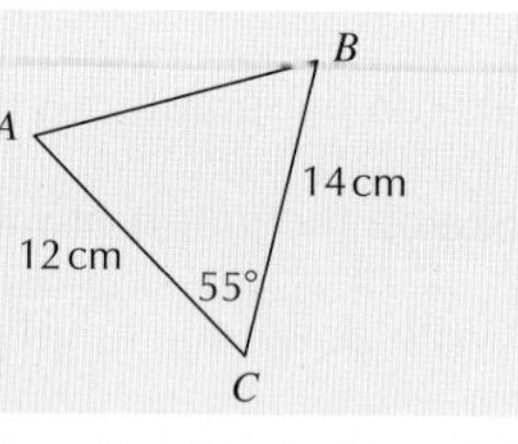

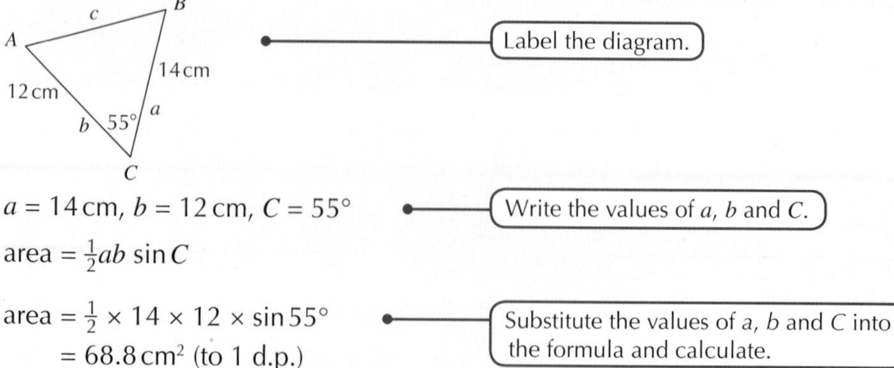

$a = 14\,\text{cm}$, $b = 12\,\text{cm}$, $C = 55°$ — Write the values of a, b and C.

area = $\frac{1}{2}ab \sin C$

area = $\frac{1}{2} \times 14 \times 12 \times \sin 55°$
= $68.8\,\text{cm}^2$ (to 1 d.p.) — Substitute the values of a, b and C into the formula and calculate.

Exercise 25A

1 Use the area formula to calculate the area of each triangle giving your answers to 1 decimal place.

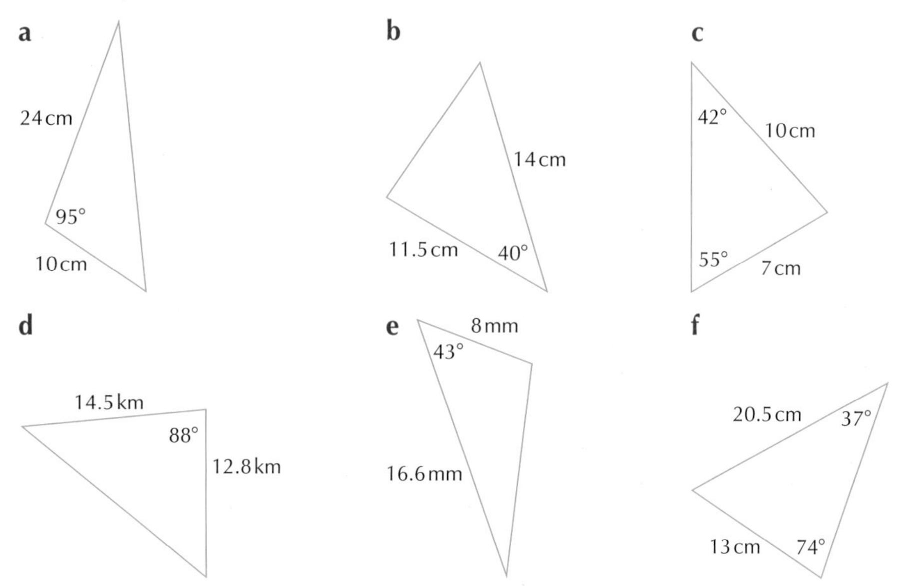

2 A banner for a maths club is being designed in the shape of a triangle. The cost of material for the banner is £3.50 per m^2. Which of the two designs shown would be cheaper and by how much?

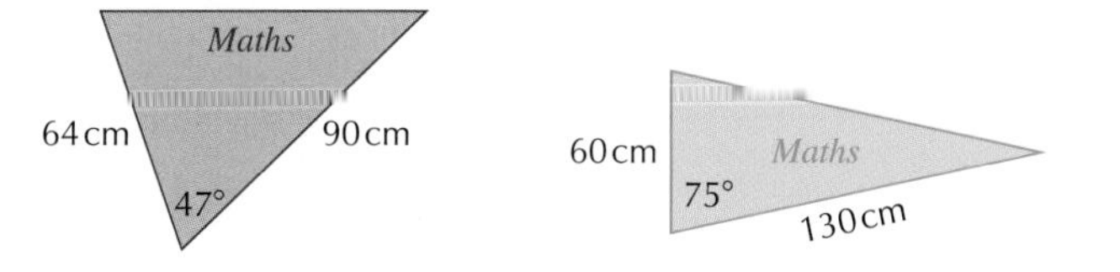

3 An artist has created a pyramid as part of an installation and has to paint all the triangular faces. Each tin of paint he buys will cover 8 m^2. How many tins of paint will he require to complete the job?

★ **4** Find the area of the rhombus shown giving your answer to 3 significant figures.

5 Ship A sails from dock on a bearing of 070° for 3.5 km.
Ship B sails from the same dock on a bearing of 120° for 5 km.
The area between the two boats is being scanned in a search for a sunken treasure ship and forms a triangle as shown.
What is the area of the seabed being scanned?
Give your answer to 1 decimal place.

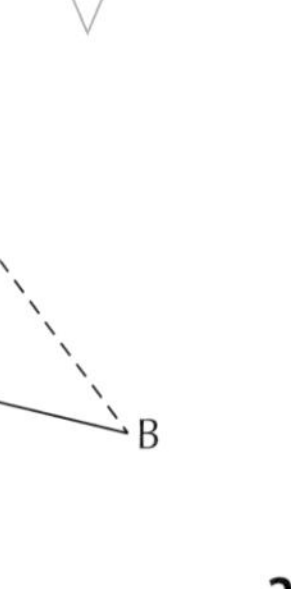

6 The sail for a boat is to be designed as shown.
Calculate the area of the sail to 3 significant figures.

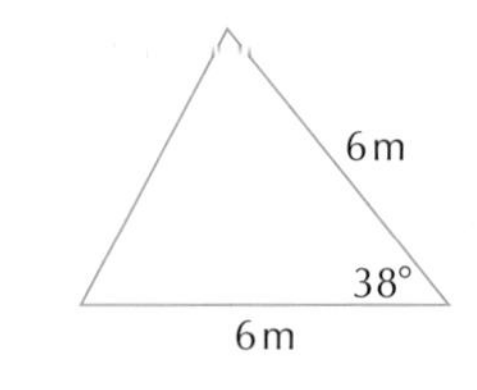

7 A manufacturing company needs to makes triangular tiles for a bathroom design. Two possible samples are shown below. Which one would be cheaper assuming both are made using the same material?

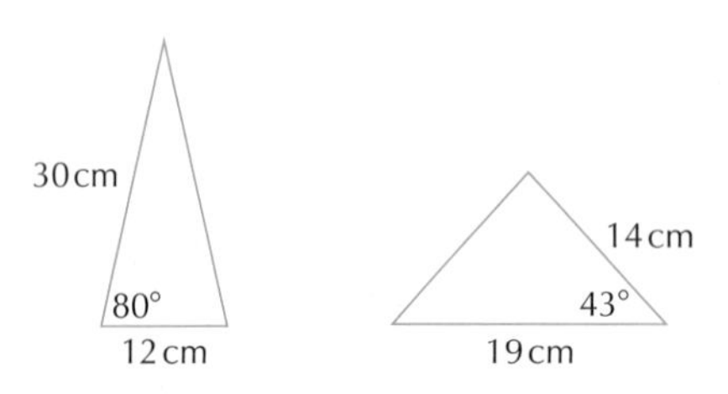

8 Calculate the area of the regular pentagon shown.

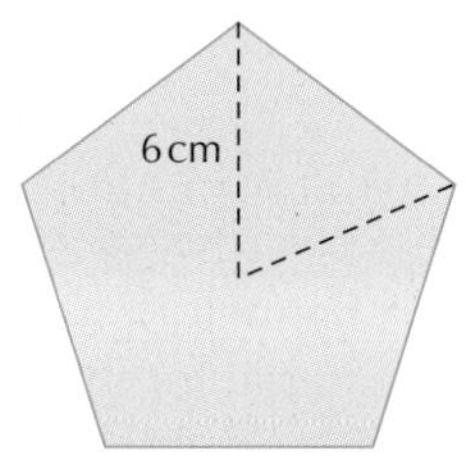

9 Using the same measurements from the centre to a vertex as in Question 8, calculate the area of:

a a regular hexagon
b a regular octagon
c a regular decagon

Using the area formula to find a missing side or angle

Sometimes you are given the area and must use the formula to calculate either a missing side or angle. To do this you need to rearrange the formula appropriately after substituting the given values.

Example 25.2

The triangle shown has an area of $45\,\text{cm}^2$. Calculate the size of the acute angle PQR. Give your answer to 1 decimal place.

P
R
10 cm
11 cm
Q

(continued)

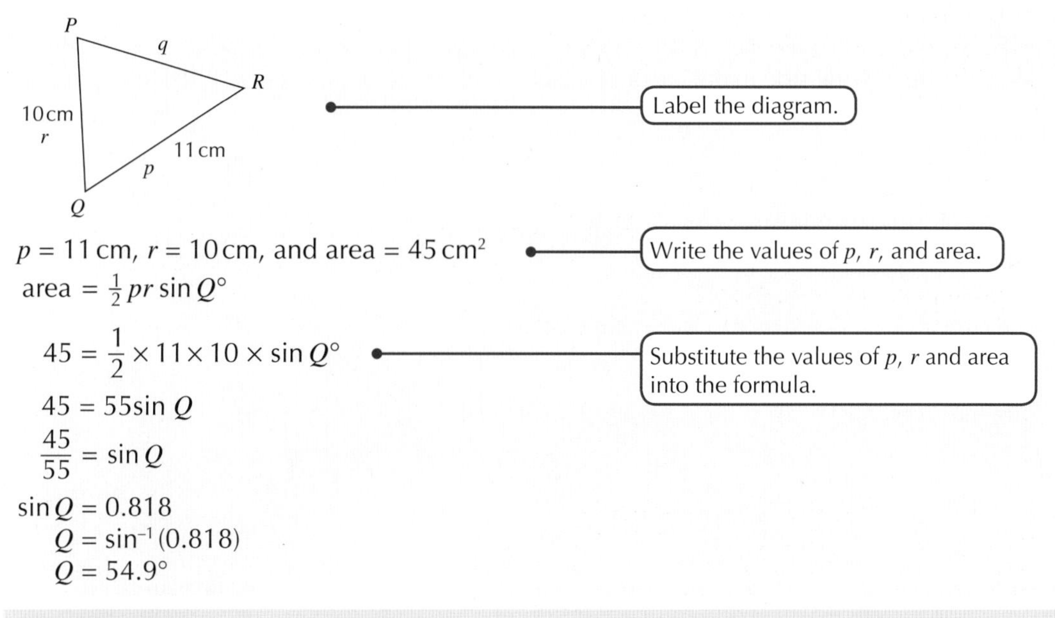
P
q
R
10 cm
r
11 cm
p
Q
Label the diagram.
Write the values of p, r, and area.
Substitute the values of p, r and area into the formula.

$p = 11\text{ cm}$, $r = 10\text{ cm}$, and area $= 45\text{ cm}^2$

area $= \frac{1}{2}pr\sin Q°$

$$45 = \frac{1}{2} \times 11 \times 10 \times \sin Q°$$

$$45 = 55\sin Q$$

$$\frac{45}{55} = \sin Q$$

$$\sin Q = 0.818$$

$$Q = \sin^{-1}(0.818)$$

$$Q = 54.9°$$

Example 25.3

Calculate the size of the missing side AB in the triangle shown.
Give your answer to 1 decimal place.

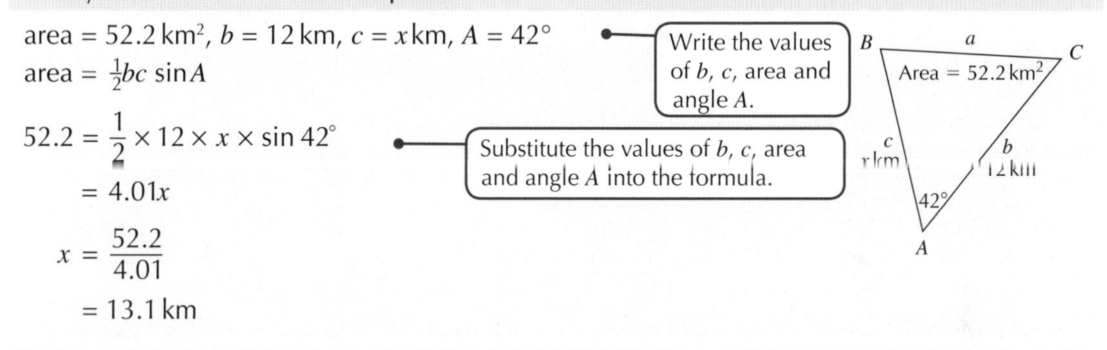
Write the values of b, c, area and angle A.
Substitute the values of b, c, area and angle A into the formula.
B
a
C
Area = 52.2 km²
c
x km
b
12 km
42°
A

area $= 52.2\text{ km}^2$, $b = 12\text{ km}$, $c = x\text{ km}$, $A = 42°$

area $= \frac{1}{2}bc\sin A$

$$52.2 = \frac{1}{2} \times 12 \times x \times \sin 42°$$

$$= 4.01x$$

$$x = \frac{52.2}{4.01}$$

$$= 13.1\text{ km}$$

Example 25.4

A triangular skateboard ramp is shown below.
Calculate the length of the smallest side of the ramp.
Give your answer to 2 decimal places.

3 m
105°
Area = 2.8 m²

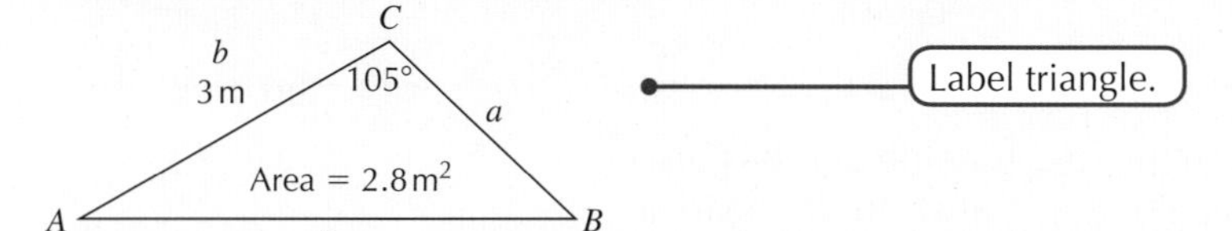
C
b
3 m
105°
a
Area = 2.8 m²
A
c
B
Label triangle.

a is the shortest side, as it is opposite the smallest angle, so we must calculate a.

$a = x\text{ m}$, $b = 3\text{ m}$, $C = 105°$, area $= 2.8\text{ m}^2$ — Write down values for area, a, b and angle C.

area $= \frac{1}{2}ab\sin C$

$$2.8 = \frac{1}{2} \times x \times 3 \times \sin 105°$$ — Substitute values into equation.

$$= 1.45x$$

$$x = \frac{2.8}{1.45}$$

$$= 1.93\text{ m (to 2 d.p.)}$$

Exercise 25B

1 Calculate the missing side in each example shown giving your answers to 2 decimal places.

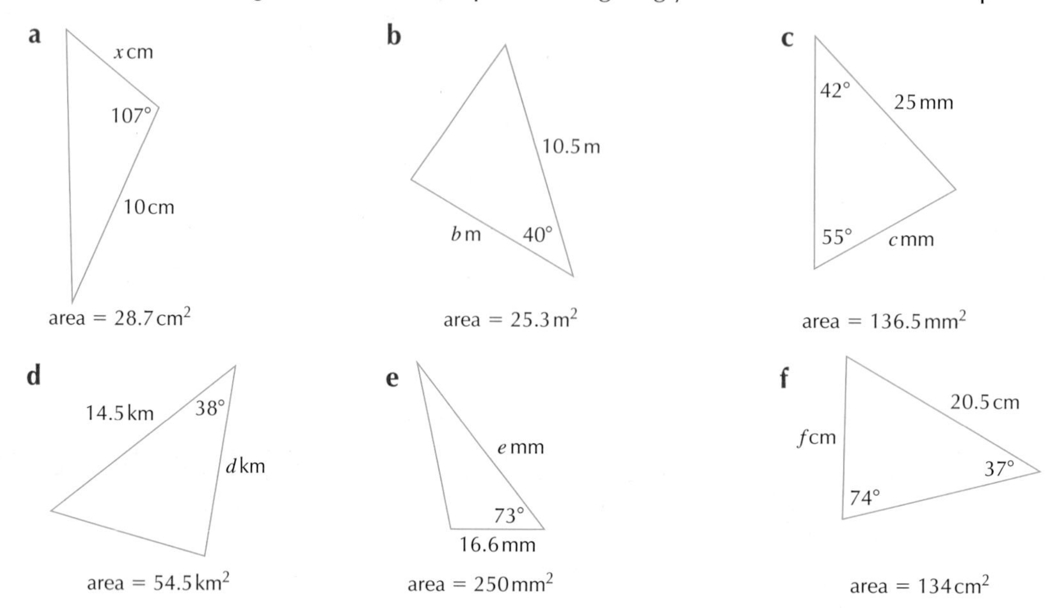

2 Calculate the missing angle in each triangle giving your answers to 1 decimal place.

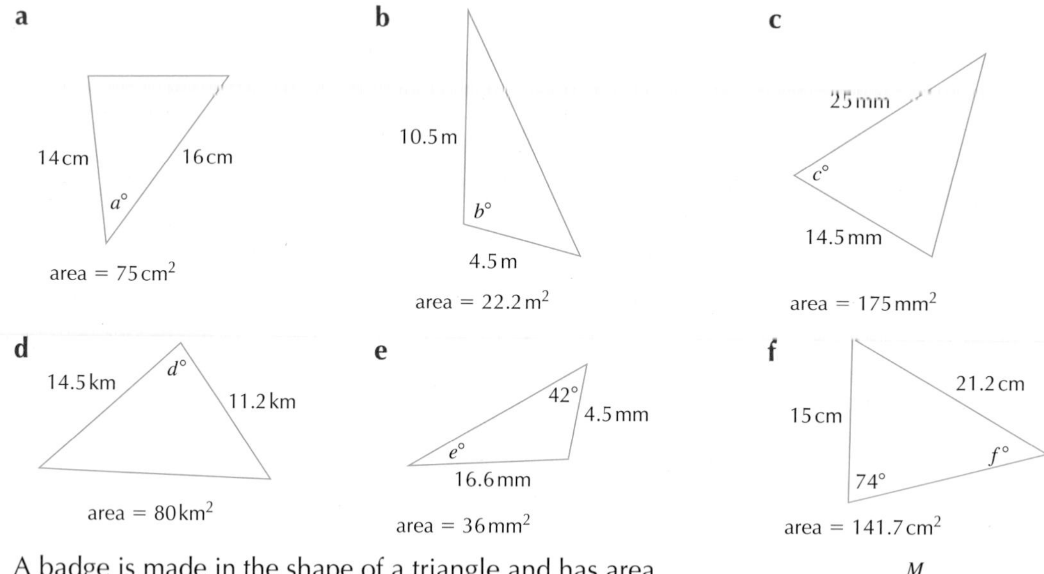

3 A badge is made in the shape of a triangle and has area 46 mm². Calculate the size of angle *MNP* to the nearest degree.

M
10 mm
P
13 mm
N

4 The roof of a barn is made of a triangle as shown. Calculate the size of the missing side of the roof x if the cross-sectional area of the roof is 4.3 m². Give your answer to 1 decimal place.

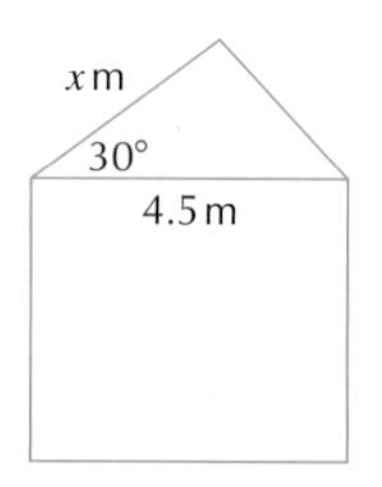

★ **5** The two triangles have the same area. Calculate the length of the missing side shown giving your answer to 2 significant figures.

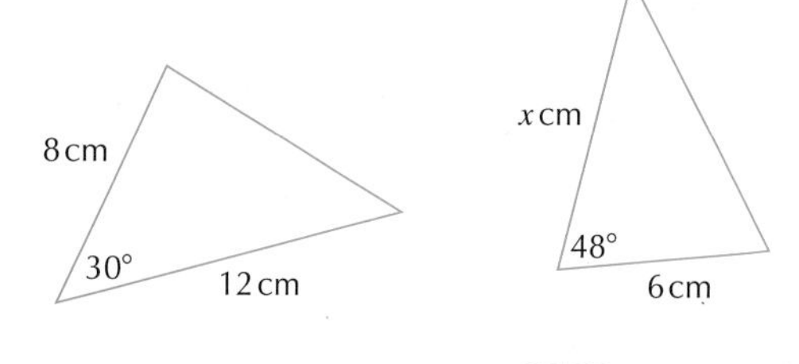

6 Calculate the area of the unshaded segment in the diagram. Give your answer to 2 decimal places.

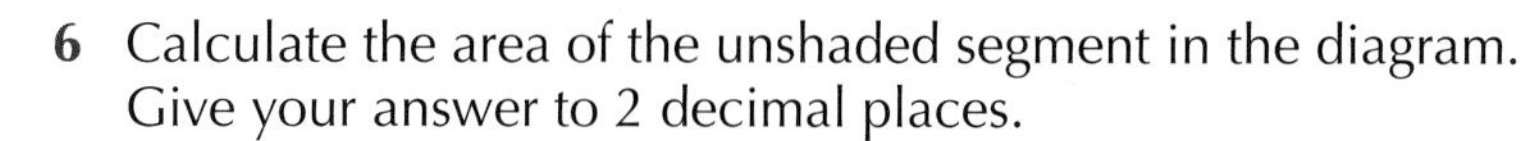

12 cm
68°

Activity

1 Try to draw two different triangles that have the same area. Check your solution by measuring two sides, the included angle and using the formula area = $\frac{1}{2}ab\sin C$.

2 Where does the formula area = $\frac{1}{2}ab\sin C$ come from? Is it related to the formula area = $\frac{1}{2}bh$?

- I can use trigonometry to work out the area of non-right-angled triangles using the formula area = $\frac{1}{2}ab\sin C$. ★ Exercise 25A Q4
- I can carry out calculations in context and calculate missing sides or angles given the area. ★ Exercise 25B Q5

For further assessment opportunities, see the Preparation for Assessment for Unit 3 on pages 373–376.

26 Using the sine and cosine rules to find a side or angle

In this chapter you will learn how to:

- use the **sine rule** to calculate a missing side or angle in a triangle
- use the **cosine rule** to find a missing side or angle in a triangle
- choose the correct formula to solve problems.

You should already know:

- how to calculate the missing side of a right-angled triangle using trigonometry
- how to label a triangle appropriately.

Using the sine rule

In trigonometry, the **sine rule** is an equation relating the lengths of sides of any triangle to the sines of its angles. The sine rule states that:

$$\frac{a}{\sin A} = \frac{b}{\sin B} = \frac{c}{\sin C}$$

where the angles and sides are related as shown in the diagram.

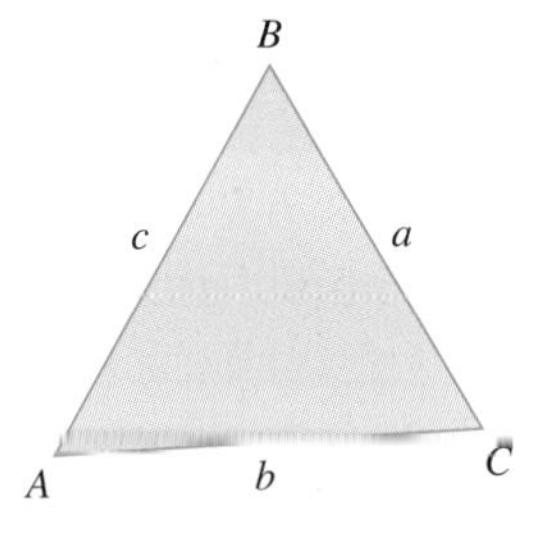
B
c
a
A
b
C

You can use this rule to calculate the size of missing angles and sides in non-right-angled triangles.

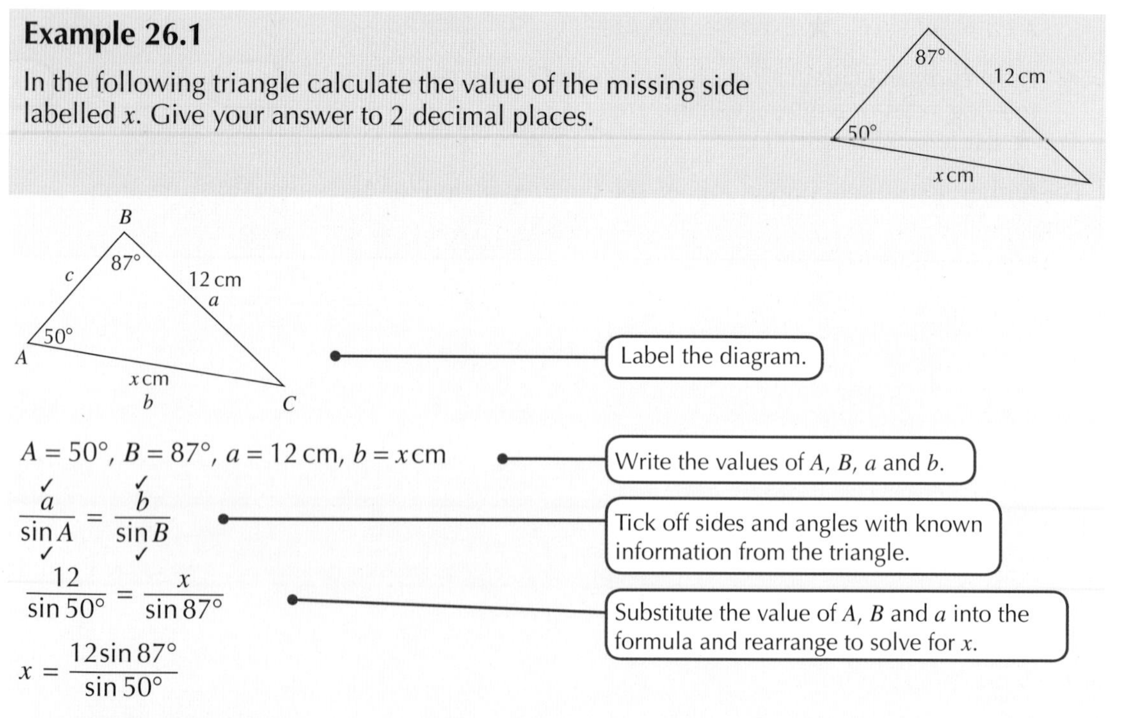

Example 26.1

In the following triangle calculate the value of the missing side labelled x. Give your answer to 2 decimal places.

87°
12 cm
50°
x cm
B
c
12 cm
a
A
x cm
b
C
Label the diagram.
Write the values of A, B, a and b.
Tick off sides and angles with known information from the triangle.
Substitute the value of A, B and a into the formula and rearrange to solve for x.

$A = 50°$, $B = 87°$, $a = 12$ cm, $b = x$ cm

$$\frac{a}{\sin A} = \frac{b}{\sin B}$$

$$\frac{12}{\sin 50°} = \frac{x}{\sin 87°}$$

$$x = \frac{12 \sin 87°}{\sin 50°}$$

$= 15.64$ cm (to 2 d.p.)

Example 26.2

Calculate the size of the missing angle x in the triangle to 1 decimal place.

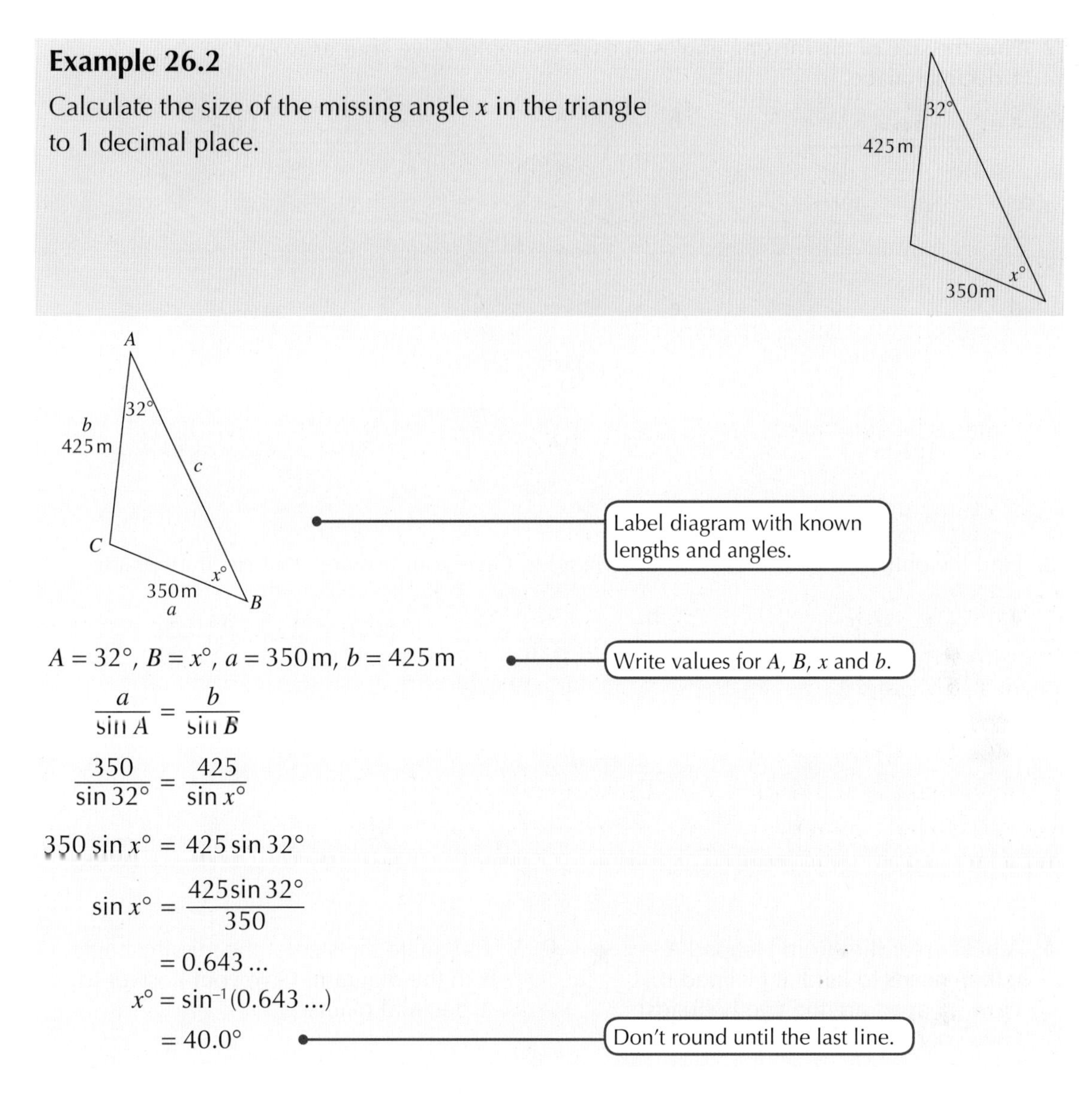

$A = 32°$, $B = x°$, $a = 350\text{ m}$, $b = 425\text{ m}$

$$\frac{a}{\sin A} = \frac{b}{\sin B}$$

$$\frac{350}{\sin 32°} = \frac{425}{\sin x°}$$

$$350 \sin x° = 425 \sin 32°$$

$$\sin x° = \frac{425 \sin 32°}{350}$$

$$= 0.643\ldots$$

$$x° = \sin^{-1}(0.643\ldots)$$

$$= 40.0°$$

Exercise 26A

1 Find the size of the missing side in each triangle. Give your answers to 3 significant figures.

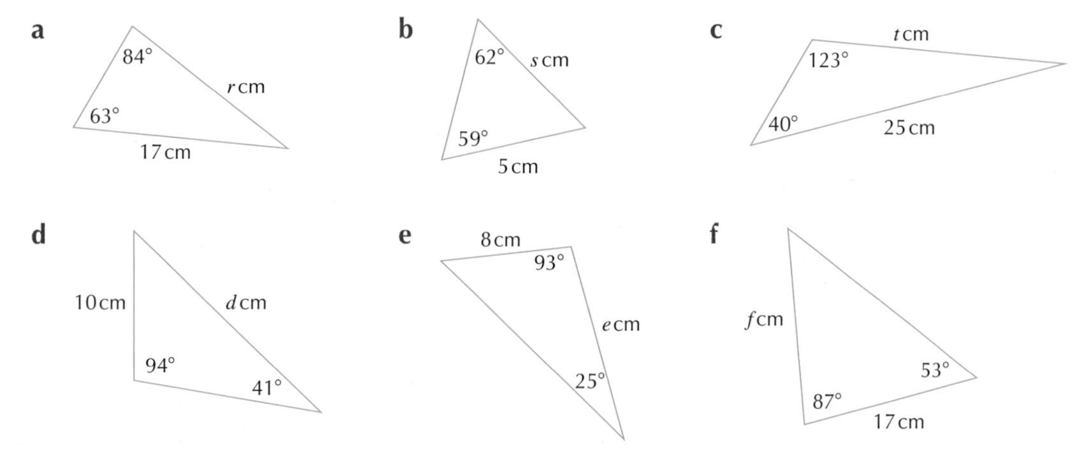

2 Find the size of the missing angles in each example. Give your answers to 1 decimal place.

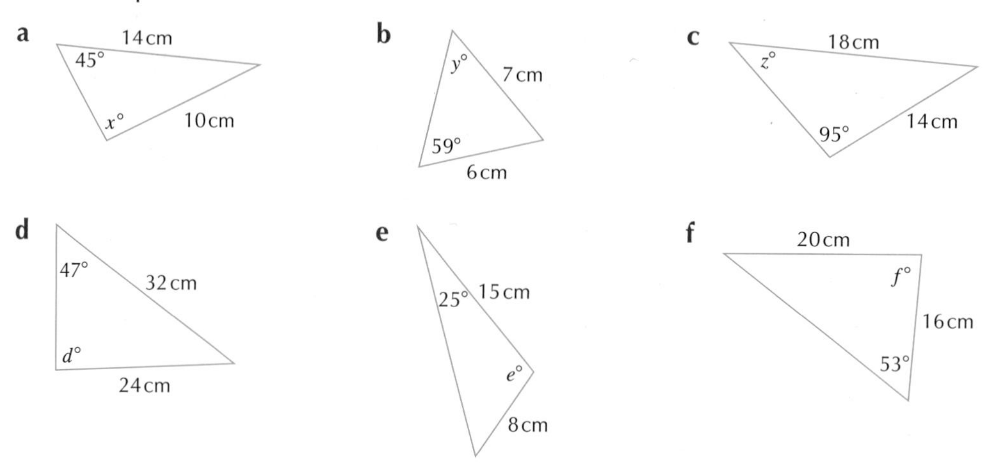

3 Find the obtuse angles in each of the triangles. Give your answers to 1 decimal place.

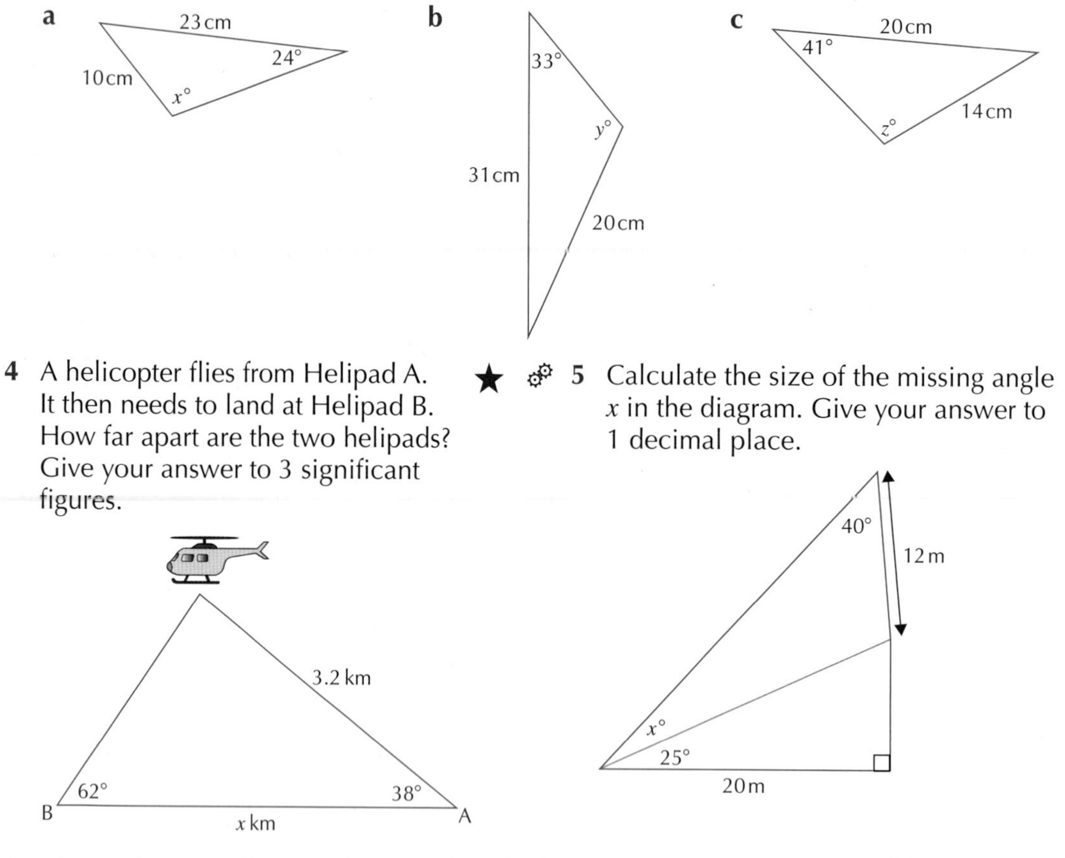

4 A helicopter flies from Helipad A. It then needs to land at Helipad B. How far apart are the two helipads? Give your answer to 3 significant figures.

★ **5** Calculate the size of the missing angle x in the diagram. Give your answer to 1 decimal place.

6 A metal support for a roof at a stadium is shown. Calculate the size of the beam labelled x to 2 significant figures.

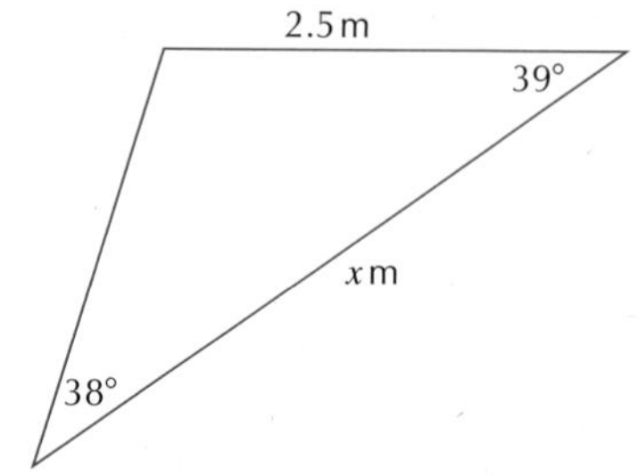

7 Two surveyors measure the distance across a river in order to build a bridge. They measure the information shown. Calculate the width W of the river giving your answer to 3 significant figures.

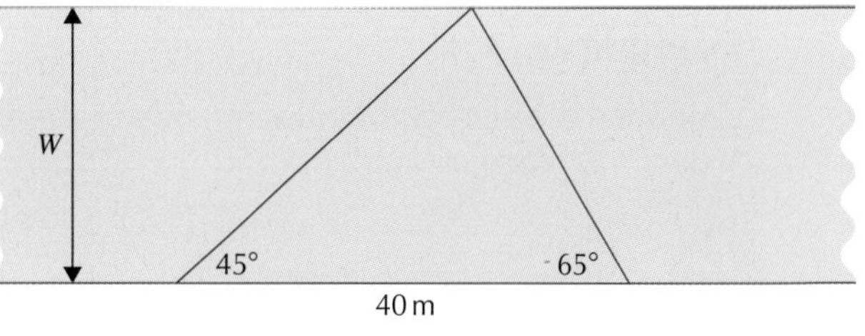

8 Jane uses a pulley system to lift a weight. The pulley is mounted to the ceiling as shown.

a Calculate the length of rope marked x giving your answer to 2 significant figures.

b The weight has to be attached to the ceiling. What distance must the weight be lifted?

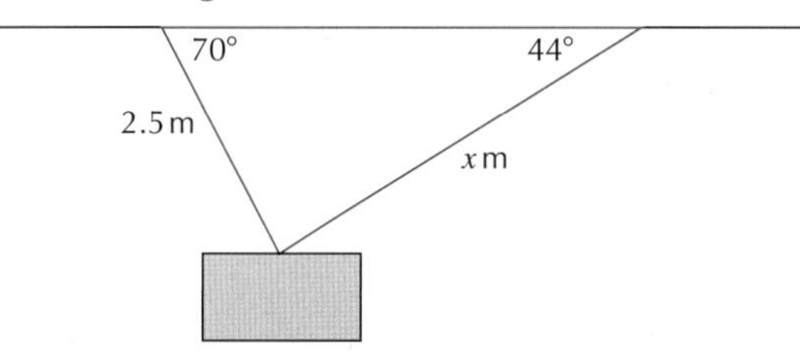

9 The cross-section of a roof truss is shown.

a Calculate the size of the missing angles. Give your answers to 1 decimal place.

b Calculate the vertical height of the roof. Give your answer to 1 decimal place.

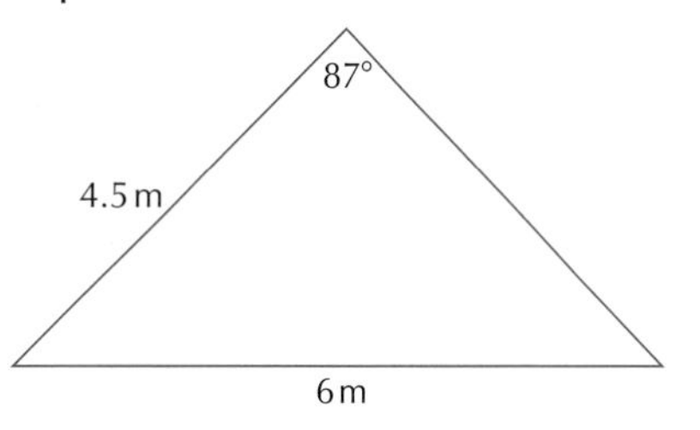

10 Two ships A and B are 500 m apart. Their navigators measure the angles from the sea level to the top of the cliff. Find the height of the cliff. Give your answer to 3 significant figures.

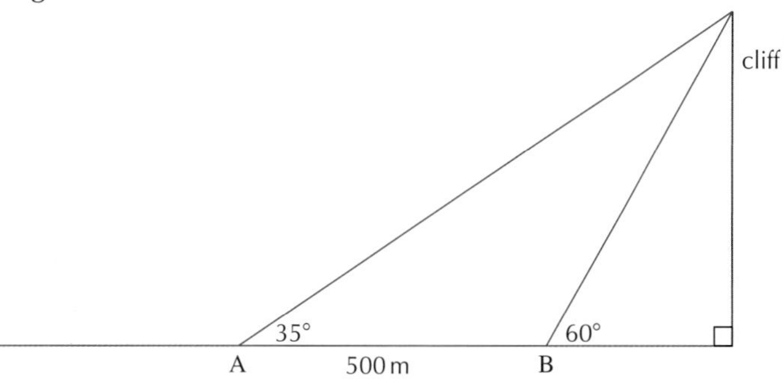

Using the cosine rule

The **cosine rule** relates the lengths of sides of any triangles to the cosine of one of its angles. The cosine rule is useful for calculating the missing side of a triangle when you know two sides and the included angle.

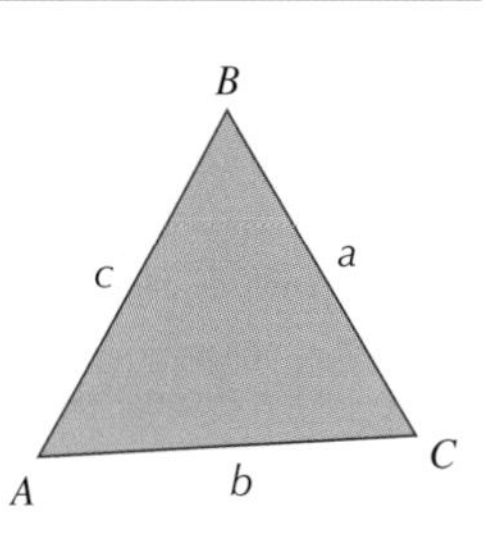

The cosine rule states:

$$a^2 = b^2 + c^2 - 2bc \cos A$$

or

$$b^2 = a^2 + c^2 - 2ac \cos B$$

or

$$c^2 = a^2 + b^2 - 2ab \cos C$$

where the angles and sides are related as shown in the diagram.

Example 26.3

Calculate the size of the missing side in the triangle to 1 decimal place.

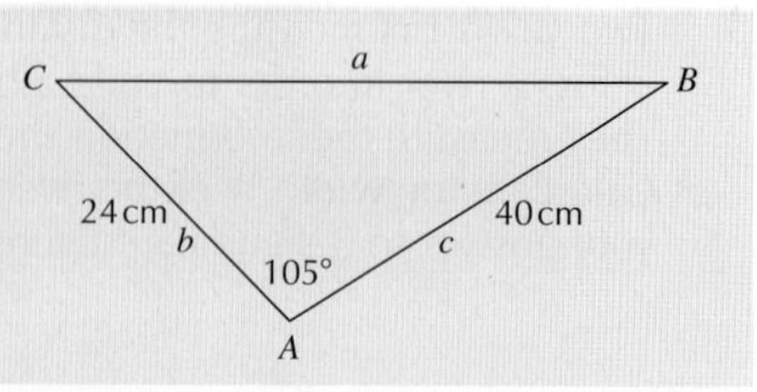

$a^2 = b^2 + c^2 - 2bc\cos A$

$a^2 = 24^2 + 40^2 - 2 \times 24 \times 40 \times \cos 105^\circ$

$= 2672.93$

$a = \sqrt{2672.93}$

$= 51.7$ cm to 1 d.p.

Exercise 26B

1 Calculate the length of the missing side in each triangle, giving your answer to 2 decimal places.

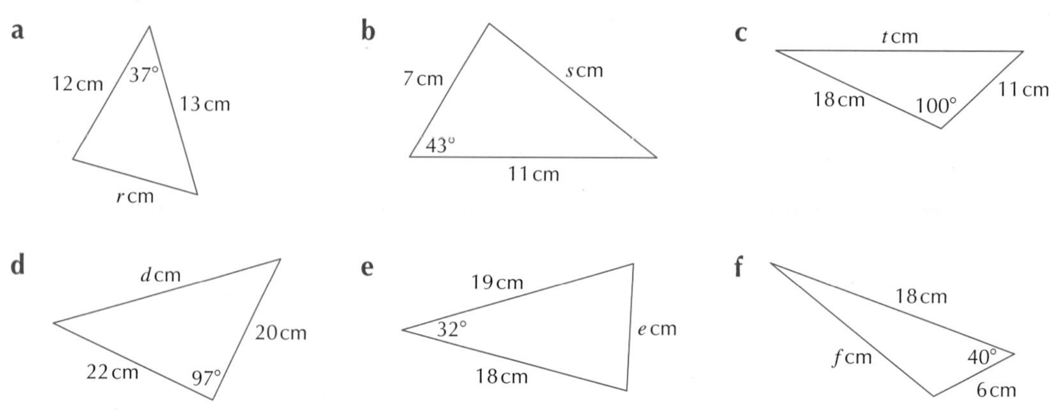

2 The equal sides of an isosceles triangle measure 6 cm. The angle between them is 40°. Calculate the size of the third side giving your answer to 1 decimal place.

★ **3** A pair of scissors is shown. Calculate the distance between the points of the scissors.

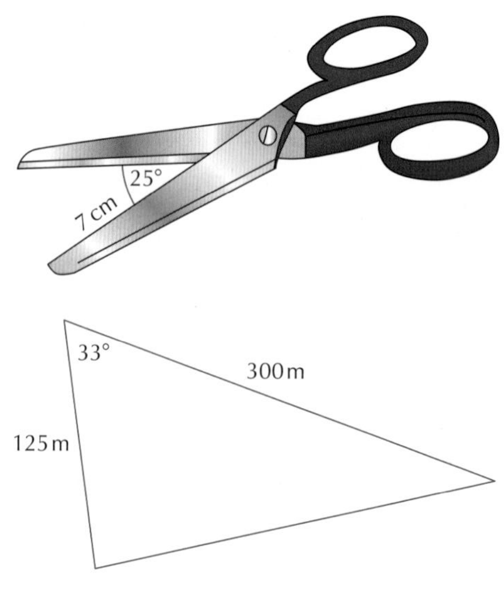

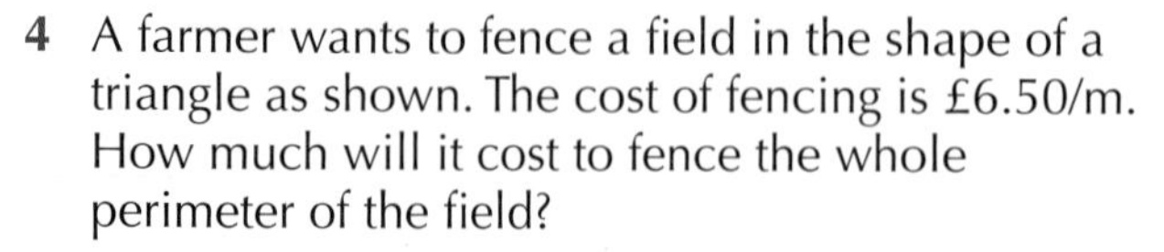

4 A farmer wants to fence a field in the shape of a triangle as shown. The cost of fencing is £6.50/m. How much will it cost to fence the whole perimeter of the field?

5 The diagram shows a trapezium.

  a Calculate the length of side AC to 1 decimal place.

  b Use your answer in **a** to calculate the length of AD to 1 decimal place.

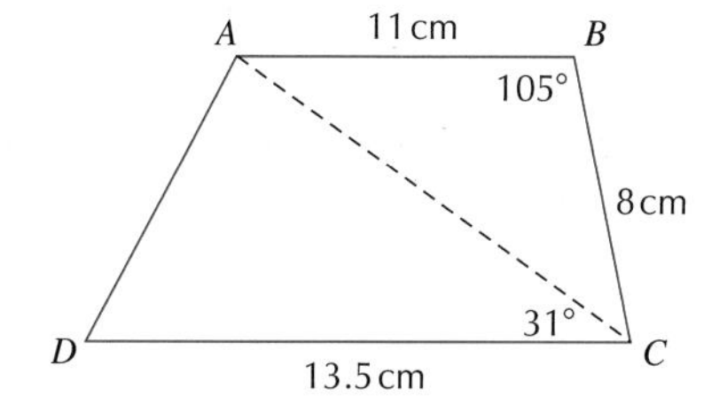

The cosine rule for an angle with all three sides known

You can rearrange the cosine rule for a side to give a formula to calculate a missing angle inside a triangle when you know all three sides.

For a triangle with sides a, b and c, with angles A, B and C, the cosine rule for an angle states:

$$\cos A = \frac{b^2 + c^2 - a^2}{2bc}$$

or

$$\cos B = \frac{a^2 + c^2 - b^2}{2ac}$$

or

$$\cos C = \frac{a^2 + b^2 - c^2}{2ab}$$

You can use this formula to find any angle in a triangle when all three sides are known.

Example 26.4

Find the size of the unknown angle $x°$ to 1 decimal place.

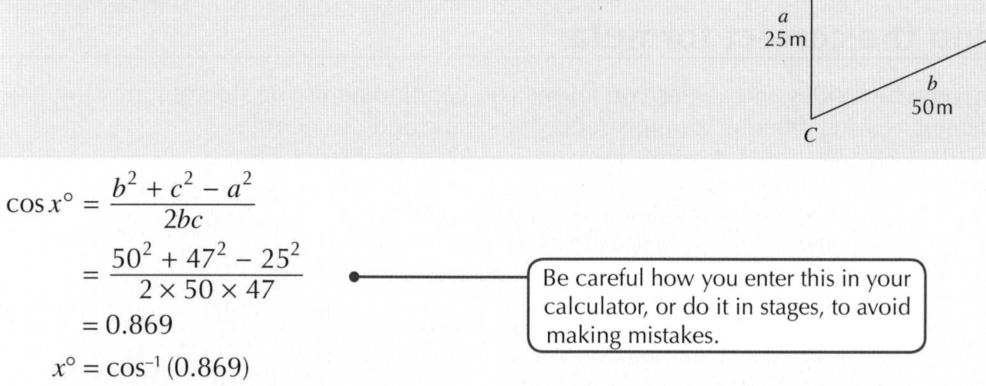

$$\cos x° = \frac{b^2 + c^2 - a^2}{2bc}$$

$$= \frac{50^2 + 47^2 - 25^2}{2 \times 50 \times 47}$$

Be careful how you enter this in your calculator, or do it in stages, to avoid making mistakes.

$$= 0.869$$

$$x° = \cos^{-1}(0.869)$$

$$= 29.7° \text{ (to 1 d.p.)}$$

Exercise 26C

1 Calculate the size of the missing angle in each triangle. Give your answers to 1 decimal place.

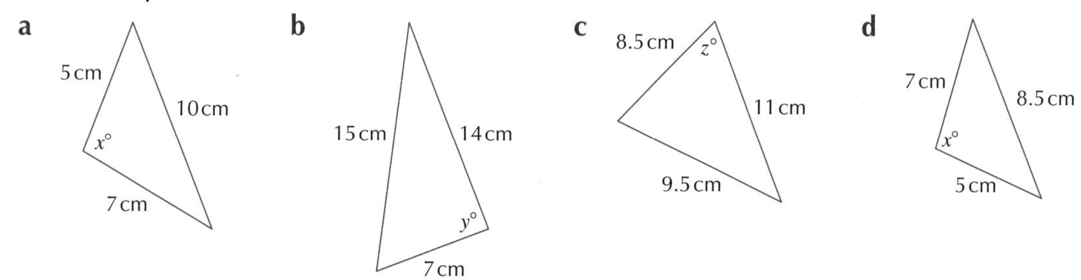

★ **2** A swimming pool is designed in the shape of a triangle as shown. Calculate the size of the largest angle. Give your answer to 1 decimal place.

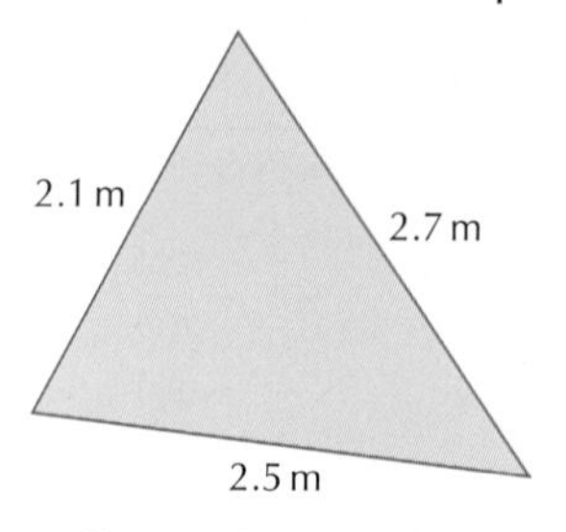

3 The perimeter of the triangle shown is 28 m. Calculate the size of angle $x°$ to 1 decimal place.

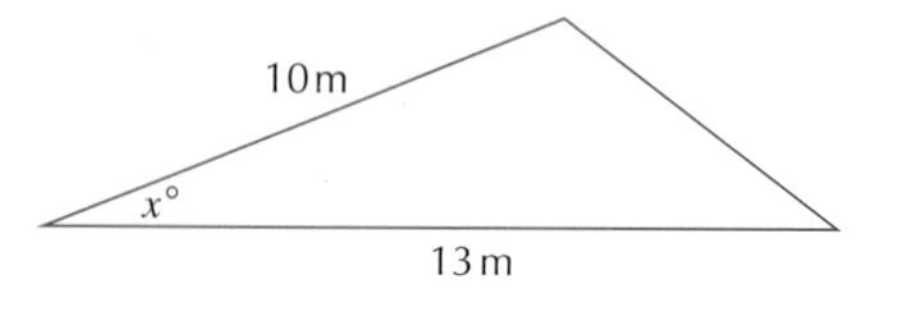

4 A flag is designed in the shape of a triangle. The edges measure 40 cm, 37cm and 20 cm. Calculate the size of the smallest angle to 1 decimal place.

5 Show that for an equilateral triangle, with sides of length x cm, $\cos A = \frac{1}{2}$.

6 Calculate the height of the balloon above the ground.

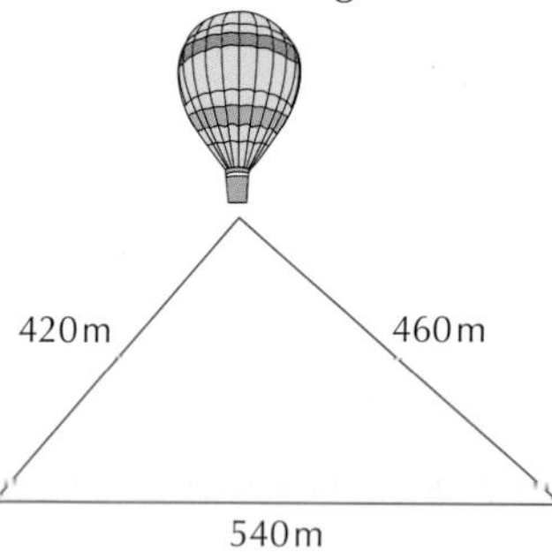

Choosing the correct formula

Depending on which sides and angles you know, you can find a missing side or a missing angle of a triangle using different arrangements of the sine and cosine rule.

B, a, A, b	**Sine rule** Find a missing side when two angles are known and one side is known in any two pairs of sides and opposite angles. *or* Find a missing angle when two sides are known and one angles is known in any two pairs of sides and opposite angles.	$\frac{a}{\sin A} = \frac{b}{\sin B}$
c, a, A, b	**Cosine rule** Find a side with two sides and the included angle known.	$a^2 = b^2 + c^2 - 2bc \cos A$
c, a, A, b	**Cosine rule** Find an angle with all three sides known.	$\cos A = \frac{b^2 + c^2 - a^2}{2bc}$

Exercise 26D

★ **1** In each triangle shown calculate the value of the letter by selecting the correct formula. Give your answers to 2 decimal places.

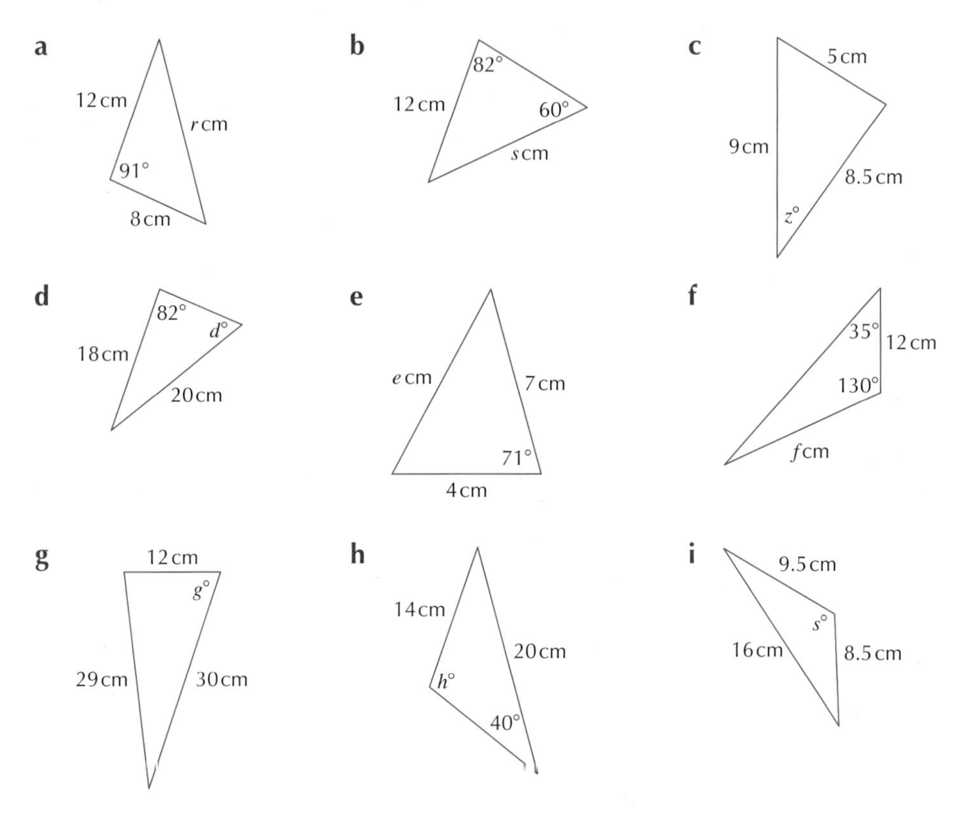

2 The legs of a pair of compasses are set 30° apart. The length of the legs are 7cm.

a Calculate the distance across the tips of the compasses to 2 decimal places.

b What diameter circle would the compasses draw at this width?

3 The hands on a clock are 4 cm and 6 cm long. What is the distance between the points when the time is 4 o'clock?

4 A picnic table is designed as shown. Calculate the length of the support bar marked x.

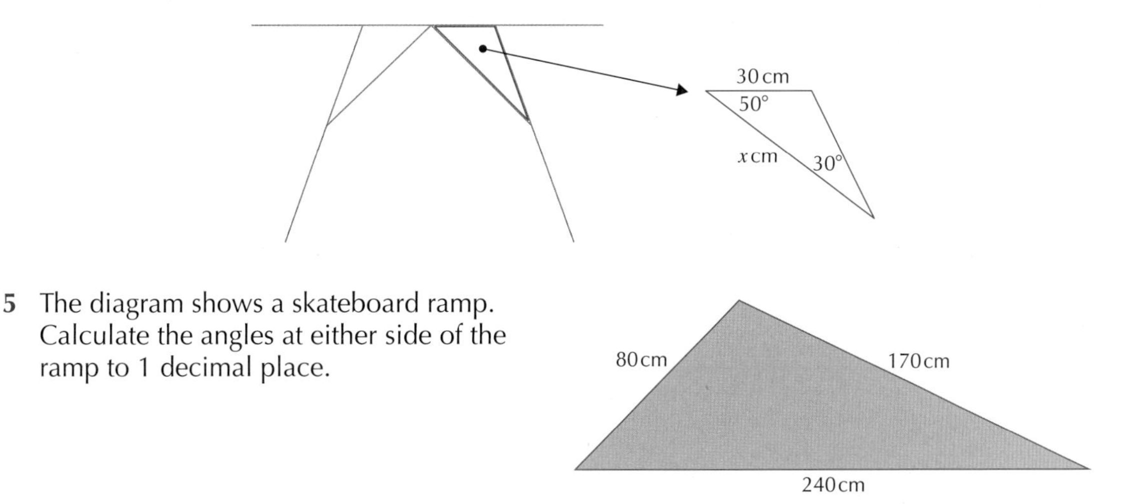

5 The diagram shows a skateboard ramp. Calculate the angles at either side of the ramp to 1 decimal place.

6 Calculate the value of x to 2 significant figures.

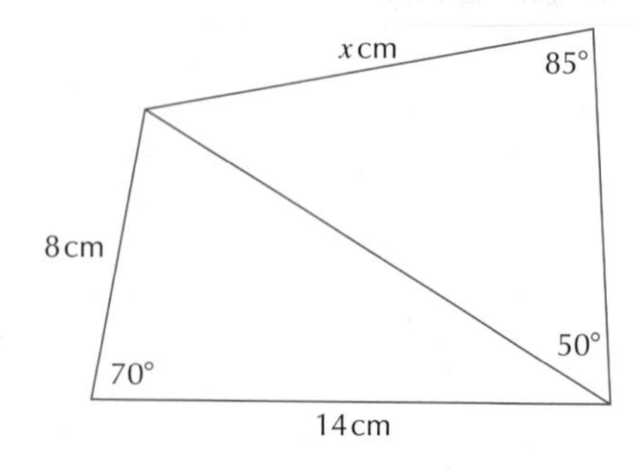

Activity

What happens when you use the trigonometric formulae on right-angled triangles? They should give you the same answers. Can you explain why this might be the case?

- I can use the sine rule to calculate a missing side or angle in a triangle. ★ Exercise 26A Q5
- I can use the cosine rules to find a missing side or angle in a triangle. ★ Exercise 26B Q3 ★ Exercise 26C Q2
- I can choose the correct formula to use given specific information. ★ Exercise 26D Q1

For further assessment opportunities, see the Preparation for Assessment for Unit 3 on pages 373–376.

27 Using bearings with trigonometry

In this chapter you will learn how to:

- find **bearings** by solving angle problems
- use your trigonometry formulae to solve problems involving bearings.

You should already know:

- that a bearing is measured clockwise from the North bearing
- how to use trigonometry formulae to find missing angles and distances
- how to find angles with intersecting and parallel lines
- how to solve angle problems with triangles and quadrilaterals.

Bearings

A **bearing** is a term used for navigation. It is a measurement of angle from a defined point, usually taken to be due North. Bearings are used in sailing, aircraft and in search and rescue. They are vital for maintaining safe navigation and planning journeys. A long journey will often involve legs of differing distances over different bearings, and a mistake made with one bearing can have very significant consequences for the whole journey.

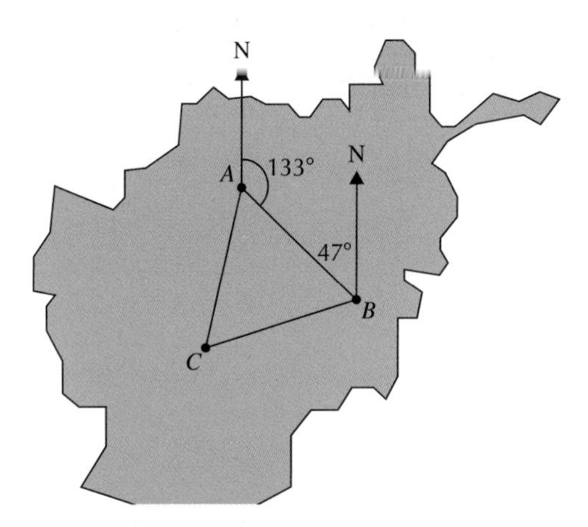

Bearings are measured from a North point so it is always useful to add North to any drawings you use in bearings problems.

Bearings are always measured in a clockwise direction from North, so in the diagram, the bearing of B from A is 133°. The bearing of A from B is $360° - 47° = 313°$.

Trigonometry and bearings are often used together in triangle problems. It is most important to be able to find angles inside the triangle.

Example 27.1

Point B is on a bearing of 320° from C and A is on a bearing of 290° from C. Using this information calculate the size of the 3 missing angles inside the triangle.

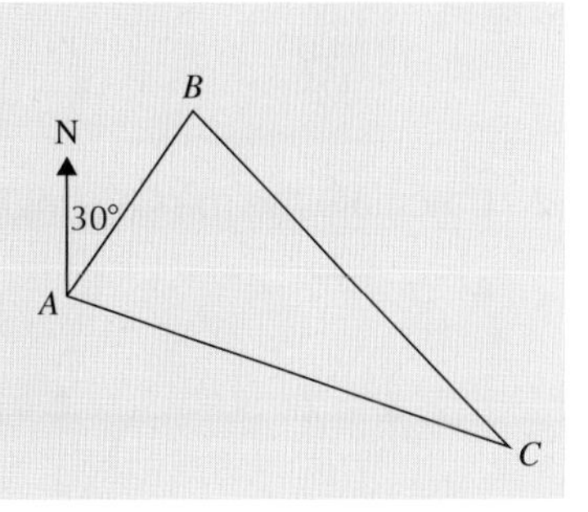

(continued)

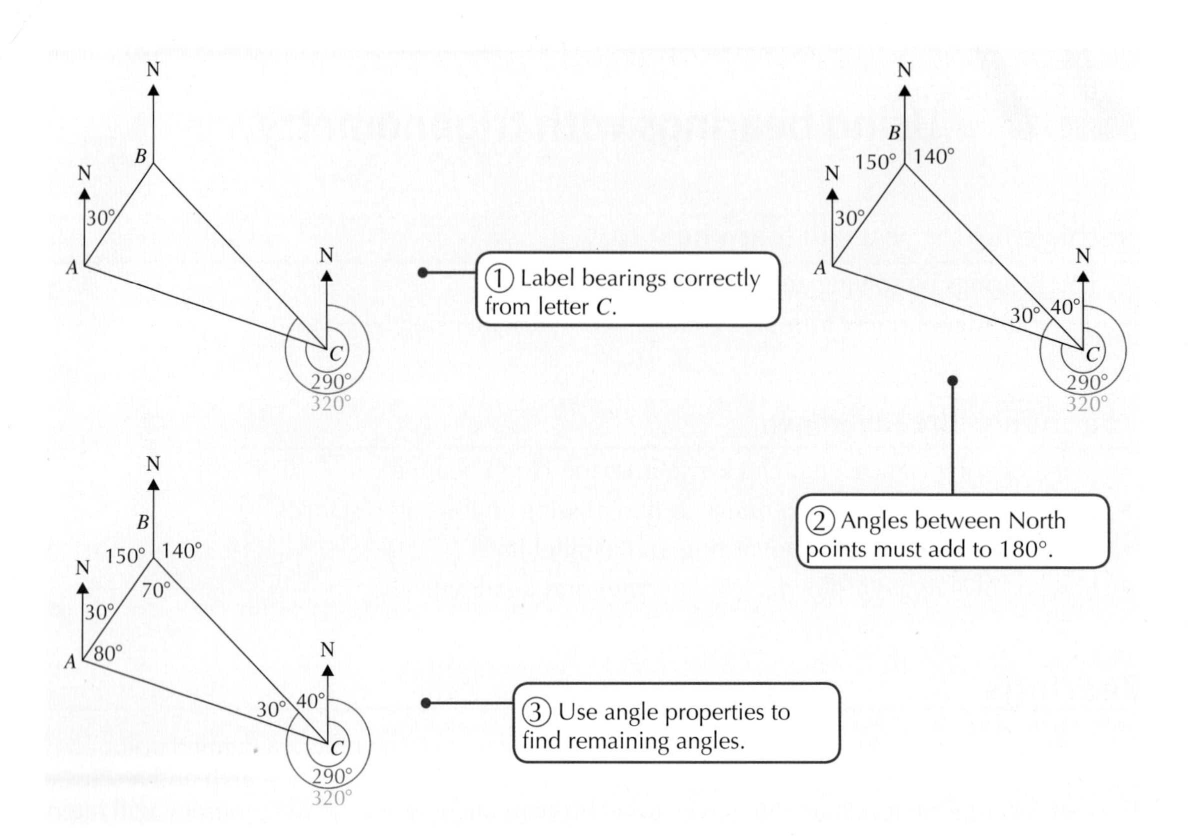

Exercise 27A

★ **1** In each triangle shown find the missing angles labelled.

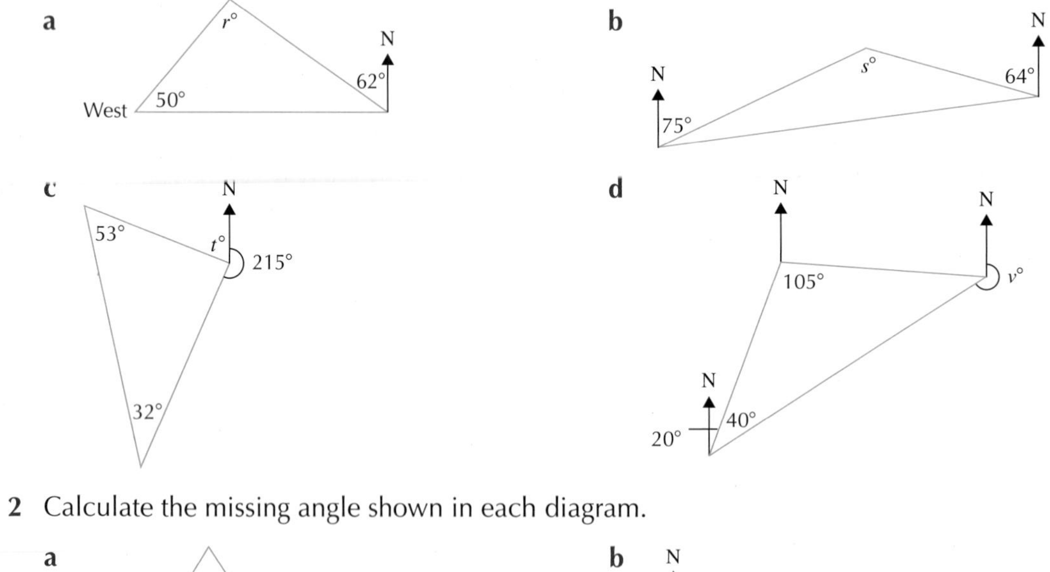

2 Calculate the missing angle shown in each diagram.

a

55°
N
N
32°
r°
t°
s°
v°
253°

b

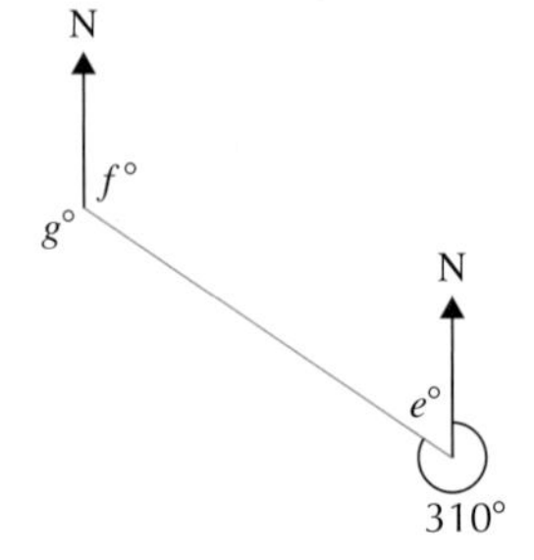

3 Boat A is on a bearing of 076° from port, boat B is on a bearing of 099°.
Calculate the bearing of boat A from boat B.

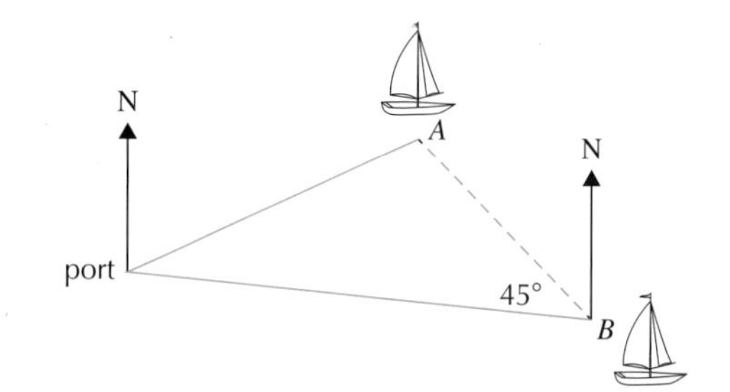

Bearings and trigonometry

Example 27.2

Given that side $AC = 12$ km, calculate the length of AB.

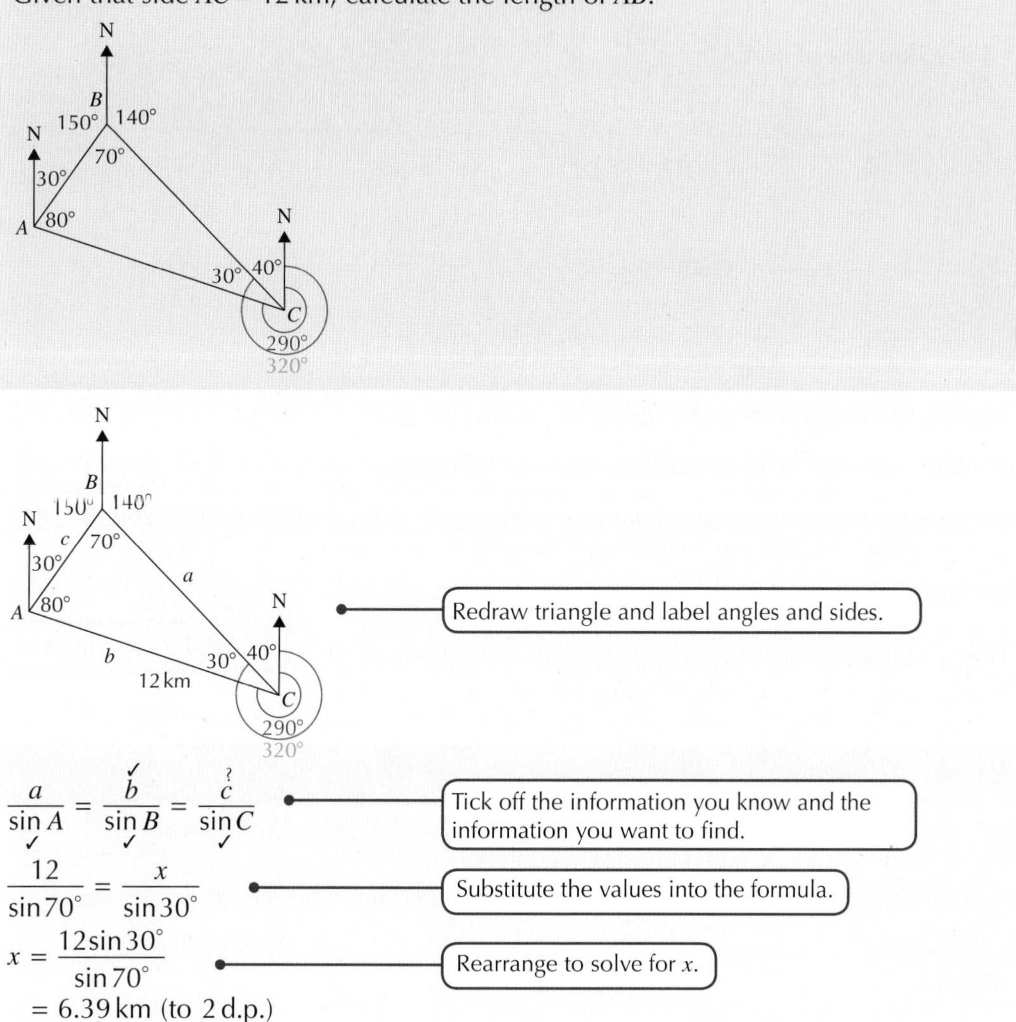

$$\frac{a}{\sin A} = \frac{b}{\sin B} = \frac{c}{\sin C}$$

$$\frac{12}{\sin 70^\circ} = \frac{x}{\sin 30^\circ}$$

$$x = \frac{12 \sin 30^\circ}{\sin 70^\circ}$$

$= 6.39$ km (to 2 d.p.)

Exercise 27B

For each question, give your answer to 2 decimal places.

1 Calculate the length of the missing side AB.

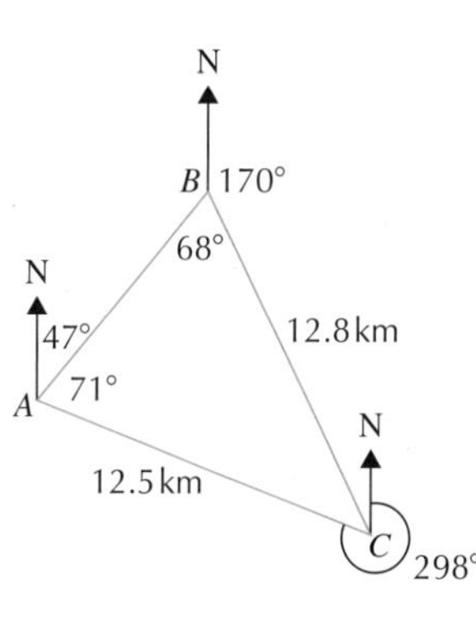

2 A ship sails east for 14.5 km then on a bearing of 130° for 11 km. Calculate the distance the boat has sailed.

Drawing a sketch might be helpful.

3 Two yachts set off from the same port as shown in the diagram. Yacht A sails 10.7 km on a bearing of 053°, while yacht B sails 11.2 km on a bearing of 112°. Calculate the distance between the two yachts.

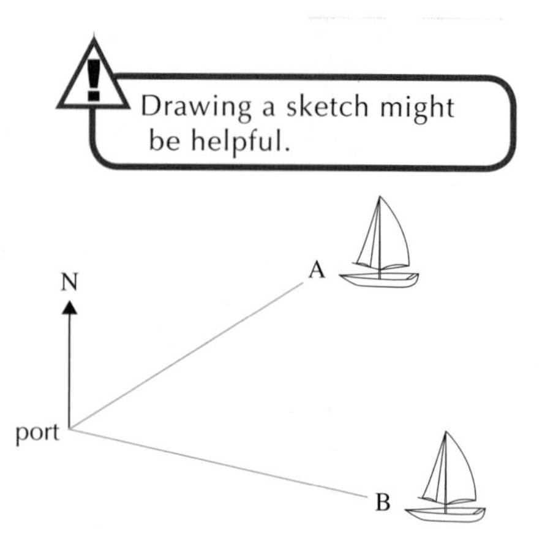

★ 4 Port B is 25 km east of port A.

port A
43°
39°
port B
25 km

a Calculate the distance of the ship to the nearer port.

b Find the bearing the ship must sail to reach this port.

5 An airplane flies 170 miles from point X at a bearing of 125°, and then turns and flies at a bearing of 230° for 90 miles. How far is the plane from point X?

6 A ship is being followed by two submarines, A and B, 3.8 km apart, with A due east of B. If A is on a bearing of 165° from the ship and B is on the bearing of 205° from the ship, find the distance from the ship to both submarines.

Look for 'Z'-angles.

Activity

1 Draw a 16-point compass rose (i.e. involving N, NNE, NE, ENE, ...) and label the directions with their associated bearings.

2 There are different types of bearings used in real life. Research the difference between:

- true bearings
- magnetic bearings
- grid bearings
- compass bearings
- relative bearings.

- I can find missing angles in a triangle using bearings. ★ Exercise 27A Q1
- I can use trigonometry formulae to solve problems involving bearings. ★ Exercise 27B Q4

For further assessment opportunities, see the Preparation for Assessment for Unit 3 on pages 373–376.

28 Working with 2D vectors

In this chapter you will learn how to:

- draw and write **vectors** using **component** form
- add or subtract 2D vectors using **directed line segments**.

You should already know

- how to plot 2D coordinates.

Drawing and writing 2D vectors

A **scalar** is a quantity which has size only, and does not depend on direction. A **vector** is a quantity which has **magnitude** (size) and direction.

Scalar	Vector
time	force applied to a snooker ball
distance	displacement
speed	velocity
volume	acceleration
temperature	

The properties of a vector are shown as:

- magnitude $|AB|$
- sense (—▸—)
- direction $\overrightarrow{AB}$ or $\overrightarrow{BA}$

Vectors can be expressed two dimensions and in three dimensions. We will start by considering movement in two dimensions. For a vector, we want to know how far something has moved from its start point to end point (**displacement**) and we want information about the direction. We do this using **directed line segments**.

A directed line segment is a line joining any two points on a coordinate diagram. For instance, the diagram shows the directed line segment from A to B as denoted by the arrow. This is written as $\overrightarrow{AB}$. This should be read as 'vector AB', and it indicates the displacement from A to B.

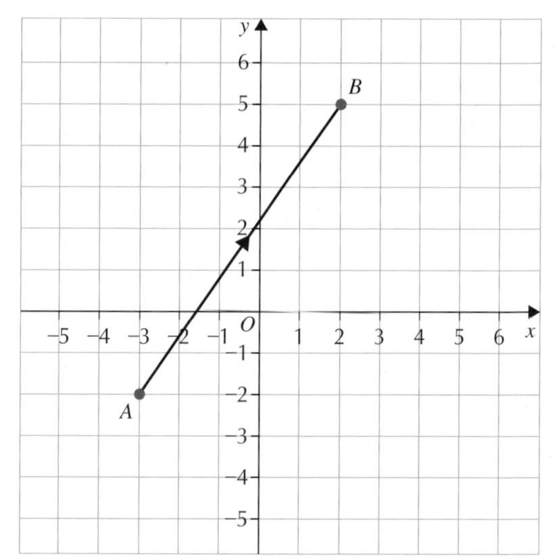

2D vectors work similarly to coordinates. We can express a vector as the movement horizontally then vertically. In the diagram on the right, vector $\overrightarrow{AB}$ is written as $\begin{pmatrix} 3 \\ 5 \end{pmatrix}$ known as **a column vector**, where 3 is the x-component and 5 is the y-component. The major difference between coordinate measures and vectors is that vectors do not have to start at the origin.

The vector $\overrightarrow{BA}$ describes the direction from B to A and is written as $\begin{pmatrix} -3 \\ -5 \end{pmatrix}$.

The same vector can be used to describe a number of different lines. The diagram shows three vectors of the same magnitude and direction $\begin{pmatrix} 3 \\ 5 \end{pmatrix}$. This shows an important vector property that parallel lines of equal magnitude can be described with the same column vector.

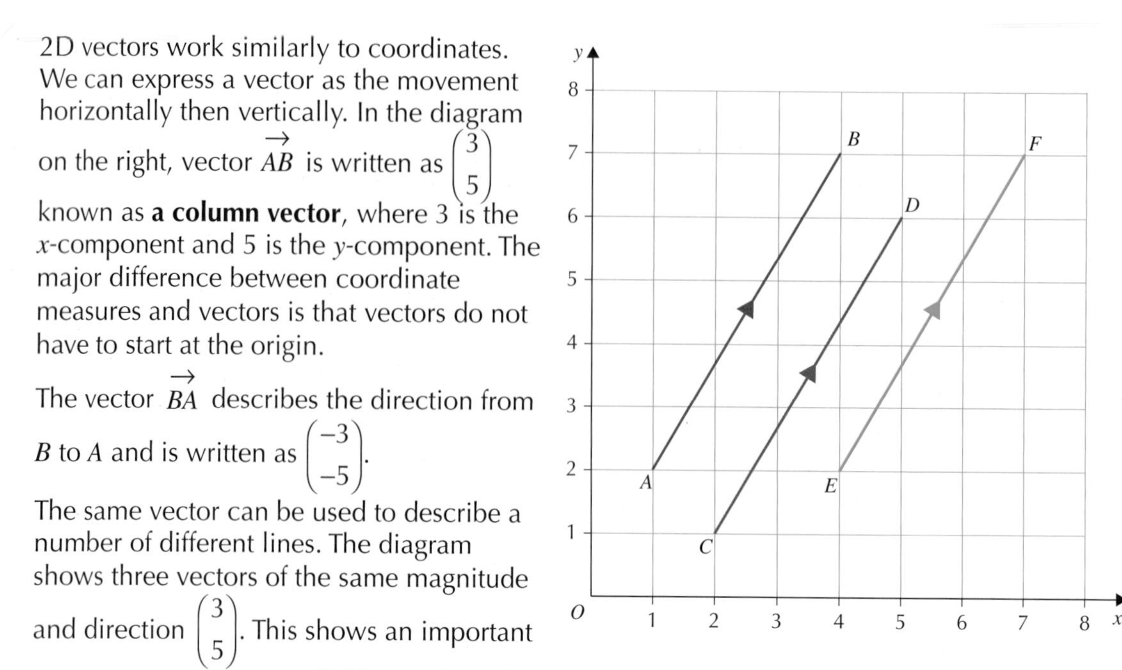

Example 28.1

Write down directed line segments $\overrightarrow{AB}$ and $\overrightarrow{CD}$ in component form.

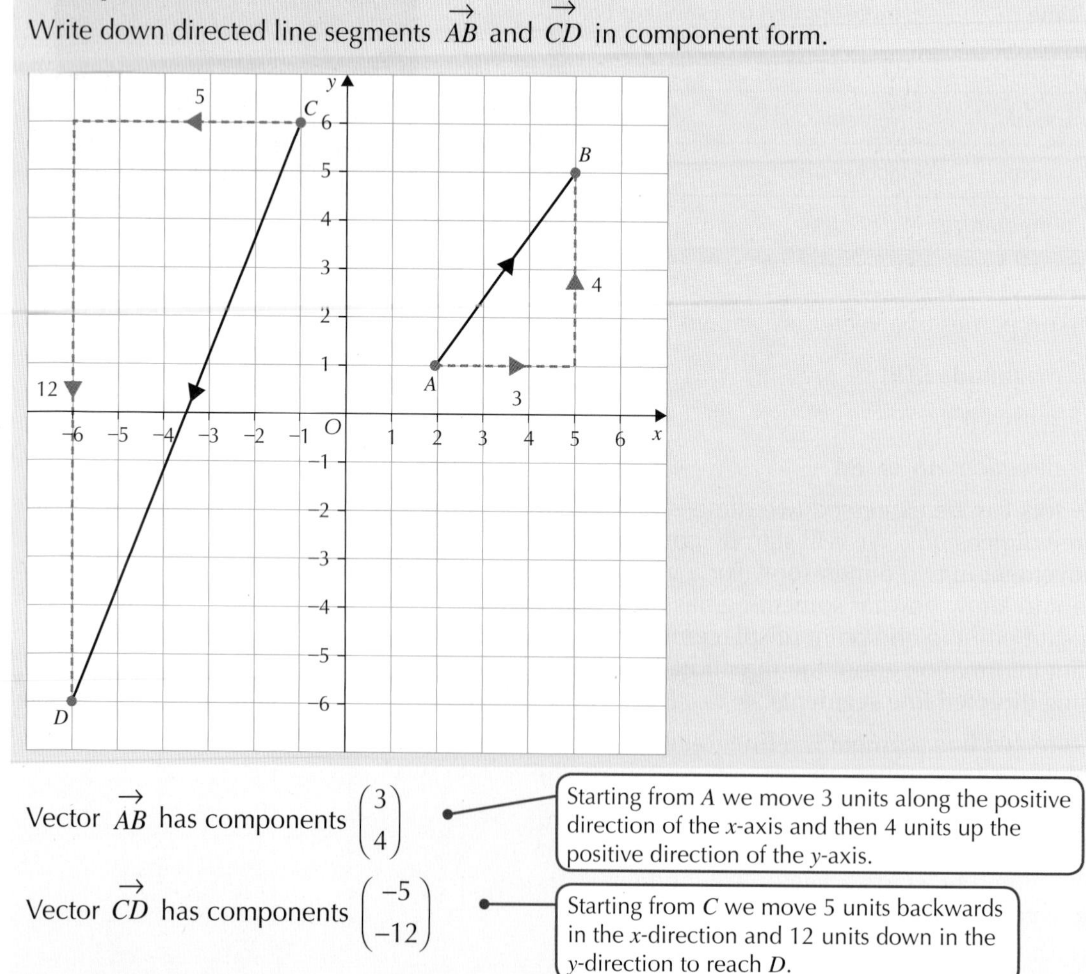

Vector $\overrightarrow{AB}$ has components $\begin{pmatrix} 3 \\ 4 \end{pmatrix}$

Starting from A we move 3 units along the positive direction of the x-axis and then 4 units up the positive direction of the y-axis.

Vector $\overrightarrow{CD}$ has components $\begin{pmatrix} -5 \\ -12 \end{pmatrix}$

Starting from C we move 5 units backwards in the x-direction and 12 units down in the y-direction to reach D.

Exercise 28A

1 Write each vector in component form.

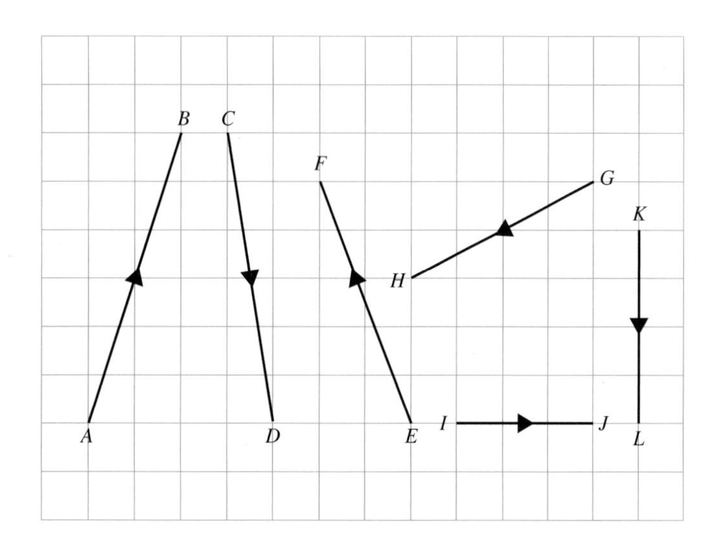

★ **2** Sketch each of these vectors.

a $\overrightarrow{AB} = \begin{pmatrix} 2 \\ 5 \end{pmatrix}$ **b** $\overrightarrow{CD} = \begin{pmatrix} 3 \\ 4 \end{pmatrix}$ **c** $\overrightarrow{EF} = \begin{pmatrix} 3 \\ -4 \end{pmatrix}$ **d** $\overrightarrow{GH} = \begin{pmatrix} 4 \\ -5 \end{pmatrix}$

e $\overrightarrow{IJ} = \begin{pmatrix} -2 \\ 5 \end{pmatrix}$ **f** $\overrightarrow{KL} = \begin{pmatrix} -3 \\ 2 \end{pmatrix}$ **g** $\overrightarrow{MN} = \begin{pmatrix} -4 \\ -1 \end{pmatrix}$ **h** $\overrightarrow{PQ} = \begin{pmatrix} -3 \\ -7 \end{pmatrix}$

Adding and subtracting vectors using directed line segments

Vectors can be added together using directed line segments by joining the first vector to the second tip to tail. This produces a **resultant vector**. This is the vector produced by adding the components of the two initial vectors.

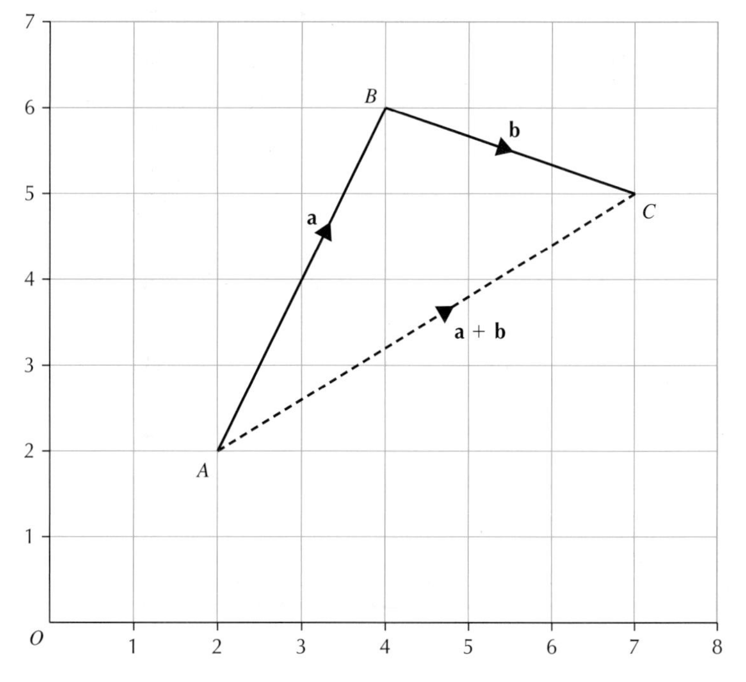

Vectors can also be multiplied and divided. Multiplying a vector by a scalar quantity (such as a number) multiplies the magnitude of the vector. The direction remains the same.

Example 28.2

Write down in component form the resultant vector $\overrightarrow{AC}$.

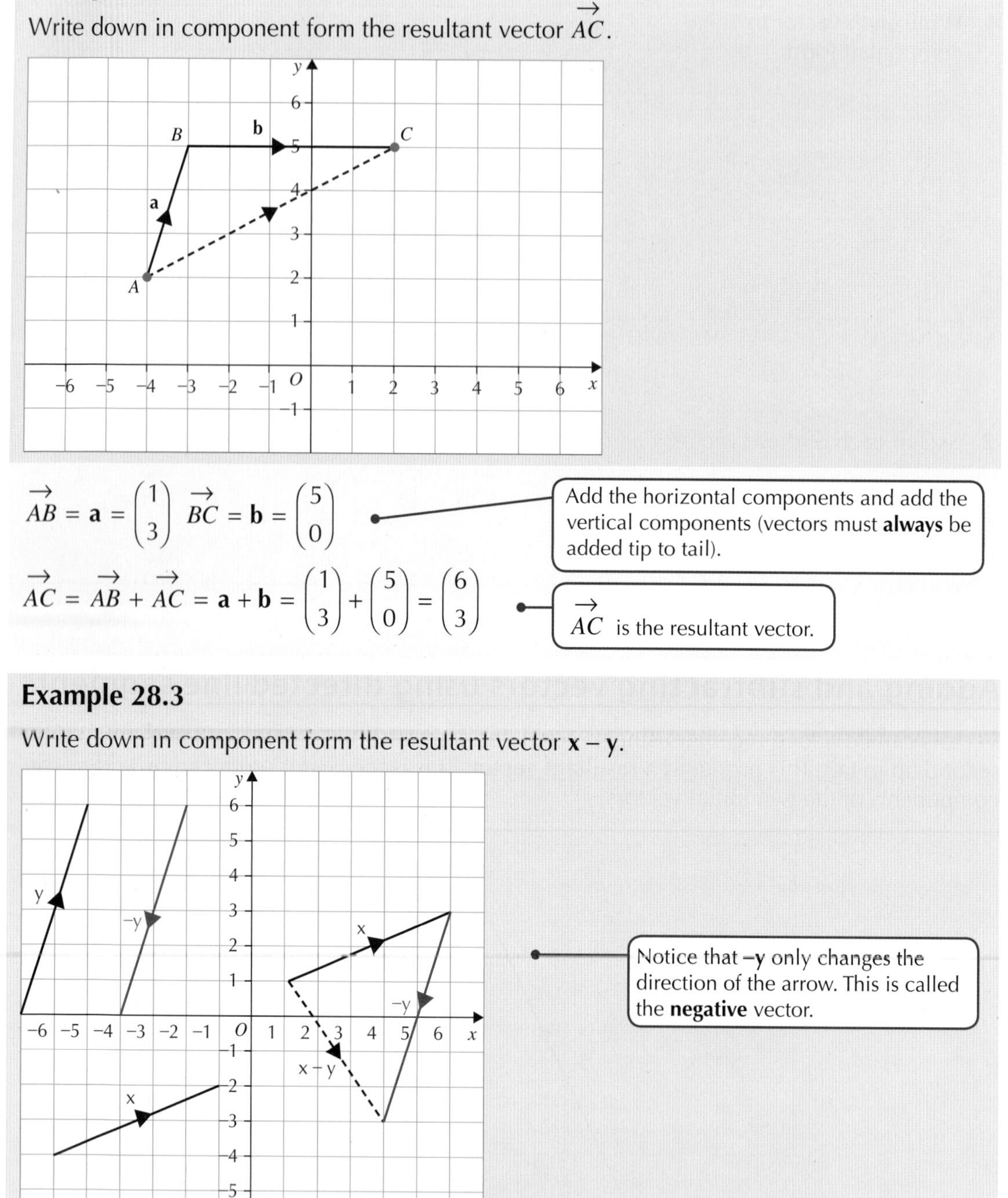

$\overrightarrow{AB} = \mathbf{a} = \begin{pmatrix} 1 \\ 3 \end{pmatrix}$ $\overrightarrow{BC} = \mathbf{b} = \begin{pmatrix} 5 \\ 0 \end{pmatrix}$

Add the horizontal components and add the vertical components (vectors must **always** be added tip to tail).

$\overrightarrow{AC} = \overrightarrow{AB} + \overrightarrow{AC} = \mathbf{a} + \mathbf{b} = \begin{pmatrix} 1 \\ 3 \end{pmatrix} + \begin{pmatrix} 5 \\ 0 \end{pmatrix} = \begin{pmatrix} 6 \\ 3 \end{pmatrix}$

$\overrightarrow{AC}$ is the resultant vector.

Example 28.3

Write down in component form the resultant vector **x** – **y**.

Notice that **–y** only changes the direction of the arrow. This is called the **negative** vector.

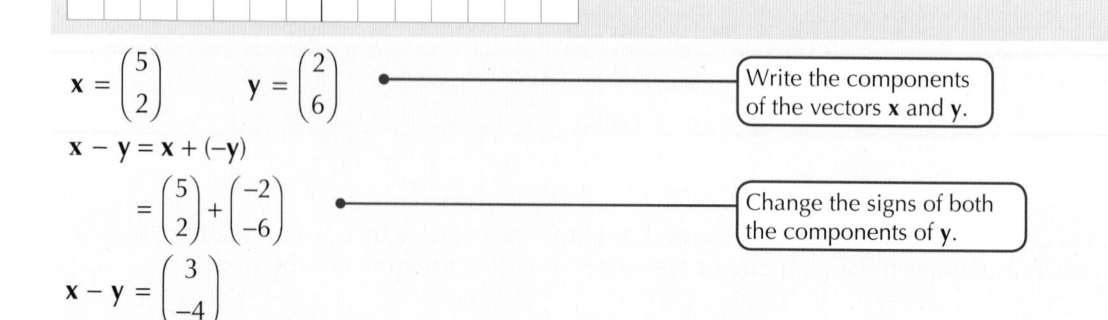

$\mathbf{x} = \begin{pmatrix} 5 \\ 2 \end{pmatrix}$ $\mathbf{y} = \begin{pmatrix} 2 \\ 6 \end{pmatrix}$

Write the components of the vectors **x** and **y**.

$\mathbf{x} - \mathbf{y} = \mathbf{x} + (-\mathbf{y})$

$= \begin{pmatrix} 5 \\ 2 \end{pmatrix} + \begin{pmatrix} -2 \\ -6 \end{pmatrix}$

Change the signs of both the components of **y**.

$\mathbf{x} - \mathbf{y} = \begin{pmatrix} 3 \\ -4 \end{pmatrix}$

Example 28.4

Using the vectors **a** and **b**, find the resultant vector 2**a** + 3**b**.

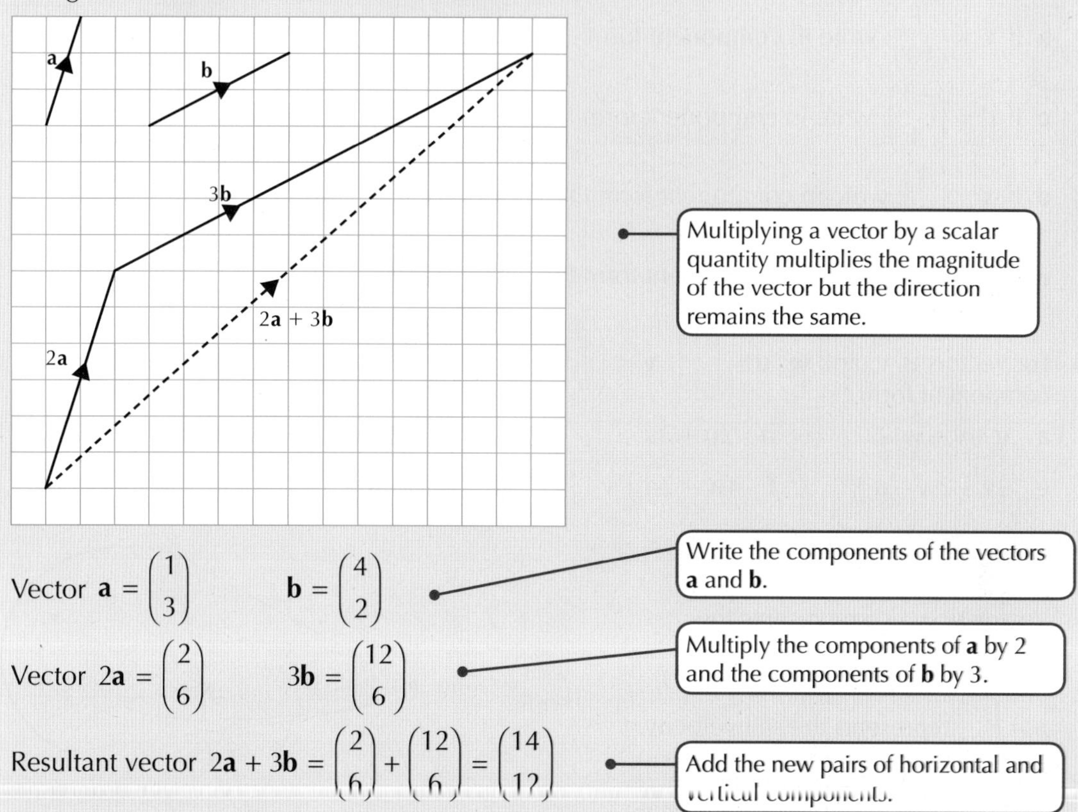
a
b
3b
2a + 3b
2a
Multiplying a vector by a scalar quantity multiplies the magnitude of the vector but the direction remains the same.

Vector **a** = $\begin{pmatrix} 1 \\ 3 \end{pmatrix}$ $\quad$ **b** = $\begin{pmatrix} 4 \\ 2 \end{pmatrix}$ — Write the components of the vectors **a** and **b**.

Vector 2**a** = $\begin{pmatrix} 2 \\ 6 \end{pmatrix}$ $\quad$ 3**b** = $\begin{pmatrix} 12 \\ 6 \end{pmatrix}$ — Multiply the components of **a** by 2 and the components of **b** by 3.

Resultant vector 2**a** + 3**b** = $\begin{pmatrix} 2 \\ 6 \end{pmatrix} + \begin{pmatrix} 12 \\ 6 \end{pmatrix} = \begin{pmatrix} 14 \\ 12 \end{pmatrix}$ — Add the new pairs of horizontal and vertical components.

Exercise 28B

1 Draw a vector diagram for **m** and **n** and the resultant vector **m** + **n**. State the components of **m** + **n**.

a $\mathbf{m} = \begin{pmatrix} 2 \\ 4 \end{pmatrix}, \mathbf{n} = \begin{pmatrix} 3 \\ 1 \end{pmatrix}$ $\quad$ **b** $\mathbf{m} = \begin{pmatrix} -1 \\ 3 \end{pmatrix}, \mathbf{n} = \begin{pmatrix} 0 \\ 1 \end{pmatrix}$ $\quad$ **c** $\mathbf{m} = \begin{pmatrix} 5 \\ -2 \end{pmatrix}, \mathbf{n} = \begin{pmatrix} -5 \\ 4 \end{pmatrix}$

2 Calculate the resultant vector **x** – **y**.

a $\mathbf{x} = \begin{pmatrix} 3 \\ 4 \end{pmatrix}, \mathbf{y} = \begin{pmatrix} 2 \\ 2 \end{pmatrix}$ $\quad$ **b** $\mathbf{x} = \begin{pmatrix} 6 \\ 2 \end{pmatrix}, \mathbf{y} = \begin{pmatrix} 3 \\ -5 \end{pmatrix}$ $\quad$ **c** $\mathbf{x} = \begin{pmatrix} -2 \\ -3 \end{pmatrix}, \mathbf{y} = \begin{pmatrix} -4 \\ -5 \end{pmatrix}$

3 Write the negative vector and sketch a diagram of the positive and the negative vector.

a $\mathbf{p} = \begin{pmatrix} 2 \\ 5 \end{pmatrix}$ $\quad$ **b** $\mathbf{q} = \begin{pmatrix} 4 \\ 7 \end{pmatrix}$ $\quad$ **c** $\mathbf{r} = \begin{pmatrix} 0 \\ 3 \end{pmatrix}$ $\quad$ **d** $\mathbf{s} = \begin{pmatrix} -2 \\ 5 \end{pmatrix}$

e $\mathbf{t} = \begin{pmatrix} 3 \\ -4 \end{pmatrix}$ $\quad$ **f** $\mathbf{u} = \begin{pmatrix} -3 \\ -5 \end{pmatrix}$ $\quad$ **g** $\mathbf{v} = \begin{pmatrix} -1 \\ -6 \end{pmatrix}$ $\quad$ **h** $\mathbf{w} = \begin{pmatrix} -5 \\ 0 \end{pmatrix}$

★ **4** Calculate the resultant vectors.

$\mathbf{v} = \begin{pmatrix} -2 \\ 5 \end{pmatrix}, \mathbf{w} = \begin{pmatrix} 0 \\ 2 \end{pmatrix}, \mathbf{x} = \begin{pmatrix} 6 \\ 4 \end{pmatrix}, \mathbf{y} = \begin{pmatrix} -3 \\ -7 \end{pmatrix}$

a **v** + **w** $\quad$ **b** **x** – **w** $\quad$ **c** **v** + **w** + **x** $\quad$ **d** **x** – **y** – **v**

5 **a** If $\mathbf{a} = \begin{pmatrix} 2 \\ 5 \end{pmatrix}$, write in component form the vector $3\mathbf{a}$.

b If $\mathbf{v} = \begin{pmatrix} 4 \\ 10 \end{pmatrix}$, write in component form the vector $\frac{1}{2}\mathbf{v}$.

c If $\mathbf{d} = \begin{pmatrix} -3 \\ 4 \end{pmatrix}$, write in component form the vector $4\mathbf{d}$.

d If $\mathbf{x} = \begin{pmatrix} 4 \\ 5 \end{pmatrix}$, write in component form the vector $-2\mathbf{x}$.

e If $\mathbf{y} = \begin{pmatrix} 3 \\ -5 \end{pmatrix}$, write in component form the vector $-4\mathbf{y}$.

6 For vectors **u**, **v**, and **w**: $\mathbf{u} = \begin{pmatrix} 3 \\ 1 \end{pmatrix}$, $\mathbf{v} = \begin{pmatrix} 2 \\ 6 \end{pmatrix}$, $\mathbf{w} = \begin{pmatrix} -2 \\ 1 \end{pmatrix}$, calculate the resultant vector in component form.

a $\mathbf{u} + \mathbf{v} + \mathbf{w}$ **b** $2\mathbf{u} + 3\mathbf{v}$ **c** $3\mathbf{u} - 2\mathbf{v}$ **d** $5\mathbf{u} + \mathbf{v} - \mathbf{w}$

e $3\mathbf{v} + 2\mathbf{w} - \mathbf{u}$ **f** $4\mathbf{w} - 2\mathbf{u} + \mathbf{v}$ **g** $\frac{1}{2}\mathbf{v} + 2\mathbf{u} + 3\mathbf{w}$ **h** $-\mathbf{v} - 2\mathbf{u} - 3\mathbf{w}$

7 A sailing dinghy is shown.

Vector $\mathbf{w} = \begin{pmatrix} 4 \\ 0 \end{pmatrix}$ represents the normal wind velocity and vector $\mathbf{s} = \begin{pmatrix} 5 \\ 2 \end{pmatrix}$ represents the sail velocity.

Both forces act together to create a resultant vector force. Calculate the resultant vector.

wind velocity $3w$

$2s$ sail velocity

8 A white snooker ball is hit in a direction represented by the vector $\mathbf{w}_1 = \begin{pmatrix} 0 \\ 4 \end{pmatrix}$. The white ball then hits a red ball. After it hits the red ball, the movement of the white ball is described by the vector $\mathbf{w}_2 = \begin{pmatrix} 3 \\ 7 \end{pmatrix}$.

a What is the resultant vector that represents the movement of the white ball?

b The same shot is repeated but the white ball is hit with three times the force and reflects off the red ball with double the original force. What is the resultant vector in this shot?

9 A swimmer is swimming across a river in the direction described by the vector. The river flow has the direction of the vector as shown.

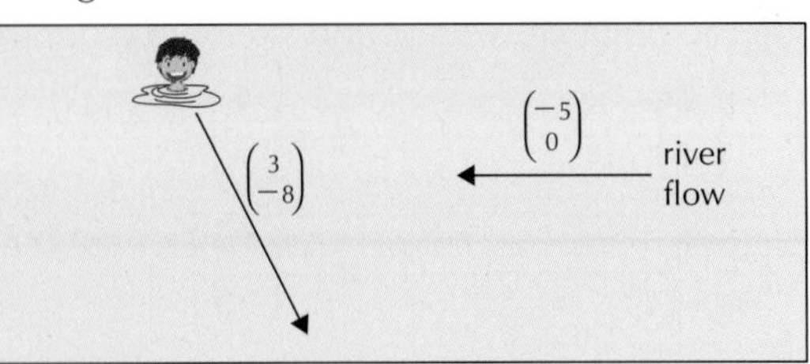

a What is the resultant vector of the swimmer?

b What would be the effect if the river flow doubled?

GO! Activity

1 Write the directed line segments for each individual vector $\overrightarrow{AB}$, $\overrightarrow{BC}$, $\overrightarrow{CD}$, $\overrightarrow{AM}$, and $\overrightarrow{MD}$ and show that the resultant $\overrightarrow{AD}$ can be achieved in two different ways.

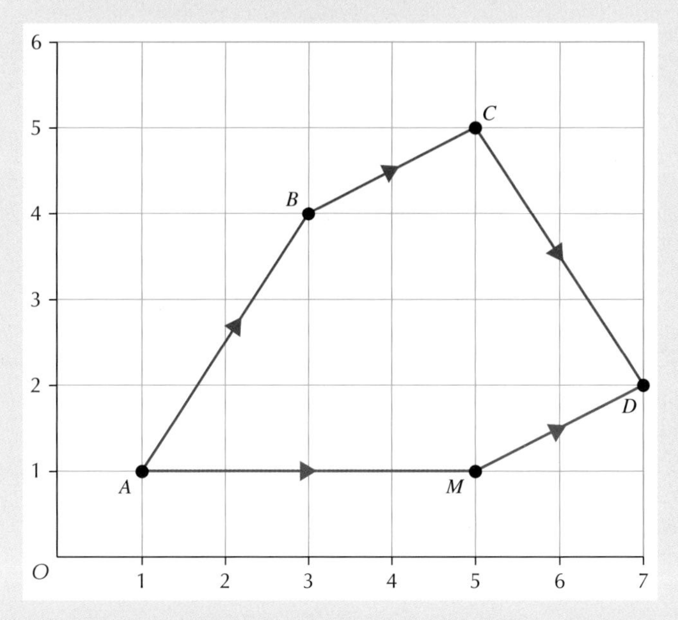

2 By adding vectors to make a vector path can you draw a diagram?
Here's one to try for yourself:

$$\begin{pmatrix}6\\0\end{pmatrix}+\begin{pmatrix}-2\\-3\end{pmatrix}+\begin{pmatrix}-8\\0\end{pmatrix}+\begin{pmatrix}-2\\3\end{pmatrix}+\begin{pmatrix}6\\0\end{pmatrix}+\begin{pmatrix}0\\8\end{pmatrix}+\begin{pmatrix}5\\-7\end{pmatrix}+\begin{pmatrix}-10\\0\end{pmatrix}+\begin{pmatrix}5\\7\end{pmatrix}$$

- I can draw 2D vectors given the vector in component form. ★ Exercise 28A Q2
- I can add and subtract 2D vectors. ★ Exercise 28B Q4

For further assessment opportunities, see the Preparation for Assessment for Unit 3 on pages 373–376.

29 Working with 3D coordinates

In this chapter you will learn how to:

- determine coordinates of a point from a diagram representing a 3D object.

You should already know:

- how to plot coordinates in 2 dimensions.

Determining coordinates in 3 dimensions

3D coordinates work in a similar way to 2D coordinates, with the addition of an extra axis. As well as having an x- and y-axis, 3D coordinates also use a z-axis. Just as 2D coordinates are plotted in terms of (x, y) and 2D shapes can be plotted using 2D coordinates, 3D objects can be plotted using 3D coordinates in terms of (x, y, z).

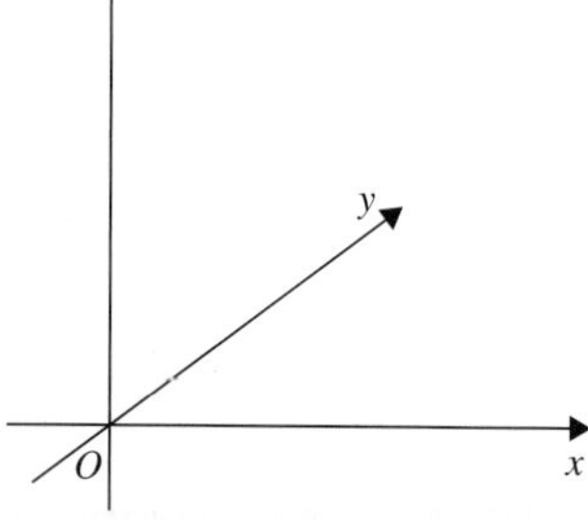

z
y
O
x

Example 29.1

Label the vertices of the cuboid shown using 3D coordinates.

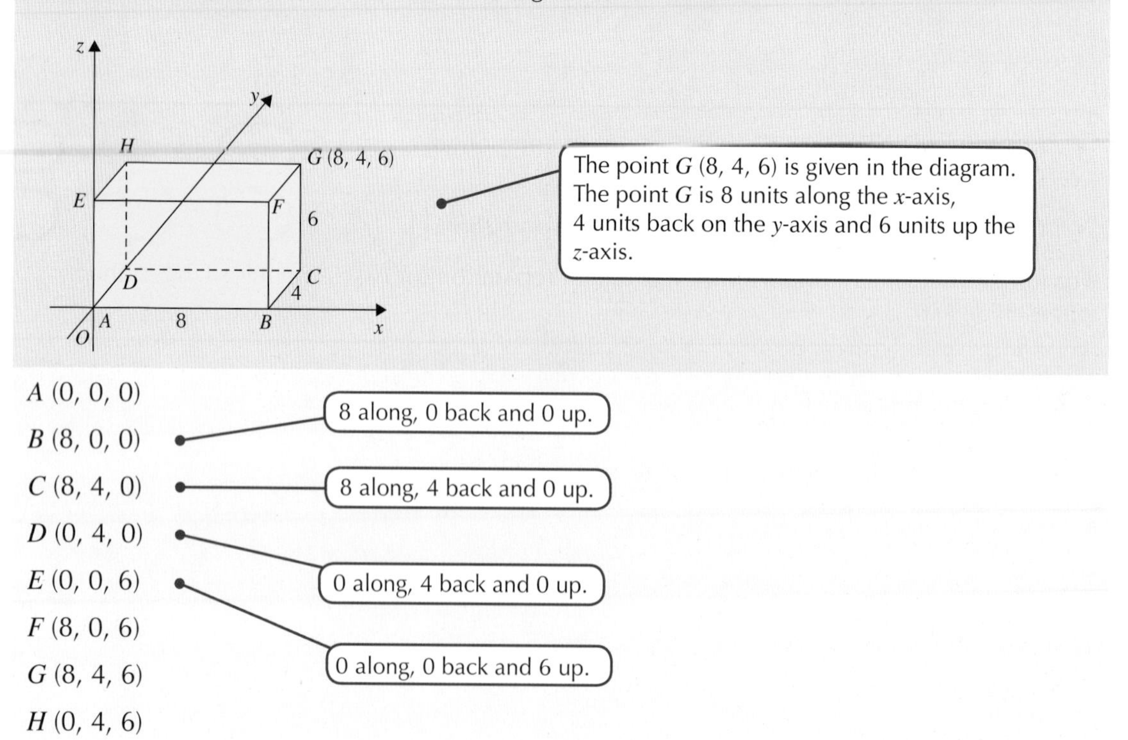

z
y
H
G (8, 4, 6)
E
F
6
D
C
4
A
8
B
x
O
The point G (8, 4, 6) is given in the diagram. The point G is 8 units along the x-axis, 4 units back on the y-axis and 6 units up the z-axis.

A (0, 0, 0)

B (8, 0, 0) — 8 along, 0 back and 0 up.

C (8, 4, 0) — 8 along, 4 back and 0 up.

D (0, 4, 0) — 0 along, 4 back and 0 up.

E (0, 0, 6) — 0 along, 0 back and 6 up.

F (8, 0, 6)

G (8, 4, 6)

H (0, 4, 6)

Exercise 29A

★ **1** For each diagram, write down the coordinates of the vertices shown.

a

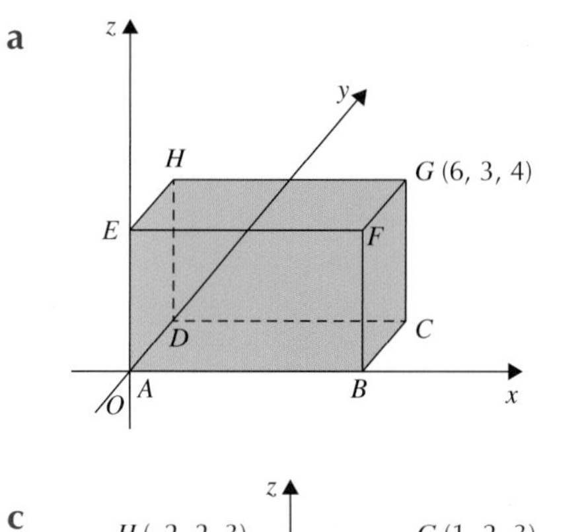

b

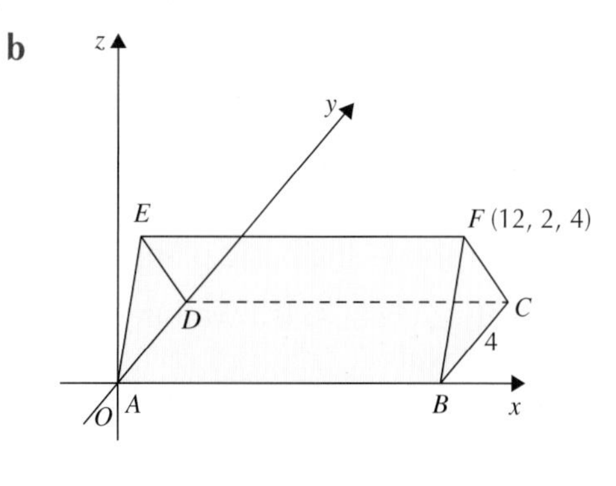

c

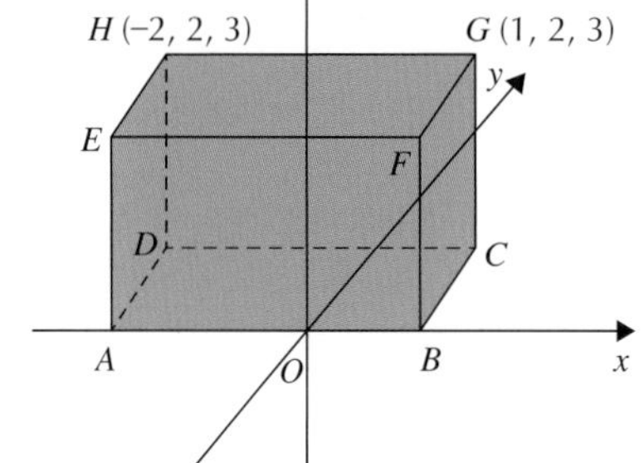

d

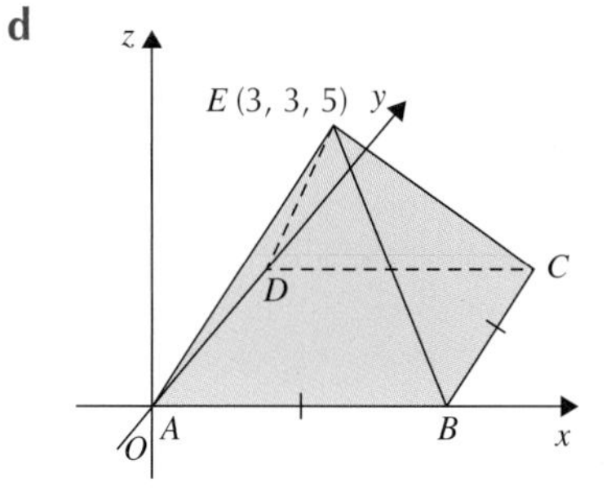

e

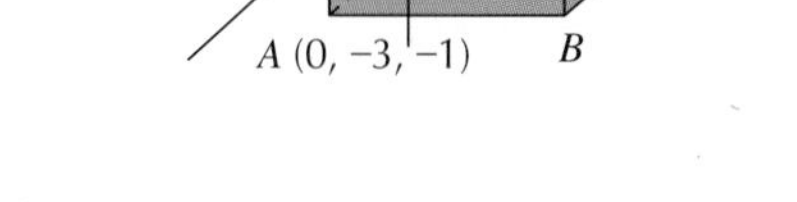

2 Calculate the coordinates of:

a the midpoint of EF

b the midpoint of BF

c the midpoint of FG.

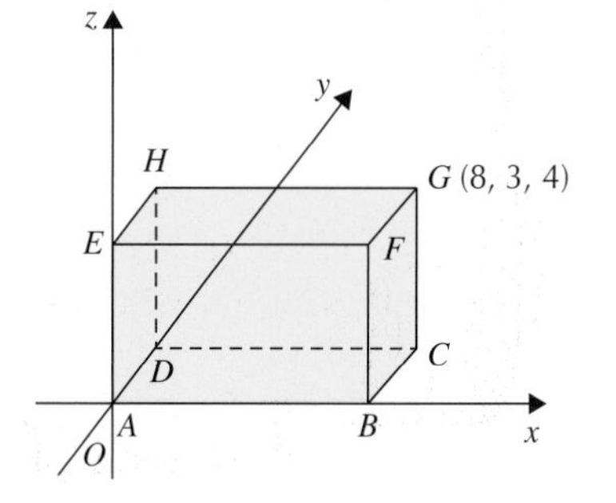

3 For the triangular prism shown:

a find the coordinates of the midpoint of AE

b find the coordinates of the midpoint of BF

c using your answers from **a** and **b**, calculate the centre point of face $ABFE$.

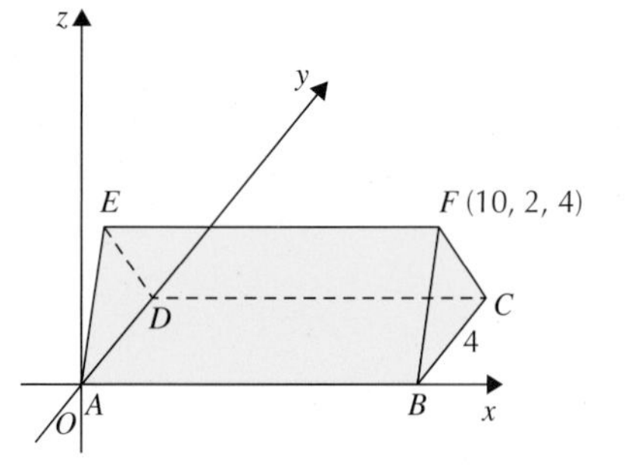

4 **a** Write down the length of the line AB.

b Write down the length of the line BF.

c Calculate the length of the line AF shown to 1 decimal place.

d Calculate the size of angle BAF to the nearest degree.

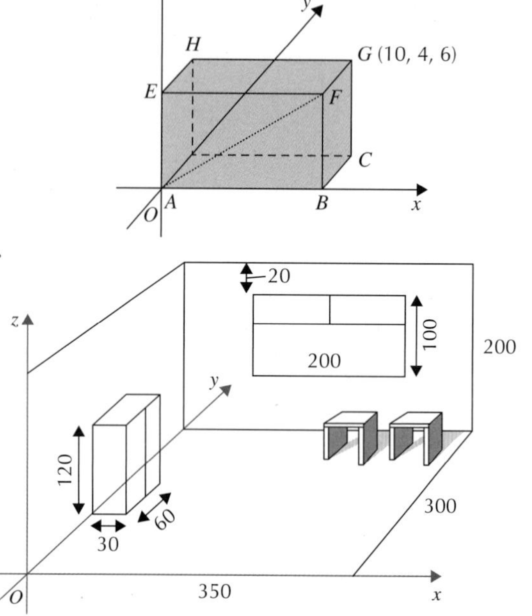

5 The diagram shows the basic plans for a room.

The cabinet against the left wall is in the centre of the wall.

a Write down the coordinates of the vertices of the cabinet.

b The window has equal lengths of wall on either side. Write down the coordinates of the corners of the window.

c A light fitting is to be placed on the centre of the ceiling. Write down the coordinates of the point where the fitting should be placed.

d Calculate the volume of the room assuming sizes are in centimetres. Give your answer in cubic metres.

GO! Activity

1 In Question 1**d**, use your knowledge of Pythagoras and trigonometry to calculate the size of angle AEB.

2 Practise drawing your own shapes using 3D coordinates. It can be quite challenging. Start off simple: draw a cuboid that is 3 by 5 by 8. Try drawing and labelling some other shapes. What is the most complicated shape you can manage?

- I can accurately label coordinates on a 3D diagram using (x, y, z) notation. ★ Exercise 29A Q1

For further assessment opportunities, see the Preparation for Assessment for Unit 3 on pages 373–376.

30 Using vector components

In this chapter you will learn how to:

- find the **resultant vector** by adding and subtracting 2D and 3D vectors in component form
- calculate the **magnitude** of a vector.

You should already know:

- what a vector is
- how to add and subtract 2D vectors
- how to determine points from a 3D diagram.

Adding and subtracting vectors using component form

2D and 3D vectors can be added and subtracted using their component forms to find the **resultant** vector. The resultant vector is a single vector used to represent the combination of a number of separate vectors.

Example 30.1

a Show that the vector $\overrightarrow{AG}$ is the same as $\overrightarrow{AB} + \overrightarrow{BC} + \overrightarrow{CG}$.

b Use the position vectors **b** and **c** to find the vector $\overrightarrow{BC}$.

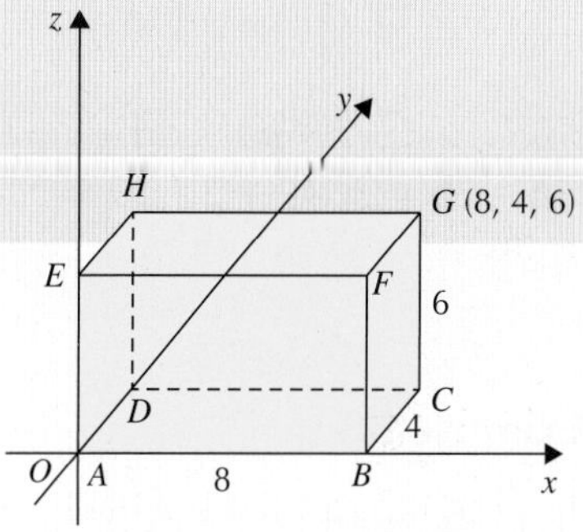

a Vector $\overrightarrow{AG} = \begin{pmatrix} 8 \\ 4 \\ 6 \end{pmatrix}$

$$\overrightarrow{AB} + \overrightarrow{BC} + \overrightarrow{CG} = \begin{pmatrix} 8 \\ 0 \\ 0 \end{pmatrix} + \begin{pmatrix} 0 \\ 4 \\ 0 \end{pmatrix} + \begin{pmatrix} 0 \\ 0 \\ 6 \end{pmatrix}$$

$$= \begin{pmatrix} 8 \\ 4 \\ 6 \end{pmatrix}$$

The vector components of $\overrightarrow{AG}$ are given by the (x, y, z) coordinates of point G because point A is at the origin.

From the diagram, write the coordinates of each vector.

Add the x-components, the y-components and the z-components.

b position vector $\mathbf{b} = \overrightarrow{OB} = \begin{pmatrix} 8 \\ 0 \\ 0 \end{pmatrix}$

position vector $\mathbf{c} = \overrightarrow{OC} = \begin{pmatrix} 8 \\ 4 \\ 0 \end{pmatrix}$

$$\overrightarrow{BC} = \mathbf{c} - \mathbf{b} = \begin{pmatrix} 8 \\ 4 \\ 0 \end{pmatrix} - \begin{pmatrix} 8 \\ 0 \\ 0 \end{pmatrix} = \begin{pmatrix} 0 \\ 4 \\ 0 \end{pmatrix}$$

Example 30.2

The vector $\mathbf{a} = \begin{pmatrix} 3 \\ 4 \end{pmatrix}$ and the vector $\mathbf{b} = \begin{pmatrix} -2 \\ 5 \end{pmatrix}$. What is the value of $\mathbf{a} + \mathbf{b}$?

$$\mathbf{a} + \mathbf{b} = \begin{pmatrix} 3 \\ 4 \end{pmatrix} + \begin{pmatrix} -2 \\ 5 \end{pmatrix} = \begin{pmatrix} 3 + (-2) \\ 4 + 5 \end{pmatrix} = \begin{pmatrix} 1 \\ 9 \end{pmatrix}$$

Example 30.3

The vector $\mathbf{v} = \begin{pmatrix} 2 \\ 5 \\ -1 \end{pmatrix}$ and the vector $\mathbf{u} = \begin{pmatrix} -2 \\ 2 \\ 4 \end{pmatrix}$. Find the resultant vector $\mathbf{v} - \mathbf{u}$.

$$\mathbf{v} - \mathbf{u} = \begin{pmatrix} 2 \\ 5 \\ -1 \end{pmatrix} - \begin{pmatrix} -2 \\ 2 \\ 4 \end{pmatrix} = \begin{pmatrix} 2 - (-2) \\ 5 - 2 \\ -1 - 4 \end{pmatrix} = \begin{pmatrix} 4 \\ 3 \\ -5 \end{pmatrix}$$

Example 30.4

If $\mathbf{a} = \begin{pmatrix} 2 \\ 4 \\ 5 \end{pmatrix}$ and $\mathbf{b} = \begin{pmatrix} -4 \\ 2 \\ 0 \end{pmatrix}$, calculate the resultant vector $3\mathbf{a} + 2\mathbf{b}$.

$$3\mathbf{a} + 2\mathbf{b} = 3\begin{pmatrix} 2 \\ 4 \\ 5 \end{pmatrix} + 2\begin{pmatrix} -4 \\ 2 \\ 0 \end{pmatrix} = \begin{pmatrix} 6 \\ 12 \\ 15 \end{pmatrix} + \begin{pmatrix} -8 \\ 4 \\ 0 \end{pmatrix} = \begin{pmatrix} -2 \\ 16 \\ 15 \end{pmatrix}$$

Exercise 30A

1 Find the resultant of each set of vectors.

a $\begin{pmatrix} 4 \\ 5 \end{pmatrix} + \begin{pmatrix} 3 \\ 7 \end{pmatrix}$ **b** $\begin{pmatrix} 2 \\ 9 \end{pmatrix} + \begin{pmatrix} -1 \\ 2 \end{pmatrix}$ **c** $\begin{pmatrix} 4 \\ 2 \\ 3 \end{pmatrix} + \begin{pmatrix} 5 \\ 1 \\ 1 \end{pmatrix}$

d $\begin{pmatrix} 2 \\ 7 \\ -1 \end{pmatrix} + \begin{pmatrix} -2 \\ 4 \\ 4 \end{pmatrix}$ **e** $\begin{pmatrix} -2 \\ -1 \\ -6 \end{pmatrix} + \begin{pmatrix} -1 \\ 2 \\ 2 \end{pmatrix} + \begin{pmatrix} 1 \\ 1.5 \\ 3 \end{pmatrix}$ **f** $\begin{pmatrix} 6 \\ 0 \\ 3 \end{pmatrix} + \begin{pmatrix} -3 \\ -2 \\ 1 \end{pmatrix} + \begin{pmatrix} -3 \\ 0 \\ -4 \end{pmatrix}$

2 Find the resultant of each set of vectors.

a $\begin{pmatrix} 3 \\ 2 \end{pmatrix} - \begin{pmatrix} 4 \\ 1 \end{pmatrix}$ **b** $\begin{pmatrix} 5 \\ -2 \end{pmatrix} - \begin{pmatrix} 3 \\ -1 \end{pmatrix}$ **c** $\begin{pmatrix} -1 \\ -5 \end{pmatrix} - \begin{pmatrix} 6 \\ -3 \end{pmatrix}$

d $\begin{pmatrix} 5 \\ 4 \\ 2 \end{pmatrix} - \begin{pmatrix} 3 \\ 3 \\ 3 \end{pmatrix}$ **e** $\begin{pmatrix} 5 \\ 1 \\ 4 \end{pmatrix} - \begin{pmatrix} 12 \\ 2 \\ 2 \end{pmatrix}$ **f** $\begin{pmatrix} 21 \\ 10 \\ 8 \end{pmatrix} - \begin{pmatrix} 9 \\ 2 \\ 4 \end{pmatrix} - \begin{pmatrix} 5 \\ 5 \\ 5 \end{pmatrix}$

★ **3** For each vector:

a write down the vector in component form

b calculate the resultant vector **a** + **b** in component form.

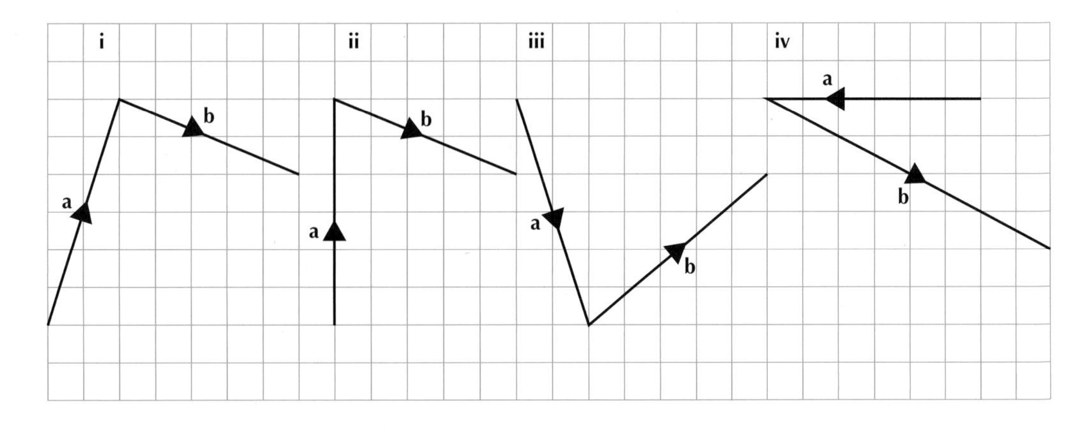

4 For each of the vectors **a** and **b** in Question 3, calculate the resultant vector:

a 2**a** + 4**b** **b** 3**a** – 2**b** **c** –**a** – 3**b** **d** 5**a** – 5**b**

5 Calculate the missing values.

a $\begin{pmatrix} 4 \\ 2 \\ a \end{pmatrix} - \begin{pmatrix} 3 \\ 5 \\ 1 \end{pmatrix} = \begin{pmatrix} 1 \\ -3 \\ 5 \end{pmatrix}$ **b** $\begin{pmatrix} x \\ y \\ z \end{pmatrix} + \begin{pmatrix} 3 \\ 4 \\ 2 \end{pmatrix} = \begin{pmatrix} 5 \\ -2 \\ 2 \end{pmatrix}$ **c** $\begin{pmatrix} -6 \\ 1 \\ 4 \end{pmatrix} + \begin{pmatrix} a \\ 3 \\ 2b \end{pmatrix} = \begin{pmatrix} -b \\ 4 \\ 2a \end{pmatrix}$

Calculating the magnitude of a vector

The coordinates of a vector are used to describe the direction of the vector, and can also be used to find the size (**magnitude**) of the vector. The magnitude of a vector is an important quantity of vectors. The diagram shows the path taken by a walker.

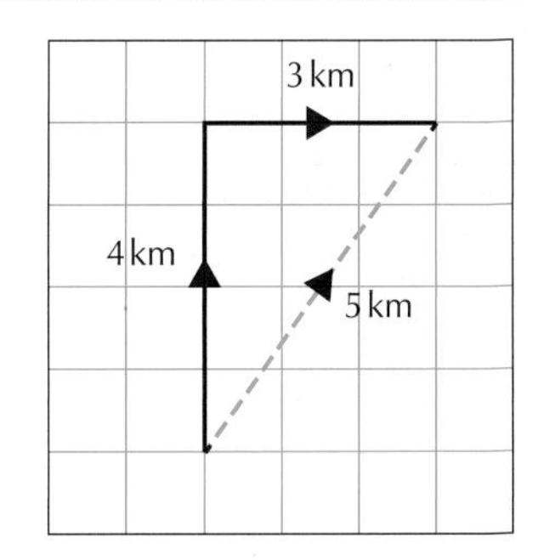

He walks north for 4 km and then east for 3 km. The total distance walked is 4 km + 3 km = 7 km.

The **displacement** is shown by the dotted line. This is the distance he has moved from the start point directly to the end point. The magnitude of his displacement is 5 km because he is only 5 km away from where he started.

The magnitude is calculated using Pythagoras' theorem.

Example 30.5

Calculate the magnitude of the vector $\overrightarrow{AB}$.

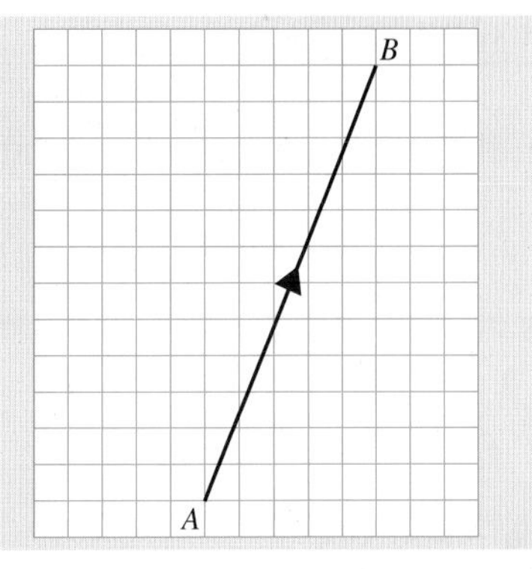

(continued)

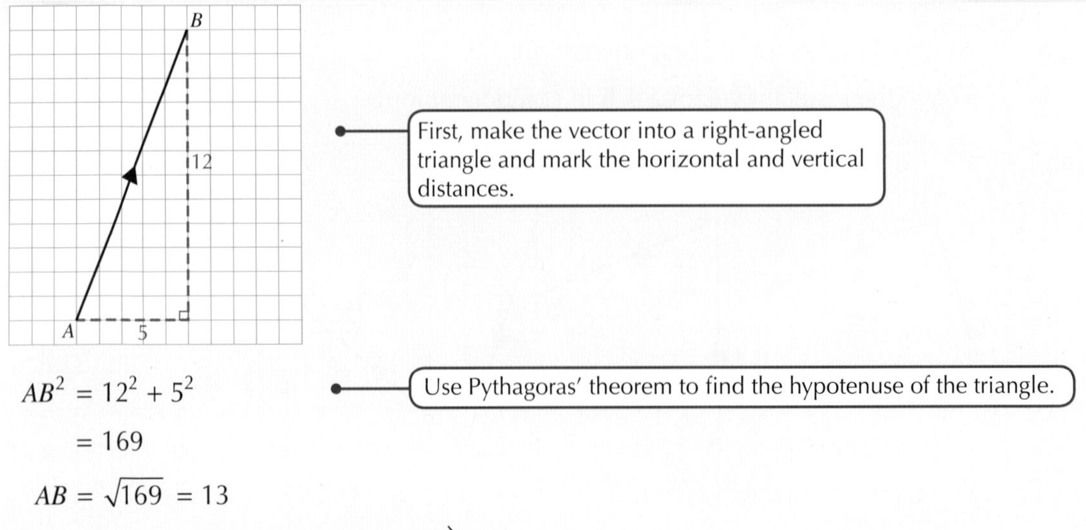

B
12
A
5
First, make the vector into a right-angled triangle and mark the horizontal and vertical distances.
Use Pythagoras' theorem to find the hypotenuse of the triangle.

$AB^2 = 12^2 + 5^2$

$= 169$

$AB = \sqrt{169} = 13$

This means the magnitude of vector $\overrightarrow{AB}$ is 13 units.

We write the magnitude of $\overrightarrow{AB}$ as $|\overrightarrow{AB}| = 13$.

Notice that in Example 30.5, the components of vector $\overrightarrow{AB}$ are $\begin{pmatrix} 5 \\ 12 \end{pmatrix}$.

In general $|\overrightarrow{AB}| = \sqrt{(x^2 + y^2)}$ where $\overrightarrow{AB} = \begin{pmatrix} x \\ y \end{pmatrix}$.

For 3D vectors, if $\overrightarrow{AB} = \begin{pmatrix} x \\ y \\ z \end{pmatrix}$, then $|\overrightarrow{AB}| = \sqrt{(x^2 + y^2 + z^2)}$.

Example 30.6

Calculate the magnitude of the 3D vector $\overrightarrow{VW} = \begin{pmatrix} 2 \\ 4 \\ -2 \end{pmatrix}$ giving your answer as a surd in simplest form.

$|\overrightarrow{VW}| = \sqrt{(2^2 + 4^2 + (-2)^2)}$

$|\overrightarrow{VW}| = 2\sqrt{6}$

Example 30.7

Calculate the magnitude of the resultant vector $\mathbf{a} + \mathbf{b}$ where $\mathbf{a} = \begin{pmatrix} 1 \\ 4 \\ 6 \end{pmatrix}$ and $\mathbf{b} = \begin{pmatrix} 3 \\ -2 \\ 4 \end{pmatrix}$.

$\mathbf{a} + \mathbf{b} = \begin{pmatrix} 1 \\ 4 \\ 6 \end{pmatrix} + \begin{pmatrix} 3 \\ -2 \\ 4 \end{pmatrix} = \begin{pmatrix} 4 \\ 2 \\ 10 \end{pmatrix}$

$|\mathbf{a} + \mathbf{b}| = \sqrt{(4^2 + 2^2 + 10^2)}$

$= \sqrt{120} = 2\sqrt{30}$

Exercise 30B

1 Calculate the magnitude of each vector leaving your answer as a surd in simplest form where appropriate.

a $\mathbf{m} = \begin{pmatrix} 3 \\ 4 \end{pmatrix}$ **b** $\mathbf{n} = \begin{pmatrix} 4 \\ -7 \end{pmatrix}$ **c** $\mathbf{o} = \begin{pmatrix} 24 \\ -7 \end{pmatrix}$ **d** $\mathbf{p} = \begin{pmatrix} -7 \\ -12 \end{pmatrix}$

e $\mathbf{q} = \begin{pmatrix} 2 \\ 3 \\ 4 \end{pmatrix}$ **f** $\mathbf{r} = \begin{pmatrix} -2 \\ 7 \\ 11 \end{pmatrix}$ **g** $\mathbf{s} = \begin{pmatrix} 8 \\ -1 \\ 0 \end{pmatrix}$ **h** $\mathbf{t} = \begin{pmatrix} 5 \\ -3 \\ 4 \end{pmatrix}$

★ **2** Calculate the magnitude of the resultant vectors leaving your answer as a surd in simplest form where appropriate.

a $\begin{pmatrix} 4 \\ 5 \end{pmatrix} + \begin{pmatrix} 3 \\ 7 \end{pmatrix}$ **b** $\begin{pmatrix} 2 \\ 9 \end{pmatrix} + \begin{pmatrix} -1 \\ 2 \end{pmatrix}$ **c** $\begin{pmatrix} 4 \\ -6 \end{pmatrix} + \begin{pmatrix} -5 \\ 7 \end{pmatrix}$

d $\begin{pmatrix} 3 \\ 5 \end{pmatrix} + \begin{pmatrix} -12 \\ -7 \end{pmatrix}$ **e** $\begin{pmatrix} 4 \\ 2 \\ 3 \end{pmatrix} + \begin{pmatrix} 5 \\ 1 \\ 1 \end{pmatrix}$ **f** $\begin{pmatrix} 2 \\ 7 \\ -1 \end{pmatrix} + \begin{pmatrix} -2 \\ 4 \\ 4 \end{pmatrix}$

g $\begin{pmatrix} -2 \\ -1 \\ -6 \end{pmatrix} + \begin{pmatrix} -1 \\ 2 \\ 2 \end{pmatrix} + \begin{pmatrix} 1 \\ 1.5 \\ 3 \end{pmatrix}$ **h** $\begin{pmatrix} 3 \\ 5 \end{pmatrix} - \begin{pmatrix} 2 \\ 2 \end{pmatrix}$ **i** $\begin{pmatrix} 3 \\ 2 \end{pmatrix} - \begin{pmatrix} 4 \\ 1 \end{pmatrix}$

j $\begin{pmatrix} 5 \\ -2 \end{pmatrix} - \begin{pmatrix} 3 \\ -1 \end{pmatrix}$ **k** $\begin{pmatrix} -1 \\ -5 \end{pmatrix} - \begin{pmatrix} 6 \\ -3 \end{pmatrix}$ **l** $\begin{pmatrix} 5 \\ 4 \\ 2 \end{pmatrix} - \begin{pmatrix} 3 \\ 3 \\ 3 \end{pmatrix}$

m $\begin{pmatrix} 5 \\ 1 \\ 4 \end{pmatrix} - \begin{pmatrix} 12 \\ 2 \\ 2 \end{pmatrix}$ **n** $\begin{pmatrix} 21 \\ 10 \\ 8 \end{pmatrix} - \begin{pmatrix} 9 \\ 2 \\ 4 \end{pmatrix} - \begin{pmatrix} 5 \\ 5 \\ 5 \end{pmatrix}$

3 Calculate the length of side AC of triangle ABC where $\overrightarrow{AB} = \begin{pmatrix} 2 \\ 7 \end{pmatrix}$ and $\overrightarrow{BC} = \begin{pmatrix} 4 \\ -3 \end{pmatrix}$.

4 If vector $\mathbf{v} = \begin{pmatrix} 2 \\ -2 \\ 4 \end{pmatrix}$ and $\mathbf{u} = \begin{pmatrix} 3 \\ 5 \\ 1 \end{pmatrix}$, calculate the magnitude of:

a $\mathbf{v}$ **b** $\mathbf{u}$ **c** $\mathbf{v} + \mathbf{u}$ **d** $2\mathbf{v}$

e $3\mathbf{u}$ **f** $2\mathbf{v} + 3\mathbf{u}$ **g** $3\mathbf{v} - 2\mathbf{u}$ **h** $5\mathbf{v} - \mathbf{u}$

5 A ship's journey can be represented by the vectors $\mathbf{a} = \begin{pmatrix} 12 \\ 0 \end{pmatrix}$ then $\mathbf{b} = \begin{pmatrix} 0 \\ 5 \end{pmatrix}$.

Calculate the displacement of the ship assuming distances are in kilometres.

Activity

1 Using vector diagrams, investigate whether **a** + **b** = **b** + **a**.

2 Does **a** – **b** = **b** – **a**?

3 If you add two of the same vector what happens? For example **a** + **a** = ? Use some of the examples and draw out the vectors. Is there a faster way than adding each component together?

- I can add and subtract 2D and 3D vectors in component form. ★ Exercise 30A Q3
- I can calculate the magnitude of a vector. ★ Exercise 30B Q2

For further assessment opportunities, see the Preparation for Assessment for Unit 3 on pages 373–376.

31 Working with percentages

In this chapter you will learn how to:

- calculate the result of a percentage increase or decrease over time intervals while knowing what is meant by the terms **appreciation** and **depreciation**
- calculate **compound interest**
- use **direct proportion** and **reverse percentages** to calculate an original quantity.

You should already know:

- how to convert between percentages, fractions and decimal fractions
- how to find a percentage of a quantity with or without a calculator
- how to solve problems involving direct proportion
- how to calculate the result of a percentage increase or decrease
- how to express one quantity as a percentage of another
- how to express an increase or decrease as a percentage.

Calculating percentage increases and decreases over periods of time

In maths, science, business and finance, it is very common to have to calculate increases or decreases in values over periods of time, or intervals. For example, in biology, it is common to look at the growth rate of bacteria over a period of hours, while in finance, it is usual to look at the decline in the value of a piece of equipment over a period of years.

Increases and decreases are often given in terms of percentage changes. For a single increase or decrease, the percentage is expressed as a **multiplier**, and a single calculation is used. For changes over a number of intervals, the calculations involving multipliers need to be repeated a number of times.

Calculating percentage increases and decreases using a multiplier

A **multiplier** is the number used to calculate the percentage increase or decrease.

$35\% = \frac{35}{100} = 0.35$

so: 35% of £240 = 0.35 × £240 = £84

0.35 is the multiplier in this calculation.

In examples involving a percentage increase or decrease, the first step is to calculate the appropriate multiplier. The multiplier is found by adding or subtracting the percentage increase or decrease to 100% and converting the percentage to a decimal by dividing by 100.

Increase by 15%	Decrease by 15%
Add to 100%: 100% + 15% = 115%	Subtract from 100%: 100% – 15% = 85%
Divide by 100: $\frac{115}{100} = 1.15$	Divide by 100: $\frac{85}{100} = 0.85$
Multiplier = 1.15	Multiplier = 0.85
New value = 1.15 × original value	New value = 0.85 × original value

Remember that if the question is in problem form, you should round your answer appropriately as required and the answer should be given in the form of a short sentence.

Example 31.1

A dress costing £90 is reduced by 20% in a sale.

Calculate the sale price.

Multiplier $= \frac{(100 - 20)}{100} = 0.80$

Find the multiplier by subtracting % decrease from 100% and convert to a decimal (÷ 100).

Remember order of operations.

Sale price = 0.80 × £90 = £72

Use the multiplier to calculate the new value.

The dress will cost £72 in the sale.

Example 31.2

A garage adds a 12% fitting fee to the cost of a set of tyres.

A set of tyres costs £135.99.

How much does it cost to buy a set of tyres and have them fitted?

Multiplier $= \frac{100 + 12}{100} = 1.12$

Find the multiplier by adding % increase to 100% and convert to a decimal (÷ 100).

Total cost = 1.12 × £135.99 ≈ £152.31

Use the multiplier to calculate the total cost.

It will cost £152.31 to have the tyres fitted.

Exercise 31A

1 A golf professional gives 5% of his winnings to his caddy. How much does the professional keep for himself if he wins £120 000?

2 The cost of a washing machine before VAT is £240. Calculate the price after VAT at 20% is added.

★ 3 A TV is reduced by 15% in a sale. It originally cost £600. How much does it cost now?

4 The population of a small town grew by 6.5% over the course of a year. At the start of the year there were 12 400 people. How many were there at the end of the year? Give your answer to the nearest hundred people.

Calculating percentage changes over several periods of time

In Exercise 31A Question 4, the population of a town grew by 6.5% over the course of one year. For planning purposes, it would be useful to know what the population would be if the population grew at the same annual rate for a period of five years.

The general method of using the multiplier remains the same but it can be applied in two ways. You can repeat the calculation for the specified number of time periods, or you can raise the multiplier to the power of the number of time periods. Multiplying by 1.065 five times is the same as multiplying once by 1.065^5.

In the special case of increases and decreases in the value of equipment and physical objects such as houses and equipment (often called **assets**), the terms appreciation and depreciation are used:

- **appreciation** is when a value increases
- **depreciation** is when a value decreases.

Example 31.3

A population of bacteria increases by 15% each hour. If the population at 10 am was 120, what would it be at 2 pm?

Method 1

Multiplier $= \dfrac{100 + 15}{100} = 1.15$

Population at 10 am = 120

Population at 11 am = 120 × 1.15 = 138

Population at 12 am = 138 × 1.15 = 158.7

Population at 1 pm = 158.7 × 1.15 = 182.5

Population at 2 pm = 182.5 × 1.15 = 209.9

Method 2

Multiplier $= \dfrac{100 + 15}{100} = 1.15$

Multiplier for 1 hour = 1.15

So multiplier for 4 hours = 1.15^4

Population at 10 am = 120

Population at 2 pm = $120 \times 1.15^4 = 209.9$

By 2 pm the number of bacteria would have increased to approximately 210.

Calculating percentage changes with different rates of change

In Example 31.3 you had a choice of method because the percentage increase, and hence the multiplier, was the same at each stage. If the percentage is different for different time intervals you cannot use a shortcut of raising the multiplier to the power of the number of time periods. Each increase or decrease must be calculated one step at a time as shown in Example 31.4.

Example 31.4

Over a 3-year period, the value of a house appreciated in value by 3% in the first year, depreciated by 2% in the second year and appreciated by 2.3% in the third year.

If the original value of the house was £240 000, what was its final value at the end of the three years?

(continued)

$$\text{Year 1 multiplier} = \frac{100 + 3}{100} = 1.03$$

The value appreciated, so find the multiplier by adding % appreciation to 100% and convert to a decimal.

$$\text{Year 2 multiplier} = \frac{100 - 2}{100} = 0.98$$

The value depreciated, so find the multiplier by subtracting % depreciation from 100% and convert to a decimal.

$$\text{Year 3 multiplier} = \frac{100 + 2.3}{100} = 1.023$$

The value appreciated, so find the multiplier by adding % appreciation to 100 % and convert to a decimal.

Original value = £240 000

Value after 1 year = £240 000 × 1.03 = £247 200

Apply calculations in three stages.

Value after 2 years = £247 200 × 0.98 = £242 256

Value after 3 years = £242 256 × 1.023 = £247 828

At the end of three years the house was valued at approximately £247 800.

Example 31.5

Over time prices tend to increase. A percentage, known as the **rate of inflation**, measures the rate of increase. State pensions rise each year in line with inflation. How much would a pension of £6400 rise by in a 3-year period when the inflation rate was calculated at 2.8%, 3.2% and 4.9% over the period?

$$\text{Year 1 multiplier} = \frac{100 + 2.8}{100} = 1.028$$

$$\text{Year 2 multiplier} = \frac{100 + 3.2}{100} = 1.032$$

$$\text{Year 3 multiplier} = \frac{100 + 4.9}{100} = 1.049$$

Original pension value = £6400

Pension after 1 year = £6400 × 1.028 = 6579.20

Pension after 2 years = 6579.20 × 1.032 = 6789.73

Pension after 3 years = 6789.73 × 1.049 = 7122.43

Pension rise = £7122.43 – £6400 = £722.43

The pension would rise by £722.43 over the 3-year period.

Exercise 31B

★ **1** City police aim to reduce crime figures by 8% each year for the next three years. If the current number of crimes committed each year is 25 000, how many do they hope for in three years time? Give your answer to the nearest hundred.

★ **2** The number of bacteria in a Petri dish is increasing at a rate of 3% every hour. If there are 12 000 bacteria at the start then how many will there be in 4 hours? Give your answer to the nearest hundred.

3 The number of subscribers to a news magazine is expected to increase by 24% every year. This year there are 16 500 subscribers.

How many subscribers are there expected to be in 3 years time? Give your answer to 3 significant figures.

4 The amount of serum in a patient's bloodstream decreases by 20% every hour. A patient is injected with 6 mg of the serum at 9 am.

How many milligrams will remain in his bloodstream at 1 pm? Answer to the nearest tenth of a milligram.

5 A bouncy ball, when dropped, loses 20% of its height each time it bounces. How high will a ball dropped from a height of 1.5 m reach after three bounces?

6 Jamie bought a house for £120 000. It appreciated in value by 3% for each of the next four years.

How much was the house worth after four years?

7 The population of a rural town was 23 400 at the end of 2009. In 2010 it increased by 4%, followed by decreases of 6% in 2011 and 3% in 2012.
How many people were living in the town at the end of 2012? Give your answer to the nearest hundred.

8 A car which was bought for £8000 depreciated annually by 13%, 9% and 6% respectively over the course of the next three years.

How much was it worth at the end of the three years?

9 A city council introduced a campaign to reduce litter levels by 5% each year.

At the start of the campaign, 250 tonnes of litter was being dropped annually. At the end of three years the amount of litter being dropped had reduced to 220 tonnes.

Were the council successful in meeting their target for that time? Give a reason for your answer.

10 A computer lost 20% of its value in the first year followed by a depreciation of 6% for each subsequent year. It originally cost £560.

How many years did it take to fall below half of its original value?

11 The value of a flat increased from £56 000 to £58 000 over the course of one year.

a Calculate the percentage increase.

b The flat is expected to increase at the same rate for each of the next three years. How much should it be worth at the end of this time?

12 Over 3 consecutive years the rate of inflation was calculated at 3.1%, 5.2% and 2.2%, respectively.

a If Alastair's wage of £24 900 rose in line with inflation over the 3-year period, what would he be earning at the end of the three years?

b Unfortunately his wage was frozen for the first two years (no increase) and he was awarded only 1% in the final year. How much less is he earning than he would have been if his wage had gone up in line with inflation?

13 Between 1960 and 2012, the world population increased by more than it had during the previous 2 million years. It more than doubled, growing from about 3 billion to about 7 billion. If, in the future, it were to grow by 1.8% each year, how many years would it take to reach 10 billion?

Activity

Arctic Ice Shrinks 18% Against Record

The newspaper headline above referred to the fact that in September of 2012 the area of ice in the Arctic was at a record low of 3.41 million km^2. This was 18.2% lower than the previous record set five years earlier in 2007.

a What area of ice would be expected in the Arctic in another 20 years if it were to continue to shrink at a rate of 18.2% every five years?

b What would some of the effects of such a serious melt of ice be?

There is no doubt that the result is dramatic, but dramatic headlines sell newspapers. If the same article had been written one year earlier in September 2011 it would have shown an area of 4.33 million km^2, 3.8% higher than in 2007. Instead of a shrinkage of Arctic ice over a 5-year period you would have been looking at a growth of 3.8% over a 4-year period!

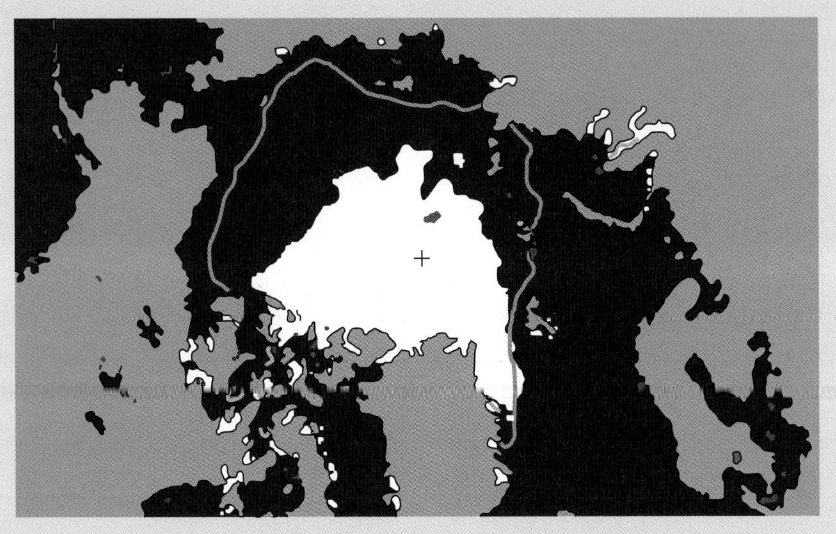

Arctic sea ice extent on 12 September 2012, in white, compared with the 1979–2000 median, marked with a red line. Photograph: NSIDC

The 2011 question would have been: What area of ice would be expected in the Arctic in another 20 years if it were to continue to grow at 3.8% every four years? What is the difference between the two scenarios?

This activity gives a clear indication of the dangers of making predictions with insufficient evidence and evidence based on a relatively short period of time. While long-term research would indicate that the trend is downwards for late summer ice area, it is actually less dramatic than the newspaper headline would have predicted. You can learn more by accessing the website of the National Snow and Ice Data Centre based in the University of Colorado.

Compare what is happening in September with other times of the year.

Calculating compound interest

Money which is invested usually increases in value. The money invested is called the **principal** and the extra value is the **interest** on the investment.

Interest is calculated as a percentage of the principal invested. The percentage used is referred to as the **interest rate**. To allow for simple comparisons, the interest rate is usually quoted for a 1-year period (known as 'per annum' which means 'each year'). 'Per annum' is often abbreviated to 'pa'.

When the interest is not withdrawn it starts to gain interest in its own right. It is added to the principal before interest is calculated for the next stage. Interest calculated in this way is called **compound interest**.

Example 31.6

David invested £8000 for four years at a rate of 2.3% per annum.

a How much money did he have at the end of the 4 years?

b What was the compound interest?

a Multiplier $= \frac{100 + 2.3}{100} = 1.023$ — Find the multiplier for 1 year.

Original investment = £8000

Investment after 4 years = £8000 × 1.023^4 = 8761.78 — Multiply the original investment by the multiplier for 4 years.

After 4 years he had £8761.78.

b Compound interest = £8761.78 – £8000 — Subtract the original investment from the total value to calculate the compound interest.

= £761.78.

Example 31.7

Interest rates for savers change from year to year. Joan invested £5000 in an account which guaranteed an interest rate of 3.5% pa for the first three years. Unfortunately it then dropped to 1.6% pa for the next two years. How much money did she have at the end of the 5-year period and what was the compound interest?

Over the first 3 years:

Multiplier for 1 year $= \frac{100 + 3.5}{100} = 1.035$ — Find the multiplier for 1 year.

Multiplier for 3 years = 1.035^3 — Find the multiplier for 3 years.

Over the next 2 years:

Multiplier for 1 year $= \frac{100 + 1.6}{100} = 1.016$

Multiplier for 2 years = 1.016^2 — Find the multiplier for the next 2 years.

Original investment = £5000

Value after 3 years = £5000 × 1.035^3 = £5543.59

Value after 5 years = £5543.59 × 1.016^2 = £5722.40

After 5 years Joan had £5722.40 in her account.

Her compound interest was £5722.40 – £5000 = £722.40

Exercise 31C

1 Sarah invested £3000 in a savings account offering a rate of 3% per annum.

a How much would she have in her account after four years?

b Calculate the compound interest.

★ **2** A bank offers interest at a rate of 4.2% per annum on savings. Calculate the compound interest on an investment of £5000 over three years.

Danny invested £4500 at a rate of 4.8% per annum. At the same time, Michael invested £5000 with a different bank at a rate of 2.6% per annum.

a Calculate the compound interest for both after two years.

b How many years will it take until Danny has more in his account than Michael?

4 A bank offers interest at a rate of 3.6% per annum in the first year followed by a rate of 2.4% per annum thereafter.

Calculate the compound interest on an investment of £8500 over:

a 3 years b 4 years.

5 Iain invested a lump sum of £25 000 in a savings account originally offering an interest rate of 3.8% per annum.

a What would he have had in his account at the end of five years if the interest rate remained at that level?

After two years the interest rate dropped to 2.2% per annum.

b i How much did he actually have at the end of the five years?

ii How much less is this than if the interest rate had remained the same?

GO! Activity

Suppose you are short of funds and need a quick short-term loan of £500 which you mean to pay back in 10 days when you are paid. One option open to you is to apply for a Payday Loan.

A typical Payday Loan Company would lend you £100 for 10 days for a charge of £0.30, which may at first glance seem quite reasonable. What would it cost to borrow £500 for 10 days?

The name 'Payday Loan' is meant to reinforce the fact that they should only be treated as short-term loans to tide a person over until pay day. The danger comes if the borrower cannot pay it back.

The **daily** interest rate for the loan advertised above is 1.008%. How much would be owed by the end of a year if the borrower was not in the position to pay anything back?

What is this the annual rate of interest?

Finding original values

If you are given the result of a percentage increase or decrease, you sometimes have to find the original value. For example, if a business is charged 20% VAT for a piece of equipment, they need to know the initial price without VAT.

Original values can be calculated in two ways:

- using reverse percentages
- using direct proportion.

Finding original values using reverse percentages

If 15% has been **added** you now have **115%** of the original value.

If 15% has been **deducted** you now have **85%** of the original value.

After calculating the resultant percentage you can work back to the original value of 100% using what we will term **reverse percentages**.

Suppose a quantity has risen in value by 15%.

You now have 115% of the original value and can use a multiplier of 1.15 to calculate the new value:

Original value × 1.15 = new value.

Rearranging to solve for the original value gives:

$$\text{Original value} = \frac{\text{new value}}{1.15}$$

Example 31.8

A cottage has appreciated in value by 3% over the past year. If it is now worth £154 000, what was its value at the start of the year? Answer to the nearest hundred pounds.

$$\text{Multiplier} = \frac{100 + 3}{100} = 1.03$$ Find the multiplier.

$$\text{Original value} = \frac{£154\ 000}{1.03} = £149\ 515$$ Divide by the multiplier to find the original value.

The value at the start of the year was £149 500.

Example 31.9

An electricity bill for the six months from May to November was £436.27 including VAT at 5%. What was the cost before VAT had been added?

$$\text{Multiplier} = \frac{100 + 5}{100} = 1.05$$ Find the multiplier.

$$\text{Cost before VAT} = \frac{£436.27}{1.05} = £415.50$$ Divide by the multiplier to find the original value.

Exercise 31D

Use a calculator for this Exercise

1. A painting has increased in value by 5% since it was bought in 2005. At auction in 2013 it was sold for £13 650. How much did it cost in 2005?
2. The average attendance at a football ground has decreased by 10% since last year. The average attendance this year is 36 800. What was the average attendance last year?
3. A TV costs £900 after VAT is added at 20%. What was the cost before VAT?
4. A house appreciated in value by 2% over the course of a year. At the end of the year it was worth £117 300. How much was it worth at the start of the year?
5. A car has depreciated by 40% since it was bought. It is currently worth £1800. How much did the car originally cost?
6. ★ Michael received a 3% pay rise, increasing his annual wage to £26 780. What was his wage before the increase?
7. ★ A 540 g cereal box claims that it has 20% extra free. What does a normal box weigh?
8. The value of a car has decreased by 12%. It is currently worth £7480. What was the original value?

The population of a Scottish town has risen by 3.5% over the past 10 years. If the population now stands at 22 300, what was it 10 years ago? Answer to the nearest hundred.

10 The rate of inflation in a particular year was 2.8%. At the end of the year a TV was priced at £635.99. If the cost had risen in line with inflation, what did it cost at the start of the year?

11 John put money in an account offering compound interest at 3.2% per annum. After three years it was worth £604.51. How much did he initially place in the account?

Finding original values using direct proportion

Using reverse percentages is the most efficient way of finding an original value when you have a calculator handy. There is another method using **direct proportion** which is also valid and which may be easier if no calculator is allowed or available.

Example 31.10

The number of students on a course fell by 20% over the course of a year. The final number of students was 240.

How many students started the course?

The final number of students was 80% of the original number (100% – 20%).

80% of original number = 240 — Calculate the final percentage.

1% of original number $= \frac{240}{80} = 3$

100% of original number = 100 × 1% = 100 × 3 = 300 — Use direct proportion to work back to the original value (100%).

300 people started the course.

As this is non-calculator work, it could be simplified further by working from 80% to 10% to 100% as shown below.

80% of original number = 240

10% of original number $= \frac{240}{8} = 30$

100% of original number = 10 × 10% = 10 × 30 = 300

300 people started the course.

Example 31.11

The price of a washing machine was £420 after a 5% installation fee was added. Calculate the original price of the washing machine.

Final price = 105% of original price (100% + 5%)

105% of original price = £420

1% of original price $= \frac{420}{105} = £4$ — If you don't have a calculator, dividing by 105 could seem daunting but you should notice that 420 = 105 × 4.

100% of original price = 100 × £4 = £400

The washing machine cost £400.

Exercise 31E

Do not use a calculator for this Exercise.

1 A computer cost £240 after VAT at 20% was added. What was the cost before VAT?

2 Mike took his family out for dinner. He paid £132 for the meal including a 10% tip. What was the price of the meal without the tip?

3 A dress cost £72 in a sale where all items were reduced by 20%. What was the price of the dress before the sale?

★ 4 Pensioners get 40% off of the price of a full adult train ticket. A pensioner pays £21 for a particular journey. What would be the cost of a full adult ticket for the same journey?

5 The population of a small village decreased by 25% last year. There are now 225 people. What was the population last year?

6 Jamie bought a computer last year. It depreciated by 12.5% over the course of the year and is now valued at £315. How much was it worth when he bought it?

7 The cost of LPG has increased by 15% since last year. This year it cost Heather £460 for a full tank of gas. How much would it have cost her last year?

GO! Activity

A property developer has two houses for sale in different areas of town. He sold them both for £120 000 each.

Based on the original prices, one house made him a profit of 15%, while the other made him a loss of 15%.

Without doing any calculations, explain whether you think that the owner has made an overall profit or a loss.

- I can solve simple problems involving percentage increase or decrease. ★ Exercise 31A Q3
- I have learned how to find the result of a percentage increase or decrease over several time intervals. ★ Exercise 31B Q1, Q2
- I know that compound interest is the profit made on an investment and can calculate it over several years. ★ Exercise 31C Q2
- I can work backwards to find an original value when given the result of a percentage increase or decrease either with or without a calculator. ★ Exercise 31D Q6, Q7 ★ Exercise 31E Q4

For further assessment opportunities, see the Preparation for Assessment for Unit 3 on pages 373–376.

Working with fractions

In this chapter you will learn how to:

- divide by a fraction
- extend calculations with mixed numbers to multiplication and division
- deal with more complex calculations involving any combination of the four operations and brackets.

You should already know:

- how to find a fraction of a quantity
- how to create equivalent fractions
- how to convert between mixed numbers and improper fractions
- how to add, subtract and multiply simple fractions
- how to add and subtract mixed numbers
- how to multiply a mixed number by a whole number or fraction
- how to decide the order of arithmetic operations.

Dividing by a fraction

The arithmetic process of dividing by fractions makes use of the fact that the value of a fraction remains unaltered if the numerator and denominator are multiplied by the same number.

Consider the following calculation where $\frac{a}{b}$ is divided by $\frac{c}{d}$.

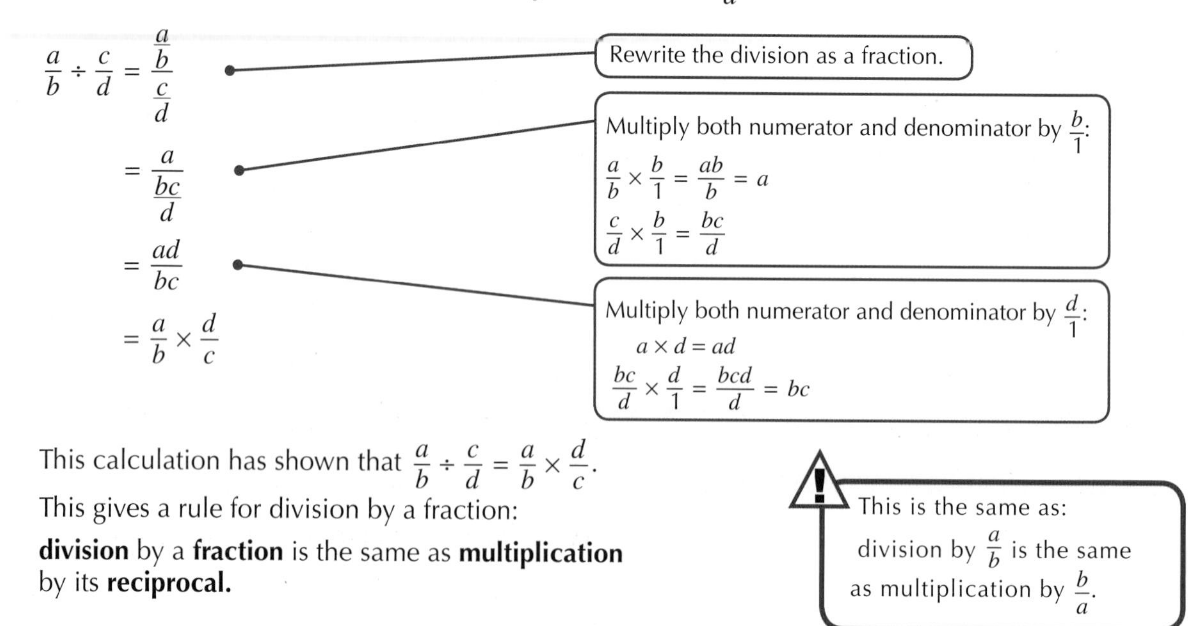

$\frac{a}{b} \div \frac{c}{d} = \frac{\frac{a}{b}}{\frac{c}{d}}$
$= \frac{a}{\frac{bc}{d}}$
$= \frac{ad}{bc}$
$= \frac{a}{b} \times \frac{d}{c}$
Rewrite the division as a fraction.
Multiply both numerator and denominator by $\frac{b}{1}$:
$\frac{a}{b} \times \frac{b}{1} = \frac{ab}{b} = a$
$\frac{c}{d} \times \frac{b}{1} = \frac{bc}{d}$
Multiply both numerator and denominator by $\frac{d}{1}$:
$a \times d = ad$
$\frac{bc}{d} \times \frac{d}{1} = \frac{bcd}{d} = bc$

This calculation has shown that $\frac{a}{b} \div \frac{c}{d} = \frac{a}{b} \times \frac{d}{c}$.

This gives a rule for division by a fraction:

division by a **fraction** is the same as **multiplication** by its **reciprocal.**

This is the same as: division by $\frac{a}{b}$ is the same as multiplication by $\frac{b}{a}$.

Example 32.1

Calculate the following without the use of a calculator.

a $\frac{2}{3} \div \frac{3}{4}$ b $\frac{4}{5} \div \frac{2}{7}$ c $8 \div \frac{2}{3}$ d $\frac{2}{3} \div 8$

a $\frac{2}{3} \div \frac{3}{4} = \frac{2}{3} \times \frac{4}{3}$ — Dividing by $\frac{3}{4}$ is the same as multiplying by $\frac{4}{3}$.

$= \frac{8}{9}$ — Multiply numerators; multiply denominators.

b $\frac{4}{5} \div \frac{2}{7} = \frac{4}{5} = \frac{\cancel{4}^2}{5} \times \frac{7}{\cancel{2}_1}$ — Dividing by $\frac{2}{7}$ is the same as multiplying by $\frac{7}{2}$. Cancel where possible.

$= \frac{14}{5}$ — Multiply numerators; multiply denominators.

$= 2\frac{4}{5}$ — Convert to a mixed number.

c $8 \div \frac{2}{3} = \frac{\cancel{8}^4}{1} \times \frac{3}{\cancel{2}_1}$ — Cancel where possible.

$= \frac{12}{1}$

$= 12$

d $\frac{2}{3} \div 8 = \frac{\cancel{2}^1}{3} \times \frac{1}{\cancel{8}_4}$ — Remember that $8 = \frac{8}{1}$. Dividing by $\frac{8}{1}$ is the same as multiplying by $\frac{1}{8}$. Cancel where possible.

$= \frac{2}{24} = \frac{1}{12}$

Exercise 32A

★ **1** Calculate.

a $\frac{3}{4} \div \frac{3}{5}$ b $\frac{4}{5} \div \frac{3}{7}$ c $\frac{3}{8} \div \frac{1}{4}$ d $\frac{2}{3} \div \frac{4}{5}$ e $\frac{5}{8} \div \frac{3}{4}$ f $\frac{4}{5} \div \frac{3}{4}$

g $\frac{5}{6} \div \frac{2}{3}$ h $\frac{5}{12} \div \frac{3}{4}$ i $\frac{3}{4} \div \frac{5}{6}$ j $\frac{2}{3} \div \frac{8}{9}$ k $\frac{7}{12} \div 7$ l $\frac{3}{10} \div \frac{12}{15}$

2 Calculate.

a $15 \div \frac{2}{3}$ b $6 \div \frac{4}{5}$ c $14 \div \frac{3}{4}$ d $3 \div \frac{2}{9}$

e $16 \div \frac{5}{8}$ f $15 \div \frac{5}{6}$ g $4 \div \frac{2}{3}$ h $10 \div \frac{4}{5}$

3 A container can hold 8 kg of wheat. Bags of wheat come in $\frac{2}{5}$ kg bags. How many bags does it take to fill the container?

4 A tank of water contains 12 litres of water. Water bottles can hold $\frac{5}{6}$ litre. How many bottles can be filled from the tank?

...e four operations with mixed numbers

Calculations using the four arithmetic operations involving mixed numbers follow the same rules as similar calculations involving whole numbers.

Adding and subtracting mixed numbers

Adding and subtracting mixed numbers uses the usual rules for adding and subtracting fractions.

Example 32.2

Calculate.

a $2\frac{2}{3} + 3\frac{4}{5}$ **b** $9\frac{2}{3} - 2\frac{1}{4}$ **c** $6\frac{2}{5} - 2\frac{3}{4}$

a $2\frac{2}{3} + 3\frac{4}{5}$ — Split into two calculations: add the whole numbers, then add the fractions.

$2 + 3 = 5$ — Add the whole numbers.

$\frac{2}{3} + \frac{4}{5} = \frac{10}{15} + \frac{12}{15}$ — Add the fractions. LCM of 3 and 5 is 15.

$= \frac{22}{15} = 1\frac{7}{15}$

Combining answers gives:

$5 + 1\frac{7}{15} = 6\frac{7}{15}$

b $9\frac{2}{3} - 2\frac{1}{4}$ — Split into two calculations: subtract the whole numbers, then subtract the fractions.

$9 - 2 = 7$ — Subtract the whole numbers.

$\frac{2}{3} - \frac{1}{4} = \frac{8}{12} - \frac{3}{12}$ — Subtract the fractions. LCM of 3 and 4 is 12.

$= \frac{5}{12}$

Combining answers gives:

$7 + \frac{5}{12} = 7\frac{5}{12}$

c $6\frac{2}{5} - 2\frac{3}{4}$

$6 - 2 = 4$

$\frac{2}{5} - \frac{3}{4} = \frac{8}{20} - \frac{15}{20}$

$= \frac{-7}{20}$

Combining answers gives:

$4 + \left(\frac{-7}{20}\right) = 4 - \frac{7}{20} = 3\frac{13}{20}$

An alternative method is to change the mixed numbers to improper fractions at the begining, as shown in Example 32.3.

Example 32.3

Calculate $3\frac{5}{8} + 7\frac{2}{3}$.

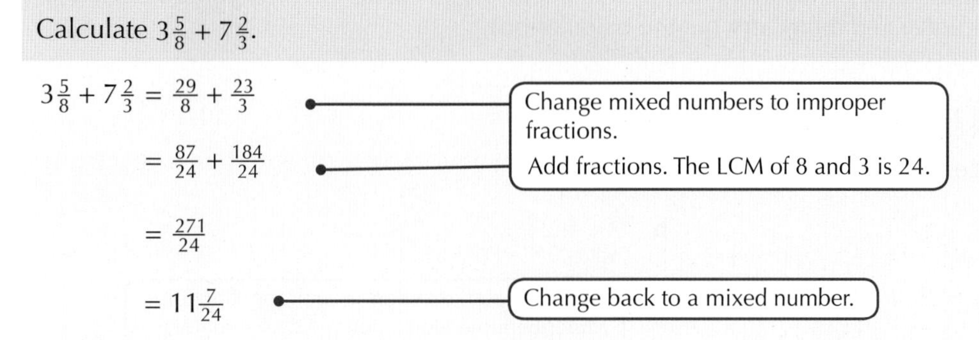

$3\frac{5}{8} + 7\frac{2}{3} = \frac{29}{8} + \frac{23}{3}$
Change mixed numbers to improper fractions.
$= \frac{87}{24} + \frac{184}{24}$
Add fractions. The LCM of 8 and 3 is 24.
$= \frac{271}{24}$
$= 11\frac{7}{24}$
Change back to a mixed number.

Exercise 32B

1 Calculate.

a $2\frac{1}{3} + 3\frac{1}{7}$ **b** $5\frac{2}{3} + 3\frac{3}{4}$ **c** $7\frac{4}{5} + 1\frac{1}{2}$ **d** $3\frac{5}{6} + 2\frac{1}{3}$

e $6\frac{5}{6} + 2\frac{2}{5}$ **f** $7\frac{3}{7} + 1\frac{1}{2}$ **g** $4\frac{7}{9} + 5\frac{1}{5}$ **h** $3\frac{3}{4} + 2\frac{5}{6}$

2 Calculate.

a $7\frac{5}{6} - 3\frac{1}{3}$ **b** $8\frac{3}{4} - 3\frac{1}{2}$ **c** $4\frac{4}{5} - 1\frac{1}{2}$ **d** $7\frac{5}{6} - 4\frac{1}{4}$

e $7\frac{1}{7} - 5\frac{1}{5}$ **f** $12\frac{3}{4} - 6\frac{7}{8}$ **g** $4\frac{1}{9} - 1\frac{1}{5}$ **h** $8\frac{1}{4} - 5\frac{5}{8}$

★ **3** Jim is building a wall. He needs the same amount of sand as cement mix. He has two bags of cement mix weighing $3\frac{1}{6}$ kg and $4\frac{2}{3}$ kg. How much sand does he need?

★ **4** Danielle lives $3\frac{1}{2}$ miles from school. Laura lives $1\frac{1}{3}$ miles from school. How much closer to school does Laura live?

5 Maria has given birth to triplets, weighing $6\frac{1}{4}$ lbs, $8\frac{1}{2}$ lbs and $7\frac{2}{3}$ lbs. Calculate their total weight.

6 Iain runs for $1\frac{3}{4}$ miles, jogs for $2\frac{5}{8}$ miles and walks for $\frac{2}{3}$ of a mile. How far has he travelled?

ultiplying and dividing mixed numbers

o multiply and divide by mixed numbers, it is necessary to convert the mixed numbers to improper fractions, and then carry out the calculations.

Example 32.4

Calculate the following without using a calculator.

a $3\frac{1}{5} \times 2\frac{2}{3}$ b $4\frac{1}{3} \div 2\frac{3}{4}$

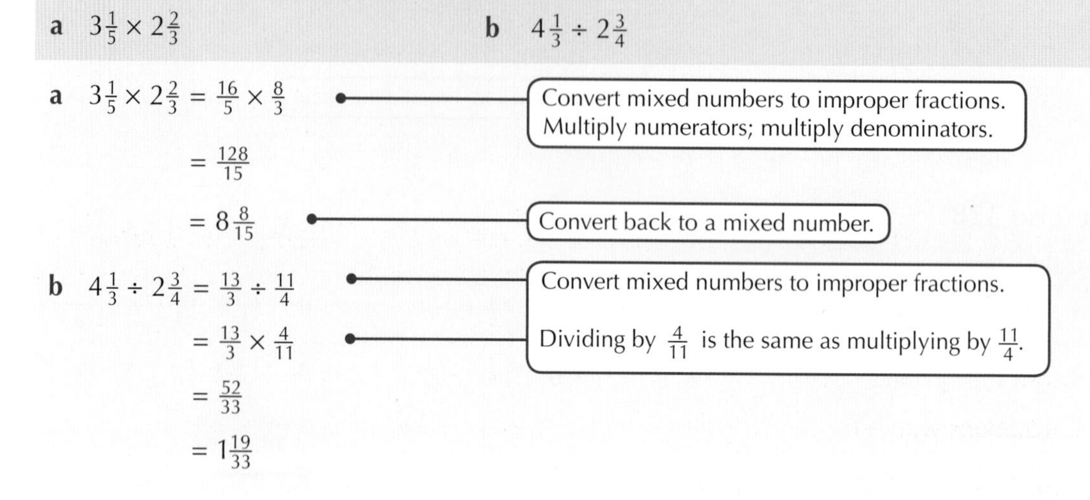

a $3\frac{1}{5} \times 2\frac{2}{3} = \frac{16}{5} \times \frac{8}{3}$
Convert mixed numbers to improper fractions. Multiply numerators; multiply denominators.
$= \frac{128}{15}$
$= 8\frac{8}{15}$
Convert back to a mixed number.
b $4\frac{1}{3} \div 2\frac{3}{4} = \frac{13}{3} \div \frac{11}{4}$
Convert mixed numbers to improper fractions.
$= \frac{13}{3} \times \frac{4}{11}$
Dividing by $\frac{4}{11}$ is the same as multiplying by $\frac{11}{4}$.
$= \frac{52}{33}$
$= 1\frac{19}{33}$

Exercise 32C

1 Calculate.

a $2\frac{1}{6} \times 1\frac{1}{3}$ b $3\frac{3}{5} \times 2\frac{1}{2}$ c $4\frac{3}{4} \times 2\frac{1}{4}$ d $1\frac{1}{2} \times 2\frac{1}{3}$

e $2\frac{2}{3} \times 2\frac{4}{5}$ f $5\frac{3}{4} \times 3\frac{5}{6}$ g $\left(2\frac{1}{4}\right)^2$ h $\left(2\frac{1}{3}\right)^3$

2 Calculate.

a $4\frac{5}{6} \div 3\frac{1}{2}$ b $5\frac{1}{4} \div 3\frac{2}{3}$ c $3\frac{2}{5} \div 1\frac{1}{2}$ d $9\frac{5}{6} \div 4\frac{3}{4}$

e $3\frac{1}{7} \div 5\frac{2}{5}$ f $4\frac{3}{4} \div 6\frac{2}{3}$ g $4\frac{1}{5} \div 1\frac{3}{5}$ h $8\frac{3}{4} \div 5\frac{5}{8}$

3 Calculate the area of the rectangles.

★ 4 A cycle route is $3\frac{1}{2}$ km long. Ally cycled it $4\frac{3}{4}$ times. How far did he cycle?

★ 5 The area of a rectangle with length $2\frac{5}{7}$ cm is $7\frac{1}{4}$ cm². Calculate its breadth.

Activity

Imagine yourself back in a 1960s classroom with no calculator available and see how fractions can be used to simplify calculations.

1 Answer the following questions without a calculator, using $\frac{22}{7}$ for π.

a Calculate the circumference and area of a circle with diameter 14 cm.

b Calculate the volume of a cylindrical can with base radius $2\frac{1}{3}$ cm and height $4\frac{1}{2}$ cm. ($V = \pi r^2 h$)

2 Use fractions to answer the following questions without a calculator.

a 1.25×8.4 **b** 75% of £345 **c** 37.5% of 152

Calculations involving combinations of the four operations with fractions and mixed numbers

Often a calculation requires the use of more than one operation, and the order in which the operations are applied will affect the answer. Operations are applied in this order:

- calculations within brackets
- powers and roots
- multiplication and division (as they appear working from left to right)
- addition and subtraction (as they appear working from left to right).

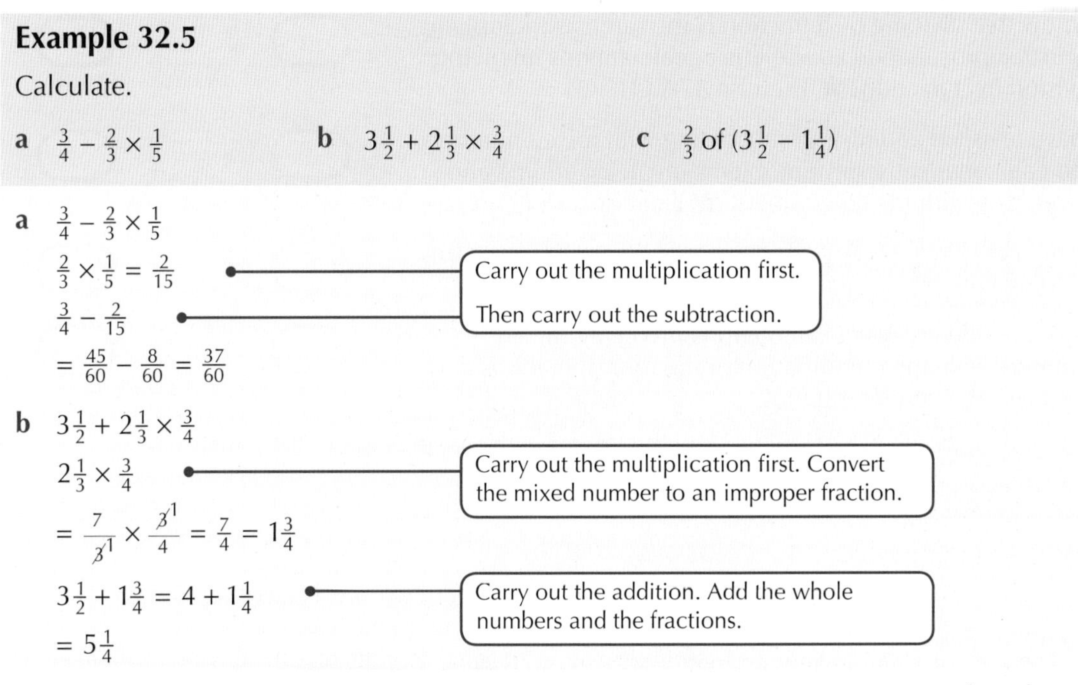

Example 32.5

Calculate.

a $\frac{3}{4} - \frac{2}{3} \times \frac{1}{5}$ **b** $3\frac{1}{2} + 2\frac{1}{3} \times \frac{3}{4}$ **c** $\frac{2}{3}$ of $(3\frac{1}{2} - 1\frac{1}{4})$

a $\frac{3}{4} - \frac{2}{3} \times \frac{1}{5}$

$\frac{2}{3} \times \frac{1}{5} = \frac{2}{15}$ — Carry out the multiplication first.

$\frac{3}{4} - \frac{2}{15}$ — Then carry out the subtraction.

$= \frac{45}{60} - \frac{8}{60} = \frac{37}{60}$

b $3\frac{1}{2} + 2\frac{1}{3} \times \frac{3}{4}$

$2\frac{1}{3} \times \frac{3}{4}$ — Carry out the multiplication first. Convert the mixed number to an improper fraction.

$= \frac{7}{\cancel{3}_1} \times \frac{\cancel{3}^1}{4} = \frac{7}{4} = 1\frac{3}{4}$

$3\frac{1}{2} + 1\frac{3}{4} = 4 + 1\frac{1}{4}$ — Carry out the addition. Add the whole numbers and the fractions.

$= 5\frac{1}{4}$

(continued)

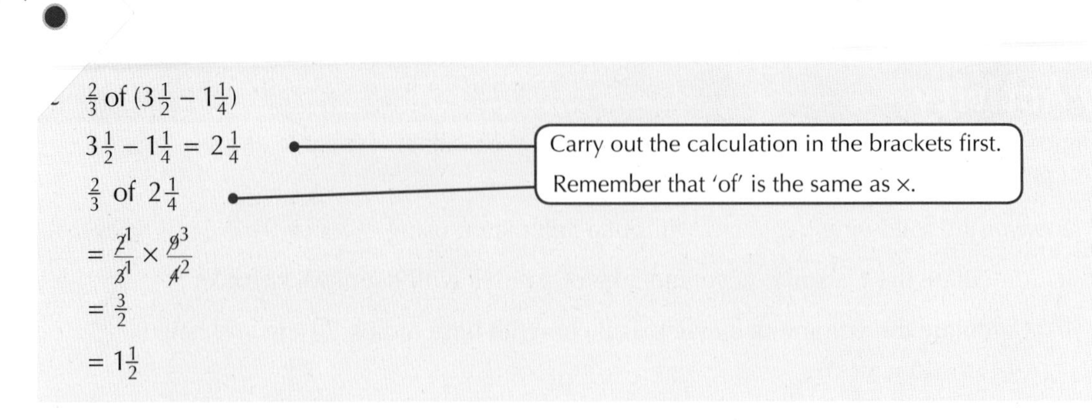

$\frac{2}{3}$ of $(3\frac{1}{2} - 1\frac{1}{4})$
$3\frac{1}{2} - 1\frac{1}{4} = 2\frac{1}{4}$
$\frac{2}{3}$ of $2\frac{1}{4}$
$= \frac{2^1}{3^1} \times \frac{9^3}{4^2}$
$= \frac{3}{2}$
$= 1\frac{1}{2}$
Carry out the calculation in the brackets first.
Remember that 'of' is the same as ×.

Exercise 32D

1 Calculate.

a $\frac{1}{2}$ of $(\frac{1}{3} + \frac{1}{5})$ **b** $(\frac{3}{4} + \frac{2}{3}) \times \frac{1}{5}$ **c** $\frac{3}{4} + \frac{3}{5} \div \frac{5}{6}$ **d** $\frac{5}{6} \div (\frac{2}{5} - \frac{1}{7})$

e $\frac{1}{5}$ of $(\frac{1}{2} + \frac{1}{3})$ **f** $\frac{1}{5}$ of $\frac{1}{2} + \frac{1}{3}$ **g** $\frac{2}{3} \times \frac{3}{4} + \frac{3}{4} \div \frac{1}{2}$

★ **2** Calculate.

a $2\frac{3}{4} + \frac{1}{2} \times 3\frac{1}{4}$ **b** $(5\frac{2}{3} + 3\frac{1}{2}) \times \frac{3}{4}$ **c** $4\frac{1}{3} - 2\frac{1}{2} \div \frac{5}{6}$

d $\frac{1}{4}$ of $(3\frac{1}{2} - 1\frac{2}{5})$ **e** $2\frac{1}{3} + (\frac{1}{2} + \frac{2}{5} \times \frac{3}{4})$ **f** $2\frac{1}{4} + (\frac{1}{3} \times 4\frac{3}{4})$

- I know that dividing by a fraction is the same as multiplying by its reciprocal and can use this in calculations involving division by a fraction. ★ Exercise 32A Q1
- I have revisited adding and subtracting with mixed numbers and can confidently solve problems involving adding and subtracting with mixed numbers. ★ Exercise 32B Q3, Q4
- I can confidently solve problems involving multiplying and dividing with mixed numbers. ★ Exercise 32C Q4, Q5
- I can confidently solve calculations involving mixed numbers with any combination of the four operations and brackets. ★ Exercise 32D Q2

For further assessment opportunities, see the Preparation for Assessment for Unit 3 on pages 373–376.

CHAPTER

33 Comparing data using statistics

In this chapter you will learn how to:

- calculate **standard deviation** from a set of data
- calculate **quartiles** and **interquartile range** from a set of data
- compare data sets using standard deviation
- compare data sets using interquartile range.

You should already know:

- how to calculate mean, median, mode and range
- how to use mean, median, mode and range to compare data sets
- how to describe, discuss and summarise data including dot plots, stem and leaf diagrams, bar graphs, line graphs and pie charts.

Standard deviation

The **standard deviation** is a measure of spread of data using all the data in a sample.

It measures the deviation of each value from the mean, to show how much the data in the sample varies from the mean average.

The dot plot below shows the results out of 10 for a mental maths test in a maths class.

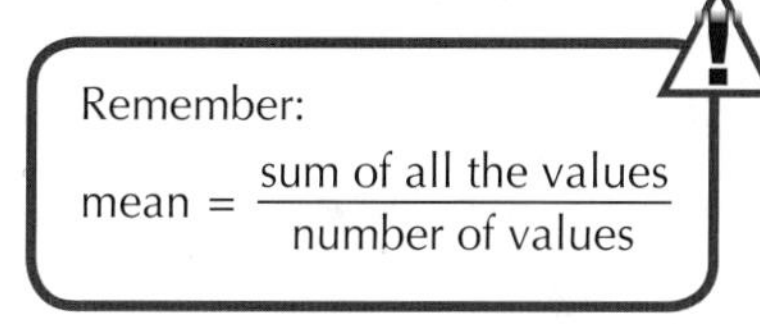

Remember:
mean = sum of all the values / number of values

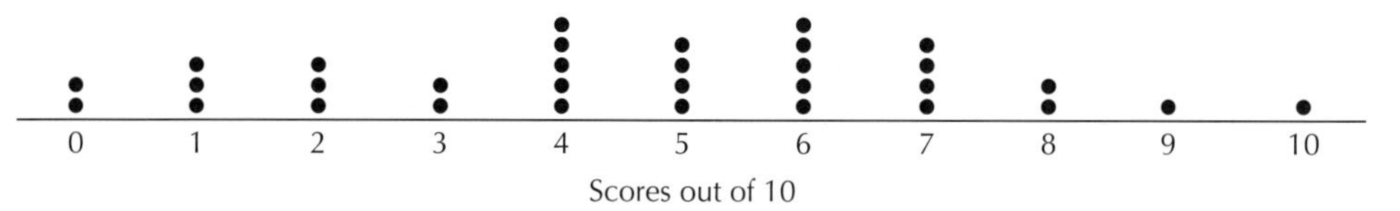

The standard deviation of a set of numbers is the average distance of each dot from the mean.

- The smaller the standard deviation, the closer the values are to the mean (the more consistent the values are).
- The larger the standard deviation, the more spread out the values are from the mean (the less consistent the values are).

The standard deviation is given by the formula: $\text{SD} = \sqrt{\dfrac{\Sigma(x - \overline{x})^2}{n - 1}}$

This formula is explained on page 347.

If you are using a **population** rather than **sample** of data, you would divide by the number of values at this stage. For example, if we had the data for the heights of all the men in Scotland we could calculate the standard deviation for the whole population. In real life, you would probably not use a full population for this kind of data, but would choose to use a sample of the height of men in Scotland to produce an estimate of the standard deviation.

.xample 33.1

The price of a bar of chocolate in five different shops is:

44p, 49p, 50p, 52p, 55p

Find the mean and standard deviation of this sample.

$$\text{Mean} = \frac{44 + 49 + 50 + 52 + 55}{5}$$

$$= 50\text{p}$$

Calculate the deviation for each value.

The deviation is the value – mean.

$44 - 50 = -6 \quad 49 - 50 = -1 \quad 50 - 50 = 0$

$52 - 50 = 2 \quad 55 - 50 = 5$

Subtract the mean value (**50**) from each individual value.

Notice that if you add all the deviations together ($-6 + -1 + 0 + 2 + 5$) you get 0. This is a helpful reminder that you have not made any careless mistakes.

Next, calculate the square of each deviation.

$(-6)^2 = 36 \quad (-1^2) = 1 \quad 0^2 = 0 \quad 2^2 = 4 \quad 5^2 = 25$

Add the deviations together and divide the total by **one less** than the number of values.

There are 5 values, so divide by 4.

$$\frac{36 + 1 + 0 + 4 + 25}{4} = \frac{66}{4} = 16.5$$

Finally take the square root to give the standard deviation:

$\text{SD} = \sqrt{16.5} = 4.1\text{p}$ (to 1 d.p.).

You may find it helpful to put these steps into a table.

Value	Deviation	$(\text{Deviation})^2$
44	–6	36
49	–1	1
50	0	0
52	2	4
55	5	25
Total	**0**	**66**

In this chapter all the data will always be a sample. In a sample always divide by one less than the number of values to take into account that the standard deviation of a sample is an estimate for an underlying population.

Calculating the standard deviation 1

There are two versions of the standard deviation formula. It is useful to be able to use both versions and to know when to use the different versions. Both forms of the formula can be used to gain full marks in examination questions.

One form of the formula is:

$$\textbf{SD} = \sqrt{\frac{\Sigma(x - \bar{x})^2}{n - 1}}$$

This connects to the steps previously mentioned in Example 33.1, provided you understand the notation given in the formula.

Symbol	Means
Σ	sum of
x	each value
$\bar{x}$	mean (pronounced 'x bar')
n	number of values of data

So: $x - \bar{x}$ represents each deviation

and: $\Sigma(x - \bar{x})^2$ represents the total of the squared deviations.

Example 33.2

The marks (out of 100) of six pupils in a maths test are:

52 60 77 82 88 91

Find the mean and standard deviation.

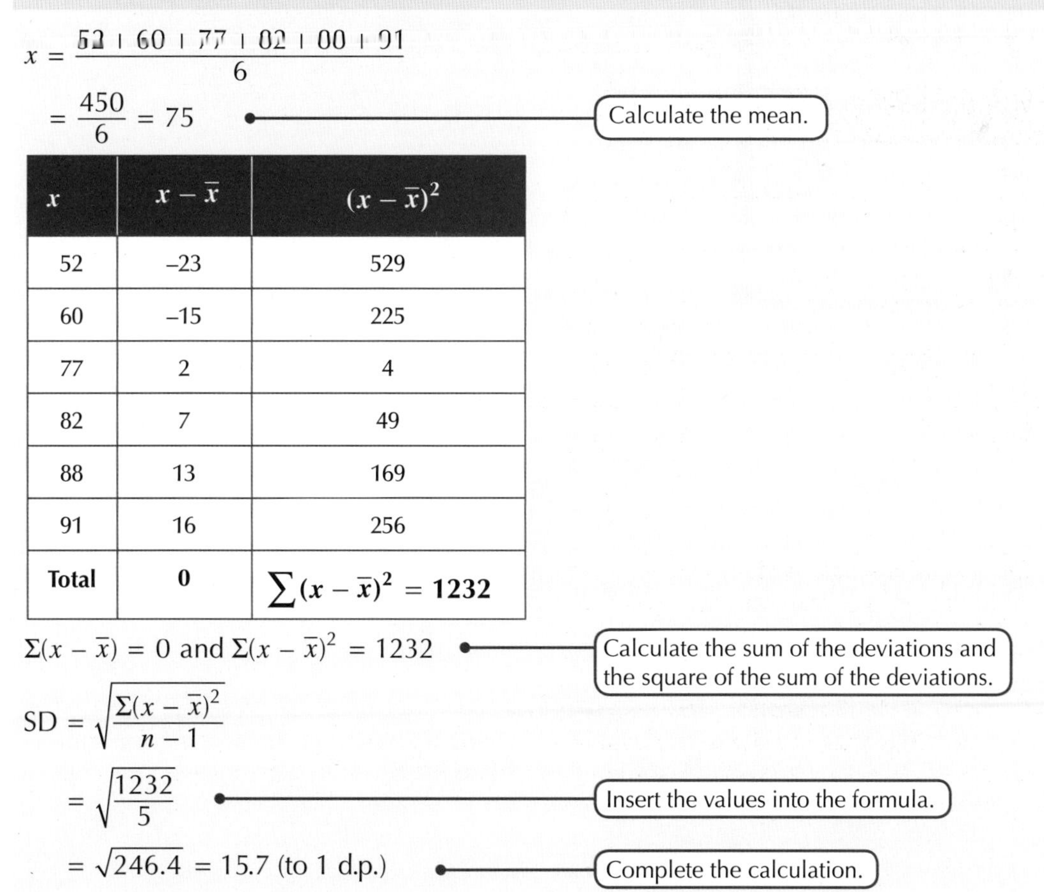

$$x = \frac{52 + 60 + 77 + 82 + 88 + 91}{6}$$

$$= \frac{450}{6} = 75$$

x	$x - \bar{x}$	$(x - \bar{x})^2$
52	–23	529
60	–15	225
77	2	4
82	7	49
88	13	169
91	16	256
Total	**0**	$\sum(x - \bar{x})^2 = \mathbf{1232}$

$\Sigma(x - \bar{x}) = 0$ and $\Sigma(x - \bar{x})^2 = 1232$

$$\text{SD} = \sqrt{\frac{\Sigma(x - \bar{x})^2}{n - 1}}$$

$$= \sqrt{\frac{1232}{5}}$$

$$= \sqrt{246.4} = 15.7 \text{ (to 1 d.p.)}$$

Calculate the mean.
Calculate the sum of the deviations and the square of the sum of the deviations.
Insert the values into the formula.
Complete the calculation.

Calculating the standard deviation 2

An alternative form of the formula is:

$$\text{SD} = \sqrt{\frac{\Sigma x^2 - \frac{(\Sigma x)^2}{n}}{n - 1}}$$

Symbol	Means
Σx^2	square each value then add together
$(\Sigma x)^2$	add all the values together then square the total

This is a rearranged form of the original and is a more useful form to use if:

- you are not required to calculate the mean, or
- you are asked to calculate the standard deviation of a large sample of numbers.

This version of the formula reduces rounding errors.

Using this formula, we can see that we obtain the same answer as above in Example 33.2.

x	x^2
52	2704
60	3600
77	5929
82	6724
88	7744
91	8281
$\Sigma x = 450$	$\Sigma x^2 = 34\,982$

$$\text{SD} = \sqrt{\frac{\Sigma x^2 - \frac{(\Sigma x)^2}{n}}{n - 1}}$$

$$= \sqrt{\frac{34\,982 - \frac{450^2}{6}}{5}}$$

$$= \sqrt{\frac{34\,982 - 33\,750}{5}}$$

$$= \sqrt{\frac{1232}{5}}$$

$$= 15.7 \text{ (to 1 d.p.)}$$

Example 33.3

a The maximum daytime temperature (in °C) was recorded in Forres on seven consecutive days in April. The results were as follows:

7 9 12 10 14 12 6

Calculate the mean and standard deviation of these maximum daytime temperatures, correct to 2 decimal places.

b The maximum daytime temperature was recorded in Menorca's capital city, Mahon, on the same seven consecutive days in April. The mean of these temperatures was 17°C and the standard deviation was 1.68.

Make two valid comparisons between the maximum daytime temperatures in Forres and Mahon in April.

a $\bar{x} = \dfrac{7+9+12+10+14+12+6}{7} = \dfrac{70}{7} = 10°C$

x	$x - \bar{x}$	$(x - \bar{x})^2$
7	–3	9
9	–1	1
12	2	4
10	0	0
14	4	16
12	2	4
6	–4	16
Total	**0**	$\Sigma(x - \bar{x})^2 = 50$

$\Sigma(x - \bar{x}) = 0$ and $\Sigma(x - \bar{x})^2 = 50$

$$\text{SD} = \sqrt{\frac{\Sigma(x-\bar{x})^2}{n-1}}$$

$$= \sqrt{\frac{50}{6}} = \sqrt{8.333} = 2.89°C \text{ (to 2 d.p.)}$$

b On average the temperatures in Mahon were higher and were less spread out (or more consistent) than the temperatures in Forres.

In your answer, make sure that you connect the mean to the word 'average' and that you connect the standard deviation to the spread of the data. You will not be given marks in an exam for simply stating that the mean or standard deviation is higher or lower.

Exercise 33A

1 Find the mean and standard deviation of:

a 4 6 6 9 10

b 12 17 20 21 25 31

2 The figures below show the wingspan, in cm, of a sample of five buzzards captured in Norfolk:

121 152 134 142 131

Calculate the mean and standard deviation for these measurements, giving your answers correct to 1 decimal place.

3 The fat content (per 100 g) of different soups in a supermarket were recorded as shown in the table.

Soup	Tomato	Leek	Cream of Chicken	Chicken Broth	Crab	Oxtail
Fat per 100 g	2.6 g	1.7 g	2.5 g	1.1 g	1.5 g	2.6 g

Calculate the mean and standard deviation of the fat content in soup based on this sample.

★ 4 a During the monsoon season in Mumbai (India) the average rainfall (in mm) is as shown in the table.

Month	June	July	August	September
Rainfall in mm	495	531	737	349

Calculate the mean and standard deviation for the rainfall, giving your answers correct to 1 decimal place.

b During the same monsoon season in New Delhi (also in India) the average rainfall (in mm) is as follows.

Month	June	July	August	September
Rainfall in mm	97	201	190	136

Calculate the mean and standard deviation for the rainfall giving your answers correct to 1 decimal place.

c Using your answers to **a** and **b** above, compare the average rainfall of the two cities and make two valid comparisons.

> ! You should see that there is a difference in the mean and standard deviation of the rainfall during monsoon season in Mumbai and New Delhi. What factors contribute to this difference?

5 Nine men who work for the same company were asked how much (on average) they earn in a month.

The data gave these summary totals:

$\Sigma x = 29\,319$ and $(\Sigma x)^2 = 101\,364\,919$

a Calculate the sample mean and standard deviation, giving your answers correct to the nearest pound.

b A group of women working in the same company doing equivalent jobs to the men in the survey were also asked how much they earned in a month. The mean of this group was £2670 and the standard deviation was £982.

Based on this sample, how does the rate of pay of women compare to that of the pay of men?

> ! Can you think of the factors that contribute to the difference in rate of pay in men? Why do you think that the standard deviation of women is higher?

6 A company claims that there are 60 matches in each box of matches.

a Find the mean and standard deviation of this sample of boxes correct to 1 decimal place:

61 62 59 59 58 60 61

b For the company's claim to be valid, the Advertising Standards Agency specifies that the mean of a sample must be between 59 and 61, and the standard deviation of a sample should be less that 2. Based on this sample, is the company's claim valid?

7 The age of a sample of eight men living in the same town in the UK who had heart attacks were recorded as follows:

54 56 58 61 53 47 80 55

a Calculate the mean and standard deviation of age, giving your answers correct to 1 decimal place.

b The age of a sample of eight men living in a town in the USA were also recorded. The mean of this sample was 63 and the standard deviation was 7.2.

Make two valid comparisons between the ages of men having heart attacks in these two samples.

c Why can you not generalise the comparisons made in **b** to reflect a comparison between the ages of men having heart attacks in the whole population of the UK and USA?

How does the sample size compare with the total population?

8 According to the Office for National Statistics, in 2009 the homicide rates per million people of population in six countries were as follows:

Country	Germany	Spain	Italy	Netherlands	France	Australia
Homicide rate per million people	8.6	9.0	10.4	10.9	11.2	13.3

a Find the mean and standard deviation of the homicide rate per million people of these six countries.

b The homicide rate per million people in the USA is 49.7.

Outlying points have a big impact on both the mean and standard deviation of a sample and can skew results significantly.

i Calculate the mean and standard deviation of the homicide rate per million people of all seven countries.

ii What impact has the addition of the USA's statistics to the group of countries had on the mean and standard deviation?

You can see that the homicide rate per million people in the USA is significantly higher than other countries. What factors contribute to the cause of this homicide rate?

GO! Activity

Ask each member of your class for one of the following: shoe size, height (to nearest cm), or number of people living in each pupil's home. (You could also make up an example of your own.)

a Use your data to calculate the mean and standard deviation.

b Make a comparison of your results.

Interquartile range

As you can see in Exercise 33A Question 8, an outlying piece of data can skew results and give an average and measure of spread which do not really reflect the data.

The **interquartile range** is a measure of the spread of data when compared with the median of the data. It only takes into account the middle 50% of the data, therefore it is not affected by outlying pieces of data and is easier to calculate than standard deviation, especially with large pieces of data.

> Remember that the median average is the middle value of a set of data.

To find the median, firstly list the numbers in numerical order (usually smallest to largest).

If the list has an odd number of values the median will be the middle value. If the list has an even number of values the median will be the mean of the two middle values.

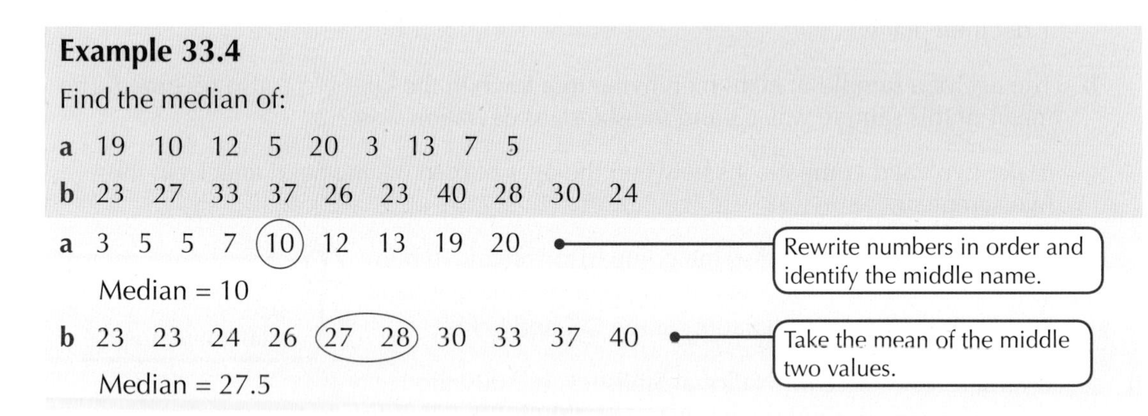

Example 33.4

Find the median of:

a 19 10 12 5 20 3 13 7 5

b 23 27 33 37 26 23 40 28 30 24

a 3 5 5 7 (10) 12 13 19 20 — Rewrite numbers in order and identify the middle name.

Median = 10

b 23 23 24 26 (27 28) 30 33 37 40 — Take the mean of the middle two values.

Median = 27.5

Quartiles

Before we can calculate the interquartile range first we need to calculate the quartiles of our data. Quartiles are the three values that divide the data set into four equal groups.

As above, order the numbers in numerical value.

The **median** (**Q_2**, the **second quartile**) is the first to be calculated.

The median splits the data set into two equal groups:

- the **lower quartile** (**Q_1**) is the **median** of the **lower half** of the numerical data
- the **upper quartile** (**Q_3**) is the **median** of the **upper half** of the numerical data.

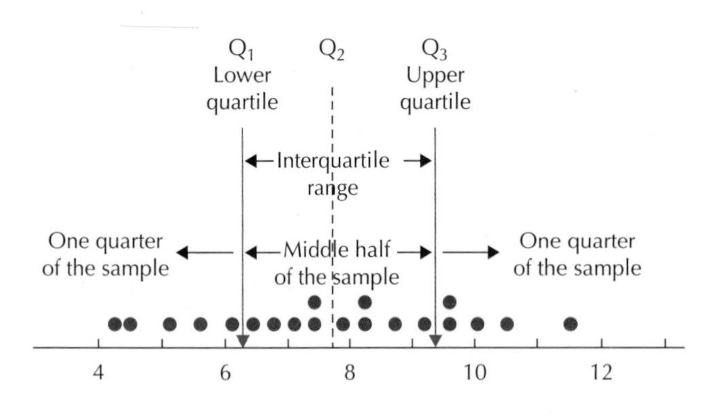

Example 33.5

Find the median and quartiles of this list of values:

3 4 5 5 7 9 11 11 12

There are 9 numbers in the list so Q_2 is the 5th value:

3 4 5 5 (7) 9 11 11 12

$Q_2 = 7$

There are 4 numbers in the lower half so Q_1 is between the 2nd and 3rd values in this list:

3 (4 5) 5

$Q_1 = 4.5$

There are 4 numbers in the upper half so Q_3 is between the 2nd and 3rd values in this list:

9 (11 11) 12

$Q_3 = 11$

3 (4 5) 5 (7) 9 (11 11) 12

↑ ↑ ↑

4.5 7 11

Q_1 Q_2 Q_3

So: $Q_1 = 4.5$, $Q_2 = 7$, $Q_3 = 11$

Example 33.6

Find the median and quartiles of this list of values:

5 6 9 10 12 13 13 15 15 21

There are 10 numbers in the list so Q_2 is between the 5th and 6th values.

5 6 9 10 (12 13) 13 15 15 21

$Q_2 = 12.5$

There are 5 numbers in the lower half so Q_1 is the 3rd value in this list:

5 6 (9) 10 12

$Q_1 = 9$

There are 5 numbers in the upper half so Q_3 is the 3rd value in this list:

13 13 (15) 15 21

$Q_3 = 15$

5 6 (9) 10 (12 13) 13 (15) 15 21

↑ ↑ ↑

9 12.5 15

Q_1 Q_2 Q_3

So: $Q_1 = 9$, $Q_2 = 12.5$, $Q_3 = 15$

If the median is between the two middle values (if there is an even number of values) make sure you don't include the calculated median in the calculation of Q_1 and Q_3.

Interquartile range (IQR)

Using Q_1 and Q_3, the **interquartile range** is calculated as follows:

$$\mathbf{IQR = Q_3 - Q_1}$$

This is a measure of spread giving the range of the middle 50% of the data.

- A small IQR represents the middle values being closer together and indicates greater consistency.
- A larger IQR represents the middle values being further apart and indicates less consistency.

Showing median and quartiles using a boxplot

If you want to show the median and quartiles in a diagram you can draw a **boxplot** to represent this information.

For this you need to state a five-figure summary. This consists of:

- lowest number **L**
- lower quartile $\mathbf{Q_1}$
- median $\mathbf{Q_2}$
- upper quartile $\mathbf{Q_3}$
- highest Number **H**

To draw the diagram:

- draw a graduated number line with a label clearly shown under the number line
- five vertical lines above the number line represent the five-figure summary
- draw a box around the upper and lower quartiles
- join the box to the lowest and highest numbers with a horizontal line.

Example 33.7

The weight of 12 15-year-old boys was recorded and is shown below.

60 kg, 60 kg, 62 kg, 64 kg, 65 kg, 65 kg, 66 kg, 69 kg, 70 kg, 71 kg, 72 kg, 83 kg

a Find the interquartile range for this data set and draw a boxplot to represent the data.

A second group of 12 boys had their weights recorded. This time all 12 boys were 14 years old. Their median was 61 kg and their interquartile range was 9 kg.

b Make two valid comparisons about the weights of the two groups.

a Middle two values: 65, 66

So $Q_2 = \frac{65 + 66}{2} = 65.5$ kg — Find the median

Lower half of numbers: 60, 60, 62, 64, 65, 65

$Q_1 = \frac{62 + 64}{2} = 63$ kg — Find Q_1 and Q_3.

Upper half of numbers: 66, 69, 70, 71, 72, 83

$Q_3 = \frac{70 + 71}{2} = 70.5$ kg

(*continued*)

$\text{IQR} = Q_3 - Q_1$
$= 70.5 - 63$
$= 7.5\text{ kg}$

Calculate the interquartile range.

The boxplot is drawn as shown below.

60 65 70 75 80 85

Weight of boys (kg)

Boxplots must be labelled or drawn along a number line.

b The median average of the second group was lower than that of the first group.

The weights of the second group of boys were more spread out (less consistent) as their interquartile range was larger.

In this style of question make reference to the average and connect it to the median, and connect the spread of data to the IQR.

Exercise 33B

1 Find the median and quartiles of the following data sets:

a 1, 1, 3, 3, 3, 6, 7, 7, 8, 9, 9

b 23, 25, 29, 29, 32, 36, 41, 42, 42, 51, 52, 60

c 4.5, 3.2, 1.6, 8.9, 5.3, 2.7, 8.4, 7.2, 5.6

★ **2** Biologists are studying the differences between hamsters and rats.

a The lengths (from nose to the tip of the tail) in cm of a sample of 10 Syrian hamsters are:

13 13 14 15 15 15 16 17 17 18

Calculate:

i the median **ii** the quartiles **iii** the interquartile range of the sample.

b The lengths (from nose to the tip of the tail) in cm of a sample of 10 brown rats are:

9 10 12 14 15 15 18 22 23 25

Calculate:

i the median **ii** the quartiles **iii** the interquartile range of the sample.

c Make two valid comparisons between the lengths of the hamsters and rats.

3 The price, in pence per litre, of diesel at 12 city garages is shown below:

136.9 145.4 139.2 148.2 138.1 142.5

137.5 143.6 138.2 140.8 139.8 146.7

a Calculate:

i the median **ii** the quartiles **iii** the interquartile range of these prices.

In 12 rural garages the petrol prices have a median of 146.3p and an interquartile range of 15.3p.

b How do the rural prices compare with the city prices?

4 Patients in Central Scotland waiting for hip replacements have to wait a number of days on a waiting list in order to see a consultant. The average number of days in 20 different hospitals is recorded in the stem and leaf diagram below.

Average number of days waiting to see a consultant prior to a hip replacement

Stem	Leaves
0	2 5
1	6
2	0 2 2 6 6 7 7
3	1 2 4 4 8
4	3 5
5	1 2 8

Key: 2 | 6 = 26

$n = 20$

a Find the median and quartiles.

b Calculate: **i** the range **ii** the interquartile range.

c Another hospital's data is included in the list. Its average waiting time is 76 days.

i Calculate the new range and interquartile range.

ii Which is most affected by the addition of one piece of data: the range or the interquartile range?

5 A popular city centre theatre has an audience capacity of 3039 people.

The audience figures for 12 performances of a musical were:

680 1427 2532 1793 2838 1982

717 1816 2997 2739 1407 1834

a Calculate:

i the median **ii** the quartiles **iii** the interquartile range of these figures.

> The interquartile range only calculates the range of the middle 50% of the data, so if an outlying point is included in the data, it does not affect the interquartile range.

b Draw a boxplot to represent this data.

During a Comedy Festival 12 different famous comedians were booked to perform at the same theatre on 12 separate performances. The audience figures were:

2782 2815 2976 3009 1897 2156

2419 2795 3030 2184 1982 2864

c Calculate:

i the median **ii** the quartiles **iii** the interquartile range of these figures.

d Draw a boxplot on the same scale as in **b** to represent this data.

e Based on these figures, which do you think is more popular, musicals or comedy? Give two reasons for your answer.

6 14 pupils sat a maths exam (marks out of 50). Their results were:

14 21 22 25 31 33 34 34 34 36 37 40 41 48

a Make a five-figure summary of the results.

b Draw a boxplot to illustrate the data.

7 **a** The maximum temperatures recorded in the UK in 1911 are shown in the table.

Month	Jan	Feb	Mar	Apr	May	Jun	Jul	Aug	Sept	Oct	Nov	Dec
Temp. (°C)	6	7	7	10	16	18	21	21	17	11	8	7

i Make a five-figure summary of the temperatures.

ii Calculate the interquartile range.

iii Draw a boxplot to illustrate the data.

You will need to order the temperatures from smallest to largest.

b The maximum temperatures recorded in the UK in 2011 are shown in the table.

Month	Jan	Feb	Mar	Apr	May	Jun	Jul	Aug	Sept	Oct	Nov	Dec
Temp. (°C)	6	8	10	16	15	18	19	18	15	15	12	8

i Make a five-figure summary of the temperatures.

ii Calculate the interquartile range.

iii Draw a boxplot to illustrate the data.

c Make two valid comparisons between the maximum temperatures in 1911 and the maximum temperatures in 2011.

8 A class of S4 pupils sat a maths test. Their results are shown in the boxplot.

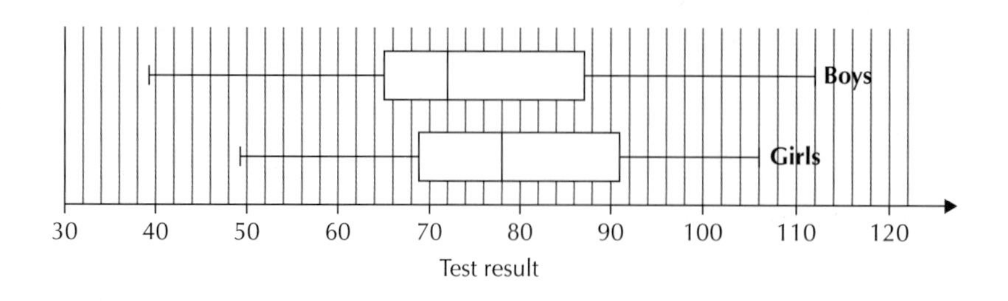

Make two valid comparisons between the boys' and girls' test results.

9 A group of students went ten-pin bowling. Their scores after 10 games each are recorded in the boxplot.

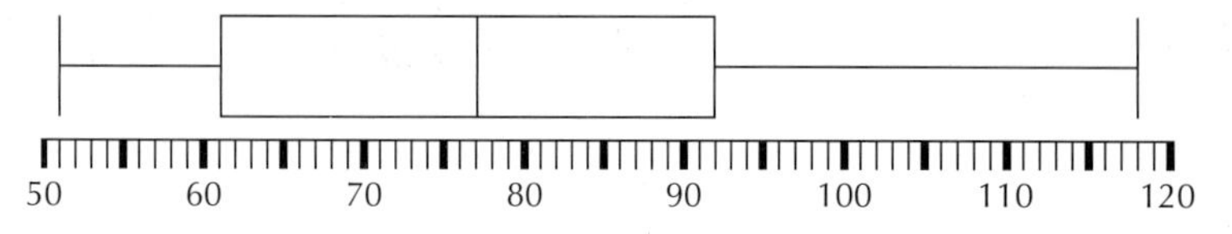

a Use the boxplot to make a five-figure summary of the scores.

b Calculate the interquartile range of the scores.

10 Two garages, Fixit and Park Auto Services, record the time taken, to the nearest minute, to repair the exhausts on 20 cars.

Their results are recorded in the boxplot.

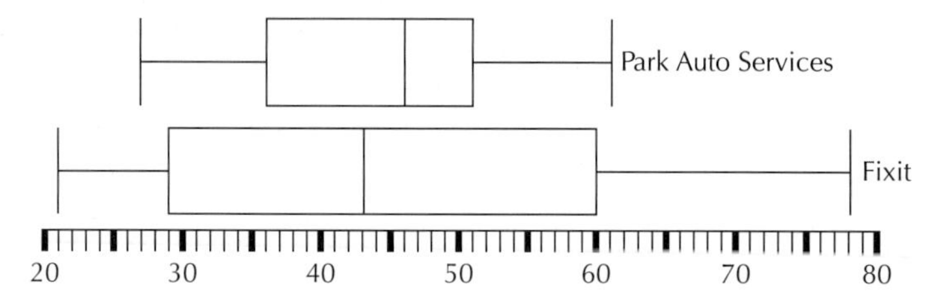

a Use the boxplots to make a five-figure summary of the times recorded for each garage.

b Calculate the interquartile range for each garage.

c Make two valid comparisons between the time taken to repair an exhaust in the two garages.

Semi-interquartile range

Sometimes the **semi-interquartile range** (SIQR) is used instead of the interquartile range. It is calculated as follows:

$$\textbf{SIQR} = \frac{Q_3 - Q_1}{2}$$

As the semi-interquartile range is half of the interquartile range, it is a measure of the average spread from the centre of the data. This can be used as a reasonable comparison with the standard deviation but tends to be used as a valid comparison only if the data set is sufficiently large.

Any comparisons in SIQR between two data sets would be answered in the same way as responses to interquartile range.

Which measure of spread should I use?

When you are answering questions in an exam it is obvious which measure of spread to use as the question will direct you. However, if you are making comparisons outwith the examination process it is important that you can recognise which would be best to use.

- Standard deviation: Use this if there are no outlying points, the data set is relatively small and you want a more accurate measure of the spread.
- Interquartile range: Use this if you have outlying points and a relatively large set of data.

Activity

1 Use the data you recorded in the activity on page 351 when you asked each member of your class for one of the following: shoe sizes, height (to nearest cm), number of people living in each pupil's home.

a Use your data to calculate the median, quartiles and interquartile range.

b Make a comparison of your results.

Make the Link

The **population** is the complete set of people, items or data used in a statistical survey.

For example, if the mean and standard deviation of the shoe sizes of your whole school was required then you would ask each pupil in the school for their shoe size then carry out the calculations above.

However, this can take a long time and require a lot of work, so it would be expected that you would take a **sample** of the population, i.e. select a cross-section of the population. Using the example above of shoe sizes in your school you would select a sample of pupils in each year group and ensure you had a fair gender representation.

$$\text{The standard deviation of a population} = \sqrt{\frac{\text{sum of squared deviations}}{\text{number of values in the population}}}$$

This represents the average distance between the mean and each value.

When a sample is used we divide by **one less** than the number of values to take into account that the standard deviation of a sample is an estimate for the underlying population and it is less accurate (usually too small).

Activity

Consider the following population of shoe sizes in an S3 class of 15 pupils:

3 4 5 5 5 5 7 7 8 8 9 9 9 10 11

a Calculate the mean and standard deviation of the above population.

b Now select a sample of 5 pupils' shoe sizes from the above sample.

- **i** Calculate the mean and **population** standard deviation of your sample (divide by n).
- **ii** Now calculate the **sample** standard deviation of your sample (divide by $n - 1$).
- **iii** Compare these answers to your answer in **a**. Is your standard deviation in **b i** smaller than the population standard deviation? Do you get a more accurate answer when you divide by $(n - 1)$ in **b ii**? Explain.

c Compare your answers in **b** to other members of your class who will have selected a different sample.

This has demonstrated the significance of dividing by $n - 1$ in the sample standard deviation. In National 5 it will always be the sample standard deviation that is used.

- I can calculate standard deviation from a set of data. ★ Exercise 33A Q4
- I can compare data sets using standard deviation. ★ Exercise 33A Q4
- I can calculate quartiles and interquartile range from a set of data. ★ Exercise 33B Q2
- I can compare data sets using interquartile range. ★ Exercise 33B Q2

For further assessment opportunities, see the Preparation for Assessment for Unit 3 on pages 373–376.

34 Forming a linear model from a given set of data

In this chapter you will learn how to:

- calculate the equation of a best-fitting straight line on a scattergraph
- use the equation of a best-fitting straight line to estimate solutions.

You should already know:

- how to construct a scattergraph
- how to draw and apply a best-fitting line
- how to draw and recognise a graph of a line or equation
- how to determine the equation of a straight line
- how to recognise the trend and correlation of a graph.

Determine the equation of a best-fitting straight line on a scattergraph

Scattergraphs are used to investigate the relationship between two variables. Corresponding values are plotted on a graph in the same way as coordinates are plotted.

When changes in one variable match changes in the other, there may be a cause-and-effect relationship between the two. This is called **correlation**.

Correlation can be described in three ways:

- **positive correlation**
- **negative correlation**
- **no correlation**

Positive correlation

If both variables are increasing you get a **positive correlation**. For example, you would expect that as a person's height increases, their weight also increases, so there is a positive correlation between height and weight.

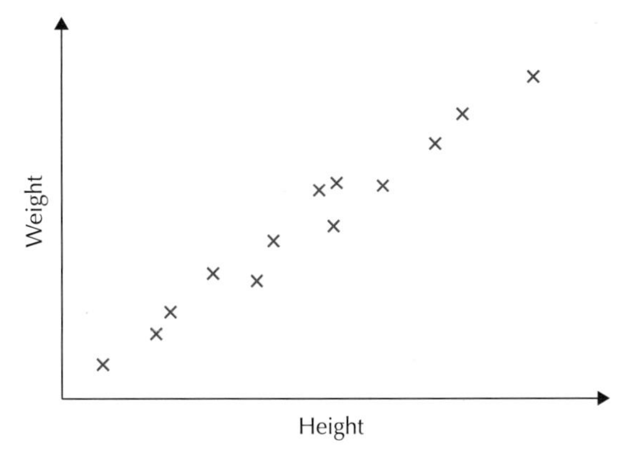

Negative correlation

If one variable increases as the other decreases you get a **negative correlation**. For example, you would expect that, as the temperature lowers, the sale of gloves increases, so there is a negative correlation between temperature and sales of gloves.

No correlation

If there is no relationship between the variables there is **no correlation**. For example, there is no correlation between shoe size and mathematical ability.

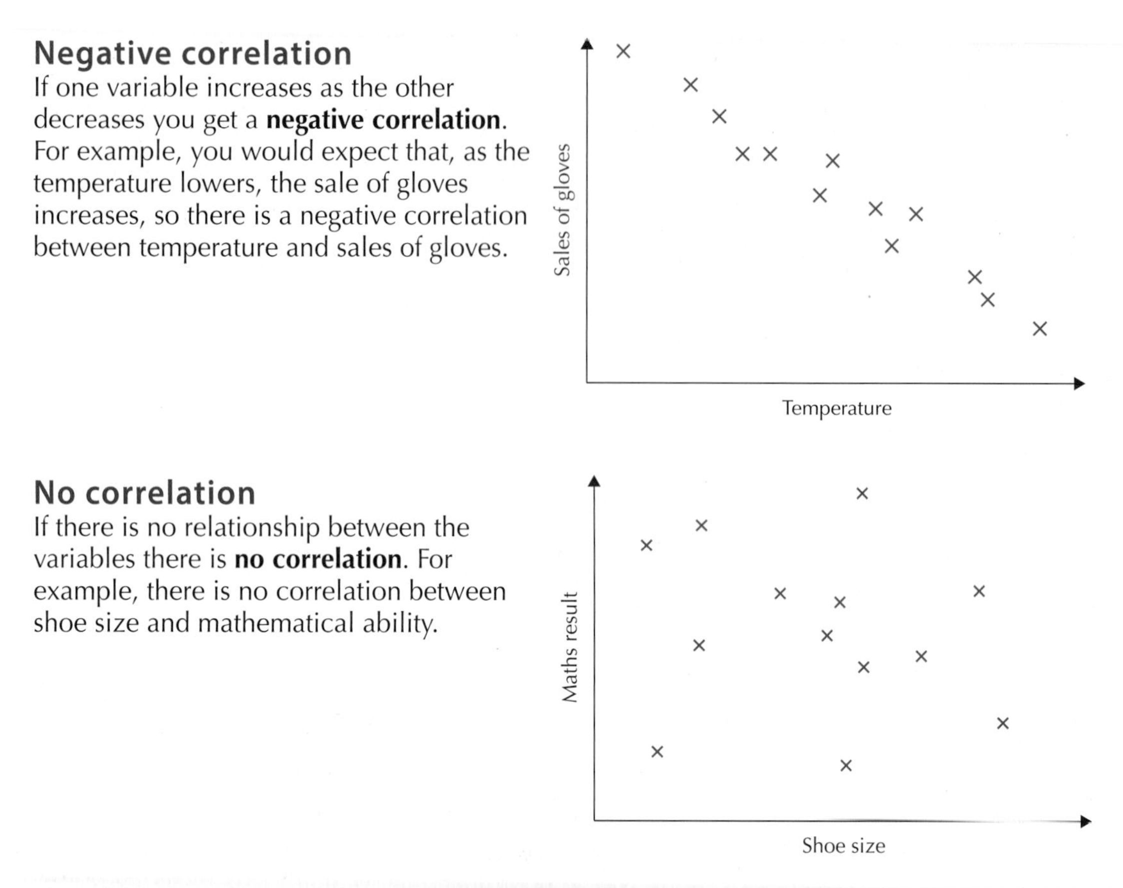

The table shows the connection between the gradient of a scattergraph and the correlation.

Gradient	Direction	Correlation
positive	graph goes **upwards** reading from left to right	positive
negative	graph goes **downwards** reading from left to right	negative

Drawing a scattergraph and line of best fit

Drawing a scattergraph and a line of best fit has a few basic guidelines:

- choose the scales carefully to show all the data
- plot points carefully. Always plot the first set of data as your horizontal axis and the second set of data as your vertical axis.
- use letters on the axes to represent the context instead of x and y.
- draw a straight line that best fits the data. Ensure you have approximately the same number of points on either side of the line and that the line is following the direction of the data. The line of best fit might not go through the first and last data points on the graph.
- calculate the equation for the line of best fit.

> The equation of a straight line can expressed as:
> $y = mx + c$
> where m = gradient and c = y-intercept (value of y when graph crosses the y-axis).

Example 34.1

1 The scattergraph shows the age of some cars and their values.

a By selecting two points on the line, calculate the gradient of the line.

b Read the intercept of the vertical axis from the graph.

c Using your answers to parts **a** and **b**, write down the equation of the line.

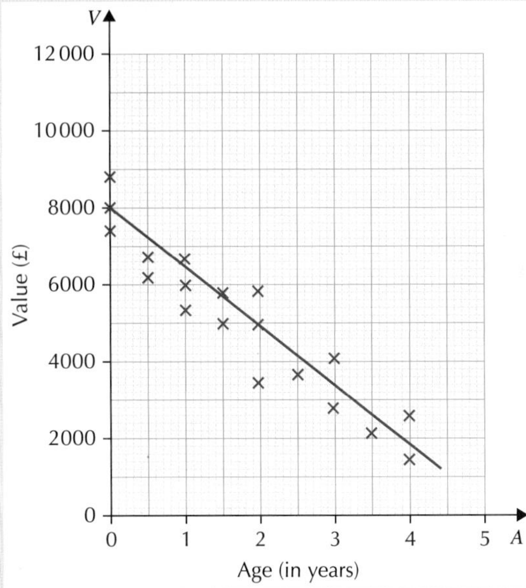

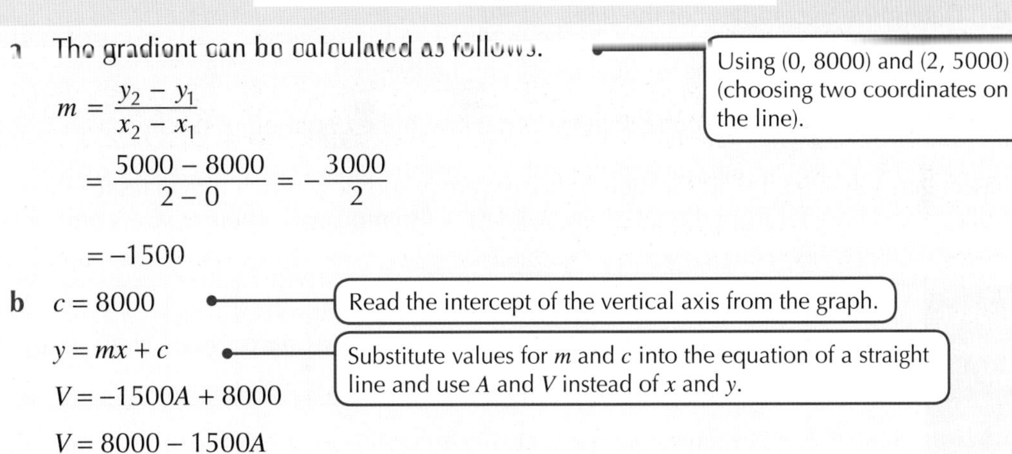

a The gradient can be calculated as follows.

Using (0, 8000) and (2, 5000) (choosing two coordinates on the line).

$$m = \frac{y_2 - y_1}{x_2 - x_1}$$

$$= \frac{5000 - 8000}{2 - 0} = -\frac{3000}{2}$$

$$= -1500$$

b $c = 8000$

Read the intercept of the vertical axis from the graph.

$y = mx + c$

Substitute values for m and c into the equation of a straight line and use A and V instead of x and y.

$V = -1500A + 8000$

$V = 8000 - 1500A$

Exercise 34A

1 The table shows the volume of ice cream sold by a shop and the daily maximum temperature outside.

Temperature (°C)	Sales of ice cream (litres)
3	150
5	190
8	220
10	235
13	230
15	261
20	285
25	304

a Draw a scattergraph to illustrate the data.

b Describe the type of correlation and what the graph tells you.

c Draw a line of best fit on the graph.

d Determine the gradient of the line.

e Read the intercept of the vertical axis from the graph.

f Using your answers to parts **d** and **e**, write down the equation of the line.

2 Passengers travelling by air for their holiday were asked how much they paid for their holiday and how far from the UK their holiday destination was. Their answers are recorded in the table.

Distance travelled (km)	600	800	1000	1100	1300	1400	1600	1700	2000	2400	2500	2600	3500	4500
Cost of holiday (£)	400	450	500	450	550	550	600	550	650	650	700	700	750	900

a Draw a scattergraph to illustrate the data.

b Describe the type of correlation and what the graph tells you.

c Draw a line of best fit on the graph.

d Calculate the gradient of the line.

e Read the intercept of the vertical axis from the graph.

f Using your answers to parts **d** and **e**, write down the equation of the line.

★ **3** Each month the average outdoor temperature was recorded together with the number of therms of gas used to heat a house. The results are shown in the table.

Temperature (°C)	1	3	5	8	9	10	12	13	15	16
Therms used per month	42	40	34	30	25	21	20	15	10	6

a Draw a scattergraph to illustrate the data.

b Describe the type of correlation and what the graph tells you.

c Draw a line of best fit on the graph.

d Calculate the equation of the line of best fit.

4 The size of engine (cm^3) and fuel consumption in miles per gallon (mpg) of 13 cars is shown in the table.

Size of engine (cm^3)	1400	1500	2000	2100	2400	2800	3000	3000	3400	3900	4000	4500	4800
Fuel consumption (mpg)	48	50	42	43	41	34	35	33	28	27	23	22	18

a Draw a scattergraph to illustrate the data.

b Describe the type of correlation and what the graph tells you.

c Draw a line of best fit on the graph

d Calculate the equation of the line of best fit.

5 The table shows the latitude of nine cities in the northern hemisphere and their average high temperatures.

City	Latitude (degrees)	Average high temperature (°C)
Casablanca	34	22
Dublin	53	13
Hong Kong	22	25
Istanbul	41	18
St Petersburg	60	7
Manila	15	32
Mumbai	19	31
Oslo	60	10
Paris	49	15

a Draw a scattergraph to illustrate the data.

b Describe the type of correlation and what the graph tells you.

c Draw a line of best fit on the graph.

d Calculate the equation of the line of best fit.

Use the equation of a best-fitting straight line to estimate solutions

The equation found can be used to analyse and interpret data and to forecast outcomes.

Example 34.2

A group of 18 S2 pupils sat a maths exam and science exam.

Their results are shown in the table.

Pupil	1	2	3	4	5	6	7	8	9	10	11	12	13	14	15	16	17	18
Maths	18	24	26	34	46	50	64	62	14	20	26	28	30	36	42	16	40	54
Science	30	30	34	38	46	54	62	68	30	34	38	40	44	46	52	34	44	60

a Draw a scattergraph and draw a line of best fit on the graph.

b Use the line of best fit to find an equation connecting the two sets of data.

c Another pupil in the same S2 class had been absent on the day of the science exam but achieved a mark of 36% in the maths exam.

Use your equation of line of best fit to predict the pupil's science mark.

a

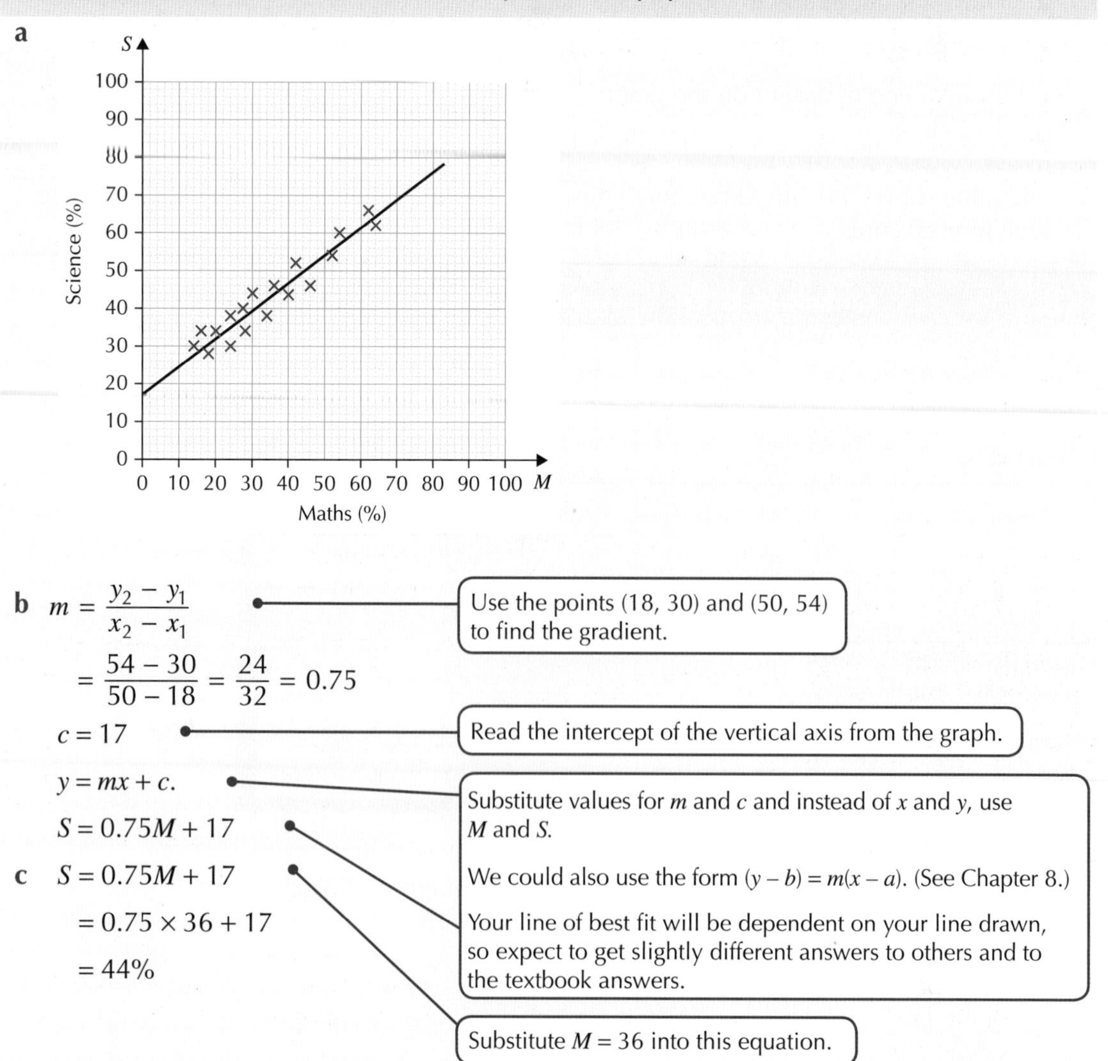

b $m = \dfrac{y_2 - y_1}{x_2 - x_1}$

$= \dfrac{54 - 30}{50 - 18} = \dfrac{24}{32} = 0.75$

$c = 17$

$y = mx + c.$

$S = 0.75M + 17$

c $S = 0.75M + 17$

$= 0.75 \times 36 + 17$

$= 44\%$

Use the points (18, 30) and (50, 54) to find the gradient.

Read the intercept of the vertical axis from the graph.

Substitute values for m and c and instead of x and y, use M and S.

We could also use the form $(y - b) = m(x - a)$. (See Chapter 8.)

Your line of best fit will be dependent on your line drawn, so expect to get slightly different answers to others and to the textbook answers.

Substitute $M = 36$ into this equation.

Exercise 34B

1 An experiment is carried out to measure the extension of a spring as weights are attached to it. The results are shown in the table.

Mass, M (g)	0	20	50	80	100	125	150	160	180	200
Extension, E (cm)	0	1.2	2	3.9	5	5.9	7.2	7.9	8.5	9.6

a Draw a scattergraph to illustrate the data.

b Draw a line of best fit on the graph.

c Calculate the equation of the line of best fit.

d Use your equation to estimate the length of the extension of the spring if the mass used is 250 g.

e Use your equation to estimate the size of mass attached to the spring if the extension is 16.4 cm.

For part **e** you need to rearrange the equation after substitution to solve.

2 The table shows the test marks for 10 S4 pupils in their English and history tests.

Student	**Ali**	**Brian**	**Clare**	**David**	**Ed**	**Fergus**	**Greg**	**Henry**	**Marie**	**Sarah**
English	35	51	62	27	34	59	66	45	52	39
History	36	38	54	25	30	55	60	45	47	35

a Plot the data on a scattergraph. Plot English (E) on the horizontal axis and History (H) on the vertical axis.

b Draw a line of best fit on the graph.

c Calculate the equation of the line of best fit.

d Kelly was absent on the day of the history exam. She got 82 in her English test. Estimate the mark she would have got if she had sat the history test.

You may be given a scattergraph with line of best fit already drawn where the line does not cross the vertical axis. If this is the case then extend the line yourself by drawing on the graph in order to read the intercept of the vertical axis from the graph. This is much easier than calculating the intercept using the alternative method later in this chapter.

★ **3** In Abruzzo, Italy, the average hours of sunshine and the average temperatures were recorded each month in 2011. This information is shown in the table.

Month	Jan	Feb	Mar	Apr	May	Jun	Jul	Aug	Sep	Oct	Nov	Dec
Hours of sunshine, H	3	4	5	6	8	9	10	9	7	6	4	3
Temperature, T (°C)	9	11	13	17	21	25	28	28	25	20	15	12

a Draw a scattergraph to illustrate the data.

b Draw a line of best fit on the graph.

c Calculate the equation of the line of best fit.

d On a particularly hot summer's day the temperature soared to 31 °C. Use the equation of the line of best fit to estimate the number of hours of sunshine (to the nearest whole hour) on that day.

4 In a PE class 30 pupils were set the following two challenges:

- throw a 5 kg medicine ball and record the distance thrown.
- do shuttle runs and record the levels (where the level is the number of shuttles completed in a set time).

The results for the class were recorded in the table.

Pupil	1	2	3	4	5	6	7	8	9	10	11	12	13	14	15
Ball throw, B (m)	6.2	8.9	9.3	7.0	10.6	6.1	6.5	9.9	7.3	11.4	5.7	5.9	7.9	6.0	8.9
Shuttle run level, S	22	45	41	30	56	20	30	44	28	60	22	23	38	23	42
Pupil	16	17	18	19	20	21	22	23	24	25	26	27	28	29	30
Ball throw, B (m)	9.3	7.9	5.8	8.9	6.8	6.3	9.4	9.6	8.4	10.2	7.3	7.9	6.2	9.5	8.3
Shuttle run level, S	40	33	2	45	26	30	48	44	38	51	33	30	24	45	35

a Draw a scattergraph to illustrate the data.

b Draw a line of best fit on the graph.

c Calculate the equation of the line of best fit.

d A boy was late for class. He completed the ball throw test but was unable to complete the shuttle run due to lack of class time. He threw the ball 12.1 m.

Use your equation to estimate his shuttle run level.

5 The lengths and widths of 12 holly leaves are shown in the table.

Length (mm)	42	38	51	32	45	61	84	76	70	88	95	69
Width (mm)	29	28	31	25	30	36	40	41	36	42	45	35

a Draw a scattergraph to illustrate the data.

b Draw a line of best fit on the graph.

c Calculate the equation of the line of best fit.

d An unusually large holly leaf is found. If its length is 104 mm, estimate its width.

Calculating the value of the intercept of the vertical axis which cannot be read from the graph

Sometimes the axes have been chosen (or you will choose your axes) to take into account the nature of the points. For example, if the values to be plotted are close together, the axes do not have to start at 0.

If this occurs then the equation line of best fit needs to be calculated in a different way.

Example 34.3

A class measured their height and handspan.

Their results are shown in the table.

Pupil	1	2	3	4	5	6	7	8	9	10	11	12	13	14	15
Span, in S (cm)	16	17	18	18	18	19	19	19	19	20	20.5	21	21	22	22
Height in H (cm)	140	145	140	150	155	145	150	160	170	150	160	155	165	165	170

a Draw a scattergraph and draw a line of best fit on the graph.

b Use the line of best fit to find an equation connecting the two sets of data.

a

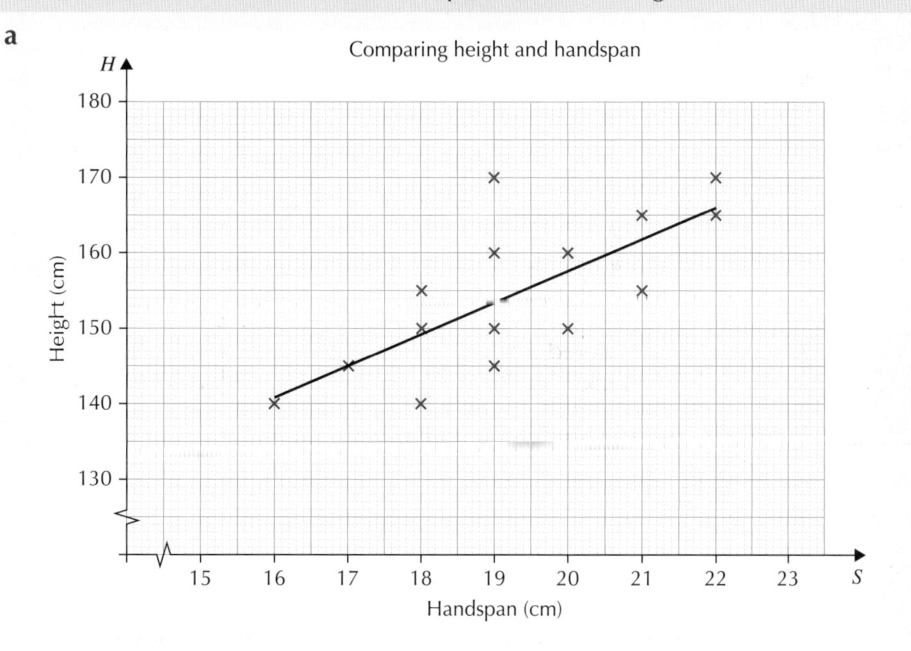

b $\frac{160 - 145}{20.5 - 17}$

$m = \frac{y_2 - y_2}{x_2 - x_1} = \frac{160 - 145}{20.5 - 17}$ — Choose (17, 145) and (20.5, 160) as two points on the line.

$= \frac{15}{3.5} = 4.3$ (to 1 d.p.)

In this case we cannot read c from the graph so we need to calculate it using the equation above.

$H = 4.3S + c$ — Substitute the gradient into the general equation and keep the intercept of the vertical axis as c.

$145 = 4.3 \times 17 + c$ — Substitute one of the points on the line into the equation. Use (17,145) where $S = 17$ and $H = 145$

$= 73.1 + c$

$c + 73.1 = 145$ — Rearrange to find the value of c.

$c = 71.9$

$H = 4.3S + 71.9$ — Finally, substitute the value of c into the equation above.

(continued)

You can also use the **alternative formulation** for the equation of a straight line:

$y - b = m(x - a)$

where m = gradient and (a, b) is a point on the line.

The gradient has already been calculated, so using $m = 4.3$ and $(a, b) = (17, 145)$:

$y - b = m(x - a)$

$y - 145 = 4.3(x - 17)$ — Substitute values.

$= 4.3x - 73.1$

$y = 4.3x - 73.1 + 145$

$y = 4.3x + 71.9$

So: $H = 4.3S + 71.9$

You can see this produces the same equation as above.

Exercise 34C

1 The maths and history results of 11 pupils are shown in the table.

Pupils	1	2	3	4	5	6	7	8	9	10	11
Maths Test % (M)	76	79	38	42	49	75	83	82	66	61	54
History Test % (H)	70	36	84	70	74	42	29	33	50	56	64

These results are shown in the scattergraph with a line of best fit drawn.

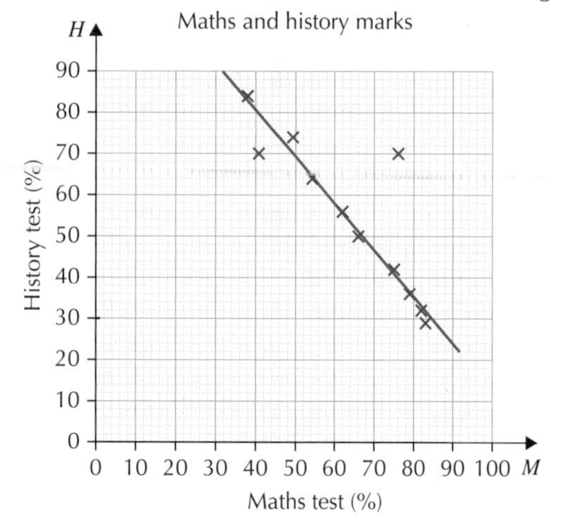

a Use the line of best fit to find an equation connecting the maths test % (M) and the history test % (H).

b A pupil was absent for his history test but achieved 28% in his maths test. Use the equation of the line of best fit to estimate his history mark.

This only gives an estimate. The pupil who achieved a maths mark of 76% and a history mark of 70% does not follow the shape of the graph therefore this pupil appears to be good in both subjects. Be careful about using the line of best fit to estimate values outside the group of plotted points. For example, a mark of 5% in the history test would not achieve 106% in maths.

★ 2 The ages and heights of 10 gifted young basketball players are shown in the table.

Age (years)	11.9	12.4	12.8	13.2	13.5	13.7	14.4	14.7	14.7	15.2
Height (metres)	1.52	1.54	1.62	1.61	1.68	1.72	1.77	1.79	1.88	1.85

a Plot the data on a scattergraph. Make sure that Age (A) is on the horizontal axis and Height (H) is on the vertical axis.

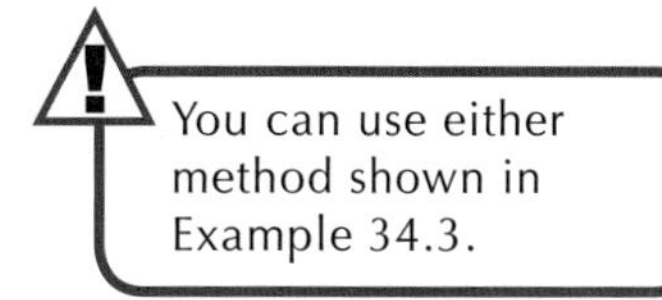

b Draw a line of best fit on the graph.

c Determine the equation of the line of best fit.

d Use the equation of the line of best fit to estimate the height of a basketball player who is aged 16.1 years. Give your answer to 2 decimal places.

e Use the equation of the line of best fit to estimate the age of a basketball player whose height is 2.2 m. Suggest why this might not be correct.

f Explain why it would not be appropriate to use this equation to estimate the height of a 21-year-old basketball player.

You can use either method shown in Example 34.3.

3 Gross domestic product (GDP) is a measure of economic growth in a country's economy.

GDP per capita is a measure of the current GDP per person, and is found simply by dividing the GDP by the population. The GDP per capita (in thousands) and average life expectancy for countries in Western Europe is recorded in the table below.

Country	GDP per capita (€000s)	Life expectancy (years)
Austria	21.4	77.48
Belgium	23.2	78.29
Finland	20.0	77.32
France	22.7	78.63
Germany	20.8	77.17
Ireland	18.6	76.39
Italy	21.5	78.51
Netherlands	22.0	78.15
Spain	23.8	78.99
United Kingdom	21.2	77.37

a Plot the data on a scattergraph.

b Draw a line of best fit on the graph.

c Calculate the equation of the line of best fit.

d Use the equation to estimate the life expectancy of a Western European country whose GDP per capita (in €000s) is recorded as 17.5.

Make the Link

Geography

Look at Question 3 in Exercise 34C. The correlation between life expectancy and GDP in Western Europe is a positive correlation.

However, the correlation does not hold for other countries and the shape of the graph looks more like the graph on the right when all countries in the world are taken into account.

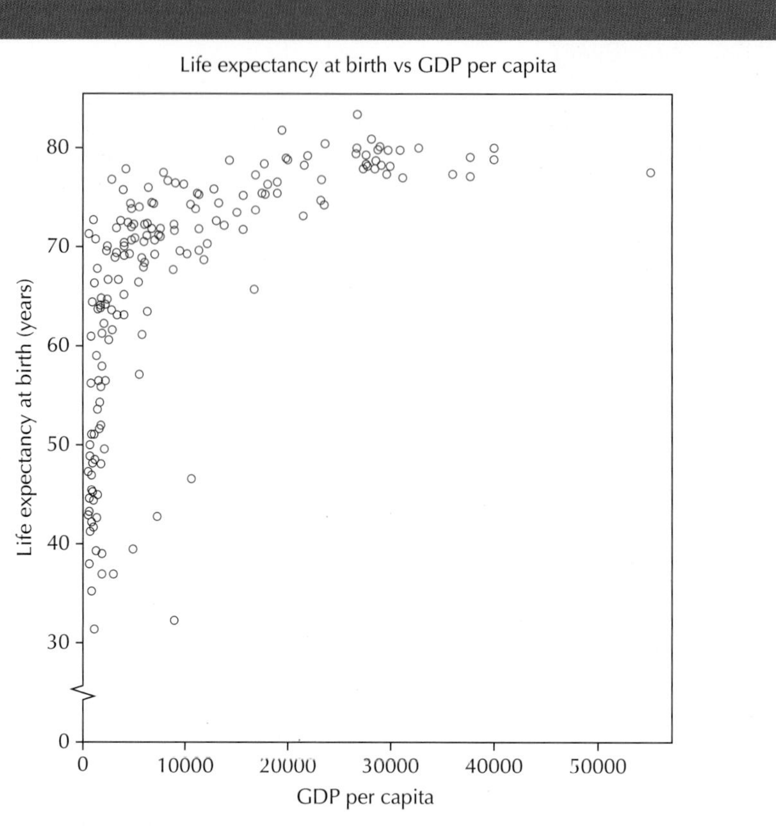

Activity

Research GDP and life expectancy of other continents.

1 Draw scattergraphs of the following continents and comment on any correlation.
  - **a** Africa
  - **b** South America
  - **c** Asia

2 **a** What are the causes of lower GDP and low life expectancy?
  - **b** What are the effects of lower GDP and low life expectancy?

- I can determine the equation of a best-fitting straight line on a scattergraph. ★ Exercise 34A Q3
- I can use the equation of a best-fitting straight line to estimate solutions. ★ Exercise 34B Q3
- I can calculate the value of the intercept of the vertical axis. ★ Exercise 34C Q2

For further assessment opportunities, see the Prepartion for Assessment for Unit 3 on pages 373–376.

Preparation for Assessment: Unit 3

The questions in this section cover the minimum competence for the content of the course in Unit 3. They are a good preparation for your unit assessment. In an assessment you will get full credit only if your solution includes the appropriate steps, so make sure you show your thinking when writing your answers.

Remember that reasoning questions marked with the ⚙ symbol expect you to interpret the situation, identify a strategy to solve the problem and then clearly explain your solution. If a context is given you must relate your answer to this context.

The use of a calculator is allowed.

Applying trigonometric skills to triangles which do not have a right angle (Chapters 25, 26 and 27)

1 The Devil's Triangle is an area of sea off the coast of Florida. The diagram gives the dimensions of the area.

Calculate the area of the Devil's Triangle giving your answer to 2 significant figures.

2 A manufacturer is creating soft play toys with a triangular cross-section as shown in the diagram.

Calculate all the missing angles and side lengths of the cross-section.

⚙ **3** An artist has created a picture in the shape of a triangle. Use the information in the diagram to calculate the length of the frame required for the picture.

⚙ **4** Part of an orienteering course is shown in the diagram. Competitors start at point A and must travel on a bearing of 100° to point B. The distance from B to C is 2.4 km and A to C is 2.6 km. Angle BAC is 30°. Calculate the bearing of C from B and the total distance around the course.

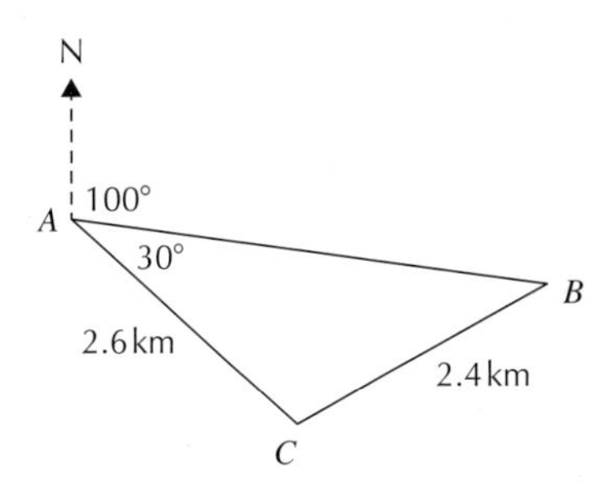

Applying geometric skills to vectors (Chapters 28, 29 and 30)

5 The diagrams below show two vectors as directed line segments **a** and **b**.

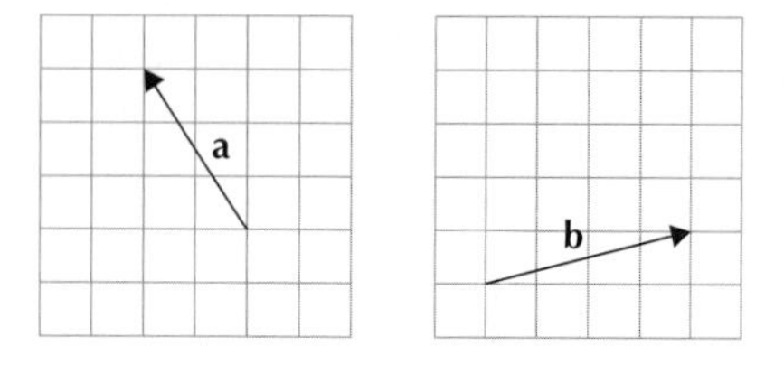

On squared paper draw the following resultants:

a **a** + **b** b **b** – **a** c 2**a** + 4**b** d 3**a** – **b**

6 The diagram shows a cuboid on a 3D coordinate grid. AB is 2 cm, AD is 4 cm and AE is 6 cm.

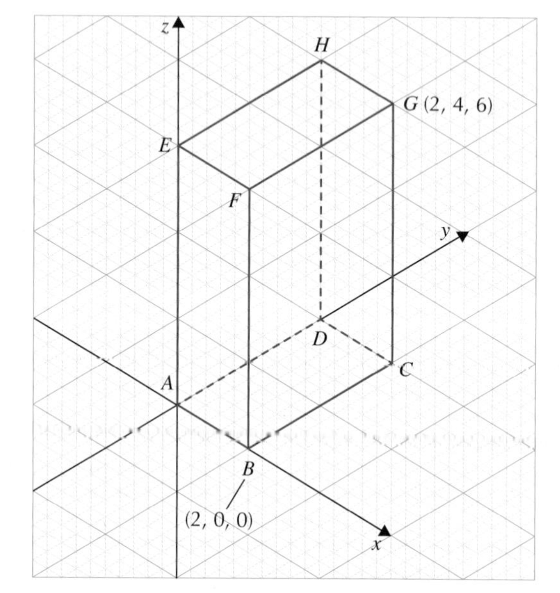

a Write down the coordinates of all the vertices.

b List the coordinates at the centre of each face of the cuboid.

7 Vectors can be used to represent the collision of two balls moving in different directions. Find the resultant force when

vector $\mathbf{a} = \begin{pmatrix} 9 \\ 3 \end{pmatrix}$ and vector $\mathbf{b} = \begin{pmatrix} -2 \\ -4 \end{pmatrix}$

8 The forces acting on a plane can be simplified as drag, thrust, gravity and lift. These forces are represented by the vectors **d**, **t**, **g**, and **l** as given below.

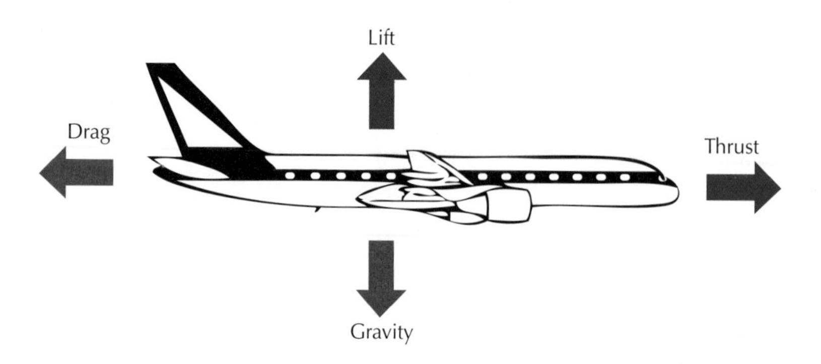

Find the resultant force when

$$\mathbf{d} = \begin{pmatrix} -4 \\ -6 \\ 0 \end{pmatrix}, \mathbf{t} = \begin{pmatrix} 8 \\ 5 \\ 2 \end{pmatrix}, \mathbf{g} = \begin{pmatrix} 0 \\ 0 \\ -6 \end{pmatrix} \text{ and } \mathbf{l} = \begin{pmatrix} 2 \\ 2 \\ 3 \end{pmatrix}$$

9 Vector $\mathbf{a} = \begin{pmatrix} 6 \\ 8 \end{pmatrix}$ and vector $\mathbf{b} = \begin{pmatrix} 4 \\ -13 \end{pmatrix}$.

- **a** Find the resultants $|\mathbf{a}|$ and $|\mathbf{b}|$.
- **b** Find $\mathbf{a} + \mathbf{b}$.
- **c** Find the resultant $|\mathbf{a} + \mathbf{b}|$.
- **d** Find the resultant $|2\mathbf{a} + 4\mathbf{b}|$.

Applying numerical skills to fractions and percentages (Chapters 31 and 32)

10 A farmer compares his yield of wheat in 2013 to his yield in 2012. In 2013 his crop weighs 350 tonnes, which is 17% less than 2012. What was the weight of his wheat crop in 2012?

11 Kiera's luxury car has devalued by 5% per year. Calculate the value of her car after 5 years if she bought it for £21 000.

12 Matthew has a box of collectable bean bag bears. The value of the collection will appreciate 2.5% per year. Calculate the value of the collection after 3 years if the initial value is £840.

13 A gardener has a plot that measures $2\frac{2}{3}$ m^2. He plants sufficient brussels sprouts to cover the whole plot and he expects to harvest $5\frac{1}{7}$ kg/m^2. Calculate the total weight of brussels sprouts that he expects to grow on the plot.

14 Calculate $2\frac{1}{5} + 2\frac{2}{3} - 3\frac{7}{8}$.

15 Calculate $\left(3\frac{1}{4}\right)^2$.

16 Calculate $11\frac{1}{8} \div 3\frac{5}{6}$.

Applying statistical skills to analysing data (Chapters 33 and 34)

17 Below is a back-to-back stem and leaf diagram showing the time spent on homework (in minutes) for 30 S1 pupils and 30 S4 pupils.

Time spent on homework

S1		S4
7 6 6 5	0	0 0
9 9 2 1 0	1	0 5 5 8
8 6 5 4 3 3 0 0	2	2 3 5 5 7 8 8
9 6 5 1 1 1	3	3 5 5 6
9 5 5 2	4	6 7 8 8 9 9
6 5 5 5 0	5	5 6 7 9

Key 1|3 means 31 minutes Key 4|6 means 46 minutes

a Find the median and interquartile range for each group.

b Compare the amount of time spent on homework by each group.

18 Two work teams at a car manufacturer collect data on the number of cars they complete every hour. The table contains the data for both teams for Friday 14 March.

Cars produced on Friday 14 March

Team	Time						
	0800–0859	0900–0959	1000–1059	1100–1159	1300–1359	1400–1459	1500–1559
Team A	15	16	14	16	17	15	14
Team B	16	17	20	14	12	20	10

a Calculate the mean and standard deviation for both teams.

b Using your answers to part **a**, compare the completion rates for both teams and make two valid comparisons.

19 A chemical engineer is conducting an experiment to measure reaction rates of a compound as she changes the temperature of the reaction chamber. The table shows her results.

Reaction time, r (s)	5	14	21	17	21	7	27	9	18	23
Temperature, T (°C)	20	45	73	50	65	28	84	34	59	77

a Draw a scatter graph to compare r and T.

b Describe the correlation.

c Draw a line of best fit onto your scatter graph.

d Find the equation of your line of best fit.

e Use your equation for the line of best fit to estimate the reaction time for the compound at a temperature of 124 °C.

Preparation for Assessment: Examination style questions (Added Value Questions)

The final course assessment will be an examination (this is sometimes called the 'added value' unit). To prepare for this examination you should practise answering questions that combine skills from several parts of the course and added value skills that go beyond the minimum competence that you met in the unit assessments. Practising on past papers questions is a good way to prepare for your examination.

In this section there are examples to show how you should write your answers and some added value questions to practise on.

When answering an examination question you should think about the following:

- *What bit of maths is this question asking for?* Identify the topic/concept, and then decide on a strategy.
- *What calculation do I need to do?* Show the process or working out that you are using.
- *What accuracy do I have to use?* As appropriate, round your answer or give it in a specific format.
- *How should I communicate my answer?* Where possible relate your answer to the context in the question.

Example 1 (Non-calculator)

A right-angled triangle has dimensions as shown. Calculate the length of AB, leaving your answer as a surd in its simplest form.

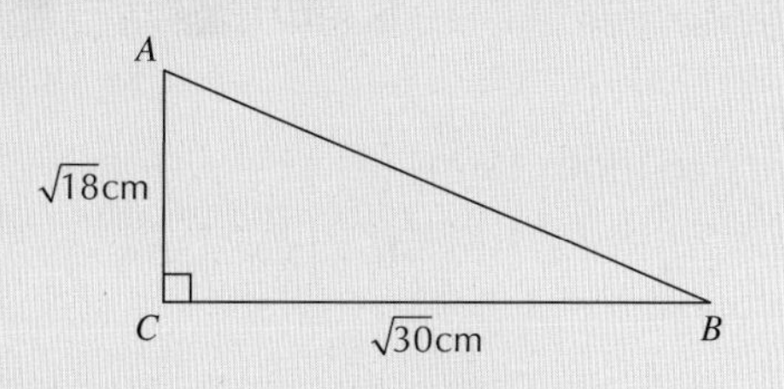

Select a strategy: Use Pythagoras' theorem: $a^2 + b^2 = c^2$

Process: $(\sqrt{18})^2 + (\sqrt{30})^2 = c^2$

$$18 + 30 = c^2$$

$$c^2 = 48$$

$$c = \sqrt{48}$$

Accuracy: Leave answer as a surd in its simplest form.

$$\sqrt{48} = \sqrt{(16 \times 3)}$$

$$= 4\sqrt{3}$$

Communication: AB is $4\sqrt{3}$ cm long.

Example 2 (Calculator allowed)

Tony is attending a class to learn how to touch type on a keyboard. He can type 20 words per minute at the beginning of his course. He wants to increase his word per minute speed by 15% every week. If he sticks to his target, how many words per minute will he be able to type by the end of week 4?

Select a strategy: Percentage increase — 1.15 represents a 15% increase.

Process: $20 \times (1.15)^4 = 34.980$ — Even though you are using a calculator you should write down the calculation.

Accuracy and communication: Round to nearest whole number.

Tony should be able to type 35 words per minute if he meets his target.

Exercise

For each question consider carefully your strategy, process, accuracy and communication.

Non-calculator questions

1 Express $\frac{3}{\sqrt{12}}$ with a rational denominator. Give your answer in its simplest form.

2 A square with side length t centimetres has a diagonal 10 cm long. Calculate the length of t giving your answer as a surd in its simplest form.

3 Express the following without brackets in their simplest form.

a $(b^3)^5$ **b** $(d^{-2})^6$

4 Rewrite the following in index form.

a $\sqrt[5]{k^3}$ **b** $\left(\sqrt[3]{x^3}\right)^{-4}$

5 Multiply out the brackets and collect like terms.

a $(3b + 2)(b^2 + 4b - 5)$ **b** $(4x + 1)(x^2 - 7x + 4)$

6 Fully factorise.

a $3x^2 - 27$ **b** $\frac{1}{2}t^2 - 32$

7 Factorise these trinomials.

a $3s^2 + 8s + 4$ **b** $6y^2 + 30y + 36$

8 A parabola can be expressed as the equation $y = x^2 - 4x + 3$.

a Express $x^2 - 4x + 3$ in the form $(x - a)^2 - b$.

b Make a sketch of the parabola, clearly labelling the y-intercept and the turning point.

c State the equation of the line of symmetry of the parabola.

9 The straight line shown on the diagram passes through the points (0, 2) and (5, 17).

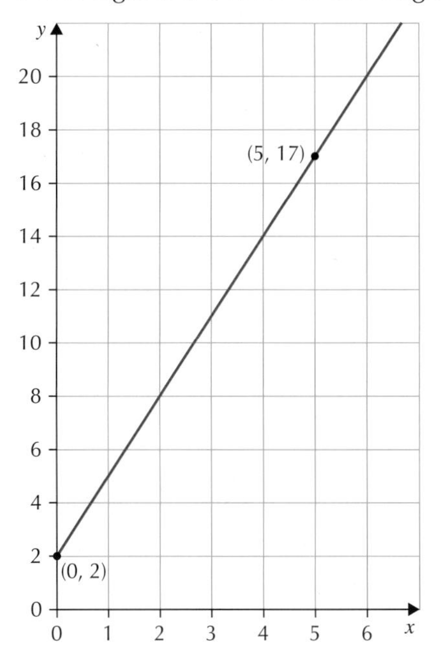

a Find the equation of the line.

b A second line with equation $x = \frac{7 - y}{2}$ is drawn onto the diagram. Rearrange this equation for y.

c Find the coordinates of the intersection of the two lines.

10 A scattergraph has been drawn to compare the sales (S) and advertising spend (A) for a manufacturing firm over the past 10 years. A line of best fit has been drawn on the scattergraph.

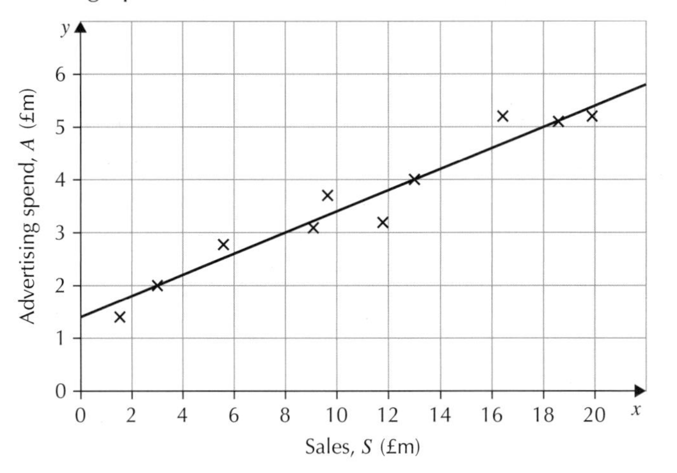

a Find the equation of the line of best fit.

b Use your equation to estimate the sales of the firm if they had spent £12million on advertising.

11 Identify the gradient and y-intercept of straight lines with these equations.

a $y = 3x + 5$

b $x + y = 2$

c $x = 2 - 3y$

d $6x + 3y - 8 = 0$

12 Show that $\frac{1-\cos^2 x}{\cos^2 x} = \tan^2 x$

13 Sketch the graph of $f(x) = 3\sin 2x^{\circ}$, $0 \leqslant x \leqslant 360$.

Calculator questions

1 The dome of St Peter's Basilica in the Vatican can be modelled as a hemisphere with a diameter of 41.47 metres. Calculate the volume of the dome. Give your answer to 4 significant figures.

2 Solve the quadratic equation $3x^2 + 11x + 9 = 0$. Give your answers to 2 decimal places.

3 The diagram below shows a cuboid with sides 5, 12 and 14. Find the magnitude of $(\overrightarrow{AG})$. Give your answer to 3 decimal places.

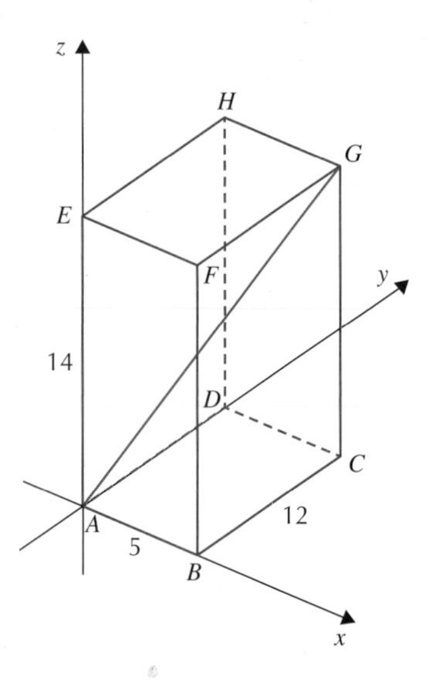

4 A millionaire movie star has bought a house with a swimming pool in the shape of a star. Her pool takes 1.23×10^5 litres of water. She decides to build a bigger pool that is mathematically similar to the original but the depth of the pool will increase from 2 metres to 4.5 metres. Calculate the volume of water that will be required to fill the new pool. Give your answer in scientific notation.

5 The diagram shows the cross-section of a railway tunnel. It consists of part of a cross-section of a circle with a horizontal base, AB. O is the centre of the circle. OB is 3.1 metres long and AB is 5.8 metres. Calculate the height of the tunnel.

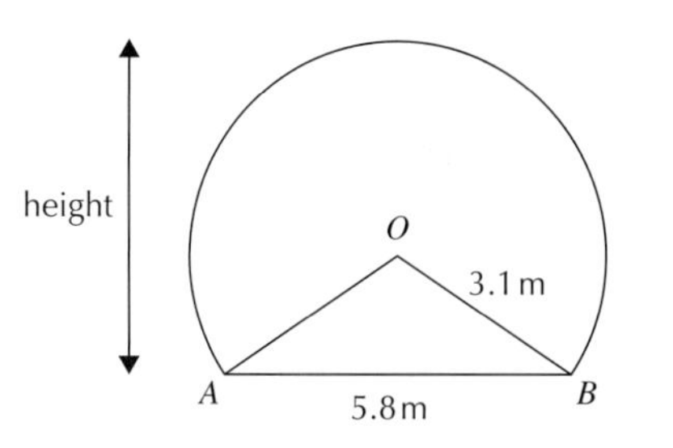

6 **a** Express the following without brackets in their simplest form.

i $x^2(x^{-\frac{1}{2}} + x^{\frac{2}{3}})$ **ii** $2x^{-3}(x^{-\frac{1}{6}} - x^{-4})$

b Find the value (to 2 decimal places) for each expression when $x = 2$.

7 A sail boat left point A on a bearing of 042°. The distances between each point are shown on the diagram. Calculate the boat's bearing from B to C.

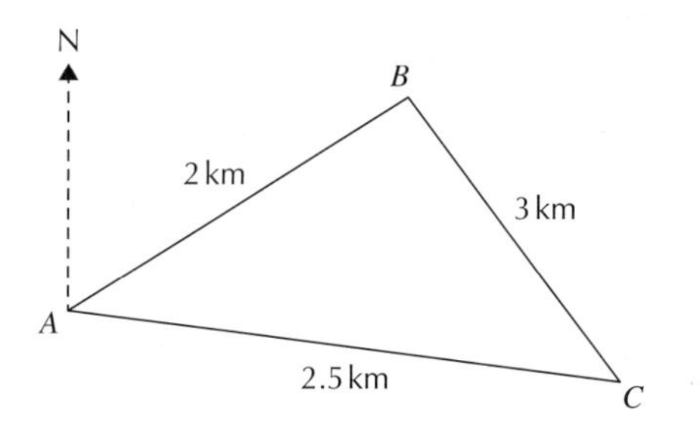

8 Peter buys a new tablet computer. He signs up for a wireless broadband contract. He is charged x pence for every megabyte (Mb) of data plus y pence for every text message he sends. In the first month he uses 230 Mb and sends 150 text messages. His bill is £60.

a Write an equation in x and y which satisfies the above conditions.

The next month he is charged £29.90 after using 120 Mb and sending 70 text messages.

b Write a second equation using this new information.

c Calculate the cost per Mb of data and the cost of sending one text message.

9 A market research company has collected data on the gross annual wage of a sample of 10 people in Glasgow and 10 people in London. The results are shown in the table below.

London	£23 000	£34 000	£18 500	£21 600	£120 000	£14 000	£56 000	£78 500	£20 000	£176 000
Glasgow	£23 000	£45 000	£12 000	£26 000	£7000	£22 000	£56 500	£12 000	£16 000	£19 000

a Find the mean wage for each group.

b Find the standard deviation for each group.

c Make two valid comparisons between the wages of these two groups.